Lecture Notes in Networks and Systems

Volume 273

The series "Lecture Notes in Networks and Systems" publishes the latest developments in Networks and Systems—quickly, informally and with high quality. Original research reported in proceedings and post-proceedings represents the core of LNNS.

Volumes published in LNNS embrace all aspects and subfields of, as well as new challenges in, Networks and Systems.

The series contains proceedings and edited volumes in systems and networks, spanning the areas of Cyber-Physical Systems, Autonomous Systems, Sensor Networks, Control Systems, Energy Systems, Automotive Systems, Biological Systems, Vehicular Networking and Connected Vehicles, Aerospace Systems, Automation, Manufacturing, Smart Grids, Nonlinear Systems, Power Systems, Robotics, Social Systems, Economic Systems and other. Of particular value to both the contributors and the readership are the short publication timeframe and the world-wide distribution and exposure which enable both a wide and rapid dissemination of research output.

The series covers the theory, applications, and perspectives on the state of the art and future developments relevant to systems and networks, decision making, control, complex processes and related areas, as embedded in the fields of interdisciplinary and applied sciences, engineering, computer science, physics, economics, social, and life sciences, as well as the paradigms and methodologies behind them.

Indexed by SCOPUS, INSPEC, WTI Frankfurt eG, zbMATH, SCImago.

All books published in the series are submitted for consideration in Web of Science.

More information about this series at http://www.springer.com/series/15179

Ravindra S. Goonetilleke ·
Shuping Xiong · Henrijs Kalkis ·
Zenija Roja · Waldemar Karwowski ·
Atsuo Murata
Editors

Advances in Physical, Social & Occupational Ergonomics

Proceedings of the AHFE 2021 Virtual Conferences on Physical Ergonomics and Human Factors, Social & Occupational Ergonomics, and Cross-Cultural Decision Making, July 25–29, 2021, USA

Springer

Editors
Ravindra S. Goonetilleke
Division of Integrative Systems
and Design
Hong Kong University of Science
and Technology
Kowloon, Hong Kong

Shuping Xiong
Department of Industrial and Systems
Engineering
Korea Advanced Institute of Science
and Technology
Daejeon, Korea (Republic of)

Henrijs Kalkis
University of Latvia
Riga, Latvia

Zenija Roja
University of Latvia
Riga, Latvia

Waldemar Karwowski
University of Central Florida
Orlando, FL, USA

Atsuo Murata
Okayama University
Okayama, Japan

ISSN 2367-3370 ISSN 2367-3389 (electronic)
Lecture Notes in Networks and Systems
ISBN 978-3-030-80712-2 ISBN 978-3-030-80713-9 (eBook)
https://doi.org/10.1007/978-3-030-80713-9

This Springer imprint is published by the registered company Springer Nature Switzerland AG
The registered company address is: Gewerbestrasse 11, 6330 Cham, Switzerland

Advances in Human Factors and Ergonomics 2021

AHFE 2021 Series Editors

Tareq Z. Ahram, Florida, USA
Waldemar Karwowski, Florida, USA

12th International Conference on Applied Human Factors and Ergonomics and the Affiliated Conferences (AHFE 2021)

Proceedings of the AHFE 2021 International Conferences on Physical Ergonomics and Human Factors, Social & Occupational Ergonomics and Cross-Cultural Decision Making, July 25–29, 2021, Manhattan, New York, USA.

Advances in Neuroergonomics and Cognitive Engineering	Hasan Ayaz, Umer Asgher and Lucas Paletta
Advances in Industrial Design	Cliff Sungsoo Shin, Giuseppe Di Bucchianico, Shuichi Fukuda, Yong-Gyun Ghim, Gianni Montagna and Cristina Carvalho
Advances in Ergonomics in Design	Francisco Rebelo
Advances in Safety Management and Human Performance	Pedro M. Arezes and Ronald L. Boring
Advances in Human Factors and Ergonomics in Healthcare and Medical Devices	Jay Kalra, Nancy J. Lightner and Redha Taiar
Advances in Simulation and Digital Human Modeling	Julia L. Wright, Daniel Barber, Sofia Scataglin and Sudhakar L. Rajulu
Advances in Human Factors and System Interactions	Isabel L. Nunes
Advances in the Human Side of Service Engineering	Christine Leitner, Walter Ganz, Debra Satterfield and Clara Bassano
Advances in Human Factors, Business Management and Leadership	Jussi Ilari Kantola, Salman Nazir and Vesa Salminen
Advances in Human Factors in Robots, Unmanned Systems and Cybersecurity	Matteo Zallio, Carlos Raymundo Ibañez and Jesus Hechavarria Hernandez
Advances in Human Factors in Training, Education, and Learning Sciences	Salman Nazir, Tareq Z. Ahram and Waldemar Karwowski

(continued)

(continued)

Advances in Human Aspects of Transportation	Neville Stanton
Advances in Artificial Intelligence, Software and Systems Engineering	Tareq Z. Ahram, Waldemar Karwowski and Jay Kalra
Advances in Human Factors in Architecture, Sustainable Urban Planning and Infrastructure	Jerzy Charytonowicz, Alicja Maciejko and Christianne S. Falcão
Advances in Physical, Social & Occupational Ergonomics	Ravindra S. Goonetilleke, Shuping Xiong, Henrijs Kalkis, Zenija Roja, Waldemar Karwowski and Atsuo Murata
Advances in Manufacturing, Production Management and Process Control	Stefan Trzcielinski, Beata Mrugalska, Waldemar Karwowski, Emilio Rossi and Massimo Di Nicolantonio
Advances in Usability, User Experience, Wearable and Assistive Technology	Tareq Z. Ahram and Christianne S. Falcão
Advances in Creativity, Innovation, Entrepreneurship and Communication of Design	Evangelos Markopoulos, Ravindra S. Goonetilleke, Amic G. Ho and Yan Luximon
Advances in Human Dynamics for the Development of Contemporary Societies	Daniel Raposo, Nuno Martins and Daniel Brandão

Preface

The discipline of human factors and ergonomics (HF/E) is concerned with the design of products, process, services, and work systems to assure their productive, safe, and satisfying use by people. Physical ergonomics involves the design of working environments to fit human physical abilities. By understanding the constraints and capabilities of the human body and mind, we can design products, services, and environments that are effective, reliable, safe, and comfortable for everyday use. A thorough understanding of the physical characteristics of a wide range of people is essential in the development of consumer products and systems. Human performance data serve as valuable information to designers and help ensure that the final products will fit the targeted population of end users. Mastering physical ergonomics and safety engineering concepts is fundamental to the creation of products and systems that people can use, avoidance of stresses, and minimization of the risk for accidents. This book focuses on the advances in the physical HF/E, which are a critical aspect in the design of any human-centered technological system.

An exploration of how ergonomics can contribute to the solution of important societal and engineering challenges; advances in social and organizational factors discuss the optimization of sociotechnical systems, including their organizational structures, policies, and processes. It includes coverage of communication, crew resource management, work design, design of working times, teamwork, participatory design, community ergonomics, cooperative work, new work paradigms, organizational culture, virtual organizations, telework, and quality management.

The book also highlights issues with special populations, detailing how to design and adapt products and work situations for these groups. In addition to exploring the challenges faced in optimizing sociotechnical systems, the book underlines themes that play a role in all the challenges and how they are linked to each other. It concludes with an exploration of emotional ergonomics and the important positive effects of making people happy and healthy. With chapter authors from around the globe, the book supplies a broad look at current challenges and possible solutions. This book contains a total of ten sections that covers the following topics.

The ideas and practical solutions described in the book are the outcomes of dedicated research by academics and practitioners aiming to advance theory and practice in this dynamic and all-encompassing discipline. A total of ten sections are presented in this book:

Social and Occupational Ergonomics

1. Management and Efficiency
2. Physical Ergonomics and Work-Related Musculoskeletal Disorders
3. Social and Occupational Ergonomics

Physical Ergonomics

4. Holistic Approach in Safety Management During the Pandemic
5. Wearable Sensing in Physical Ergonomics and Safety
6. Workload Assessment Methods and Techniques
7. Job Analysis and Ergonomic Design
8. Human Characteristics and Influencing Factors

Cross-Cultural Decision Making

9. Cross-Cultural Decision Making
10. Cross-Cultural Decision Making

Each section contains research papers that have been reviewed by members of the International Editorial Board. Our sincere thanks and appreciation to the Board members as listed below:

Physical Ergonomics

K. Kotani, Japan
Y. Kwon, Korea
M. Lehto, USA
C. Lung, Taiwan
A. Luximon, Hong Kong
L. Ma, China
S. Maly, Czech Republic
S. Muraki, Japan
M. Nasarwanji, USA
J. Niu, China
E. Occhipinti, Italy
Y. Okada, Japan
H. Pacaiova, Slovak Republic
W. Park, South Korea
G. Paul, Australia
P. Ray, India
Z. Roja, Latvia
L. Saenz, Colombia
L. Shijan, China
J. Sinay, Slovak Republic
S. Xiong, Korea
J. Yang, USA

Social and Occupational Ergonomics

J. Charytonowicz, Poland
D. Horn, USA
S. Hwang, Taiwan
J. Kantola, Finland
B. Kleiner, USA
L. Pacholski, Poland
M. Robertson, USA
S. Saito, Japan
M. Smith, USA
H. Vanharanta, Finland
Z. Wisniewski, Poland
R. Yu, China

Cross-Cultural Decision Making

M. Alama, USA
U. Asgher, Pakistan
E. Cakit, Turkey

V. Cavojova, Slovakia
A. Cybal-Michalska, Poland
A. Divakaran, USA
T. Doi, Japan
C. Fidopiastis, USA
J. Frank, USA
M. Hail, USA
C. He, China
A. Heaton, USA
M. Hoffman, USA
A. Itoh, Japan
D. King, USA
G. Klein, USA
M. Kruger, USA
S. Numrich, USA
N. Okabe, Japan
J. Pfautz, USA
P. Picucci, USA
E. Raybourn, USA
E. Reitz, USA
A. Ruvinsky, USA
L. Saner, USA
D. Scapin, France
S. Schatz, USA
J. Stodd, UK
C. Tajima, Japan
J. Urakami, Japan

We hope that this book, which is the international state of the art in physical domain of human factors, will be a valuable source of theoretical and applied knowledge enabling the human-centered design of a variety of products, services, and systems for global markets.

July 2021

Ravindra S. Goonetilleke
Shuping Xiong
Henrijs Kalkis
Zenija Roja
Waldemar Karwowski
Atsuo Murata

Contents

Management and Efficiency

Integrating Individual and Intra-organizational Learning for Calibration of Organization's Performance

Harsh Chauhan and Henrijs Kalkis(✉)

Faculty of Business, Management and Economics, University of Latvia,
Aspazijas blvd. 5, Riga 1050, Latvia

Abstract. The pragmatic literature review explores the significance of individual and intra-organizational learning for organization's performance. Literature review depicts topic actuality, conceptual contribution, research methods and scientific discussions. The relativity of learning for individuals and organization have been evaluated on behalf of selected scientific papers. The articles reviewed for developing literature review have shown similarities of subject knowledge. The comparative analysis of conceptual framework is based on examination of experiments and modelling documented in previous research. Learning is considered as a process of transferring skills and knowledge from recognized intellectual source. The perspective of individual learning is more concentrated towards self-development and efficacy. Organizational learning is complimented by individual learning. The concentration is on empowering employees and departments for cumulative growth. Continuous Individual and organizational learning and transforming contribute to sustainable long-term organization's performance.

Keywords: Learning · Individual learning · Intra-organization learning · Organization's learning · Organization's performance

1 Introduction

Organization's performance in the context of learning organization and individual learning has been thoroughly researched [1, 2]. It is a combination of one or more sub-systems carried out within the organizational boundaries [3]. The antecedents of organizational performance are referred to the application of both forms of learning. It comprises of individual and intra-organizational learning has shown significant influence on the outcome of organization's performance [4, 5].

Learning is directly related to the constant improvement that enables positive influence on creating organizational performance. It is responsible for transforming and refining knowledge sources [6]. Although individual learning and intra-organizational learning are distinct in comparison, when put together in organization's context having similar interests to grow both have common motives to achieve [7]. Constant learning process helps employees to develop new skills and acquire required knowledge for improving the quality of work [8]. Manifestation of learning in organization's culture is

R. S. Goonetilleke et al. (Eds.): AHFE 2021, LNNS 273, pp. 3–10, 2021.
https://doi.org/10.1007/978-3-030-80713-9_1

critical for sustaining competitive advantage and it successfully sustaining competitive advantage and it successfully contributes to organization's performance [9].

Organization's performance comprises of individual and organizational learning. It is comprehensive shared responsibility between individuals and business management. It is one of the most common feature in hi-tech organizations primarily based on processing knowledge into product and service delivery. Knowledge intensive organizations reliance heavily on intellectual human capital for gaining competitive advantage. Their business development structure constantly focus on harnessing knowledgeable human resource capabilities. Human resources management facilitating fast track learning programs through training and peer to peer knowledge transfer. It is evident that individual learning promote organizational learning. The focus is on optimizing the potential of employees for making substantial contribution to overall organizational learning. The aim of integrating organizational and individual learning successfully calibrate long-term positive impact on organization's performance [10].

2 Research Method

The literature review envisage on learning organization primarily focus on improving individual learning skills that can be transferred into intra-organizational learning. The synchronization between both individual and organizational learning produce positive outcomes on organization's performance.

The construct of the paper relies on previous independent researches. For the purpose of developing literature review various scientific articles have been chosen from recognised academic journal indexed in data base by using title and key words: "Integrating individual and intra-organizational learning for calibration of organization's performance", "Individual learning", "Intra-organizational learning", "Organisational learning", "Learning organization", "Knowledge management", "Knowledge transfer", "Knowledge stock", "Intellectual capital", "Training and development".

The relevant articles which has been selected in consideration to the research by covering additional studies in the subject area has the total number of 46 overall research papers including individual and organizational learning along with organization's performance.

3 Significance of Learning

Learning provides a set platform for organization's growth and performance. Learning is beneficial for both individuals and organizations. It is considered as a part of receiving knowledge based instruction [11]. Learning helps to utilize knowledge and provides an understanding to operate it for organizational purposes [12]. Workplace learning comprises of social experience with competence elements, knowledge building and collaborative efforts to transfer learning. The theory of effective learning at work place demands participation in knowledge-building process, developing creative environment for desired solutions [13].

The social participation, human actions, cognition, enabling learning progression from individual to organizational learning has gone through with many transformation

[14, 15]. The participation and social interaction of how things can be done establishes routine, skills and insights of crafting their own way of learning [16]. Learning has profound influence on career experience and formation. This includes development of key skills, expertise and social contacts from employment. Such experiential learning is relational, functional and problem solving in nature [17, 18].

The development of "experiential and social theories of learning in combination of actions, conceptualization and social practices in past decades have gone through with the process of comprehensive social and behavioral theory including dimensions of meaning, identity and community" [19, 20]. There are two most popular types of learning in context of experiential learning. Both of these types has its own significance depending on the nature and usage. Reproductive learning is associated with dealing routine problems and becoming competent for work practices [21]. Expansive learning is related to solving problems in creative ways, it requires major levels of knowledge implementation and executing competencies [22]. Mentoring, apprenticeships, one-on-one training and team building exercise are another effective and efficient aspects of learning [23].

The transfer of learning is vital for organization's performance and it is most commonly termed as transferable skills. The knowledge and experienced gained during business practices needs to flow and pass on to other individuals. Sharing the accumulated knowledge is a form of learning transfer ensures the distribution of technological skills and business know-how. The most common methods of teaching from predecessors to unexperienced is on job learning [24]. The experienced individuals pass on the expertise and knowledge to freshers. The peer to peer relationship helps closing the knowledge gap and both the parties are the beneficiary of the exercise [25]. The process of exchange aids individual from different backgrounds, age, values, beliefs and motivates them to learn together [26].

The outcome of transfer and/or exchange remains positive. It enhances knowledge and skills retention and creation of meaningful trustworthy relationship [27]. Combined learning develops the sense of belongingness and being felt valued in respect to the contribution made and it motivates individuals to perform and put forward more constructive efforts. On occasions it has been observed that freshers are exchanging latest technological skills with experienced employees [28, 29]. The senior and junior professionals learning and teaching goes hand by hand. From this point of view both organizations and individuals relish knowledge, learning and teaching exchange. This is an ongoing process transferred from one generation to another helps in maintaining knowledgeable resources.

4 Individual Learning

The cognitive paradigm of an individual learning is limited to the individual's ability of processing the information. Individual learning demonstrates the earnest desire for self-development, efficacy and career progression [30]. Well-informed, educated and willing to improve employees are proved to be an asset. Individual learning on its own cannot gain self-sufficiency. Individual learning is a group activity of interdependent people within organization that cannot be learn in isolation. Training, workshops, conferences, and seminars helps employees further in enhancing their portfolio [31]. Training and workshop provide holistic scenario of industrial trends, requirements & specifications. Group exercise helps employees to overcome shortfalls and prepare them by changing or adapting to organizational needs. Equal participation in learning exercise, training and other developmental activities is a part of important strategy for improving individual and organization's performance [32].

Employee's desire for self-development and recognition within organizational framework encourage knowledge exchange and facilitate social learning [33]. They become self-confident and ready to share their ideas with top level management. The environment created fosters open discussion through formal & informal exchange of knowledge and information helps employees to solve task related problems and assist other employees to perform individually [34]. Individual learning environment is created by transferring required skills and knowledge from one recognized source to another. The medium of transfer is from intellectual source to knowledgeable users either among individuals, in teams or in combinations. Their earnest desire to grow motivates other individuals willing to improve their skills. The combined effect of individual and group learning positively affects organization's performance in continuous progression.

5 Intra-organizational Learning

The most common objectives in intra-organizational learning are associated with managing intellectual capital, improving knowledge stock and ability to transfer knowledge and share vital information for enhancing key skills. Organizations strategically promote learning environment and acquiring new skills for sustaining competitive advantage [35, 36].

Organizational learning capability targeted more towards staying ahead of competitors, acquiring cutting edge technological know-how and dominating industrial trends. Organizations thrive on intellectual capital that are "willing to express their opinions, experiment new ideas, make trivial errors without the fear of consequences & rejections" in order to strengthen the core competencies [37].

Learning organizations have focused on the dissemination of learning as a means of remaining competitive in dynamic environment [38]. Learning of organizations are more directed towards individuals, teams, divisions, departments and organization as a whole to be involved in continuous learning and sharing learning. Over a period of time learning transforms into the creation of knowledge and it gets absorbed by organization's work culture [39]. The integration of training, mentorship and skills transfer exercise is valuable for the development of learning organization [40]. The emphasis is on effective and interactive learning techniques. They devolve on technical, managerial, operational and functional skills development. Learning organizations are well prepared and better equipped with knowledge resources. In other words learning organizations constantly evolve themselves and reciprocates organization's performance.

6 Discussion and Conclusions

The performance of an organization is an accumulation of overall organizational learning. Organization's performance in many literatures is defined as the result of constant learning, application of skills and acquiring knowledge resources. Maintaining intellectual capital is pertinent for organizations performance. It is depending on the execution and the resultant of intra-organizational learning and individual learning effectiveness. To maximize the impact of both organizational learning and individual learning on organizational performance learning programs are required to be multi-dimensional. The specifications are needed to be best suited for case to case instead of utilitarian approach. The collective focused on organizational motive of training and learning tools should not be based on the theory of one set fits for all.

The research and practice of an ongoing trend by moving away from egalitarian learning approach and training program of employee's development to be more concentrated and classified in nature will yield prospective results [41]. The effectiveness of learning program should consist of human aesthetics, personal development, not just vocational skills development [42]. Constructive paradigm shift allows flexible approach and creativity to work in organizing better workplace learning [43]. Learning should be a part of understanding learning process and procedure embedding in organizational work structure [44]. Equal participation, extensive social interaction is most relevant to learning in organizational and individual framework [45].

Based on theoretical dimensions a conceptual framework has been proposed. In Fig. 1 learning is shown as an integral part of individual and intra-organizational learning.

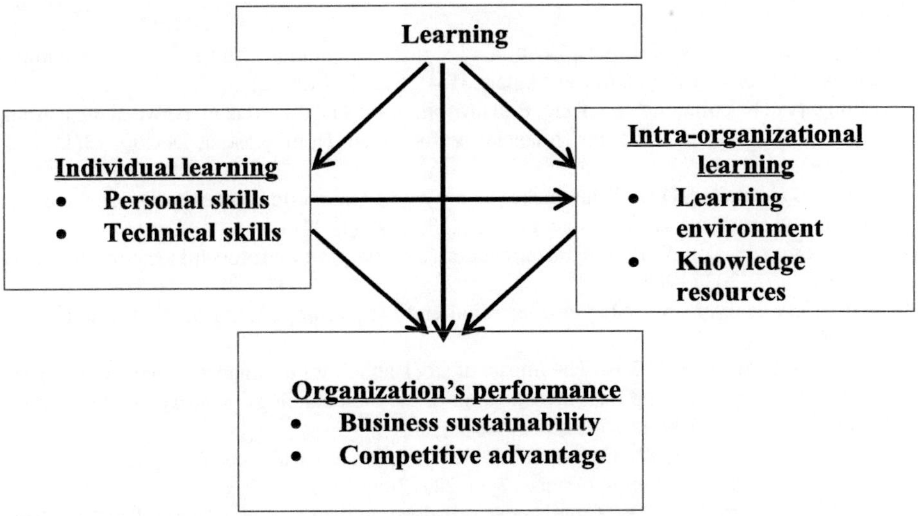

Fig. 1. Impact of Individual and intra-organizational learning on organization's performance.

Learning effects individual's personal and technical skills development. Learning in organization creates environment for become learning organization and enhances knowledge resources in the form of man and material. By integrating individual and organizational learning the joint effect on organization's performance for business sustainability and competitive advantage can be validated. The individual learning and intra-organizational learning environment cultivate new process and procedure for facilitating the organization's performance requires both the entities to work together.

The biggest barrier keeping organizations to move forward from individual learning to intra-organizational learning and transformation into learning organizations is the gap existed between individual learning transfer and absorption of knowledge in organizational management structure [46]. Intra-organizational learning occurs if individual learning take place within organizational boundaries and individuals are willing to share it with teammates, subordinates, supervisors and top-level management. Intra-organizational and individual learning helps in improving internal learning process such as fluency in communication procedure, exchanging information, and developing group and individual capabilities.

In order to facilitate organization's learning individuals must learn and transfer learning efficiently. Organizations ensure developing intellectual capital either by raising skills level of employees or by adopting additional measures for building knowledge resources. Integrating individual learning and intra-organizational learning to occur at both levels resulting in organizational performance.

References

1. Calantone, R.J., Cavusgil, S.T., Zhao, Y.: Learning orientation: firm innovation capability, and firm performance. Ind. Mark. Manag. **31**(4), 515–524 (2002)
2. Ellinger, A.D., Ellinger, A.E., Yang, B., Howton, S.W.: The relationship between the learning organization concept and firms' financial performance. Hum. Resour. Dev. Q. **13**(1), 5–21 (2002)
3. Cohen, S.G., Bailey, D.R.: What makes teams work: group effectiveness research from shop floor to the executive suite. J. Manag. **23**(3), 239–290 (1997)
4. Cangelosi, V.E., Dill, W.R.: Organizational learning observations: toward a theory. Adm. Sci. Q. **10**(2), 175–203 (1965)
5. Slater, S.F., Narver, J.C.: Market orientation and the learning organization. J. Mark. **59**(7), 63–74 (1995)
6. Decarolis, D.M., Deeds, D.L.: The impact of stock and flows of organizational knowledge on firm performance: an empirical investigation of the biotechnology industry. Strateg. Manag. J. **20**(7), 953–968 (1999)
7. Lopez, S.P., Peon, J.M.M., Ordas, C.J.V.: Organizational learning as a determining factor in business performance. Learn. Organ. **12**(3), 227–245 (2005)
8. London, M., Smither, J.W.: Empowered self-development and continuous learning. Hum. Resour. Manag. **38**, 3–15 (1999)
9. Buhler, P.M.: Managing the new millennium: building the learning organization for the 21st century: a necessary challenge. Supervision **63**(12), 20–23 (2002)
10. Barnes-Farrell, J.L.: Older Worker Issues, Encyclopaedia of Industrial and Organizational Psychology, 21. Sage, Newbury (2006)
11. Mead, G.: Mind, Self and Society. University of Chicago Press, Chicago (1934)
12. Kolb, D.: Experiential Learning: Experience as the Source of Learning and Development. Prentice-Hall, Englewood Cliffs (1984)
13. Poell, R.F., Vander Krogt, F.J.: Learning-program creation in work organizations. Hum. Resour. Dev. Rev. **2**(3), 252–272 (2003)

14. Wenger, E.: Communities of Practice: Learning, Meaning, and Identity. Cambridge University Press, Cambridge (1998)
15. Gibb, A.: The enterprise culture and education. Entrep. Theory Pract. **11**(3), 11–34 (1993)
16. Gibb, A.: Entrepreneurship and small business management: can we afford to neglect them in the twenty-first century business school? Br. J. Manag. **7**(4), 309–321 (1996)
17. Deakins, D., Freel, M.: Entrepreneurial learning and the growth process in SME's. Learn. Organ. **5**(3), 144–155 (1998)
18. Harre, R.: Language Games and Texts of Identity. Sage, London (1989)
19. Kelly, G.: The Psychology of Personal Constructs. Norton, New York (1955)
20. Vygotsky, L.: Thought and Language. MIT Press, Cambridge (1962)
21. Ellstrom, P.: Integrating learning and work: problems and prospects. Hum. Resour. Dev. Q. **12**(4), 412–435 (2001)
22. Engestrom, Y.: Expansive learning at work: toward an activity theoretical reconceptualization. J. Educ. Work. **14**(1), 133–156 (2001)
23. Mainemelis, C., Boyatzis, R.E., Kolb, D.A.: Learning styles and adaptive flexibility: Testing experiential learning theory. Manag. Learn. **33**(1), 5–33 (2002)
24. Spannring, R.: Intergenerational Learning in Organisations – Literature Report, European Commission, Brussel (2008)
25. Bransford, J., Brown, A., Cocking, R.: How People Learn. National Academy Press, Washington (2002)
26. Loch, C.H., Sting, F.J., Bauer, N., Mauermann, H.: How BMW is defusing the demographic time bomb. Harv. Bus. Rev. **88**(3), 99–102 (2010)
27. Newman, S., Hatton-Yeo, A.: Intergenerational learning and the contributions of older people. Ageing Horiz. **8**, 21–31 (2008)
28. Baily, C.: Reverse intergenerational learning: a missed opportunity? AI Soc. **23**(1), 111–115 (2009)
29. Tempest, S.: Intergenerational learning. Manag. Learn. **34**(2), 181–200 (2003)
30. Bandura, A.: Social Foundations of Thought and Action: A Social Cognitive Theory. Prentice-Hall, Englewood Cliffs (1986)
31. Torbert, W.R.: Managerial learning and organizational learning: a potentially powerful redundancy. Manag. Learn. **25**(1), 57–70 (1989)
32. Adler, P.S., Kwon, S.W.: Social capital: prospects for a new concept. Acad. Manag. Rev. **27**(1), 17–40 (2002)
33. Bandura, A., Walters, R.H.: Social Learning and Personality Development. Holt, Rinehart & Winston, New York (1963)
34. Baer, M., Frese, M.: Innovation is not enough: climates for initiative and psychological safety, process innovations, and firm performance. J. Organ. Behav. **24**(1), 45–68 (2003)
35. Grant, R.M.: Prospering in dynamically competitive environments: organizational capability as knowledge integration Organizational. Science **7**(4), 375–387 (1996)
36. Spender, J.C.: Organizational knowledge, learning and memory: three concepts in search of a theory. J. Organ. Chang. Manag. **9**(1), 63–78 (1996)
37. Oldham, G.R., Cummings, A.: Employee creativity: personal and contextual factors at work. Acad. Manag. J. **39**, 607–634 (1996)
38. Leigh, D.E., Gifford, K.D.: Workplace transformation and worker upskilling: the perspective of individual workers. Ind. Relat. **38**(2), 174–191 (1999)
39. Eisenhardt, K.M., Martin, J.A.: Dynamic capabilities: what are they? Strateg. Manag. J. **21**(10/11), 1105–1121 (2000)
40. Appelbaum, S.H., Goransson, L.: Transformational and adaptive learning. Learn. Organ. **4**(3), 115–128 (1997)

41. Doornbos, A.J., Bolhuis, S., Simons, P.R.J.: Modelling work-related learning on the basis of intentionality and developmental relatedness: a noneducational perspective. Hum. Resour. Dev. Rev. **3**(3), 250–274 (2004)
42. Gibb, S.: Imagination, creativity, and HRD: an aesthetic perspective. Hum. Resour. Dev. Rev. **3**(1), 53–74 (2004)
43. Fuller, A., Unwin, L.: Older and wiser? Workplace learning from the perspective of experienced employees. Int. J. Lifelong Educ. **24**(1), 21–39 (2005)
44. Sfard, A.: On two metaphors for learning and the dangers of choosing just one. Educ. Res. **27**(2), 4–13 (1998)
45. Illeris, K.: The Three Dimensions of Learning. Krieger, Malabar (2002)
46. Romme, A.G.L., Witteloostuijn, A.: Circular organizing and triple loop learning. J. Organ. Chang. Manag. **12**(5), 439–454 (1999)

Social Distancing, Stress and Unethical Behavior: A Study on Italian University Students in the First Period of Isolation Due to COVID-19

Oronzo Parlangeli(✉), Paola Palmitesta, Stefano Guidi, Ileana Di Pomponio, Margherita Bracci, and Enrica Marchigiani

Department of Social, Political and Cognitive Sciences, University of Siena, P.zzo san Niccolò, via Roma 56, 53100 Siena, Italy
{Oronzo.Parlangeli,Paola.Palmitesta,Stefano.Guidi, Ileana.DiPomponio,Margherita.Bracci,Enrica.Marchigiani}@unisi.it

Abstract. Some studies report high levels of perceived stress by university students in relation to academic misconduct [1]. Therefore, it seemed interesting to evaluate how and to what extent the social containment measures that have been put in place to mitigate the Covid-19 pandemic can influence the relationship between stress and ethical/unethical behaviors in the population of university students. 282 students from 3 Universities in Italy, filled in an online questionnaire aimed at detecting socio-demographic information, the level of perceived stress, the degree of interference between home life and study activities, the perception of unethical behavior. Results show a high level of stress and more serious issues with relationships with other students. In addition, problems in obtaining adequate online informational support appear to increase perceptions of misconduct during exams and in relationships with professors.

Keywords: Social distancing · Stress · Unethical behavior · COVID-19 isolation

1 Introduction

The pandemic due to coronavirus disease (Covid-19) has forced countries all over the world to take severe social isolation measures to respond to the virus and control its spread. In Europe Italy has been seriously affected, earlier than other European states. Starting from the spring 2020, Italian Schools and Universities had to transform face-to-face lessons in distance learning. Some researchers have begun to explore the psychological pressure that the pandemic and social isolation have had on general population [2], elderly [3] children and adolescents [4, 5], university students [6–12].

Cao and colleagues [6] found that 24.9% of the sample of 7143 college students in China had symptoms of anxiety. Other studies conducted in Spain [8] and Switzerland [12] have linked stress, and symptoms such as anxiety and depression, to the condition of students confined in conditions of isolation. Son and colleagues [13] reported difficulties in focusing on study activities and worries on academic performance in a sample of U.S. university students during the Covid-19 pandemic.

R. S. Goonetilleke et al. (Eds.): AHFE 2021, LNNS 273, pp. 11–18, 2021.
https://doi.org/10.1007/978-3-030-80713-9_2

In isolation for students it is more challenging to maintain adequate levels of motivation and engagement as well as adhere to certain moral behavioral norms that are more easily detectable in face-to-face interactions [5, 12].

A few studies focused their attention on a correlation between stress level and misconduct as far as academic student behavior is concerned [1, 14, 15]. In particular, one of these researches [1] on mental distress and unethical behavior, conducted with a sample of Italian students, found that a higher perception of misconduct is related to a higher perception of mental distress, where unethical behavior refers mostly to cheating, plagiarizing, copy the work of another student or allow another student to copy during an exam, using unauthorized materials, copy materials from other publications without citing the source [1, 16, 17].

The study presented here is aimed at investigating the relationship between psychological stress and the perception of unethical behaviors in a sample of university students which have been living measures aimed at social containment put in place to curb the spread of the pandemic due to COVID-19.

2 Participants and Procedure

The research involved 282 students from 3 Italian Universities, in the region of Tuscany. Participants were mostly females (74.8%), aged between 18 and 27 years (mean age: 22.9, SD: 5.8), and all were resident in Italy during the first pandemic period, March–May 2020.

The students, who took part in the study on a voluntary basis, were invited via their university e-mail address to fill in an anonymous, self-reported online questionnaire, that takes about 15–20 min to complete.

The delivery of the questionnaire began online in October 2020.

The protocol of the study was approved by the Ethics Committee of Human and Social Sciences of the University of Siena.

Approximately half the respondents had spent the March–May 2020 lockdown in the city where their university is located or nearby (49.7%). The majority of the students reported to had lived with their family (85.8%), during the lockdown period and 6.4% with other students or roommates.

2.1 Measures

All the items in the questionnaire were referring to the first lockdown in Italy, March–May 2020. The questionnaire was divided in 4 sections. The first section was aimed at detecting socio-demographic information.

The second section (6 questions) was aimed at gathering information on aspects that were perceived as most problematic such as: attending the lessons at a distance, online exams, the relations with other students, keeping in touch with professors and obtaining adequate information support from online institutional channels. Problems were rated on a 5-point Likert scale (from 0 = 'not a problem at all' to 4 = 'very much of a problem').

The third section had three scales (34 questions) related to the psychological impact of confinement due to COVID-19 pandemic. The level of perceived stress was detected with the four item version of the Perceived Stress Scale [18, 19], aiming at rating the frequency of perceived stressful situation during lockdown. The answers were collected on a frequency scale from 0 = "Never" to 4 = "Very often".

The degree of psychological well-being, was measured through the Italian version of the Warwick-Edinburgh Mental Well-Being Scale (WEMWBS) [20, 21], a 14-item scale, which contains all positively worded items on different aspects of positive mental health, that estimate the individual's ability to manage and get through problematic or difficult situations in life. Answers were collected on a frequency Likert scale, from 1 (= "Never") to 5 ("Always").

Functional and dysfunctional relationships between student life and private life, were explored through the 16 item questionnaire structured by Fisher ad colleagues [22] that is composed by four scales: Work Interference on Private Life; Work Enhancement on Private Life; Private Life Interference on Work; Private Life Enhancement on Work. The answers were collected on a 5-point Likert scale where 1 = "Never" and 5 = "Always", adapting the items to study activities.

Finally, the fourth section of the questionnaire (4 questions) was dedicated to the perception of misconducts [1, 14, 23] related to studying and taking exams, relationships with other students and both toward and from professors. An example of these questions is: "How often did your colleagues engage in incorrect behavior (i.e. cheating on exam or prompting each other to pass an exam more brilliantly or using pieces written by others for their own degree theses)?". Answers were collected on a 5-point frequency scale from 0 = "Never" to 4 = "Always".

2.2 Statistical Analysis

Descriptive statistics were computed (mean and standard deviations for numeric variables, and frequencies tables for nominal variables) for all the measures of interest. Paired sample t-tests were used to compare the average ratings about the problems in the different dimensions of study life considered. Independent sample t-test were used to compares the average scores by gender. Correlations were also computed among the different measures of stress, well-being and study-life balances, and among these measures and other variables related to living and studying during lockdown. Finally, ordinal multiple regression logistic models were used to assess stress, well-being and study-life balance measures as predictors of the perceived frequencies of misconducts.

3 Results and Discussion

3.1 Issues in Studying and Learning

In Table 1, the descriptive statistics on the ratings of the problems experienced by respondents during lockdown concerning different dimensions of student life (expressed on a scale from 0 = 'not a problem at all' to 4 = 'very much of a problem') are reported.

The highest scores were found for the dimension about the relationships with other students, for which the ratings were significantly higher than for all the other dimensions ($p < .001$). The lowest ratings were found for problems concerning online exams

Table 1. Average scores for the ratings of the difficulties experienced during lockdown concerning different dimensions of student life.

Issue	Mean	SD	S.E.	Range	Lower C.I.	Upper C.I.
… with online exams	1.9	1.3	0.1	4 (0–4)	1.8	2.1
… in relationships with students	2.8	1.3	0.1	4 (0–4)	2.6	2.9
… in online lectures	2.2	1.3	0.1	4 (0–4)	2.0	2.3
… in relationships with professors	2.4	1.3	0.1	4 (0–4)	2.2	2.5
… in administrative practices	2.1	1.4	0.1	4 (0–4)	2.0	2.3
… in receiving technical support	2.4	1.1	0.1	4 (0–4)	2.2	2.5

(significant differences: vs administrative duties: $p < .05$; vs online lectures: $p < .01$; vs other dimensions: $p < .001$). Independent samples t-tests revealed that the average ratings for the problems with online exams were significantly higher for females than for males (mean difference = 0.4, $p < .05$).

3.2 Stress, Mental Well-Being and Work-Life Balance During Lockdown

Table 2 presents descriptive statistics about the measures of stress, well-being and interference/enhancement between study and personal life. The average level of perceived stress in the sample was considerable (mean = 9.2, SD = 3.4), significantly higher than the published normative value (6.7) for a sample of English people in the 19–29 age range [20]. The average score for the WEMWBS scale was instead significantly lower than the published value (41.5) for Italian students [21]. The average scores for the levels of dysfunctional interference between study and private life were significantly higher than the average scores for the levels of functional interference.

We analyzed the correlations among the different measures of stress, well-being and study-life balances, and among these measures and other variables related to living and studying during lockdown. As it can be seen in Fig. 1, where the significant correlations are displayed as colored ovals, perceived stress was strongly and negatively correlated with mental well-being ($r = -0.59$, $p < .001$), negatively correlated with study enhancement of private life ($r = -0.27$, $p < .01$) and private life enhancement of study ($r = -0.19$, $p < .05$). Perceived stress was instead positively correlated with study interference on personal life ($r = 0.24$, $p < .01$), personal life interference on study ($r = 0.35$, $p < .001$) and receiving informative support online ($r = 0.21$, $p < .05$). Mental well-being, on the other hand, was negatively correlated with study interference on personal life ($r = -0.2$, $p < .05$), personal life interference on study ($r = -0.2$, $p < .05$). It was instead positively correlated with study enhancement of personal life ($r = 0.31$, $p < .001$), personal life enhancement of study ($r = 0.29$, $p < .001$).

Perceived stress was positively correlated with issues in receiving informative support ($r = 0.21$, $p < .05$).

Table 2. Average scores for the measures of perceived stress, well-being and study-personal life interference.

Variable	Mean	SD	S.E.	Range	Lower C.I.	Upper C.I.
Perceived stress (PSS)	9.2	3.4	0.2	16 (0–16)	8.8	9.6
Mental Well-being (WEMWBS)	35.0	8.5	0.5	48 (12–60)	34.0	36.0
Study interference with Private Life	2.7	1.1	0.1	4 (1–5)	2.6	2.9
Study enhancement of Private Life	2.4	0.9	0.1	3.67 (1–4.67)	2.3	2.5
Private Life interference with Study	2.7	1.0	0.1	4 (1–5)	2.6	2.8
Private Life enhancement of Study	2.3	0.8	0.1	4 (1–5)	2.2	2.4
N. of daily study hours	4.7	2.2	0.1	16 (0–16)	4.4	5.0

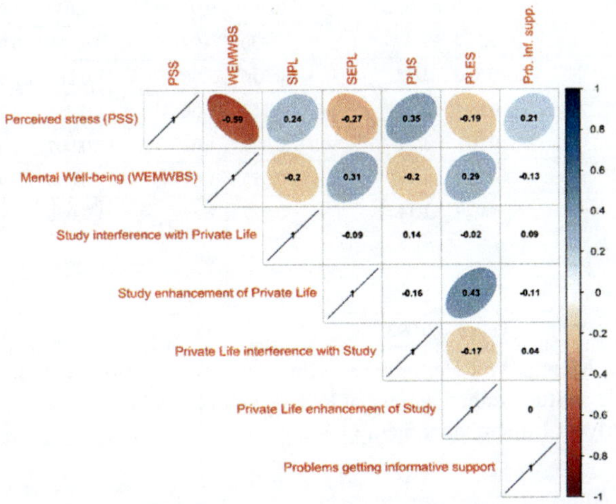

Fig. 1. Plot of the correlation among Perceived Stress (PSS), Mental Well Being (WEMWBS), Study-Life balance and a variable related to problems in getting informative support. The coefficients reported are Pearson's product momentum indices, and statistically significant correlations are represented by colored ovals.

3.3 Predictors of Misconducts During Lockdown

We investigated the predictive effects of different variables (the first column in Table 3) on respondents' perception of the frequency of some misconduct (the first row in Table 3), using ordinal logistic multiple regression models.

Results show that females ($odds_{male} = 0.56$) and student having issue with technical informative support (odds = 1.27) were more likely to perceive misconduct related to studying/taking exam ($R^2 = .084$, $p < .05$).

High levels of study interference with private life are significant are predictors of the perceptions of misconduct related to other students ($R^2 = .092$, $p < .05$). On the other side, perceptions of misconduct related to behavior towards professor are significantly predicted by younger age (odds = .94), low mental well-being (odds = 0.95) higher ratings related to receiving informative support online (odds = 1.41) ($R^2 = .145$, $p < .05$).. Finally, perceptions of misconducts acted by professor are predicted by high work interference with private life (odds = 1.40) ($R^2 = .115$, $p < .05$).

Table 3. Ordinal logistic multiple regression models: predictors of misconducts

	Misconducts related to studying/taking exams		Misconducts related to other students		Misconducts towards professor		Misconducts by professor	
Predictors	Odds Ratios	p	Odds Ratios	p	Odds Ratios	p	Odds Ratios	p
Gender [M]	0.56	**0.042**	0.62	0.127	0.73	0.351	0.79	0.449
Age	0.97	0.139	0.97	0.236	0.94	**0.014**	0.96	0.067
WEMWBS	1.01	0.455	1.00	0.953	0.95	**0.032**	1.02	0.361
PSS	1.03	0.587	1.00	0.930	0.92	0.169	0.99	0.913
WIPL	1.02	0.860	1.37	**0.016**	1.08	0.562	1.40	**0.010**
WEPL	0.96	0.794	0.91	0.554	1.06	0.753	0.73	0.056
PLIW	0.99	0.921	1.11	0.481	1.02	0.907	0.92	0.557
PLEW	0.97	0.837	1.17	0.355	1.07	0.694	0.97	0.876
Issue with information support	1.27	**0.029**	1.14	0.277	1.41	**0.008**	1.11	0.381
N	276		243		205		220	
R^2 Nagelkerke	0.084		0.092		0.145		0.115	

Note. WEMWBS = Mental Well-Being, PSS = Perceived stress, WIPL = Study interference with Private Life, WEPL = Study enhancement of Private Life, PLIW = Private Life interference with Study, PLEW = Private Life enhancement of Study.

4 Conclusions

The results of this study indicate that under conditions of isolation students' stress levels are increased, and this is related to the occurrence of problems essentially in relationships with other students. Evidently, aspects related to the deprivation of opportunities for peer relationships are experienced as particularly problematic.

Staying at home, then, seems to have negative effects on studying, and may have as a consequence that teachers' behaviors are judged as less correct, something that younger students also do. Probably the opacity of a distant and unfamiliar world produces negative effects on perceptions of a reality that thus does not seem characterized by an adequate level of integrity.

In this regard, even the usability of interfaces for obtaining information, as shown in many other domains, can contribute negatively [24, 25]. And this fact affirms once again the need to have work and interaction tools that are effective from the point of view of their usability and the quality of the information they convey.

References

1. Parlangeli, O., Palmitesta, P., Bracci, M., Marchigiani, E., Liston, P.M.: Stress misconduct and reduced ability to express dissent: a study on a sample of students at the University of Siena. In: Arezes, J.B.P.M. (ed.) Occupational Safety and Hygiene, pp. 443–446. CRC Press, Taylor & Francis Group, London (2018). eBook ISBN 9781351008884
2. Wang, C., et al.: Immediate psychological responses and associated factors during the initial stage of the 2019 coronavirus disease (COVID-19) epidemic among the general population in China. Int. J. Environ. Res. Public Health **17**(5), 1729 (2020). https://doi.org/10.3390/ijerph17051729
3. Mukhtar, S.: Psychological impact of COVID-19 on older adults. Curr. Med. Res. Pract. **10**(4), 201–202 (2020). https://doi.org/10.1016/j.cmrp.2020.07.016
4. Wagner, K.D.: Addressing the experience of children and adolescents during the COVID-19 pandemic. J. Clin. Psychiatry **81**(3) (2020). https://doi.org/10.4088/JCP.20ED13394
5. Shah, K., Mann, S., Singh, R., Bangar, R., Kulkarni, R.: Impact of COVID-19 on the mental health of children and adolescents. Cureus **12**(8) (2020). https://doi.org/10.7759/cureus.10051
6. Cao, W., et al.: The psychological impact of the COVID-19 epidemic on college students in China. Psychiatry Res. **287** (2020). https://doi.org/10.1016/j.psychres.2020.112934
7. Duan, L., Zhu, G.: Psychological interventions for people affected by the COVID-19 epidemic. Lancet Psychiatry **7**(4), 300–302 (2020). https://doi.org/10.1016/S2215-0366(20)30073-0
8. Odriozola-González, P., Planchuelo-Gómez, Á., Irurtia, M.J., de Luis-García, R.: Psychological effects of the COVID-19 outbreak and lockdown among students and workers of a Spanish university. Psychiatry Res. **290** (2020). https://doi.org/10.1016/j.psychres.2020.113108
9. Xiao, C.: A novel approach of consultation on 2019 novel coronavirus (COVID-19)-related psychological and mental problems: structured letter therapy. Psychiatry Investig. **17**(2), 175–176 (2020). https://doi.org/10.30773/pi.2020.0047
10. Rogowska, A.M., Kuśnierz, C., Bokszczanin, A.: Examining anxiety, life satisfaction, general health, stress and coping styles during COVID-19 pandemic in polish sample of university students. Psychol. Res. Behav. Manag. **13**, 797–811 (2020). https://doi.org/10.2147/PRBM.S266511
11. Shah, K., Bedi, S., Onyeaka, H., Singh, R., Chaudhari, G.: The role of psychological first aid to support public mental health in the COVID-19 pandemic. Cureus **12**(6), e8821 (2020). https://doi.org/10.7759/cureus.8821
12. Elmer, T., Mepham, K., Stadtfeld, C.: Students under lockdown: comparisons of students' social networks and mental health before and during the COVID-19 crisis in Switzerland. PLoS ONE **15**(7), e0236337 (2020). https://doi.org/10.1371/journal.pone.0236337
13. Son, C., Hegde, S., Smith, A., Wang, X., Sasangohar, F.: Effects of COVID-19 on college students' mental health in the United States: interview survey study. J. Med. Internet Res. **22**(9), e21279 (2020). https://doi.org/10.2196/21279
14. Parlangeli, O., Bracci, M., Guidi, S., Marchigiani, E., Liston, P.: Stress and unethical behaviour in the research of the untenured university researchers. In: 10th International Conference of Education Research and Innovation, ICERI, Seville, Spain (2017). https://doi.org/10.21125/iceri.2017.0585

15. Parlangeli, O., Guidi, S., Marchigiani, E., Bracci, M., Liston, P.M.: Perceptions of work-related stress and ethical misconduct amongst non-tenured researchers in Italy. Sci. Eng. Ethics **26**(1), 159–181 (2019). https://doi.org/10.1007/s11948-019-00091-6
16. Trost, K.: Psst, have you ever cheated? A study of academic dishonesty in Sweden. Assess. Eval. High. Educ. **34**(4), 367–376 (2009). https://doi.org/10.1080/02602930801956067
17. Baran, L., Jonason, P.K.: Academic dishonesty among university students: the roles of the psychopathy, motivation, and self-efficacy. PLoS ONE **15**(8), e0238141 (2020). https://doi.org/10.1371/journal.pone.0238141
18. Cohen, S., Kamarck, T., Mermelstein, R.: A Global measure of perceived stress. J. Health Soc. Behav. **24**(4), 385–396 (1983). https://doi.org/10.2307/2136404
19. Warttig, S.L., Forshaw, M.J., South, J., White, A.K.: New, normative, English-sample data for the short form perceived stress scale (PSS-4). J. Health Psychol. **18**(12), 1617–1628 (2013). https://doi.org/10.1177/1359105313508346
20. Tennant, R., et al.: The Warwick-Edinburgh mental well-being scale (WEMWBS): development and UK validation. Heal. Qual. Life Outcomes **5**(63) (2007). https://doi.org/10.1186/1477-7525-5-63
21. Gremigni, P., Stewart-Brown, S.: Una misura del benessere mentale: validazione italiana della Warwick-Edinburgh Mental Well-Being Scale (WEMWBS). G. Ital. di Psicol. **XXXVIII**(2), 485–508 (2011). https://doi.org/10.13140/2.1.2425.9203
22. Fisher, G.G., Bulger, C.A., Smith, C.S.: Beyond work and family: a measure of work/nonwork interference and enhancement. J. Occup. Health Psychol. **14**(4), 441–456 (2009). https://doi.org/10.1037/a0016737
23. Parlangeli, O., Marchigiani, E., Bracci, M., Duguid, A.M., Palmitesta, P., Marti, P.: Offensive acts and helping behavior on the internet: an analysis of the relationships between moral disengagement, empathy and use of social media in a sample of Italian students. Work **63**(3), 469–477 (2019). https://doi.org/10.3233/WOR-192935
24. Caratozzolo, M.C., Bagnara, S., Parlangeli, O.: Use of information and communication technology to supply health-care services to nomadic patients: an explorative survey. Behav. Inf. Technol. **27**(4), 345–350 (2008). https://doi.org/10.1080/01449290701760658
25. Cassano, C., et al.: Developments and problems in the man-machine relationship in computed tomography (CT). In: Bagnara, S., Tartaglia, R., Albolino, S., Alexander, T., Fujita, Y. (eds.) IEA 2018. AISC, vol. 822, pp. 488–496. Springer, Cham (2019). https://doi.org/10.1007/978-3-319-96077-7_52

Personal Gains from Materials in Social Networks

Tom Sander[1], Biruta Sloka[2], and Henrijs Kalkis[2](✉)

[1] University of Applied Sciences of Ludwigshafen, Ludwigshafen, Germany
[2] Faculty of Business, Management and Economics, Department of Management, University of Latvia, Aspazijas blvd. 5, Riga 1050, Latvia
{biruta.sloka,henrijs.kalkis}@lu.lv

Abstract. Importance of gains from social networks is analyzed in great extent in many countries by many researchers as social networks are reaching more and more importance. Research methods applied: analysis of scientific publications, survey of social network users on gains from materials in social networks. For evaluations by respondents in survey it was used evaluation scale 1–6 as it is used in grading system in Germany. Survey data results are analyzed by indicators of descriptive statistics – indicators of central tendency or location: arithmetic mean, median, mode and indicator of variability – standard deviation. In the research there were found differences by countries in use of social networks.

Keywords: Social networks · Personal gains · Survey · Evaluation scale

1 Introduction

Social networks are becoming more and more important in everyday life in many countries. Use of social networks are analyzed by many researchers world-wide by several aspects including habits of use social network sites in different countries investigating differences by countries, by network use: either for private purposes or for professional needs. Aim of the paper is to analyze how many contacts in social networks have people in Germany and United Kingdom and to find out how much time they spend daily in social network sites and for how long time they are members in social network sites. During the research such tasks were stated: for what purpose do the investigation group use social network sites; how important it is for social network users to enter dialogue with other social network users; what is the importance of social network sites for finding unique information; to what extent is possible to save money to get information in social network sites in comparison with purchasing newspapers and possibility to find unique information. Most of the aspects asked were evaluated in scale 1–6, where 1 is strongly agree and 6 is strongly disagree. The evaluation scale used in this research corresponds to evaluation scale in German speaking countries.

R. S. Goonetilleke et al. (Eds.): AHFE 2021, LNNS 273, pp. 19–24, 2021.
https://doi.org/10.1007/978-3-030-80713-9_3

2 Theoretical Findings

The recent research on social network site applications use new and innovative ideas [1–3] indicating importance of structure of social network sites as important part of the social capital [4] underlining the importance of use the social network sites as open street available for all [5].

Social network sites are good friends for higher education institution students [6] as they benefit from them and the social network sites contribute for the rise of their social capital. Online surveys are good contribution for researchers in the collection of information [7] and significantly faster data analysis and data processing. Entrepreneurs use social network sites for advertising but attitude from social network users are different towards advertising in social network sites [8] having also very negative attitude to the advertisements in social network sites. The attitude contributes to economic outcomes [9] and the new solutions is applied to find the best and optimal solutions taking into account each social network sites specifics [10], requirements and limitations.

Local media act innovatively and creatively to be part of the system [11] and benefit from the social network sites as much as possible. Some social network sites like Facebook have special experience and achievements in generating social relationships and growing persons social capital [12]. Researchers are stating the question on investment size for social network sites asking either it is too much of investment [13]. Digital media had historically big importance but in recent years other new technological solutions are becoming more and more important [14].

Trust for information by well-known people, friends and relatives have been important source for information and in recent years more and more important is becoming online word-of-mouth [15] expressions in social network sites. Social technologies are transforming the world and knowledge obtained on social network sites [16]. New theories are on discussion and development process recently and the approaches are becoming very innovative as well as very efficient [17]. Social network sites are becoming also as object in purchasing of decisions [18] taking into account cultural differences of different nations in different continents [19].

Deep analysis of social network sites selections are carried out [20] as the business development in this field is increasing speed and gaining benefits from innovative solutions [21]. Exchange of contents on social network sites are different in different countries as the cultural and historical background plays also important role [22] taking into consideration social capital aspects [23] as well as formal group settings [23]. The findings in social network applications is effectively used in e-learning becoming more and more popular and efficient [24, 25].

Researchers are discussing also problematic aspects of use of social network sites including cyberattacks and crime [26] and those aspects also are taken into serious account in development of the theory [27] of social network sites and social capital as well as application of possibilities in different science fields [28] and different countries [29]. New technological solutions and applications of social network sites will be discussed in future even more as social network sites are becoming more and more important in professional life as well as in personal life.

3 Empirical Research Results on Use of Social Network Sites in Several Countries

The research of social network users in different continents has indicated importance of cultural differences as it was mentioned above in the theoretical findings by researchers in several countries. Our research was focusing to evaluate social network use in very developed countries: United Kingdom and Germany to find how many contacts on the social network sites have social network users in both analyzed countries, how much time social network users do spend daily and how long time (in years) they are social network users. Main information characterized by indicators of central tendency or location: arithmetic mean, median and mode as well as indicators of variability: range and standard deviation is included in Table 1.

Table 1. Main information on participant in social statistical indicators of responses on question "What do you personally gain by reading materials on social network sites (SNS), Survey conducted by Tom Sander.

Questions	Country	Mode	Median	Max	Min	Mean	Standard Deviation
How many contacts do you have on the SNS on which you have the highest number of contacts?	UK	200	500	3500	2	660.22	635.42
	GER	0	250	10000	0	594.38	1171.34
How many minutes on average do you spend on it in a day?	UK	30; 60	30	300	2	59.73	66.78
	GER	60	30	360	0	45.88	59.29
How many years have you been a member of this social network?	UK	10	8	15	2	8.71	2.89
	GER	10	7	15	0	6.55	4.02

Data of the Table 1 indicate that respondents in United Kingdom have in average more than 660 contacts, respondents in Germany have less than 600 contacts but by social network users in Germany there is bigger variety indicated by bigger standard deviation. In a day for social networks spent by people in United Kingdom biggest time was 30 min and 60 min (characterised by mode) with average time a little less than 60 min but having bigger variety as in Germany (characterised by standard deviation).

Main statistical indicators on evaluations on what social network users gain from the social networks are included in Table 2.

Results of survey indicate that the most important aspect for respondents was willingness to enter the dialogue with other social network members as the average evaluation indicates on that with most often given evaluation 2 (characterized by mode) and half of respondents gave evaluation 2 or less and half of respondents gave evaluation 2 or more (characterized by median).

Table 2. Main statistical indicators of responses on question "What do you personally gain by reading materials on social network sites, Survey conducted by Tom Sander, evaluation scale 1–6, where 1- strongly agree, 6- strongly disagree.

Statistical Indicators	By sharing/reading personal updates I maintain my relationships e.g. with friends or family	I want to enter a dialogue with other members	I can easily access information and news that would be hard to find elsewhere	The expense to get the information is cheaper than other sources e.g. people have to pay for a newspaper…	I can find or share information that is specifically important to my own interests
N	157	156	157	156	156
Mean	3.52	2.66	3.49	3.16	3.04
Mode	2	2	2	2	2
Std. Dev.	1.65	1.44	1.59	1.62	1.55
Median	3	2	3	3	3

4 Conclusions

Social network sites as they are becoming more and more important have different applications in different countries: United Kingdom and Germany but having rather similar use of network sites in both countries by number of contacts having around six hundred contacts and time spent in social network sites which for most of the social network users varies from half an hour till one hour per day. The most importance of social network users have indicated that social network sites are used for social dialogue with other people than followed by possibility to share the information having importance for respective social network user and indicating importance of price paid for this information. Less importance in use of social network sites was indicated the sharing/reading personal updates and maintenance of relationships with friends or family – it means that in with more important people social network users prefer personal contacts and not communication via social network sites.

References

1. Batraga, A., Salkovska, J., Braslina, L., Legzdina, A., Kalkis, H.: New innovation identification approach development matrix. In: Kantola, J.I., Nazir, S., Barath, T. (eds.) AHFE 2018. AISC, vol. 783, pp. 261–273. Springer, Cham (2018). https://doi.org/10.1007/978-3-319-94709-9_26
2. Batraga, A., et al.: Code of breakthrough innovations. In: Karwowski, W., Ahram, T. (eds.) IHSI 2019. AISC, vol. 903, pp. 345–351. Springer, Cham (2019). https://doi.org/10.1007/978-3-030-11051-2_53
3. Bernoff, J., Schadler, T.: Empowered. Harvard Business Review Press, Boston (2010)
4. Burt, R.S.: The network structure of social capital. Res. Organ. Behav. **22**, 345–423 (2000)

5. Curran, K., Crumlish, J., Fisher, G.: OpenStreetMap. Int. J. Interact. Commun. Syst. Technol. **2**, 69–78 (2009)
6. Ellison, N.B., Steinfield, C., Lampe, C.: The benefits of Facebook "friends:" social capital and college students' use of online social network sites. J. Comput. Mediated Commun. **12**, 1143–1168 (2007)
7. Evans, J.R., Mathur, A.: The value of online surveys. Internet Res. **15**(2), 195–219 (2005)
8. Ferreira, F., Barbosa, B.: Consumers' attitude toward Facebook advertising. Int. J. Electr. Market. Retail. **8**(1), 28–38 (2017)
9. Granovetter, M.: The impact of social structure on economic outcomes. J. Econ. Perspect. **19**(1), 33–50 (2005)
10. Halaszovich, T., Nel, J.: Customer–brand engagement and Facebook fan-page "Like"-intention. J. Product Brand Manag. **26**(2), 120–134 (2017)
11. Hampton, K.N., Lee, C., Her, E.J.: How new media affords network diversity: direct and mediated access to social capital through participation in local social settings. New Media Soc. **13**(7), 1031–1049 (2011)
12. Jin, C.-H.: The role of Facebook users' self-systems in generating social relationships and social capital effects. New Media Soc. **17**(4), 501–519 (2013)
13. Kadushin, C.: Too much investment in social capital? Soc. Netw. **26**(1), 75–90 (2004)
14. Kapitan, S., Silvera, D.H.: From digital media influencers to celebrity endorsers: attributions drive endorser effectiveness. Mark. Lett. **27**(3), 553–567 (2015). https://doi.org/10.1007/s11002-015-9363-0
15. Kucukemiroglu, S., Kara, A.: Online word-of-mouth communication on social networking sites, an empirical study of Facebook users. Int. J. Commer. Manag. **25**(1), 2–20 (2015)
16. Li, C., Bernoff, J.: Groundswell, Winning in a World Transformed by Social Technologies. Harvard Business Review Press, Boston (2011)
17. Lin, N.: Building a network theory of social capital. In: Lin, N., Cook, K., Burt, R. (eds.) Social Capital Theory and Research, New Jersey, pp. 3–30 (2001)
18. Miranda, F.J., Rubio, S., Chamorro, A., Loureiro, S.M.C.: Using social networks sites in the purchasing decision process. Int. J. E-Bus. Res. **10**(3), 18–35 (2014)
19. Omori, K., Allen, M.: Cultural differences between American and Japanese self-presentation on SNSs. Int. J. Interact. Commun. Syst. Technol. **4**(1), 47–60 (2014). https://doi.org/10.4018/ijicst.2014010104
20. Rezaei, S., Ismail, W.K.W.: Examining online channel selection behaviour among social media shoppers: a PLS analysis. Int. J. Electron. Mark. Retail. **6**(1), 28 (2014). https://doi.org/10.1504/IJEMR.2014.064876
21. Salkovska, J., et al.: Four conceptual perspectives of innovation components. In: Kantola, J.I., Nazir, S. (eds.) AHFE 2019. AISC, vol. 961, pp. 72–82. Springer, Cham (2020). https://doi.org/10.1007/978-3-030-20154-8_7
22. Sander, T., Sloka, B., Puke, I.: Differences in the exchange of contents of different countries in social network sites. Int. J. E-Collaboration **13**(2), 64–80 (2017)
23. Sander, T.: Organization theory under the consideration of aspects of social capital. In: New Challenges of Economic and Business Development, Riga, pp. 545–554 (2013)
24. Thakurta, R.: Understanding digital media penetration in a formal group setting. J. Serv. Res. **16**(1), 97–117 (2016)
25. Uden, L.: Towards a new model of co-creation of value in E-learning service systems. Int. J. Interact. Commun. Syst. Technol. **1**(1), 36–49 (2011). https://doi.org/10.4018/ijicst.2011010103
26. Wegge, D., Vandebosch, H., Eggermont, S., Walrave, M.: The strong, the weak, and the unbalanced: the link between tie strength and cyberaggression on a social network site. Soc. Sci. Comput. Rev. **33**(3), 315–342 (2014)

27. Williams, K., Durrance, J.: Social networks and social capital: rethinking theory in community informatics. J. Commun. Inform. **4**(3) (2013)
28. Wilson, R.E., Gosling, S.D., Graham, L.T.: A review of Facebook research in the social sciences. Perspect. Psychol. Sci. **7**(3), 203–220 (2012)
29. Xiong, B.: Tell China's story well? Int. J. Interact. Commun. Syst. Technol. **5**(1), 26–40 (2016). National Center for Biotechnology Information. http://www.ncbi.nlm.nih.gov

Factors and Barriers of Implementing Early Warning, Support and Second Chance Support Systems for SMEs in the Baltic States

Liga Braslina, Anda Batraga, Aija Legzdina, Jelena Salkovska, Henrijs Kalkis(✉), Daina Skiltere, Girts Braslins, and Daina Saktiņa

Faculty of Business, Management and Economics, University of Latvia, Aspazijas blvd, 5, Riga 1010, Latvia
{liga.braslina,anda.batraga,jelena.salkovska, daina.skiltere}@lu.lv, girts.braslins@inbox.lv, daina.saktina@arei.lv

Abstract. COVID-19 creates insolvencies time bomb, even if economies are supported by the state. Following the liquidation or bankruptcy of a business, entrepreneurs in the EU mostly opt for a paid professional job rather than re-establishing their business [1]. Those entrepreneurs who re-establish their business after bankruptcy are experiencing faster growth than start-ups. The study points to significant current barriers and factors influencing the implementation of support, early warning and second chances in the Baltic States. There is a need to increase the competencies of both the businessmen and support providers on crisis management and the support already available in the broadest sense. By increasing support for the businesses in crisis in the Baltics, the wave of COVID-19 bankruptcies would be both reduced and used productively to create new, already stronger companies, thus providing a productive support to the Baltic business environment.

Keywords: Early warning · SMEs support · Region development · Second chance · Financial difficulties

1 Introduction

COVID-19 creates insolvencies time bomb, even if economies are supported by the state. In European business culture, closing a business is seen as a failure that leaves a negative impression on further business. Following the liquidation or bankruptcy of a business, entrepreneurs in the EU mostly opt for a paid professional job rather than re-establishing their business [1]. Insolvency analysts Allianz Research [2] and Euler Hermes [3] believe that insolvency processes are still ahead for countries in 2021, reaching an insolvency increase of up to 35% in 2021. However, the experience of liquidation is inspiring - those entrepreneurs who re-establish their business after bankruptcy are experiencing faster growth than start-ups. Research shows that entrepreneurs who have failed are more successful than start-ups [4]. This points out that an early warning and second

R. S. Goonetilleke et al. (Eds.): AHFE 2021, LNNS 273, pp. 25–32, 2021.
https://doi.org/10.1007/978-3-030-80713-9_4

chance policy that supports companies in difficulty or bankrupt entrepreneurs to start a new business as soon as possible is one of the most promising ways to strengthen existing and create new successful companies and thus jobs after COVID-19 crisis. Several early warning and second chance support organizations for SMEs are already operating successfully in Europe, while there are no such support systems in the Baltic States.

The need for such support arises as a result of several specific aspects: (1) a businessman in financial difficulty is in a high-stress situation, when a rapid action is required, which creates additional stress; (2) the businessman is acting outside its area of competence, which is already related to legal aspects; (3) in a financial crisis situation, the businessman does not have financial resources for business and legal advisers available; (4) the businessman is often already in a tax debt zone, which further closes the door to available potential support and/or potential growth. These specific situation circumstances have an extremely strong impact on the businessmen and their ability to productive action in times of crisis, when a rapid and constructive action is crucial to resolving the crisis.

The aim of the study is to increase the understanding of the factors influencing the ability of the Baltic States to implement early warning and second chance support systems, as well as to identify the barriers to insolvency prevention in the context of early warning and second chance policy making, to improve the capacity of the region's economy to improve SME support systems.

The assessment of 68 experts from industrial, academic, civic, and governmental environments in Latvia, Lithuania, and Estonia from April 2020 to November 2020, selected within the Quadruple helix model approach, has been used in the study. In scope of the study 46 in-depth expert interviews, 68 expert surveys and 6 thematic discussions have been conducted. Secondary data have been derived from official national, policy, academic and industry statistical databases. Qualitative and quantitative methods of secondary and primary data processing have been used.

2 Support, Early Warning and Second Chance System Concept

The support, early warning and second chance system is not homogeneous. It is a complex system consisting of a complex legal framework and a dynamic business environment, which is affected by both internal company factors and life cycle crises, as well as changing external conditions. This is especially true in relation to the Baltic States as small economies, directly open to external influences. In economics, an early warning system is generally interpreted as a system used to forecast the future performance of a company or country and possible deviations from the plan and to reduce the risks in the event of a crisis. The early warning system relates to the operations, systems, phenomena, people and other elements that make up an economic system. The early warning system can identify the current situation and potential risks both qualitatively and quantitatively [5, 6].

By the middle of 2020, a unified definition and early warning and second chance system of support for companies in financial difficulties, binding on all EU member states and parties concerned, has not been identified at the level of the European Commission.

The European Commission points out that insolvency regulations cover a wide range of measures - early intervention before a company faces major difficulties, and timely restructuring [7]. As other related definitions from the academic, partly administrative and civic sectors point out, legal insolvency proceedings are only a part of a much broader and a more comprehensive support, early warning and second chance system [8–14].

In academic environment, early warning systems and their efficiency have been extensively studied since the middle of the last century and cover a wide range of research, at regional, national, sectoral and enterprise level [6, 15–20], including SMEs. There are various conclusions, which are often contradictory. The studies also draw very specific conclusions on the key factors that provide early warning of potential financial difficulties and insolvency of SMEs [21–26]. In addition, the growing technological capabilities of processing large amounts of data and the latest generation of business information approaches to data mining have increased the accuracy of early warning [27]. The unifying conclusion of various academic studies is that the logic of the operation of early warning systems is based on deviations from the usual indicators, finding non-conventional and unusual behaviour in the subject. Academic studies also indicate that it is insufficient to identify increasingly efficient tools, models and subsystems in order to be able to reveal the companies that need preventive action to avoid insolvency [28, 29]. It is equally important to create support system(s) that launch these support tools on a regular basis.

There are several support, early warning and second chance organizations in the European Union. One of the largest is *Early Warning Europe*, which operates in several countries. EWE uses the conceptual framework for support, early warning and second chance defined by the European Commission and the OECD [30]. In general, all EU support, early warning and second chance organizations mainly offer mentoring consultancy support to the businessman. Support organizations are like a missing link between the experts and the specialists who, in a crisis, pre-crisis or post-crisis situation, professionally assist to the businessman to develop the most optimal roadmap for recovery and growth.

3 Challenges of Early Warning, Support and Second Chance System in the Baltic States

The support, early warning and second chance systems in the Baltic States are different. There is no unified support system in the Baltic States, but there are various separate, not interrelated support instruments for companies in financial difficulties, such as education measures pertaining to insolvency, national taxpayer rating system, information available from the credit register, guidelines for out-of-court settlements, ability to anonymously apply for a support from support organizations, legal and out-of-court protection process, consultations in business support organizations, consulting opportunities in county development centres, rural development support funds, as well as consultations available in business consulting companies. Businessmen in the Baltics have a fragmentary assessment of the current support available to a company in financial difficulties. In the Baltics, support instruments are analytical rather than mentor-oriented, like in other EU countries. In Baltic countries, the business environment has an intensive and positively stimulating support for start-ups and companies without financial difficulties. For a company that is in financial difficulties even with a minor tax debt (starting from 50 EUR), all state aid is unavailable, including the opportunity to participate in public procurement, which deprives the company of support when it is most needed, and the businessman faces a holistic problem stigma. It has been identified within the study that the stigma of problems causes a significant drop in a company's productivity - up to 90%, which further pushes it into difficulties.

3.1 The Most Significant Obstacles Faced by Businessmen in Financial Difficulties in the Baltic States

The aligned data on the main obstacles faced by the businessmen in financial difficulties in the Baltic States are related to the knowledge on the support already available and the current low efficiency of the known support instruments. It should be noted that there is a significant difference between the assessment of barriers by the businessmen and public administration representatives.

In Latvia, the low trust of the state in businessmen and the dishonest attitude of the businessmen towards the insolvency process are also mentioned as significant obstacles. In Estonia, the lack of knowledge and skills of business managers pertaining to crisis management, the late reaction of businessmen in seeking support, and economic recession are also mentioned as significant obstacles. In Lithuania, economic stagnation or recession, negative public attitudes towards business failures and the long time required for legal insolvency proceedings are also mentioned as significant obstacles. In addition, in Lithuania, the negative attitude of the society and the businessmen towards business failures has also been highlighted as a significant obstacle.

3.2 Most Influencing Factors of Implementing Early Warning, Support and Second Chance Support Systems for SMEs in the Baltic States

The aligned data on the most significant factors influencing the support, early warning and second chance environment in the Baltic States are related to the businessmen's distrust in the efficiency of the support system and the relatively low knowledge of small and medium-sized businessmen in pre-crisis and crisis management. An important existing influencing factor in the Baltic States is the currently closed support door for a company with even a minor tax debt, that, according to the study, all companies in difficulties have.

Significant influencing factors include the environment of economic recession, the fear of the businessmen to recognize the crisis in the company themselves and their inability to cope on their own, as well as the changing political and tax environment. Another important factor is the negative public perception of business failures, which contribute to the company's resistance to recognizing the crisis in a timely manner and to address it constructively through a crisis management approach. Low experience of public administration institutions and, consequently, their competence to promote successful crisis resolution for the businessmen at the early stages of a crisis, without going into insolvency proceedings, should also be mentioned.

Table 1. The most significant obstacles for the businessmen who are in financial difficulties in the Baltic States. Source: Expert survey, conducted by the authors of the research in the target group.

No	The most significant obstacles for the businessmen who are in financial difficulties in the Baltic States	Assessment in points[a]	Coefficient of variation
		X	CV
1	Lack of information on restructuring opportunities	2.88	37%
2	Insufficient efficiency of support services provided by the state and local government – the level of effectiveness of instruments, services, mechanisms, and incentives available	2.82	47%
3	Insufficient number of support services provided by the state and local government - lack of the available support instruments, services, mechanisms, incentives, etc	2.80	43%
4	Economic stagnation or recession	2.80	48%
5	Unscrupulous conduct of a cooperation partner, learning that there are difficulties	2.63	46%
6	Difficult access to venture capital for a company in difficulties	2.54	62%

[a]Grade scale for Table 1 and Table 2: The rating is given in points. 4 - means that the factor is significant and strongly influencing, 3 - the factor is significant and influencing rather than insignificant and non-influential, 2 - the factor is insignificant and little influencing rather than influencing and significant, 1 - the factor is not significant.

In Latvia, the lack of coherence of the political will - legal and economic - is also mentioned as a significant influencing factor. In Estonia, incomplete insolvency regulation (narrow view of the support, early warning and second chance process only from a legal viewpoint, not including the provision of prevention as an integrated support system in both legal and economic support schemes) and the absence of regional support to the companies in difficulties. In Lithuania, insufficient funding for the support of businessmen in difficulties from the part of non-governmental organizations, which are well aware of the need for such a support and would be ready to become more involved, has also been mentioned as a significant influencing factor.

A significant influencing factor is the negative self-attitude of the businessmen in the Baltic States towards business failures, which marks the necessary reorientation of business attitudes towards failures in all countries from a blocking experience to the

Table 2. The most significant influencing factors for the introduction of support, early warning and second chance system in the Baltic States. Source: Expert survey, conducted by the authors of the research in the target group.

No	The most significant influencing factors for the introduction of support, early warning and second chance system in the Baltic States	Assessment in points[a]	Coefficient of variation
		X	CV
1	Businessmen's distrust in the efficiency of the support system	3.20	34%
2	Low competences of the businessmen pertaining to business crisis prevention activities	3.05	33%
3	Venture capital which is difficult for the businessmen to access	2.91	52%
4	Economic stagnation and recession	2.88	38%
5	Lack of coherence of political will - legal and economic	2.82	36%
6	High fear of businessmen to recognize the inability to cope with the crisis in the company	2.79	32%
7	Stability of political regulation	2.74	37%
8	Distrust of the society that business failure is the result of an honest action	2.72	27%
9	Little experience and lack of competence to implement support on the part of the administration and civil support society - difficulties in ensuring the efficiency of the system	2.63	41%
10	Limited competence of state and local government institutions in cooperation with the businessmen	2.59	55%

[a]Grade scale for Table 1 and Table 2: The rating is given in points. 4 - means that the factor is significant and strongly influencing, 3 - the factor is significant and influencing rather than insignificant and non-influential, 2 - the factor is insignificant and little influencing rather than influencing and significant, 1 - the factor is not significant.

one that ensures further successful business in the future, giving the businessman with a self-confidence of its experience. It is positive that overall, in the Baltics the attitude of state and local government policy makers towards business failures is positive.

4 Conclusion and Discussion

The study points to significant current barriers to the introduction of support, early warning and second chance system in the Baltic States and the factors that influence it. At present, support in the field of early warning in the Baltic States is incomplete and provides little incentive to keep the business. The study points to the need to increase information among both the businessmen and support providers on the support already available, to increase the number of support mechanisms by recommending the introduction of new, already tested and efficient support mechanisms, based on the experience of other EU countries, and to promote small and medium-sized enterprise managers' knowledge in crisis management. At present, a significant limiting factor is also the negative self-attitude of the businessmen themselves towards business failures, perceiving it as a blocking experience. By facilitating access to information on crisis management for small and medium-sized enterprise managers and support providers in the Baltics, the transfer of new, efficient support mechanisms for the businessmen in crisis, approbated in other countries, the wave of COVID-19 bankruptcies would both be reduced and used to create new ones, already stronger companies, thus providing a strong and productive support to the Baltic and common EU business environment.

References

1. Bosma, N., Hill, S., Ionescu-Somers, A., Kelley, D., Levie, J., Tarnawa, A.: Global Entrepreneurship Monitor 2019/2020 Global Report. London Business School, Regents Park, London (2020)
2. Hermes, E.: Calm before the storm Covid19 and the business insolvency time bomb. Allianz Res. **7**, 1–13 (2020)
3. Euler Hermes: COVID-19 to Drive 43% Rise in UK Insolvencies. https://www.eulerhermes.co.uk/newsroom/covid-19-to-drive-43percent-rise-in-uk-insolvencies.html
4. Stam, E., Audretsch, D.B., Meijaard, J.: Renascent entrepreneurship. J. Evol. Econ. **18**(3), 493–507 (2008)
5. Ozgulbas, N., Koyuncugil, A.S.: Financial early warning system for risk detection and prevention from financial crisis. Data Min. 1559–1590 (2013)
6. Ashraf, S., GS Felix, E., Serrasqueiro, Z.: Do traditional financial distress prediction models predict the early warning signs of financial distress? J. Risk Finan. Manag. (12), 55, 1–17 (2019)
7. European Commission: Proposal for a Directive of the European Parliament and of the Council on Preventive Restructuring Regulation, Second Chance and Measures to Increase the Efficiency of Restructuring, Insolvency and Redemption Proceedings, Strasbourg (2016)
8. Chaudhuri, A.: Bankruptcy prediction using Bayesian, hazard, mixed logit and rough Bayesian models: a comparative analysis. Comput. Inf. Sci. **6**, 103–125 (2013)
9. Barniv, R., Hathorn, J.: The merger or insolvency alternative in the insurance industry. J. Risk Insur. **64**(1), 89–113 (1997)
10. Ambrose, J.M., Seward, J.A.: Best's ratings financial ratios and prior probabilities in insolvency prediction. J. Risk Insur. **55**, 229–244 (1998)
11. Barniv, R., Mcdonald, J.B.: Identifying financial distress in the insurance industry: a synthesis of methodological and empirical issues. J. Risk Insur. **59**, 543–573 (1992)
12. Laitinen, K., Chong, H.G.: Early warning system for crisis in SMEs: preliminary evidence from Finland and the UK. J. Small Bus. Enterp. Dev. **6**(1), 89–102 (1999)

13. Saksonova, S., Vilerts, K.: Measuring returns to education: the case of Latvia. Ann. Alexandru Ioan Cuza Univ.-Econ. **62**(2), 252–262 (2015)
14. Batraga, A., Salkovska, J., Braslina, L., Legzdina, A., Kalkis, H.: New innovation identification approach development matrix. In: Kantola, J.I., Nazir, S., Barath, T. (eds.) AHFE 2018. AISC, vol. 783, pp. 261–273. Springer, Cham (2019). https://doi.org/10.1007/978-3-319-94709-9_26
15. Assaad Khalil, K.: Financial statements as sole predictors of financial distress: the need for new financial ratios. SSRN Electron. J. (2010)
16. Gallego, A., Quintana, M., Pascual, M.: Forecasting statistical methods in business: a comparative study of discriminant and logit analysis in predicting business failure. Glob. Bus. Econ. Rev. **17**, 76 (2015)
17. Zizi, Y., Oudgou, M., El Moudden, A.: Determinants and predictors of SMEs' financial failure: a logistic regression approach. Risks **8**(107), 1–22 (2020)
18. Deakin, E.B.: A discriminant analysis of predictors of business failure. J. Account. Res. **10**(1), 167–179 (1972)
19. Altman, E.I., Haldeman, G., Narayanan, P.: Zeta analysis: a new model to identify bankruptcy risk of corporations. J. Bank. Finan. **6**, 29–54 (1977)
20. Zmijewski, M.E.: Methodological issues related to the estimation of financial distress prediction models. J. Account. Res. Suppl. 59–82 (1984)
21. Zavgren, C.: Assessing the vulnerability to failure of American industrial firms: a logistic analysis. J. Bus. Financ. Acc. **12**, 19–45 (2006)
22. Devi, S., Radhika, Y.: A survey on machine learning and statistical techniques in Bankruptcy prediction. Int. J. Mach. Learn. Comput. **8**, 133–139 (2018)
23. Yang, B., Ling, X.L., Hai, J., Jing, X.: An early warning system for loan risk assessment using artificial neural networks. Knowl. Based Syst. **14**(5–6), 303–306 (2001)
24. Edison, H.J.: Do indicators of financial crises work? An evaluation of an early warning system. Int. J. Financ. Econ. **8**(1), 11–53 (2003)
25. Salkovska, J., et al.: Four conceptual perspectives of innovation components. In: Kantola, J.I., Nazir, S. (eds.) AHFE 2019. AISC, vol. 961, pp. 72–82. Springer, Cham (2020). https://doi.org/10.1007/978-3-030-20154-8_7
26. Batraga, A., et al.: Code of breakthrough innovations. In: Karwowski, W., Ahram, T. (eds.) IHSI 2019. AISC, vol. 903, pp. 345–351. Springer, Cham (2019). https://doi.org/10.1007/978-3-030-11051-2_53
27. Salas, V., Saurina, J.: Credit risk in two institutional regimes: Spanish commercial and savings banks. J. Finan. Serv. Res. **22**, 203–224 (2002)
28. Saksonova, S., Kantane, I.: Mergers and acquisitions: examples of best practice in Europe and Latvia. Contemp. Stud. Econ. Finan. Anal. **98**, 95–110 (2016)
29. Saksonova, S., Koļeda, O.: Evaluating the interrelationship between actions of Latvian commercial banks and Latvian economic growth. Procedia Eng. **178**, 123–130 (2017)
30. European Commission: A Second Chance for Entrepreneurs: Prevention of Bankruptcy, Simplification of Bankruptcy Procedures and Support for a Fresh Start. EUR-OP, Luxembourg, pp. 1–13 (2011)

Retail Skills as the Craftsmanship of Liquor Retail SMEs

Myungrae Cho(✉) and Koichiro Watanabe

Faculty of Economics, Kagawa University, 1-1 Saiwai-cho, Takamatsu, Kagawa 7608523, Japan
{cho.myungrae,watanabe.koichiro}@kagawa-u.ac.jp

Abstract. Commercial theories have examined the existence of small and medium-sized retail enterprises in Japan from the viewpoint of economic efficiency and social effectiveness. However, these businesses, expected to play a role in revitalizing city communities, are declining. This study analyzes the retail skills and abilities of a small liquor business that remains successful despite the changing times. The analysis identified the following retail skills as key to the owner's success: product proposal skill, customer nurture skill, and market recognition skill.

Keywords: Retail skills · Retail technology · Liquor retail SMEs · SMEs commercial theory

1 Introduction

The overwhelming majority of Japanese retailers have traditionally been small and medium-sized enterprises (SMEs). However, since the mid-1980s, the number of small and medium-sized retailers has been steadily decreasing. The Japanese government has implemented various policies to protect retail SMEs, but with little effect. There are various attributing factors, but it is clear that the emergence and development of large-scale retailers with the advantage of innovative retail technology have had a significant impact. On the other hand, there are still retail SMEs that have carved out a niche and remain firmly rooted in their environments. In other words, retail SMEs, generally considered to be inferior in terms of economies of scale and brand power, may be able to coexist with large-scale retailers if they possess unique advantages. What are the unique advantages of retail SMEs? This study intends to answer this question.

2 Literature Review

2.1 SMEs Commercial Theory in Japan

It has long been acknowledged that Japan's retail structure is inefficient due to the abundance of retail SMEs and their low productivity [1]. The numbers of retail SMEs increased during a period of high economic growth until the mid-1980s. This raises the

R. S. Goonetilleke et al. (Eds.): AHFE 2021, LNNS 273, pp. 33–39, 2021.
https://doi.org/10.1007/978-3-030-80713-9_5

question of why retail SMEs, which have a lower competitive advantage than large-scale retailers, continue to survive even with the economic decline. Japanese researchers point out that market slack creates a gap that large-scale retailers cannot fill [1], the family employment system within the merchant family creates employment [2], and the high shopping frequency of Japanese consumers is suited to smaller convenience stores [3].

These existing studies analyze the survival factors of retail SMEs in terms of economic efficiency, which is now irrelevant given recent market changes such as the decrease in SMEs and the emergence of large, innovative retail companies. Economic efficiency does not explain the survival and co-existence of SMEs alongside larger retailers. In fact, Japanese retail SMEs are not positively evaluated in terms of economic efficiency.

The decline of Japanese retail SMEs, which formed commercial agglomerations in city centers, has led to the demise of cities. In these circumstances, research was conducted from the perspective of social effectiveness to evaluate the role of retail SMEs as communicators. As a representative study, Ishihara [4] points out the existence of "town merchants" [5] – SME owners who actively engage in community development pursuant to a "town merchant spirit"; their community development activities help maintain a more comfortable living environment for local residents. However, evaluation based on this aspect of social effectiveness is limited. In Japan, the total amount of retail consumption is stagnant due to factors such as long-term economic stagnation, declining birthrate, and aging population. Retail SMEs are reluctant to engage in time-consuming community development activities given the uncertain rewards. In addition, the trend toward individualism is increasing and fewer residents participate in community engagements. The function of the community of retail SMEs is disappearing. Against this background, social effectiveness cannot be considered as an effective axis for evaluation.

As mentioned, in Japan the existence of retail SMEs has been examined from the viewpoints of social effectiveness and economic efficiency. However, retail SME possibilities have not been fully examined in current times and with changes to the environment. It can also be said that these existing studies have focused on the external factors of retail SMEs and have not fully focused on the internal factors, that is, the inherent abilities endemic to SME retailers.

2.2 Analytical Perspective

In recent years, emerging research has begun to focus on the individual abilities of SME retailers. For example, Komiya [6] examines the relationship between the management consciousness of retail SMEs and the formation of assortments. Komiya [6] reflected merchants' personal interests and tastes in various behaviors related to store management and highlighted the self-purpose-oriented merchants who seek enjoyment from the behaviors themselves. Such merchants add diversity to the assortment of commercial clusters to which they belong. This highlights a unique image of a merchant that is different from a merchant with general management ability.

Matsuda [7] conducted participant observation of retail greengrocer SMEs and identified merchants' craftsmanship as a competitive advantage. In this study, the greengrocer's assortment, product handling, and craftsmanship skills are unique factors that produce good performance despite the small scale of the business.

Cho and Watanabe [8] conducted a comparative case analysis of merchants at major Japanese SPA-type spectacle chain stores and small and medium-sized spectacle stores and attempted to conceptualize the merchants' unique abilities. They derived the concept of "retail skill," which refers to the craftsmanship of merchants, a concept distinct from "retail technology" represented by mechanization and systematization in retail companies. "Retail technology" is the theorized and explicit knowledge and know-how of retail, which can be embodied by systematization and mechanization. On the other hand, "retail skill" is the retail knowledge and ability inherent in a merchant that cannot be converted into data or a manual.

This study uses the concept of retail skills to clarify the particular retail skills of liquor merchants.

3 Understanding the Current State of Liquor Retailers in Japan

Prior to the discussion of the case study, the current situation of Japanese liquor retailers will be described. According to the Statistics Bureau of Japan [9], there are approximately 32,000 liquor stores in Japan with about 95,000 employees. However, 25 years ago, the number of stores exceeded 100,000, and the number of employees was about 310,000. Within a quarter of a century, the number of stores and employee numbers has shrunk to one-third. In addition, although the number of small stores (1–4 employees) accounts for nearly 90% of the total number of stores, their combined annual product sales amount to just 35% of total revenue, which indicates that liquor retail SMEs operate in a harsh environment. The appearance of large-scale liquor stores is a major contributor to the shrinking number of small liquor stores. In addition, consumers are moving away from sake due to diversified needs, consumer spending being curtailed due to the prolonged economic downturn, and the competitive environment intensifying due to the deregulation of retail liquor licenses.

However, Tamura [10] points out that while the liquor industry is considered one of the industries with the highest rate of store decline, their productivity is relatively high. Relative productivity is the share of the number of stores with one or two employees in that industry in one year divided by a similar number in another year. This number is produced on the assumption that efficient stores will be screened out due to competition between stores of various sizes. If this value is one or more, productivity is high. As a result, most industries have a value of one or less, while liquor stores have a high value of one or more. This shows that even small businesses in the liquor store industry possess an element of competitiveness, and have the potential to survive.

Tamura [10] describes the characteristics of industries with high relative productivity as follows: (1) human services are the decisive factor in attracting customers, (2) consumers have a strong preference for the freshness of products, (3) customer needs are diversified and individualized. (4) labor-intensive distribution processing or small-scale manufacturing processes, and (5) strong distribution support by manufacturers.

In the liquor sales industry, fine-tuned human services, diversification, and individualization of customer needs may apply. However, not many studies have investigated and analyzed the actual phenomenon in detail. Furthermore, it is unclear what kind of ability the merchants themselves have in responding to these phenomena. Therefore, in this study, we decided to investigate a store with a history of more than 100 years, which enjoys the strong support of local residents.

4 A Case Study of a Liquor Retail SME

This study involved a free-form interview conducted with the owner of Watanabe Liquor Store located in Takamatsu City, Kagawa Prefecture, Japan on November 12, 2020.

Watanabe Liquor store is a specialty liquor store founded in 1893. The shop is located close to a railway station in the south and a pier in the north, making it a popular place in the community of Takamatsu. The store itself was rebuilt in the mid-1970s and continues to operate to this day. Since its establishment, it has been run mainly by the founder's family; the current owner is the 4th generation, and four family members manage the store. The store area is about 100 m^2, there are no branches, and it is classified as a small retail business. The product lineup consists mainly of local Japanese alcohol, such as sake and shochu, wine, and snacks.

Originally the liquor was mainly sold by weight from barrels, but over time, bottles became popular, and the demand for beer increased indicating that the selling method and products were adapted to changing requirements. In recent times, many traditional liquor sales businesses have closed down due to the decline in the consumption of sake and shochu, the revision of the liquor laws, and the emergence of large discount and convenience stores.

In response to these changes, the owner considered switching to a convenience store business model but decided to continue as a specialized liquor store and consulted with a local brewery to partner in the manufacturing of sake.

From there, he transformed the store into a local customer-oriented store and carefully guided the manufacture of liquor at the local store-approved sake brewery. Currently, about 80% of all products are purchased directly from the brewery, and the product lineup is closely aligned to the local community. Most of the customers are regulars.

In terms of sales and detailed product explanations, the owner actively supports customers' product selection by having them enter the large brewery connected to the store directly to look for products and actively incorporates tasting into the process. In recent years, online sales have also been generated, driven by the owner's son, who is the 5th generation. The son does not just sell products online but also promotes the characteristics of the sake, how to drink it, and the appropriate dishes to match the sake.

The owner's son has extensive work experience in a local liquor store in Tokyo, and has a wealth of liquor knowledge. He is trying to establish a different management style from that of the 4th generation owner, and is developing a new brewery and proposing liquor that suits his cuisine.

The store also sells original, private brand products. The owner has an arrangement with a local sake brewery and sells original products manufactured using his method.

This is possible because the local sake brewery trusts the owner's knowledge of sake. Many of the customers are local, regular customers who have maintained a long relationship with the owner. Many new customers who are looking for specialized liquor find information on the store online or by word-of-mouth. The customers are not just looking for a good deal, but become so-called fans of the store.

The owner has a wealth of knowledge about sake, and he carefully explains the characteristics of the products to individual customers, including the characteristics of the brewery, the type of rice used as the raw material, the method of preparation, the way of drinking, and the compatibility with accompanying food dishes. The shop owner proposes sake that matches the product characteristics and consumption behavior desired by the customer. There is never any pressure to make a purchase, the shop owner enjoys imparting his knowledge so that customers can appreciate the product's true value.

Regarding his future outlook for the store, the owner conveyed that he was not very keen on expanding the business. He describes his store as a "niche industry" and is aware that selling products to consumers in the same way as large liquor stores will not work for his business. In addition, the owner understands that his business style is to have customers with a certain degree of commitment to alcohol who sympathize with the product and appreciate the "story" behind the products that the owner himself endorses. In order to maintain this business style, he believes that the current business scale is appropriate and can be passed on to the 5th generation.

5 Discussion

This study summarizes three skills from the case study which can be considered retail skills for SME liquor merchants.

The first is "product proposal skill" that refers to abundant product expertise. This includes expertise that emerges from individual experience, efforts, the individuality of the store owner as well as expertise that emerges from tacit knowledge and relationships passed down from generation to generation. It is assumed that all merchants have a certain level of ability to propose products. However, in order to obtain more specialized and sophisticated product proposal capabilities, many years of experience and effort, as well as trusting human relationships, is required, which cannot be acquired in the short term. In other words, it is difficult to create a manual because the background of this ability includes the experience and efforts of merchants accumulated over many years, their relationships, and the potential of individual merchants. Of course, this also applies to the ability to accurately read the customers' requirements.

The second is "customer nurture skill" through communication with customers in order to improve understanding of the proposed product. It is related to the ability to propose products; the more specialized the product, the more difficult it is for the customer to find and understand the product. Therefore, a merchant with a high level of product proposal ability also needs the ability to communicate accurately to convey the product and information desired by the customer. Through this communication, the customer obtains new information about the product which deepens their product knowledge. As a result, the customer will engage in new consumer behaviors, such as buying new products. Simultaneously, the higher the credibility of the merchants who provide such accurate information, the higher the possibility of the customer becoming a regular. Especially in small stores, the existence of regular customers is important. In order to get repeat customers, it is necessary to build good relationships and constantly increase the satisfaction of repeat customers. For this purpose, it is essential to continually detect and propose the product lineup and what the customer wants through communication, and at the same time, to nurture customer relationships to encourage further consumption. The higher this ability, the more stable the management becomes.

The third is "market awareness skill," which recognizes and acts on the store's appropriate business scale. In general, companies tend to aim for large-scale growth when their business performance is good. With various innovations, the aim is to increase the size of the store itself, develop multiple stores, diversify, and gain the support of more consumers, resulting in the product lineup and services becoming more popular. However, the retail SME introduced in this study has not taken these actions even though their business performance is good. That is, the more specialized the assortment and customers are, the more the sector will inevitably target niche markets. At that time, the smaller retailers are consciously and unconsciously aware that if they expand their business scale, they will lose the competitive advantage of their expertise and not be able to manage their own product lineups, services, and customers. Therefore, the scale of their business is intentionally limited.

This approach has evolved into a solid and stable management system and has produced the result of being a long-established liquor store for over 100 years. Of course, it has not been without management innovation during this time, including flexible responses to changes in the social environment, consumer behavior, and culture. This market awareness also applies to the ability to adapt according to the market needs and circumstances at that time, rather than remaining attached to a particular management style just because it is a long-established store.

6 Conclusion

This study examined retail skills as the craftsmanship of SME merchants using a case study of a liquor store. Three retail skills were clarified as the source of the unique superiority of this liquor store: product proposal skill, customer nurture skill, and market awareness skill. The study concludes that a combination of these three retail skills facilitates stable management over the long term.

Additionally, the theoretical and practical implications of this research and future research topics are discussed.

First, a new theoretical research approach for analysis is suggested given that existing commercial theories on the existence of SMEs in Japan are not practical. The proposed approach is based on the specific abilities of individual merchants, a concept called retail skills.

Second, as a practical implication, retail SMEs do not have to compete directly with large retailers. Large retailers and retail SMEs play different roles. If SMEs expand in scale and aim for mass sales, they will compete directly with large-scale retailers for homogenization, which may cause price competition.

Lastly, this study considers only one liquor store, an individual in a particular industry. It is merely an intermediate attempt toward theorizing the concept of retail skills. As a future research topic, it will be necessary to analyze the kind of retail skills that exist for merchants who handle less specialized products. In other words, in order to build a theory of retail skills, its validity must be thoroughly examined with sufficient samples to achieve "theoretical saturation" [11] in which no new relationships emerge during analysis.

Acknowledgments. This work was supported by JSPS KAKENHI Grant Number 17K04016, 18K12885, 20H01553, 20K01981.

References

1. Tamura, M.: Japanese Distribution System. Chikura Publishing, Tokyo (1986)
2. Ishii, J.: The Merchant Family and the Market Society: Another Consumer Society. Yuhikaku Publishing, Tokyo (1996)
3. Maruyama, M.: Competitive Structure in Japanese Market. Sobunsha, Tokyo (1992)
4. Ishihara, T.: Retail Externalities and Community Development. Yuhikaku Publishing, Tokyo (2006)
5. Doi, T.: Tenjin's Shopping Street: Make Use of the City, Make Use of People. Tohoshuppan, Osaka (2002)
6. Komiya, K.: Assortment formation with self-purpose-oriented retailers. J. Mark. Distrib. **6**(1), 81–93 (2003)
7. Matsuda, A.: A Field Work of Small-Sized Retailers. Sekigakusha, Tokyo (2017)
8. Cho, M., Watanabe, K.: "Retail skill" as a competitive advantage for small-sized retailers. In: 2019 International Conference on Business and Economics (ICBE 2019) Proceedings (2019)
9. Statistics Bureau of Japan: 2016 Economic Census for Business Activity (2018)
10. Tamura, M.: Rise and Fall of Retail Types and Formats. Chikura Publishing, Tokyo (2008)
11. Strauss, A., Corbin, J.: Basics of Qualitative Research. Sage, Thousand Oaks (1990). Translated to Japanese by Minami, H., Igakushoin, Tokyo (1999)

Analysis of the Employment Rate of People with Disabilities in Ecuador

Hugo Arias-Flores[1(✉)], Jorge Guadalupe-Lanas[2], and Janio Jadán-Guerrero[1]

[1] Centro de Investigación en Mecatrónica y Sistemas Interactivos MIST/Carrera de Administración/Carrera de Ingeniería en Ciencias de la Computación, Universidad Tecnológica Indoamérica, Av. Machala y Sabanilla, Quito, Ecuador
{hugoarias,janiojadan}@uti.edu.ec

[2] Centro de Investigación en Empresa, Sociedad y Tecnología ESTec/Carrera de Administración, Universidad Tecnológica Indoamérica, Machala y Sabanilla, Quito, Ecuador
jorgeguadalupe@uti.edu.ec

Abstract. In Latin America and the Caribbean, approximately 12% of the population has some type of disability, 80% of them are unemployed and the 20% who have a job do not have the same conditions as the rest of the employees. To address this, public policies have been proposed and implemented with a focus on the labor insertion of people with disabilities. In this context, people living with a disability find it more difficult to find a job. They are less likely to be employed compared to people without disabilities. This article analyzes the employment of people with disabilities in Ecuador. Since Ecuador is considered a developing country, employment plays a fundamental role in the economic development of the population, particularly of people with disabilities. We want to identify the occupation rate in the formal market in Ecuador, establishing the percentage of participation by type of disability. In addition, an analysis will be carried out according to the affiliated institution in which it is active. The research is descriptive, and the data recorded in official sources were analyzed. We conclude that the employment rate of people with disabilities is only 15.3%, and that people with intellectual disabilities have the lowest employability rate, at 7.5%.

Keywords: Person with disability · Type of disability · Employment rates · Employer sector · Affiliation

1 Introduction

Maintaining a paid job generates psychological, social, economic, and health benefits for individuals [1–4]. The working-age population maintains adequate employment, has access to medical care, maintains social recognition, and contributes to the production of a state [1].

Throughout history, people with disabilities have been excluded from society and have suffered violations of their human rights in many aspects, particularly the right to paid work. Article 27 of the Convention on the Rights of Persons with Disabilities treaty, adopted in 2006, with its new comprehensive care model, has created an opportunity for

R. S. Goonetilleke et al. (Eds.): AHFE 2021, LNNS 273, pp. 40–44, 2021.
https://doi.org/10.1007/978-3-030-80713-9_6

the right to work under equal conditions to be respected through relevant policies and measures, incentives, and adjustments in the workplace [5].

Despite advances in diversity and inclusion practices in the workplace, the entry of people with disabilities into the workforce remains problematic [6]. Only one in three people with disabilities in the United States is employed (34.9%) compared to 76% of their non-disabled counterparts [7]. In Canada, the employment rate for people with disabilities is 49%, whereas it is 79% for people without disabilities [8]. And in the European Union, these figures are 47.3% for those without disabilities. people with disabilities and 66.9% for people without disabilities [9].

The existing gap in employment rates shows that people with disabilities are less likely to be employed compared to people without disabilities [10]. In fact, about two-thirds of people with disabilities who are of working age are not in the workforce. Disparities are greater for women with disabilities, as the employment rate is lower than that of men with disabilities and lower than that of women and men without disabilities, confirming the persistence of gender disparities in labor markets [11].

Finding and keeping a job is a challenge for people living with some type of disability. The objective of this article is to examine the employment rate of people with disabilities in Ecuador; as it is considered a developing country, employment plays a fundamental role in the economic development of the population, particularly of people with disabilities. We want to identify and l kind of disability that has more people formally employed in the labor market, setting the percentage share by type of disability.

This article is structured as follows: Sect. 2 presents the methodology used as a fundamental basis in the proposal. Section 3 presents the proposal results and a comparison to some related works. Section 4 presents the conclusions obtained from the results and suggests future lines of research.

2 Methodology

The data on persons with disabilities who are formally employed were obtained from the Consejo Nacional para la Igualdad de Discapacidad de Ecuador and can be viewed on its website [12].

The sample had a total of 63,047 valid data, of which 21,703 (34.4%) were identified as female, 1 (0.0%) as LGBTI, and 41,343 as male (65.6%). Regarding age groups, 61 (0.1%) were from 13 to 18 years old, 2,227 (3.5%) were from 19 to 29 years old, 14,561 (23.1%) were from 25 to 35 years old, 42,781 (67.9%) were from 36 to 65 years old, and 3,417 (5.4%) were 65 years old or older. The data were recruited from all the provinces of the country.

3 Results

To establish the employment rate of people with disabilities, an analysis of the data from Ecuador's 24 provinces was conducted. For the calculation, only people with disabilities were considered; the data of people employed as substitutes were excluded.

One of the results shown by this analysis is that the employment rate of women with disabilities (11.9%) is lower than that of men with disabilities (17.9%) and women without disabilities (26.4%), which is an evident reality in many countries worldwide [11]. In Ecuador, the occupancy rate of people with disabilities is 15.3%, which is much lower than the occupancy rate of people without disabilities (32.1%), as presented in Table 1.

Table 1. Employment rate by gender.

Gender	Occupied	18 years and over*	Occupancy rate
Female	21,703	182,414	11.9
LGBTI	1	22	4.5
Male	41,343	230,653	17.9
Total	63,047	413,089	15.3

*Economically active population

When disaggregating the employment rate by type of disability, the least employed people are people with intellectual disabilities (7.5%), followed by people with psychosocial disabilities (10.4%), as can be seen in Table 2.

Table 2. Occupancy rate by type of disability.

Disability	Occupied	18 years and over*	Occupancy rate
Hearing	10,811	61,605	17.5
Physical	35,241	202,033	17.4
Intellectual	5,684	75,627	7.5
Psychosocial	2,205	21,142	10.4
VISUAL	9,106	52,682	17.3
Total	63,047	413,089	15.3

*Economically active population

In Ecuador, there are three institutions in the formal market that are part of social security: Instituto Ecuatoriano de Seguridad Social (IESS), Instituto de Seguridad Social de las Fuerzas Armadas (ISSFA), and Instituto de Seguridad Social de la Policía (ISSPOL). The IESS is the institution with the most affiliates with disabilities in Ecuador (97.6%), as presented in Table 3.

Finally, 7 out of every 10 people with disabilities employed in Ecuador work in the private sector (71.4%). Of disabled people, people with physical disabilities occupy the most jobs in the formal sector (55.9%), and the least represented are people with psychosocial disabilities (3.5%), as can be seen in Table 4.

Table 3. Participation of people with disabilities affiliated by institution.

Institution	Female	LGBTI	Male	Total
IESS	34.2%	0.0%	63.4%	97.6%
ISSFA	0.1%	0.0%	1.0%	1.1%
ISSPOL	0.2%	0.0%	1.2%	1.3%
Total	34.4%	0.0%	65.6%	100.0%

Table 4. Representation of the employer sector by type of disability.

Sector	Hearing	Physical	Intellectual	Psychosocial	VISUAL	Total
Private	13.1%	38.0%	8.1%	2.4%	9.9%	71.4%
Public	4.1%	17.9%	1.0%	1.1%	4.5%	28.6%
Total	17.1%	55.9%	9.0%	3.5%	14.4%	100.0%

4 Conclusion

The reality of people with disabilities in Ecuador does not differ from the employability data in different countries of the world, as they are twice less likely to be employed than people without disabilities [11]. In terms of occupation by gender, women are the most affected, since it is appreciated that their participation is lower, which confirms the gender disparity in the labor market in Ecuador, in line with what happens in the world labor market. In this sense, the occupancy rate of people with disabilities in Ecuador is 15.3%, much lower than the occupancy rate of people without disabilities, which is 32.1%.

In the future, it is considered to continue with more in-depth studies of the labor market and its barriers to the participation of people with disabilities and their families. In addition, it is considered to study the unemployed population of working age, who continue looking for a job without finding it, which was not considered in this study.

References

1. Hall, J., Kurth, N., Hunt, S.: Employment as a health determinant for working-age, dually-eligible people with disabilities. Disabil Health J. **6**(2), 100–106 (2013)
2. Helliwell, J., Layard, R., Sachs, J., De Neve, J.: World happiness report 2020 (2020). https://happiness-report.s3.amazonaws.com/2020/WHR20.pdf
3. Muller, V., et al.: Association of employment and health and well-being in people with fibromyalgia. J. Rehabil. **83**(3), 7–43 (2017)
4. Kamerāde, D., Wang, S., Burchell, B., Balderson, S., Coutts, A.: A shorter working week for everyone: how much paid work is needed for mental health and well-being?. Soc. Sci. Med. **241**, 112353 (2019)

5. United Nations. United Nations treaty collection: Convention of the rights of persons with disabilities (2016). https://treaties.un.org/Pages/ViewDetails.aspx?src=TREATY&mtdsg_no=IV-15&chapter=4&lang=en
6. Bonaccio, S., Connelly, C.E., Gellatly, I.R., Jetha, A., Martin Ginis, K.A.: The participation of people with disabilities in the workplace across the employment cycle: employer concerns and research evidence. J. Bus. Psychol. **35**(2), 135–158 (2019). https://doi.org/10.1007/s10869-018-9602-5
7. KRAUS, Lewis. Disability Statistics Annual Report. A Publication of the Rehabilitation Research and Training Center on Disability Statistics and Demographics. Institute on Disability, University of New Hampshire (2016)
8. Turcotte, M.: Persons with disabilities and employment (Statistics Canada Cat. No. 75-006-X). Statistics Canada, Ottawa (2014)
9. Eurostat. Disability statistics (2017). http://ec.europa.eu/eurostat/statistics-explained/index.php?title=Disability_statistics
10. Heymann, J., Stein, M.A., de Elvira Moreno, M.R.: Disability, employment, and inclusion worldwide. In: Heymann, J., Stein, M.A., de Elvira Moreno, M.R. (eds.) Disability and Equity at Work, pp. 1–9. Oxford University Press, Oxford (2014)
11. Stoevska, V.: Día Internacional de las Personas con Discapacidades: Cómo afecta la discapacidad a los resultados del mercado laboral (2020). https://ilostat.ilo.org/es/international-day-of-persons-with-disabilities-how-disability-affects-labour-market-outcomes/
12. Consejo Nacional para la Igualdad de Discapacidades. Estadísticas de discapacidad. (2020). https://www.consejodiscapacidades.gob.ec/estadisticas-de-discapacidad/

Physical Ergonomics and Work-Related Musculoskeletal Disorders

Ergonomic Indicators and Physical Workload Risks in Food Production and Possibilities for Risk Prevention

Henrijs Kalkis[1](✉), Ingus Graveris[2], and Zenija Roja[2]

[1] Faculty of Business, Management and Economics, Department of Management, University of Latvia, Aspazijas blvd. 5, Riga 1050, Latvia
henrijs.kalkis@lu.lv

[2] The Research Centre of Ergonomics, University of Latvia, Raina blvd. 19, Riga 1586, Latvia
zenija.roja@lu.lv

Abstract. The food industry is the most important and largest manufacturing industry in Latvia, producing almost a third of all manufacturing output. Employees in a food production enterprises are exposed to a variety of ergonomic risks: monotonous work movements that can be repeated up to 1000 times a day, overloads that exceeds 30 kg in lifting and moving operations, forced working postures, fast work pace. The aim of the study was to identify ergonomic indicators related to physical load for packers in one medium-sized company producing potato starch in Latvia. When summarizing the results of the survey on burden-lifting rates, it should be noted that, in a shift, 30% of packers lift the burden from 200 to 500 times, 26% from 40 to 200 times, while 24% note that smaller products are lifted between 500 and 1000 times. Nearly all those surveyed indicated that, during work, body posture was deeply bent and the body twisted. Analysis of physical load of male packers, applying the KIM-LHC method, shows that, when a mass weighing 16 kg and 25 kg is lifted, packers are exposed to risk levels 3 and 4 respectively. According to KIM-MHO assessment, in works where frequent hand movements and monotonous working operations are observed, the results presented a risk level 3. This could be explained by a high intensity of work. The results obtained by the QEC method show that the shoulder girdle, arms, hands, and the neck of packers who lift and mix a product with a mass of 25 kg, are at very high risk at work. The essential ergonomic indicators during physical load and repeated monotonous work operations are obtained analyzing the physical load of employees with the KIM methods and the QEC method in order to determine the load of individual body parts during work.

Keywords: Ergonomics · Indicator · Food production · Physical load

1 Introduction

The food industry is the most important and largest manufacturing industry in Latvia, producing almost a third of all manufacturing output. Despite the rapid industrialization

R. S. Goonetilleke et al. (Eds.): AHFE 2021, LNNS 273, pp. 47–53, 2021.
https://doi.org/10.1007/978-3-030-80713-9_7

in various business operations, workers in developing countries are facing new conditions with a lack of relevant knowledge and skills [1]. Workers in this sector have high rates of morbidity with musculoskeletal diseases (WRMSD), which have a significant impact not only on the health of workers, but also on the economic performance of the organisation. Generally, WRMSDs are caused by ergonomic risks in the working environment, as employees in food industry are exposed to awkward postures, standing, frequent hand movements, monotonous work operations, and physically hard work. Permanently working in such conditions, employees rapidly experience fatigue and even occupational accidents, such as slipping or tripping [2–5]. WRMSD are often associated with long-term absence of employees due to diseases [6]. Scientists point out that for the development of WRMSD the length of service is more important than the age of the employees, and therefore, the higher the number of years worked in the profession, the higher is the incidence of WRMSD [7]. Scientists have also demonstrated that the WRMSD in older workers are more prolonged than in the younger ones [8]. In some cases, lower back pain can be accompanied by changes in anatomical structures, such as leg length asymmetry [9].

The prevalence of pain and discomfort in food industry is relatively high. For workers, whose length of service is 5 years and more, pain or discomfort is more common in the following parts of the body: knees, back, shoulders and arms [10]. Studies have shown that pain and discomfort are declining for employees when they retire [11] and that due to musculoskeletal pain, employees have limited daily activities [12, 13].

WRMSD can be caused not only by physical factors but also by organizational and psychosocial risks at work. Stress at work and outside it has an impact not only on health but also on the performance of work [14] and therefore changes person's attitude towards themselves: a person starts taking excessive alcohol, smoking, becoming aggressive, sleep disorders appear, as well as anxiety and depression can occur [15].

Taking into account the development of modern technologies in the food industry, there is a need to identify ergonomic indicators, which are suitable for enabling the comprehensive and sustainable management of food processes. The ergonomic indicators could be used as organisational tools for the evaluation and improvement of ergonomics management operations of organisations.

The aim of the study was to identify ergonomic indicators related to physical load for packers in one medium-sized company producing potato starch in Latvia.

2 Materials and Methods

The company has two lines for potato starch packaging. It is packed manually in packs of different sizes, standing, in bent body position. Packaging includes the following manual handling: laying out the product on a pallet, lifting the product from scales, mixing, packaging, arranging packs in boxes.

Participants. The survey involved 64 packers - 25 males and 39 females (see Table 1). The inclusion criteria were as follows: age, length of service, health status after mandatory examination, and full consent to participate in the study. The exclusion criteria were acute pain in different body parts, undergone any musculoskeletal surgery, having not been to mandatory medical examination.

Table 1. Background factors for the packers of potato starch packing lines

Profession/Length of service (years)	N	Mean age ± SD	Range	Mean length of service ± SD
Packers/0–5	19	38.05 ± 7.13	26–51	3.21 ± 0.98
Packers/6–10	19	42.25 ± 7.45	34–59	6.80 ± 1.24
Packers/11–20	17	44.33 ± 5.09	34–55	14.83 ± 3.03
Packers/21–35	6	55.83 ± 3.19	51–60	29.00 ± 4.60
Packers/35+	3	58.33 ± 2.52	56–61	35.49 ± 1.15
Packers (total)	64	43.58 ± 8.46	26–61	11.29 ± 9.48

The physical load at work was analyzed for 15 potato starch packers basing on the following manual work:

(1) Placing of a 16 kg heavy product unit on the pallet. One pallet contains 45 units. In the shift, 7 pallets must be loaded.
(2) Transfer of a 25 kg heavy product unit from scales to a sliding tape and placing on a pallet.
(3) Mixing the product in a mixer: every 30 min the packer pours the raw materials into the mixer from sacks weighing 25 kg, together approximately 5400 kg in a shift.
(4) Placing of small packages on a pallet. Products with a mass of between 0.5 and 1.5 kg are placed on the pallet. During a shift, 8 pallets with 1296 product units are packed up.

The following research methods were used:

Survey. The survey provided information on: gender, age, length of service of employees in the current position, questions about the compliance/non-compliance of the workplace to comfortable working conditions, discomfort or pain in certain parts of the body after work, freedom to take a decision in the performance of work duties, support from colleagues and employers, harmful habits (alcohol use, smoking). Respondents were employees of food-producing company employed in food packaging. A total of 90 questionnaires were distributed, of which 64 questionnaires were returned.

The Key Indicator Method [16] was used for assessing and designing physical workloads with respect to manual Lifting, Holding and Carrying of loads (KIM-LHC). This Key Indicator Method considers the manual lifting, holding and carrying of loads ≥3 kg and serves to record relocating, holding and the mere transport of loads. This is determined by allocating a rating point to the individual key indicators according to their intensity.

The KIM-MHO is an important modular supplement to the practical methods assessing the risk from the manual handling of loads in the form of a risks core [17]. Their purpose is the recognition and removal of job design deficits. The risk assessment is carried out in two stages. The first stage is the ordinal scaled description of workload items. The second stage is the evaluation of the degree of probability of physical overload.

The quick exposure check (QEC) method was applied in the research in order to carry out quick identification and assessment of the influence of the workload on the body parts of packers [18]. The method evaluates the back position, back movement, shoulder/hand position, hand base position/movement, neck motion. Characteristics such as driving, vibration, pace of work, and stress are also taken into account. The exposure level was determined according to the results obtained. The QEC Calculator was used for the calculation.

3 Results and Discussion

The survey showed that there are more females in packaging than males, representing 61% (n = 39) and 25% (n = 25) respectively. The age groups were divided as follows: between 26 and 35 years, 39% (n = 25), between 36 and 50 years 39% (n = 25), and between 51 and 61 year 20% (n = 13). More employees have the length of service of 0 to 5 years and between 6 and 10 years, representing 60% of those surveyed, and the length of service of 17 employees is between 11 and 20 years. Packers noted that during the work the highest workload is experienced in the lower back (54%), 43% points to the leg, 25% to the shoulder, 15% all over the arm, and 35% to the hands and fingers. Packers – males (61%) admit that the mass to be lifted more frequently is between 20 and 30 kg, 28% points out that it is between 10 and 20 kg and 5% - lift 40 kg and more, but a mass of less than 10 kg is lifted by only 6% of employees. Female packers (32%) indicate that daily they lift from 15 to 25 kg, 29% - from 10 to 15 kg, between 5 and 10 kg - 14%, and up to 5 kg - 25%.

When summarizing the results of the survey on burden-lifting rates, it should be noted that, in a shift, 30% of packers lift the burden from 200 to 500 times, 26% from 40 to 200 times, while 24% note that smaller products are lifted between 500 and 1000 times. Nearly all those surveyed indicated that, during work, body posture was deeply bent and the body twisted. Only 13% noted that lifting a burden, body posture is straight and not bent. Survey data show that packers whose length of service is more than 5 years are subjected to physical load, awkward postures, and frequent hand movements at work. The prevalence of pain and discomfort in individual parts of the body after work is relatively high. This is also in line with the results of other scientists' studies on the prevalence of WRMSD in food industry. A study of pain or discomfort in workers in an Indian nut processing company shows that employees have a high incidence of pain (28.5%). Complaints from employees are mainly about pain in the lower back, knees and shoulder girdle, and the length of work in the profession is more than 5 years [7, 10]. Other studies have indicated that employees in food industry are exposed to long-term awkward postures and physical loads [19]. The results of the physical load analysis for packers are shown in Table 2.

Analysis of physical load of male packers, applying the KIM-LHC method, shows that, when a mass weighing 16 kg and 25 kg is lifted, packers are exposed to risk levels 3 and 4 respectively. During work, packers work in a bent-body posture, which is very burdensome. However, in the case of females lifting small packs, the load corresponds to risk level 2 since the product is easy to grip and not heavy. Lifting of a product if the mass indicator and position indicators are the same but the intensity indicator is variable, gives a higher score and therefore a higher level of risk.

Table 2. KIM-LHC method risk scores (RS), standard deviation (SD), risk range (R) for packers (n = 15)

Work process	N	RS ± SD	R I – V
1. Placing of a product unit of 16 kg mass on a pallet. One pallet contains 45 units. During a shift, 7 pallets must be loaded. Body posture – deep bending. The work is done by males	3	36.9 ± 3.6	III
2. Moving of a 25 kg heavy product unit from scales to a sliding tape and placing on a pallet. During a shift, 8 pallets, each containing 40 units must be loaded. Body posture – deep bending. The work is done by males	4	51.4 ± 2.1	IV
3. Mixing the product in a mixer: the packer pours the bag of raw materials weighing 25 kg in the mixer every 30 min, together approximately 5400 kg in a shift. Body posture – bent. The work is done by males	4	60.7 ± 3.4	IV
4. Placing of small packs (0.5 to 1.5 kg) on the pallet. During a shift, 8 pallets are loaded, 1296 product units on each pallet. Body position – slightly bent. The work is done by females	4	24.8 ± 2.4	II

According to KIM-MHO assessment, in works where frequent hand movements and monotonous working operations are observed, the results presented a risk level 3 (Table 3). This could be explained by a high intensity of work. The workload of individual body parts during the work process is shown in Table 4.

Table 3. KIM-MHO method risk scores (RS), standard deviation (SD), risk range (R) for female packers who load small packs (n = 4)

Work process	RS ± SD	R I – V
Placing of small packs (0.5 to 1.5 kg) on the pallet. During a shift, 8 pallets are loaded, 1296 product units on each pallet	45 ± 1.0	III

The results obtained by the QEC method show that the shoulder girdle, arms, hands, and the neck of packers who lift and mix a product with a mass of 25 kg, are at very high risk at work. On the other hand, the body parts mentioned above are less loaded when lifting lighter packs.

This indicates that physical load and awkward body postures during work, as well as work intensity, have a negative impact on the individual body parts of employees. Long-term working with such a load and high working intensity in the course of time can develop WRMSD. The results are consistent with studies by other scientists on WRMSD in relation to monotonous work operations, awkward body postures, heavy manual work, etc. [20].

Table 4. QEC method exposure scores for packers (n = 15)

	Placing of 16 kg mass on the pallet	Moving of 25 kg mass from scales to tape	Product mixing in a mixer	Placing of small packs on the pallet
N	3	4	4	4
Back (moving)	23	30	29	28
Shoulder/ arm	32	40	43	29
Wrist/ hand	44	45	44	36
Neck	12	14	16	14
Work pace	1	1	1	1
Stress	1	1	1	1
Total score	113	131	134	109
Exposure level	High	Very High	Very High	High
Risk range (R), I – V	**II**	**III**	**III**	**II**

4 Conclusions

The essential ergonomic indicators during physical load and repeated monotonous work operations are obtained analyzing the physical load of employees with the KIM methods and the QEC method in order to determine the load of individual body parts during the work. The results of these methods are measurable and easily understandable. The study will continue to focus on psychosocial indicators related to ergonomic risks at work.

Acknowledgments. This research has been supported by the project "Ergonomic stress indicators in contemporary technological work environment and possibilities of its improvement in social-technical system "Human-Machine-Environment"", Agreement No. Nr.1.1.1.2/VIAA/3/19/546.

References

1. Hämäläinen, P.: Systems Intelligence in Leadership and Everyday Life. Systems Analysis Laboratory, p. 320 (2007)
2. Deros, B.M., Daruis, D.D.I., Basir, I.M.: A study on ergonomic awareness among workers performing manual material handling activities. Procedia. Soc. Behav. Sci. **195**, 1666–1673 (2015)
3. Kumar, S.: The back compressive forces during maximal push-pull activities in the sagittal plane. J. Hum. Ergol. **23**, 133–150 (1994)
4. Vieira, E.R., Kumar, S.: Working postures - a literature review. J. Occup. Rehabil. **14**, 143–159 (2004)
5. Syed Ali, S.A., Kamat, S.R., Mohamed, S.B.: Analysis awkward posture at food production activity using RULA assessment. J. Fundam. Appl. Sci. **10**(1S), 619–632 (2018)

6. Laaksonen, M., Pitkäniemi, J., Rahkonen, O., Lahelma, E.: Work arrangements, physical working conditions, and psychosocial working conditions as risk factors for sickness absence: Bayesian analysis of prospective data. Ann. Epidemiol. **20**, 332–338 (2010)
7. Intranuovo, G., De Maria, L., Facchini, F., et al.: Risk assessment of upper limbs repetitive movements in a fish industry. BMC. Res. Notes **12**, 354 (2019)
8. Siukola, A., Virtanen, P., Huhtala, H., Nygard, C.H.: Absenteeism following a workplace intervention for food industry workers. Occup. Med. **61**, 583–585 (2011)
9. Rannisto, S., Okuloff, A., Uitti, J.: Leg-lenght discrepancy is associated with low back pain among those who must stand while working. BMC Musculoskelet. Disord. **16**, 110 (2015)
10. Girish, N., Ramachandra, K., Maiya, A.G.: Prevalence of musculoskeletal disorders among cashew factory workers. Arch. Environ. Occup. Health **67**(1), 37–42 (2012)
11. Neupane, S., Virtanen, P., Luukkaala, T.: A four-year follow-up study of physical working conditions and perceived mental and physical strain among food industry workers. Appl. Ergon. **45**, 586–591 (2014)
12. Kamaleri, Y., Natvig, B., Ihlebaek, C.M., Benth, J.S., Bruusgaard, D.: Number of pain sites is associated with demographic, lifestyle, and health-related factors in the general population. Eur. J. Pain **12**(6), 742–748 (2008)
13. Saastamoinen, P., Leino-Arjas, P., Laaksonen, M., Martikainen, P., Lahelma, E.: Pain and health related functioning among employees. J. Epidemiol. Community Health **60**(9), 793–798 (2006)
14. Alam, I., Sharmin, S.A., Kim, K.H., et al.: Proteome analysis of soybean roots subjected to short-term drought stress. Plant Soil **333**, 491–505 (2010)
15. Cartwright, S., Cooper, C.L.: Managing Workplace Stress, p. 185. Sage Publications, Inc. (1997)
16. Steinberg, U., Liebers, F., Klußmann, A., et al.: Bericht über die Erprobung, Validierung und Revision: Leitmerkmalmethode Manuelle Arbeitsprozesse. Bundesanstalt für Arbeitsschutz und Arbeitsmedizin: Dortmund (2012)
17. Klussmann, A., Steinberg, U., Liebers, F., et al.: The Key Indicator Method for Manual Handling Operations (KIM-MHO) - evaluation of a new method for the assessment of working conditions within a cross-sectional study. BMC Musculoskelet. Disord. **11**, 272 (2010)
18. David, G., Woods, V., Li, G., Buckle, P.: The development of the Quick Exposure Check (QEC) for assessing exposure to risk factors for work-related musculoskeletal disorders. Appl. Ergon. **39**(1), 57–69 (2008)
19. Soytas, U.: Physical and ergonomic hazards in the textile, chemical, foodl, metal products, and woodworking industries in Turkey. Int. J. Occup. Environ. Health **12**(1), 35–41 (2006)
20. Costa, B.R., Viera, E.R.: Risk factors for work-related musculoskeletal disorders: a systematic review of recent longitudinal studies. Am. J. Ind. Med. **53**(3), 285–323 (2010)

Assessment of Muscle Fatigue and Potential Health Risk of Low Back Pain Among Call Center Workers

Sunisa Chaiklieng(✉) and Worawan Poochada

Department of Environmental Health, Occupational Health and Safety, Faculty of Public Health, Khon Kaen University, Khon Kaen, Thailand
{csunis,worapooc}@kku.ac.th

Abstract. The aims of this study were to investigate muscle fatigue and the risk of low back pain (LBP) development among call center workers. Cornell Musculoskeletal Discomfort Questionnaires (CMDQ), and used with Rapid Office Strain Assessment (ROSA) for prediction of LBP development in a risk matrix. Muscle fatigue was measured by electromyography. Workers had a moderate to high level of ergonomic risk according to ROSA. 43.3% of workers had muscular discomfort. The potential risk level of LBP development was found to be moderate for 23.4% and high for 20.0%. The risk scores of LBP development significantly correlated with discomfort levels ($r = 0.914$). Electromyogram measurements indicated that over 80% of workers had muscle fatigue on the lower back. Workers at high ergonomic risk were more likely to have muscle fatigue than those at lower risk. These methods of risk assessment could be applied to identify muscle fatigue and LBP development among similar groups.

Keywords: Electromyography · Fatigue · Health risk assessment · Low back pain

1 Introduction

A call center is an organization that handles customer enquiries both telephonically and by email, by searching databases and recording customer information through a computer. Workers spend most of their time working in a workspace that includes a computer desk and chair. Therefore, management of shift work and interactive voice response (IVR) are used. They may also have eye-, ear-, and throat-related health problems [1] along with musculoskeletal disorders (MSDs). MSDs are caused by awkward posture (e.g. repeatedly or frequently using fingers, wrists and arms in a twisted posture and always looking at a screen for working with a computer) [2] or inappropriate workstation (e.g. insufficient working space and lack of forearm support) [3].

Workers who have a mainly static posture, such as prolonged sitting or standing, are at risk of MSDs [4]. Several studies have reported high prevalence of MSDs among call center workers. Poochada and Chaiklieng [5] showed that prevalence of neck, shoulder and back pain in the last three months among call center workers was 83.8% (95%CI:

R. S. Goonetilleke et al. (Eds.): AHFE 2021, LNNS 273, pp. 54–61, 2021.
https://doi.org/10.1007/978-3-030-80713-9_8

78.8–88.7), of which back pain was the highest at 35.2%. Likewise, the study of Subbarayalu [1] showed the highest prevalence of back pain in last 12 months among call center workers at 40.0%. This study showed that back pain was a health impact found among call center workers.

MSDs can be controlled by using an ergonomic risk assessment that considers posture, force, frequency and duration [6]. Ergonomic risk assessment has two approaches [7, 8]; 1) subjective assessment by self-reported questionnaire, such as Cornell Musculoskeletal Discomfort Questionnaires (CMDQ) for identification of severity of discomfort or pain, and 2) objective assessment by 2.1) Observation assessment as objective assessment, such as Rapid Office Strain Assessment (ROSA) of ergonomic risk factors, and 2.2) Direct measurement techniques, such as electromyography (EMG) to measure muscle fatigue [9].

Health risk assessment (HRA) considers the probability and severity of heath impact from exposure to health hazards. MSDs are known to mostly be caused by ergonomic problems and a health risk matrix could be developed from observation and self-reported questionnaire [10]. Previous studies regarding risk assessment had guidelines for only one approach to using ergonomics for estimating risk of LBP development; in fact, there were no methods on the two approaches of subjective and objective assessment for LBP prediction. Thus, the aims of this study were to investigate the risk of low back pain development and muscle fatigue among call center workers.

2 Methods

This was a cross-sectional descriptive study conducted among 30 call center workers from a number of companies in Khon Kaen province, Thailand. The survey was based on call center workers who had worked in their jobs for more than six months, with a working time of at least 32 h per week and at least four hours spent at a computer per day. Exclusion criteria were workers who had a musculoskeletal disorder at the present time and skin irritation.

2.1 Research Tools and Data Collection

1) ROSA was applied from Sonne et al. [11] to be used in Thai version validated by the previous study [12], which analyzed the sitting posture, work station (chair height, pan depth, armrest and back support), computer (monitor, mouse and keyboard), telephone and duration of time spent in each posture or activity. ROSA was used to identify the chance of ergonomic factor exposure. There were four risk levels, which were low, moderate, high and very high. The scores of each risk level were as follows; low risk level = score of 1–2 points, moderate risk level = score of 3–4 points, high risk level = score of 5–7 points and very high risk level = score of 8–10 points.
2) CMDQ was applied from Hedge et al. [13] which identified the severity of discomfort or pain from ergonomic factors. The discomfort level was classified by how often the discomfort occurred (never, 1–2 times last week, 3–4 times last week, once every day, several times every day), how uncomfortable it was (slightly, moderately,

or very uncomfortable) and how it interfered with the ability to work (not at all, slightly interfered, substantially interfered) among workers. There were five levels of discomfort, which were no pain (score of 0 points), low (score of 1–5 points), moderate (score of 6–13.5 points), high (score of 14–39.5 points) and very high (score 40–90 points).

3) The health risks of low back pain were assessed by using the matrix of back pain prediction from the previous findings [14] which considered the ROSA levels and discomfort levels, which were classified into five levels; acceptable risk (score of 0–2 points), low risk (score of 3–4 points), medium risk (score of 6–8 points), high risk (score of 9–12 points) and very high risk (score of 15–20 points)
4) The electromyography (EMG) model was an ME6000 8-channel model which included three parts: recorder, amplifier and surface electrode. MegaWin program version 2.0 was used for EMG signal recording for the median frequency (MF) at the erector spinae muscle (left and right) of the call center worker during the working period. The continuous measurement was performed for 10 min in every two hours of the working period (8.00 a.m.–05.00 p.m.). The MF/time slope was calculated by linear regression of median frequency between four and eight working hours. A negative MF/time slope represented muscle fatigue in this study, following the study of Hui et al. [15].

2.2 Data Analysis

Data were analyzed by STATA program version 10.1. MegaWin program version 2.0 was used to analyze muscle fatigue. Pearson Product Moment Correlation Coefficient was used for correlations between the health risk score and CMDQ level, health risk score and ROSA level, and muscle fatigue (MF/time slope) and ROSA level. This study obtained ethical approval from Khon Kaen University Ethics Committee, Thailand, no. HE572131.

3 Results

3.1 ROSA and CMDQ

Call center workers had only two ergonomic risk levels measured by ROSA, which were the moderate level (56.7%) and high level (43.3%). The score of ROSA ranged from 3 points to 7 points, as shown in Table 1.

Workers had complaints of discomfort at medium (26.7%) and low levels (16.7%) of both the left and right lower back according to CMDQ, as shown in Table 2.

3.2 Health Risk of Low Back Pain

From the matrix of health risk on low back pain development, one group of workers had an acceptable health risk (30.0%) or a low risk level (26.6%). Another group of workers had a medium risk level (23.4%) or a high risk level (20.0%) (as shown in Table 3).

The correlations of the health risk score of LBP development with the ROSA level and with the CMDQ of low back level were presented as coefficient (r) = 0.516 and 0.914, respectively. A high correlation was found from the risk matrix between the health risk score and the CMDQ of low back level, as shown in Fig. 1.

Table 1. The opportunity of ergonomic factor exposure was assessed by ROSA (n = 30).

Risk level	Number (%)	Mean score (95% CI)
Low (1–2 points)	0 (0.0)	0
Moderate	17 (56.7)	3.5 (3.3–3.8)
3 points	8 (26.7)	
4 points	9 (30.0)	
High	13 (43.3)	5.3 (4.9–5.7)
5 points	10 (33.3)	
6 points	2 (6.7)	
7 points	1 (3.3)	
Very high (8–9 points)	0 (0.0)	0

Table 2. The severity of low back muscular discomfort was assessed by CMDQ (n = 30).

Low back discomfort level	Left low back (%)	Right low back (%)
No pain	19 (63.3)	17 (56.7)
Low	2 (6.7)	5 (16.7)
Moderate	7 (23.3)	8 (26.7)
High	2 (6.7)	0 (0.0)
Very high	0 (0.0)	0 (0.0)

Table 3. Call center workers had a health risk of low back pain at different levels (n = 30).

Health risk level of LBP		**ROSA level**			
		Low	Moderate	High	Very high
Low back discomfort level	Very high	0 (0.0)	0 (0.0)	0 (0.0)	0 (0.0)
	High	0 (0.0)	2 (6.7)	0 (0.0)	0 (0.0)
	Moderate	0 (0.0)	2 (6.7)	6 (20.0)	0 (0.0)
	Low	0 (0.0)	4 (13.3)	3 (10.0)	0 (0.0)
	No pain	0 (0.0)	9 (30.0)	4 (13.3)	0 (0.0)

Remark: Red colour is very high risk level, dark orange is high risk, orange is medium risk, yellow is low risk, green is acceptable risk [14].

3.3 Electromyography on the Elector Spinae Muscle

Median frequencies (MF) of electromyography on the elector spinae muscle, which were measured from 8.00 a.m. to 05.00 p.m., were found to be highest at the beginning on the left lower back (39.4 Hz.) and decreased to 32.8 Hz. (lowest) at the end. The average

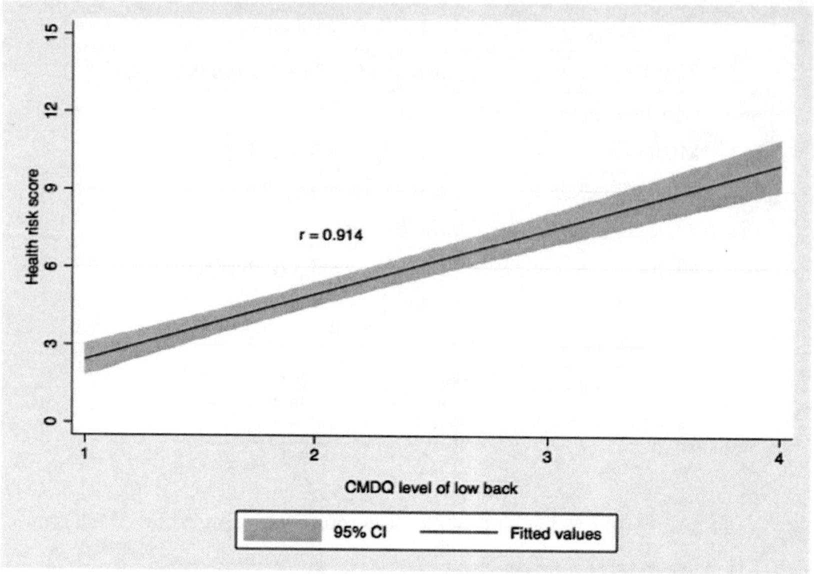

Fig. 1. Correlation between health risk score and CMDQ level of low back pain (n = 30)

MF values of the right lower back were 35.2 Hz. and 34.4 Hz. at 8.00 a.m. and at 05.00 p.m., respectively.

Thus, muscle fatigue values (MF/time slope) throughout the eight hours of work (8.00 a.m.–05.00 p.m.) of the left and right lower back were −0.648 and −0.027, respectively. For muscle fatigue of left and right lower back in the four hours of morning work (8.00 a.m. to 12.00 a.m.) MF/time slope = −2.917 and −1.725, respectively (see Table 4).

Table 4. Median frequency (MF) of electromyogram on the lower back in each duration of measurement (n = 30)

MF (Hz.)	Time					MF/time slope	
	8.00 a.m	10.00 a.m	12.00 a.m	03.00 p.m	05.00 p.m	8.00 a.m. to 12.00 a.m	8.00 a.m. to 05.00 p.m
Left	39.4	34.8	32.7	34.9	32.8	−2.917	−0.648
Right	35.2	35.4	34.1	36.5	34.4	−1.725	−0.027

There were 80% and 60% of workers who had muscle fatigue of the left and right lower back from 8.00 a.m. to 05.00 p.m. and 86.7% and 80.0% of workers who had muscle fatigue of the l MF/time eft and right lower back from 8.00 a.m. to 12.00 a.m.

4 Discussion

4.1 Health Risk of Back Pain

Objective assessment for ergonomic risk identification by ROSA indicated that all call center workers in this study were exposed to ergonomic risk from a moderate to high level. Due to the nature of office work, call centers have similar activities, which involve prolonged posture and repetitive work with a computer, and a duration of working with a computer of up to four hours, and insufficient resting time during the workday [16]. These workers are exposed to an inappropriate work environment, such as high chairs, non-armrest or nonadjustable chairs, and lack of back support or having to lean the body forward [17]. A previous study found that a non-changing posture during work and in the workplace setting was correlated with low back pain [2]. CMDQ showed that the muscle discomfort seen in this study could also confirm musculoskeletal health problems among call center workers. By using the matrix of back pain prediction from the previous finding [14], it could be shown that 43% of workers were at a medium to high risk level of LBP development with a good correlation with the perceived level of low back discomfort.

4.2 Electromyography on Elector Spinae Muscle

This study showed that workers had muscle fatigue from eight working hours. Particularly, the peak of fatigue was found at the end of the morning period before lunch. More workers had muscle fatigue of the left lower back (80.0%) than the right lower back (60.0%). According to the CMDQ, more workers complained about muscle discomfort of the right lower back (21.3%) than the left lower back (19.5%). This study did not find a statistically significant correlation between muscle fatigue by EMG and discomfort level by CMDQ. The reasons might be explained by a perception of pain or discomfort that correlated with pain tolerance, which differed among workers and depended on their fitness background [18].

The MF/time slope values of the left and right lower back in the first four working hours (8.00 a.m. to 12.00 a.m.), which were −2.917 and −1.725, were more negative than the average MF/time slope values of the left and right lower back for eight working hours (8.00 a.m. to 05.00 p.m.), which were −0.648 Hz. And −0.027 Hz., respectively. There was a one-hour lunch break from 12.00 p.m.–01.00 p.m. The maximum fatigue was found in this study after four hours of working before the lunch break. It is possible that if there were no lunch break or longer hours without a break, more fatigue would appear. That is the reason why fatigue per eight working hours was lower than the average MF per four working hours in the morning. This result could also be explained by the reduction in muscle discomfort from the program of muscle stretching during break time [19].

4.3 Correlations Between Potential Health Risk of Low Back Pain and EMG

Regarding the health risk score in the developed matrix of low back pain development, a high correlation ($r = 0.914$) between the CMDQ discomfort level and health risk score was found. Moreover, the trend of correlation between muscle fatigue from four working hours measured by EMG and the ROSA level confirms the usefulness of the health risk matrix when considering an ergonomic assessment with ROSA among call center workers. The previous health risk matrix had been proved by a new case development after following-up on the incidence of pain development [10]. Therefore, this basic risk assessment on low back pain is useful for the surveillance program and could confirm the guidance for the implementation stage of back pain prevention as per the previous report [19].

5 Conclusion

The potential risk of low back pain development (from the risk matrix) was found to be moderate for 23.4% of call center workers and high for 20.0% of them. This health risk score had the highest correlation with discomfort (CMDQ) levels ($r = 0.914$). Most call center workers (80.0%) had muscle fatigue of the left lower back, which showed that workers at high ergonomic risk were more likely to have muscle fatigue than those of lower risk. These methods of risk assessment could be applied in the prediction of low back pain development, which would be useful for health screening to prevent low back pain among call center workers. An implementation stage incorporating muscle stretching during break time is suggested for workers with similar work characteristics with regard to the health risk reduction of muscle fatigue and LBP development.

Disclosure Statement. The author declares that they have no competing interests.

References

1. Subbarayalu, A.V.: Occupational health problems of call center workers in India: a cross sectional study focusing on gender differences. J. Manag. Sci. Pract. **2**(1), 63–70 (2013)
2. Chaiklieng, S., Suggaravetsiri, P., Stewart, J.: Incidence and risk factors associated with low back pain among university office workers. Int. J. Occup. Saf. Ergon. 1–7 (2020). https://doi.org/10.1080/10803548.2019.1706827
3. Errico, A.D., Caputo, P., Falcone, U., Fubini, L., Gilardi, L., Mamo, C., et al.: Risk factors for upper extremity musculoskeletal symptoms among call center employees. J. Occup. Health **52**, 115–124 (2010)
4. Keawduangdee, P., Puntumetakul, R., Siritaratiwat, W., Boonprakob, Y., Wanpen, S., Thavornpitak, Y.: The prevalence and associated factors of working posture of low back pain in the textile occupation (fishing net) in Khon Kaen province. Srinagarind Med. J. **26**(4), 317–324 (2011)
5. Poochada, W., Chaiklieng, S.: Prevalence and neck, shoulder and back discomfort among call center workers. Srinagarind Med. J. **30**(4), 282–289 (2015)
6. Chutiwatpongstron, S., Chaiklieng, S.: Ergonomics risk assessment and ergonomic factors among computer users in Tambon Health Promoting hospital, Nongkhai province. KKU J. Public Health Res. **8**(3), 64–72 (2015)
7. Dempsey, P.G., McGorry, R.W., Maynard, W.S.: A survey of tools and methods used by certified professional ergonomists. Appl. Ergon. **36**, 489–503 (2005)
8. David, G.C.: Ergonomic methods for assessing exposure to risk factors for work-related musculoskeletal disorders. Occup. Med. **55**, 190–199 (2005)

9. Cifrek, M., Medved, V., Tonkovic, S., Ostojic, S.: Surface EMG based muscle fatigue evaluation in biomechanics. Clin. Biomech. **24**, 327–340 (2009)
10. Chaiklieng, S., Krusan, M.: Health risk assessment on neck pain among office workers. Procedia Manuf. **3**, 4941–4947 (2015)
11. Sonne, M., Villalta, D.L., Andrews, D.M.: Development and evaluation of an office ergonomics risk checklist: ROSA rapid office strain assessment. Appl Ergon. **43**, 98–108 (2012)
12. Krusan, M., Chaiklieng, S.: Ergonomics risk assessment in university office workers. KKU Res. J. **19**(5), 696–707 (2014)
13. Hedge, A., Morimoto, S., McCrobie, D.: Effects of keyboard tray geometry on upper body posture and comfort. Ergon. **42**(10), 1333–1349 (1999)
14. Chaiklieng, S., Suggaravetsiri, P.: Health risk assessment matrix for back pain prediction among call center workers. In: Ahram, T., Karwowski, W., Vergnano, A., Leali, F., Taiar, R. (eds.) IHSI 2020. AISC, vol. 1131, pp. 372–378. Springer, Cham (2020). https://doi.org/10.1007/978-3-030-39512-4_58
15. Hui, L., Gabriel, Y.F., Yeung, S.S.M., Hui-Chan, C.W.Y.: Evaluation of physical work demands and low back neuromuscular fatigue on nurses working in geriatric wards. Appl. Ergon. **32**, 479–483 (2001)
16. Chaiklieng, S., Suggaravetsiri, P., Boonprakob, Y.: Work ergonomic hazards for musculoskeletal pain among university office workers. Walailak J. Sci. Tech. **7**(2), 169–176 (2010)
17. Poochada, W., Chaiklieng, S.: Ergonomic risk assessment among call center workers. Procedia Manuf. **3**, 4613–4620 (2015)
18. Homsombat, T., Chaiklieng, S.: Physical fitness and muscular discomfort among informal garment female Workers in Udon Thani province. Thailand J. Med. Assoc. Thai. **100**(2), 230–238 (2017)
19. Chaiklieng, S., Poochada, W.: The effectiveness of ergonomics implemented program among center workers. Srinagarind Med. J. **31**(6), 325–331 (2016)

The Effects of the Physical Environment on Employee Wellbeing and Performance: A Case Study on Healthy Architecture in Call Center Interiors

Salih Ceylan(✉)

Department of Architecture, Faculty of Architecture and Design, Bahçeşehir University, North Campus, Ihlamur Yıldız Street No:10, Bl: B Fl: 8, Istanbul, Turkey
salih.ceylan@arc.bau.edu.tr

Abstract. Healthy built environment is an emerging field of interest in architecture. It possesses various approaches regarding the scale and function of the architectural products. Even though health and safety issues such as emergency situations, fire codes, accessibility, and job security have been a part of spatial organization in architectural design for a long time, current understanding of design reveals that healthy architecture is not only about the absence of disease or infirmity, but also about a state of complete physical, mental and social well-being. It consists of various characteristics like possibility-driven, balanced, user fit, and long-term effective environments. This paper tackles with the characteristics of healthy architecture on an interior scale through a survey on call center interiors. Hypothesis of the paper is that human wellbeing centered design of the built environment in call center interiors contribute to employee performance and commitment. To measure the accuracy of the hypothesis, a survey has been conducted among the employees of a call center in Istanbul, Turkey. The outcomes of the survey are discussed in the discussion and conclusion section of the paper.

Keywords: Healthy architecture · Healthy built environment · Call center interior spaces · Wellbeing

1 Introduction

Architecture as the practice responsible for the design of all built environment, is also responsible for the health and wellbeing of its occupants. Although health and safety issues such as emergencies, fire codes, accessibility, and job security have been among the elements of architectural design for a long time, creating healthy built environments is an emerging field of interest in architecture. It possesses many different approaches in terms of scale and function of design. From urban to interior scale and from residential to public buildings, the perspective of health and wellbeing has become one of the predominant determinants in design. Regional decisions for healthy urban areas, a building's location within the urban environment, building orientation, shell design, indoor

R. S. Goonetilleke et al. (Eds.): AHFE 2021, LNNS 273, pp. 62–69, 2021.
https://doi.org/10.1007/978-3-030-80713-9_9

environmental quality, and ergonomics of the furniture and equipment are among the design actions for the creation of healthy environments. Current understanding of design reveals that healthy architecture is not only about the absence of disease or infirmity, but also about a state of complete physical, mental and social well-being [1]. It consists of various characteristics like possibility-driven, balanced, user fit, and long-term effective environments [2].

This paper aims to investigate the characteristics of healthy architecture on an interior scale through a case study on call center workspaces. Call centers are special kinds of workspaces with special characteristics such as being active 24 h a day, having an intensive and flexible use of furniture and equipment, and acoustic challenges. Hypothesis of the paper is that human wellbeing centered design of the built environment in call center interiors contribute to employee performance and commitment. Existing literature shows that healthy environment in the workspaces also contribute to employee performance [3–5]. The reflection of this statement on call center interiors is inspected in the scope of this paper examining following criteria: Connection with the exterior domain, natural daylight use, artificial light use, acoustic quality, indoor air quality, layout of the workspace, finishing materials, and ergonomics of the furniture and equipment. To measure the accuracy of the hypothesis, a survey has been conducted among the employees of a call center in Istanbul, Turkey. The outcomes of the survey are discussed in the conclusion section of the paper.

2 Healthy Architecture

The relationship between the built environment and user health and wellbeing is not a new idea. It has been one of the concerns for architects during the design process. However, as an architectural concept, it has become a popular theme in the last decades, under the influence of the common interest on wellbeing and lifestyle diseases like asthma or Alzheimer's disease. According to Global Wellness Institute, the term wellness architecture defines the concept better, as architecture that *relies on the art and science of designing built environments with socially conscious systems and materials to promote the harmonious balance between physical, emotional, cognitive, and spiritual wellbeing while regenerating the natural environment* [6]. The effects of healthy architecture concept can be seen in different scales and different stages of design. It starts at urban scale through the design of healthy cities for a healthy society. WHO states that *the way neighborhoods and streets are built and designed, the way cities are planned and expanded, how effectively transport provides opportunities for easy and active mobility, are all aspects of healthy urban planning and design that can make a significant difference to the health of individuals and communities* [7]. At building scale, the concerns about healthy architecture starts from the decisions about the building location. Buildings in less densely populated and greener areas of the cities are advantageous regarding the health of the built environment, as they have to deal with less problems than the buildings in densely populated urban areas. Either way, there are strategies for building design that promote healthy architecture. Among those strategies are open space organization and interior-exterior relations, form and orientation of the building mass, design of the building envelope, and building maintenance regarding energy, water and waste management. These strategies match with the criteria of sustainable architecture so that it is possible to state that the principles of healthy architecture also serve for sustainable design. Sustainable planning and design require a holistic perspective based on a comprehensive and integrated approach, and an understanding of the different parts including healthy design principles are important for the whole [8]. Another important criterion for sustainable design is the indoor environmental quality, which is at the same time the most significant factor for healthy architecture. The health and wellbeing of the interior spaces is discussed in the next chapter of the paper.

3 Healthy Interiors - Workspaces

The most visible indicators of healthy architecture are seen in the interior spaces. The healthy interior spaces have been studied from its various aspects by different researchers [9–11]. Studies prove that the most important principles are daylight use, artificial light use, acclimatization, acoustics, materials, and ergonomics. These principles are also among the criteria of green building accreditation systems like LEED, BREEAM, or DGNB. Therefore, one might assume that healthy interior spaces are significant indicators of a healthy built environment.

The indicators of a healthy interior environment mostly reflect on workspaces. They need to have high environmental qualities because employees mostly spend one third of their day in workspaces. Health and safety issues in workspaces have long been studied and they are already a part of the design workflow [12]. However, emerging studies show that long-term health effects are as important as short-term effects, so that they must be handled with serious considerations. Emerging literature also shows that healthy environment in the workspaces also contributes to employee performance and affects wellbeing and productivity [10, 13]. This study focuses on call center interior spaces for the investigation of health and well-being issues in workspaces regarding employee satisfaction and productivity.

3.1 Healthy Workspaces at Call Center Interiors

Call center interior spaces have certain characteristics and requirements that make them a special type of workspaces. Firstly, call centers as workspaces are active 24 h a day. Consequently, health of the workspace is more important than any other workspace as it potentially has more effect on the employee performance. Additionally, call center employees work with shifts which means there is no workstations specified for each employee. That makes the principle of flexibility very important. Workstation design and ergonomics must be flexible for the use of different people. Ergonomics of the furniture and equipment in the call centers are highly important also because employees are occupying them for long times. Furthermore, success of a call center company mostly depends on the employee performance. Therefore, it is important to make the employees feel well at work to be productive. Moreover, as the work of call center agents is based on phone conversations, the acoustic quality in the call center environments is also very important. Accordingly, the use of technology might be helpful to protect the privacy of the customers and the acoustic quality of the workspace in the same time. As a result of these specific features of call centers, some strategies must be developed to design better and more productive workspaces for the call center agents. These strategies differentiate in scale and design fields, from the façade of the building to technological equipment. The strategies are proposed as follows:

Use of Daylight and Artificial Light. Proper use of daylight has a positive effect on the occupants, especially in workspaces. Workspaces with high occupants' satisfaction for spatial comfort prioritize access to daylight [3]. Even though the job of a call center agent is not related to daylight, its effect on the employee motivation is important. Additionally, providing visual connection for the employee to exterior space by providing wall openings in proper locations may improve their performance. Van Esch et al. state that offices with view possibilities to the outside contribute to the employee wellbeing [14]. Therefore, a call center interior must have enough connection with daylighting opportunities.

Use of Natural and Artificial Acclimatization. A good quality of air is one of the most important requirements of healthy indoor environments. At call center workspace it is even more important because employees spend long hours in there. Fresh air and natural ventilation are preferred, but for the regulation of temperature, proper artificial acclimatization is also a good way to achieve indoor air quality.

Layout and Spatial Organization. Call centers are workspaces where the customers have no visual access. For that reason, the layout and spatial organization are considered as secondary issues. However, for employee performance and satisfaction a good layout design and spatial organization is required. Well-planned personal spaces, sufficient circulation areas, and proper socialization spaces for the employees are important features a call center workspace needs to include.

Materials, Colors, and Furniture. Finishing materials and fixtures are the visible interface of workspaces. Their quality might reflect the overall quality of the workspace. Additionally, the suitable use of colors and textures in an interior space might contribute to the performance of the employees. Especially if they are included in the decision-making process of the materials and colors used in the workspace, as it would create a sense of belonging among the employees. Generating higher levels of employee commitment in call centers is a critical factor to their successful operation [15]. In call centers, the turnover rates of the employees are relatively high, and the design phase of the workspace can be useful as an opportunity in this regard.

Acoustic Quality. Acoustics is a prominent feature of workspaces, as mentioned by Jablonska and Trocka-Leszczynska [16]. It is extremely important for call centers because the work goes on through verbal conversations mostly on the phone. For acoustic quality, sound absorbing materials become significant and they must be used properly within the well- designed interior space. Additionally, technological devices might become useful for maintaining acoustic quality and keep the noise level on the desired level. That would also help to isolate the background noise.

4 The Questionnaire

The effects of the health of the physical environment on employee well-being and productivity were investigated through a short questionnaire among call center employees. The total number of participants in the questionnaire was 358, 216 (60.7%) of them being male, and 137 (38.5%) female employees. 309 (86.3%) of all participants had been working in a call center for less than 3 years, as 49 (13.6%) were working there for a longer time. The average age on the participants was 27.3 years, the youngest being 22, and the oldest 36 years old. The majority of the participants, 306 out of 358 people (85.4%), were between 25 and 30 years old. Following the general demographic information, the questionnaire examined the opinions of the participants on the physical environment of the call center they were working in. The results of the questionnaire are presented in the next chapter.

4.1 Results and Outcomes

The study on the employee opinions on the physical environment revealed important results and outcomes. The questionnaire followed a 5-point Likert scale for all questions except the last one, which was a multiple-choice question about the preference of the participants. Figure 1 shows the summary of the results for all opinion questions with the Likert scale. The first part investigated whether the employees think that the physical environment in their workspace has an effect on their performance at work. The majority, 256 people thought that it has a significant effect. 71 people stated that it has an effect, as 22 people were neutral about it. Only 2 people disagreed, and 5 people strongly disagreed with this statement. With a mean value of 4.60 out of 5.00 points, the call center agents clearly agree on the effects of the physical environment on their work performance.

quest no	topic	5-important effect	4- somewhat effect	3-neutral	2-no effect	1-no effect at all	mean value
1	overall evaluation	256	71	22	2	5	4.60
2	natural lighting	253	71	24	5	4	4.58
3	artificial lighting	185	64	68	26	13	4.07
4	natural ventilation	276	59	19	1	2	4.70
5	artificial acclimatization	226	67	52	9	3	4.41
6	acoustic quality	283	49	22	2	1	4.71
7	layout design	294	39	19	3	0	4.75
8	materials and furniture	247	71	29	10	0	4.55

Fig. 1. Summary of the results for the questionnaire (source: author).

Second part of the questionnaire was about the use of light in the workspace. First question investigated the natural daylight, and the second question the artificial lighting. The participants evaluated both lighting methods to have an important effect on them, but the emphasis was mostly given on natural daylight. The mean value was 4.58/5.00

for natural daylight and 4.07 for artificial lighting. 253 people stated that daylight has an extremely important effect on their performance, as only 185 people had the same statement for artificial lighting. 9 people thought that daylight has no effect on their performance, and 39 people had the same opinion on artificial lighting.

Third part of the questionnaire investigated the opinions of the participants about effect of acclimatization on their performance at work. The results indicate that the participants give greater value to natural ventilation than artificial air conditioning. 276 people stated that they thought natural ventilation has an important effect on their performance at work, but only 226 people had the same opinion about artificial air conditioning. The question about the use of natural ventilation had a mean value of 4.70/5.00, as the one about artificial air conditioning had 4.41/5.00 as the mean value.

Fourth part of the questionnaire was interested in the acoustic quality of the physical environment in the call centers and its effect on the employee wellbeing and performance. There was a significant awareness among the participants that acoustic quality in the physical environment is extremely effective on their work performances. The mean value for this question was 4.71/5.00, with 283 people saying that it has an important effect on their performance. 49 participants stated that it has an effect, as 22 people were neutral about it, and 3 people had negative opinions about the effect of acoustic quality on their performance.

Fifth part of the questionnaire was dealing with the effectiveness of plan layout and finishing materials in the workspace on employee performance. The proper layout design of the workspace with sufficient personal space, socialization areas, and circulation corridors was the mostly recognized feature of the call center interiors by the participants.294 people claimed that it has an important effect on their performance, and the mean value for that question was 4.76/5.00. The effectiveness of finishing materials and colors in the workspace was also highly anticipated by the participants, but not as high as the layout design of the space. It had a mean value of 4.55/5.00, and 247 participants thought that it has an important effect on their performance.

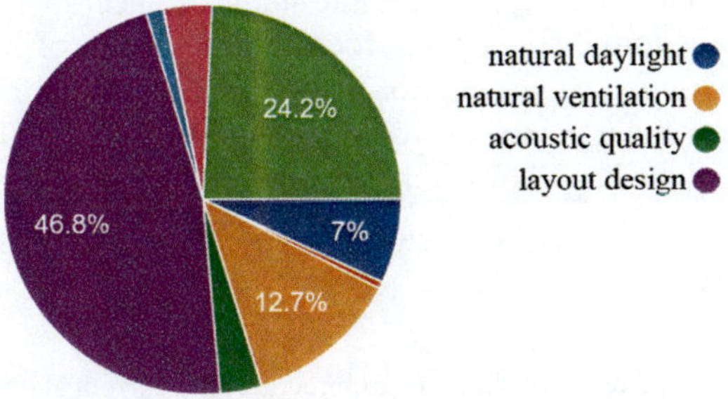

Fig. 2. Results for the last stage of the questionnaire (source: author).

The last stage of the questionnaire was a multiple-choice question about the preference of the participants for improving the physical environment of the workspace. In response, 46.8% (166 people) responded that they would improve the layout design for better personal space and circulation areas. 24.2% (86 people) said that they would improve the acoustic quality. 45 people (12.7%) voted for effective natural ventilation, and 25 people (7%) chose improving the natural daylight (see Fig. 2).

5 Conclusion

This study shows some aspects of the effects of the physical environment on the wellbeing and productivity of call center employees. As the first and the strongest conclusion, it can be asserted that the employees are aware on the effects of the physical environment on their performances, and they demand healthier and better physical environments in their workspaces. Secondly, they give greater value to natural elements like daylight or passive ventilation than artificial systems. Additionally, even though the study is focused on call center workspaces, acoustic qualities appear to be an important aspect of healthy indoor environments overall. As the last step for concluding the study, the importance of architectural details come into prominence with their effects on the occupant perception.

A big advantage of architecture is that it can order people's relationships with each other and their environment by creating interactive settings for life [17]. This gives the profession an important responsibility at the same time. Creating healthy and livable environments for the users must be one of the biggest concerns of architecture. However, there are also some setbacks preventing the creation of healthy environments. Rice and Drane summarize these setbacks as: The lack of knowledge on the relationship between health and architecture; the lack of focus on health issues in architectural education curricula; contemporary lifestyles that interrelate the design of the built environment with various phenomena; and financial limitations that present healthy architecture features costly [18]. Additionally, indoorization attracts attention as a challenge about healthy spaces. The stronger is the relationship of people with natural elements, the healthier they become. This is a valid statement for the built environment as well. With the societies spending more time in the indoor environments, it becomes harder to maintain their health [19]. Architecture must take responsibility for making indoor environments healthier. That is the reason for the necessity of academic studies on healthy indoor environments. This paper contributes to the studies that focus on the health of indoor environments through the example of call center interiors as workspaces. With the increased number of studies with similar focal points, the role of architecture for healthy societies will be recognized better, and the health of the indoor environments will improve.

References

1. Rice, L.: The nature and extent of healthy architecture: the current state of progress. Archnet-IJAR Int. J. Architect. Res. **13**(2), 244–259 (2019)
2. Desmet, P.M.A., Pohlmeyer, A.E.: Positive design: an introduction to design for subjective well-being. Int. J. Des. **7**(3), 5–19 (2013)
3. Candido, C., Marzban, S., Haddad, S., Mackey, M., Loder, A.: Designing healthy workspaces: results from Australian certified open-plan offices. Facilities **39**(5/6), 411–433 (2020). https://doi.org/10.1108/F-02-2020-0018
4. Kwon, M., Remøy, H., Van Den Dobbelsteen, A.: User-focused office renovation: a review into user satisfaction and the potential for improvement. Prop. Manag. **37**(4), 470–489 (2019)

5. Benammar, A., Anouche, K., Lesgaa, H., Hamza Cherif, Y.: Open-plan office post-occupancy evaluation: inquiring Algerian gender specificity. Facilities **36**(11/12), 546–570 (2018)
6. Global Wellness Institute. https://globalwellnessinstitute.org/initiatives/wellness-architecture-initiative/wellness-architecture-resources/. Accessed 27 Jan 2021
7. World Health Organization: Implementation framework for phase vii (2019–2024) of the WHO European Healthy Cities Network: Goals, requirements, and strategic approaches. WHO Regional Office for Europe, Copenhagen (2019)
8. Bokalders, V., Block, M.: The Whole Building Handbook: How to Design Healthy, Efficient and Sustainable Buildings, Earthscan, London (2010)
9. Fanger, P.O., Valbjorn, O.: Indoor Climate: Effects on Human Comfort, Performance, and Health in Residential, Commercial, and Light-Industry Buildings. Danish Building Research Institute, Copenhagen, Denmark (1979)
10. Samet, J.M., Spengler, J.D.: Indoor environments and health: moving into the 21st century. Am. J. Public Health **93**(9), 1489–1493 (2003)
11. Mendell, M.J., Fisk, W.J., Kreiss, K., et al.: Improving the health of workers in indoor environments: priority research needs for a national occupational research agenda. Am. J. Public Health **92**, 1430–1440 (2003)
12. Holt, A.S.J., Allen, J.: Principles of Health and Safety at Work. Routledge, London (2015)
13. Delmaz, M.A., Pekovic, S.: Environmental standards and labor productivity: understanding the mechanisms that sustain sustainability. J. Organ. Behav. **34**, 230–252 (2013)
14. van Esch, E., Minjock, R., Colarelli, S.M., Hirsch, S.: Office window views: view features trump nature in predicting employee well-being. J. Environ. Psychol. **64**, 56–64 (2019)
15. McGuire, D., McLaren, L.: The impact of physical environment on employee commitment in call centres: the mediating role of employee well-being. Team Perform. Manag. Int. J. **15**(1/2), 35–48 (2009). https://doi.org/10.1108/13527590910937702
16. Jablonska, J., Trocka-Leszczynska, E.: Ergonomics of architectural-acoustic solutions in contemporary offices. In: Charytonowicz, J. (ed.) AHFE 2020. AISC, vol. 1214, pp. 89–96. Springer, Cham (2020). https://doi.org/10.1007/978-3-030-51566-9_13
17. The Daylight Site. http://thedaylightsite.com/architecture-for-well-being-and-health/. Accessed 27 Jan 2021
18. Rice, L., Drane, M.: Indicators of healthy architecture—a systematic literature review. J. Urban Health **97**(6), 899–911 (2020). https://doi.org/10.1007/s11524-020-00469-z
19. Samet, J.M., Spengler, J.D.: Indoor environments, and health: moving into the 21st century. Am. J. Public Health **93**(9), 1489–1493 (2003)

Sustainable Work Opportunities for Drivers' Well-Being: A Case of Careem as Transportation Network Company

Amna Javed(✉) and Youji Kohda

School of Knowledge Science, Japan Advanced Institute of Science and Technology, Nomi, Japan
{amna,kohda}@jaist.ac.jp

Abstract. This study examines the services provided by the transportation network company in a developing country from the drivers' perspective. This study is set up to investigate two substantial concerns: drivers' satisfaction and well-being. By adopting a purposeful sampling practice, twelve semi-structured interviews were conducted in Lahore, Pakistan. After analyzing the facts, several results are extracted, as motivation is the main factor for the drivers to join the company. The drivers are not the workers at all—they are the users of Careem technology services, Careem defines workers as working-class entrepreneurs instead of service providers only, and finally, Careem treats its drivers as customers by re-defining the role of workers and keeping safety measures for drivers. The created values for drivers because of working conditions beyond the algorithm are indeed the effective value propositions that help Careem to retain its drivers for the betterment of the company and society's well-being.

Keywords: Sharing services · Transportation Network Company (TNC) · Careem · Sustainable work opportunities · Drivers' well-being

1 Introduction

Sharing economy has a worldwide effect on sharing businesses, therefore sharing services have channeled in the new era of work through its practice of managing works, service providers, and service recipients. However, sharing practices may vary subject to the properties and productive business models [1]. In developed countries the arrangement and the network to provide work opportunities are well established, therefore the working environment is sustainable even under fixed working conditions. On the other hand, in developing countries, there are no proper regulations of generating enough earning sources [2], because of limited resources and infrastructure. Therefore, to boost the economic situation of a developing country, sharing economy has played a vital role by considering four main factors: economic, social, practical, and environmental [3], only sharing services have the potential to create a sustainable environment by enabling flexible working situations [4]. It allows people to share their resources by their own choices rather than it is necessary [5]. It does not matter whatever the purpose is behind

R. S. Goonetilleke et al. (Eds.): AHFE 2021, LNNS 273, pp. 70–77, 2021.
https://doi.org/10.1007/978-3-030-80713-9_10

sharing, to support the expansion and success of sharing economy the only key is trust [6, 7]. And TNC is an example of transportation sharing, in which they expand their identifiable transportation facility in reaction to new competition that emerged in sharing economy [8].

In the current situation of developing countries, the launching of transportation network companies (TNCs) has competed for a vital role in the progress of the pooled transportation system and economic situation [9]. TNCs are transportation organizations that are dedicatedly associated with arranging trips using online portals or app. with the purpose to create a linkage between drivers and passengers using their vehicles for sharing reasons [10]. In transportation, the role of drivers and customers is important to get a sustainable ride-sharing service. According to Green [11], in a car-sharing business like TNC, the credibility and deliverability of ride service are more important than its execution. For example, Uber has accompanied in a new era of work through its practice of managing its drivers with algorithms, where Uber wants its drivers to work sincerely and forces them to follow designated rules, which results in a hard-working environment for drivers [12]. Regional contexts, drivers' motivation, and experiences of work, and their level of investment in the job all affect the ways they perceive its benefits or drawbacks [13]. Uber is managing almost 3 million drivers all over the world by managing the kinds of workers who decide to drive with the platform. There are several categories of workers including full-timers, part-timers, and hobbyists. Each group has its motivation to drive with Uber [12].

However, among many countries, Pakistan is a country where Careem-a TNC [14] has shown great progress by considering the service re-shaping and localization to extend the working experience of drivers beyond the algorithmic management. Careem is managing almost 1 million drivers who decide to drive with the platform spread across 14 countries and has acquired the attention from the whole world because of introducing excellent services for the well-being of the society [15]. Careem is a car-sharing service that offers reliable, secure, and convenient traveling means to the citizens in Pakistan.

The purpose of this study is to assess the basis of a sustainable working environment created by Careem along with the impressions of drivers. This research investigates the noticeable differences between the algorithm-based working environments (Uber) to the environment beyond the algorithm (Careem). Moreover, this research will investigate the motivational factors that influence drivers to work with Careem.

To extract the findings, this research has adopted a qualitative research approach. Whereas the thematic analysis technique has been used to analyze the data systematically. By analyzing the facts and results, several findings are extracted: motivation is the main factor for the drivers to work with Careem, drivers are not the employees only—they are the users of the Careem platform. Additionally, Careem defines workers first as working-class entrepreneurs with female captain prioritization. Lastly, Careem treats its drivers as customers by re-defining the role of workers and keeping safety measures for drivers.

The organization of this paper follows mainly four sections including background study and introduction with the explanation of Careem and its role as a TNC in sharing economy. Detailed research methodology and analysis technique is explained in Sect. 2. The discussion of research findings and the significance of the study is included in Sect. 3.

The final section of the paper presents the conclusion with a description of theoretical and practical contributions.

2 Research Methodology

This study has adopted a qualitative research approach with the utilization of semi-structured in-depth interview techniques. The data is collected from ten drivers (D1-10) and two management bodies (M1-2) of the most progressive transportation company named Careem in Pakistan. The selection of the study participants was made by using the purposeful sampling technique that ensured the voluntary participation of all the members; no gender discrimination; having experience with Careem. The interview guide was devised in the local language with a time duration of 30 min for each interview. The mode of the interview was face-to-face and audio-recorded with the purpose to investigate the motivation of drivers and selection reason of Careem by raising the question as; how interviewees have considered the aspects of satisfaction and well-being while working with Careem.

The collected data were analyzed using a thematic analysis technique which followed several steps as it began with the assignment of codes to the data fragments; followed up with searching for the main concepts/themes; identifying patterns and relationships to create categories and finally refining themes and perform reconfiguration of the categories for testing propositions to satisfy research aim.

3 Research Findings

A framework is established that summarized all the research findings as shown in Fig. 1.

This framework has been developed from the interview records that resulted in drivers' motivation and well-being by considering flexible working conditions to satisfy drivers with Careem services. Also examined the effect of Careem services on the lives of drivers in terms of quality and meeting everyday needs.

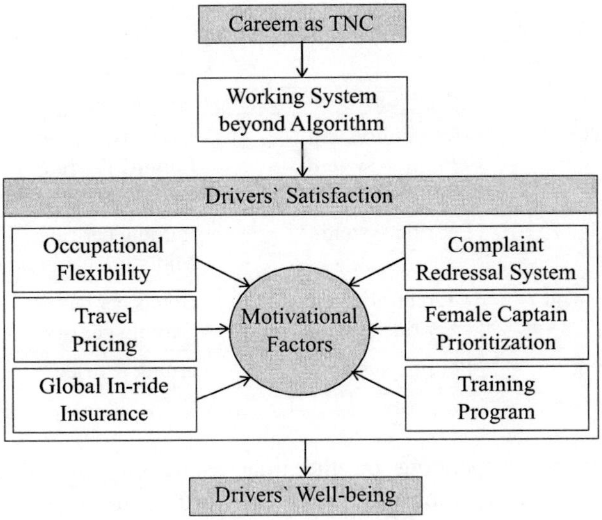

Fig. 1. Research framework to explain the factors and values that are involved in achieving the drivers' well-being in Careem.

3.1 Working System Beyond Algorithm

Career Lines. Uber is a pioneer of ride-hailing services that offers algorithm based working environment to the drivers. All the drivers are bound to follow the instructions and boundaries already determined. Three types of drivers are associated with Uber (full-timers, part-timers, and hobbyists). Uber drivers have less flexibility to select the working conditions that result in unhappiness [12]. However, Careem is a progressive ride service that meets the local needs of any region; therefore, it has introduced several career lines for the drivers as a full-time driver, *premium driver*, part-time driver, and term-time driver (see Table 1).

"Full-time employment is a most popular career among drivers because they can generate regular fixed income based on the target assigned per month. The drivers can earn their income based on per ride calculations along with bonus plus (special points) on per target exceeded rides [...] Part-time drivers follow either target-oriented line or per ride system. They are flexible to select according to their availability [...] term time drivers can only drive following per ride system" (M1).

The most special type of opportunity provided by Careem to its drivers is the status of a premium driver that is designated mainly for the full-time drivers. On each ride platform (Careem) charges its commission, but in the case of full-time drivers, after meeting their assigned target, they can still drive and charge per ride with the minimum deduction of commission by Careem as shown in Table 1.

"I am working with Careem for two years and since last year I am keeping the status of the premium driver. I am successful to earn this status after maintaining my targets every month. Now I can earn more money than before because less commission is charged by Careem on exceeded rides after meeting my target" (D1).

Table 1. Working system beyond algorithm management towards career lines.

Career lines	Working conditions	Incentives system
Full-time employment	- Target-oriented - Per ride system	- Bonus per target - Earning per ride - Bonus plus
Premium driver	Per ride system	- Earning per ride - Minimum commission charged
Part-time employment	- Target-oriented - Per ride system	- Bonus per target - Earning per ride
Term-time working	Per ride system	Earning per ride

Working Conditions. Depending on the time availability and health conditions, Careem proposes two conditions as target-oriented work, and per ride system. Drivers have the freedom to choose either of them based on their choices and convenience. The assigned target conditions and criteria are as follows:

- Daily rides (estimates 8–10)
- Weekly rides (estimates 60–75)
- Monthly rides (estimates 200–240)

"I am a graduate student at university [...] as I do not have enough time in the morning, so I work for Careem in the evening and on weekends [...] sometimes I work as part-time and occasionally as term-time" (D2).

Incentive System. Each career line has a big association with the incentive system. It offers several incentives as bonus per target (full time and part-time), earning per ride (full time, premium, part-time, and term time), *bonus plus (full time only),* and *minimum commission charged (premium only).*

"Careem is working to enhance its incentive system for the betterment of its drivers as they are the vital resources and without their role, the success of the platform is not possible especially in a developing country" (M2).

3.2 Drivers' Satisfaction

From the analysis of interviews, it is excluded that Careem drivers are satisfied with service. Also, drivers' satisfaction is associated with a wide-ranging classification of motivational factors that are comprised of drivers' emotional commitments, platform satisfaction, and overall judgment of life happiness.

Occupational Flexibility. A high degree of occupational flexibility is the core concern of Careem that determines drivers' independent choice for working. They can work as long as they want and whenever they want. This flexibility is not only provided to boost the working environment of the platform, but drivers also acknowledge the flexible conditions. Careem has a strong motive to be your own boss, make your own money, and control your own life that is mentioned on its website [16].

"I worked for Uber last year, but I have joined Careem recently because of the availability of flexible career opportunities [...]. I can work when I have free time and able to earn more money than I used to earn with Uber" (D3).

Travel Pricing. A distinctive feature of Careem from other TNCs is that travel pricing in Careem is calculated based on distance in kilometer and the associated time with that specific ride.

"I usually drive-in peak hours [...] the difference I observed from other rideshare services is fair pricing because it is calculated based on kilometers covered and minutes calculated" (D4).

Global In-ride Insurance. Careem has launched the concept of a global in-ride insurance program to facilitate free in-ride insurance for both passengers and drivers. It will cover the serious injury and death expenses in an accident occurs during the ride.

"Safety is our top priority. As Careem service is getting expansion across the borders, so we have introduced new free global insurance plans [...] Global insurance provides secured policies that cover around $20,000, if serious injury or death causalities occurred during a Careem ride in any country or region" (M1).

The coverage for drivers begins when their app is online and available for passengers to ensure the booking and continued when they wait at the pick-up location, collection of fare and end after safely drive for the passengers to their destination. By introducing successful plans of insurance, Careem has made a huge commitment to insured ride that reflects the importance of safety and well-being of both drivers and passengers.

Complaint Redressal System. Careem focuses on the situational based complaint redressal system. In case of negligence or complaint, a penalty is charged by the SSIG (safety & security investigation group) based on careful judgment.

"We investigate the complaint thoroughly and reach to some decision [...] we avoid charging instant penalty to the drivers against a single complaint" (M2).

Female Driver Prioritization. Introducing female drivers to the platform is a big success factor of Careem as it is commenced in a response to the local needs of the female passengers because they prefer to take a ride with a female driver due to security and religious reasons. In case of some problem incurred, female drivers receive an added layer of security for being prioritized with their calls during the ride.

It's been one year working for Careem [...] its app. is user friendly and being a female driver, I have an opportunity to accept or reject any request based on gender or customer rating with no effect on my profile" (D5).

Training Program. Careem has succeeded to establish a driver-training program to guarantee safe travel. Under this program, each driver learns the use of the Careem app., measures for safe driving operations and traffic authoritative rules and regulations, and TNCs policies.

"We arrange driver training programs on weekly basis to provide basic knowledge about evolving technology platform of Careem to ensure safety and security for passengers, drivers, and pedestrians" (M1).

3.3 Drivers' Well-Being

Ko and Kuo [17] have stated user well-being in their research that defined well-being as consciousness and feeling about the pleasing perception of ones' life, positive emotions, commitments, and higher satisfaction. Well-being can be achieved after maintaining good health, by building trust-based social relationships and cooperation to earn better livelihood [18]. It is generally associated with the life flourishment in ones' life with higher satisfaction.

"Careem has a worthy repute for controlling and ranking of its drivers [D6]. Careem offers a platform for drivers' opinions and feedback information [D7]. Many social campaigns and discount opportunities are initiated by Careem to boost our interest and livelihood [D8].

4 Conclusion

This research contributes significantly to the development of society because Careem promises drivers freedom, flexibility, and independence by introducing several social campaigns even in the current pandemic situation (Covid'19). There are several categories of workers including full-time drivers, premium drivers, part-time drivers, and term-time working drivers. Each group has its motivation to drive with Careem. Moreover, to attain sustainable working conditions, premium driver status is the key by which drivers gain the opportunity to decide their working hours by themselves and Careem charges a minimum commission after meeting the assigned target for making a relief to them.

The core of Careem drivers is *"Be your own boss, make your own money and control your own life"*. Careem drivers are the consumers of its software followed by a highly developed incentive system as it rewards the drivers in terms of bonuses, earnings, and bonus plus on per target exceeded rides. The created values for drivers because of working conditions are indeed the effective value propositions that help Careem to retain its drivers for the betterment of the company and society's well-being.

References

1. Davidson, A., Habibi, M.R., Laroche, M.: Materialism and the sharing economy: a cross-cultural study of American and Indian consumers. J. Bus. Res. **82**, 364–372 (2018)
2. Wallenstein, J., Shelat, U.: What's next for the sharing economy? (2017). https://www.bcg.com/en-nor/publications/2017/strategy-technology-digital-whats-next-for-sharing-economy.aspx. Accessed 20 Dec 2020
3. Lea, R.: Four keys to a successful sharing economy business model (2015). https://www.text100.com/articles/technology/sucessful-sharing-economy-business-model/
4. Parente, R.C., Geleilate, J.M.G., Rong, K.: The sharing economy globalization phenomenon: a research agenda. J. Int. Manag. **24**(1), 52–64 (2017)

5. Nica, E., Potcovaru, A.M.: The social sustainability of the sharing economy. Econ. Manag. Finan. Markets **10**(4), 69–75 (2015)
6. Botsman, R., Rogers, R.: What's Mine is Yours: The Rise of Collaborative Consumption. Harper Collins Publishers (2011)
7. Belk, R.: Sharing. J. Consum. Res. **36**(5), 715–734 (2010)
8. Wallsten, S., Wallsten, S.: The competitive effects of the sharing economy: how is uber changing taxis? (2015). https://www.ftc.gov/system/files/documents/public_comments/2015/06/01912-96334.pdf. Accessed 10 Jan 2021
9. Javaid, A., Javed, A., Kohda, Y.: Cross-business collaboration as a core of service ecosystem expansion: a case of ride-hailing service in a developing country. Sustain. (MDPI) **11**(15), 1–27 (2019)
10. Azevedo, F., Maciejewski, M.: Social, economic, and legal consequences of Uber and similar transportation network companies (TNCs). Forward Planning Policy. European Parliament, Brussels, Belgium (2015)
11. Green, H.: Trust and the sharing economy: a new business model (2015). http://trustedadvisor.com/public/White-Paper-Trust-and-the-Sharing-Economy.pdf
12. Rosenblat, A.: Uberland: How Algorithms are Rewriting the Rules of Work. University of California Press, Oakland (2018)
13. Kim, K., Baek, C., Lee, J.D.: Creative destruction of the sharing economy in action: the case of Uber. Transp. Res. Part A: Policy Pract. **110**, 118–127 (2018)
14. Anjum, D.: Top ride-hailing companies in Pakistan (2017). https://www.researchsnipers.com/top-ride-hailing-companies-pakistan/. Accessed 28 Dec 2020
15. Javaid, A., Javed, A., Kohda, Y.: Exploring the role of learning organization to improve sharing economy in developing countries: a case study of ride-hailing service in Pakistan. Int. J. Knowl. Syst. Sci. (IJKSS) **10**(4), 35–60 (2019)
16. Careem. Our mission to simplify and improve the lives of people and build an awesome organization that inspires. https://www.careem.com/en-ae. Accessed 12 Dec 2020
17. Ko, H.C., Kuo, F.Y.: Can blogging enhance subjective well-being through self-disclosure? CyberPsychol. Behav. **12**(1), 75–79 (2009)
18. Ayala-Azcarraga, C., Diaz, D., Zambrano, L.: Characteristics of urban parks and their relation to user well-being. Landsc. Urban Plan. **189**, 27–35 (2019)

Prevalence of Post-work Musculoskeletal Disorders in Social Workers and Secretaries of Second-Level Hospital

Gilberto Perea[1], Corina Flores Hernández[1](✉), Víctor Rodríguez[2], Daniel Páramo[1], and Guadalupe de los Auxilios Díaz Cisneros[1]

[1] Universidad De Guanajuato, Blvrd Cañaveral 1001, Fracciones de los Aguirre, 37672 León, Gto., México
fhcorina@ugto.mx
[2] Seguridad y Salud Ocupacional, Región Bajío Grupo OXXO, León, México

Abstract. The purpose was to identify post-work musculoskeletal alterations in social workers and secretaries of second-level hospitals. Cross-sectional study, with a sample of 185 subjects being calculated in Epidat by proportions, 20% of the population with damages at a 95% CI. The modified NORDICO questionnaire was applied to collect information on pain, fatigue or discomfort in different areas of the body. The reliability test was carried out, obtaining a 65% correlation between the items and inferential statistical analysis. The 42.8% (150) were aged $\leq$35 years and 58.2 between 36 and 50 years (35), in the seniority in the position 45.9% (80) had between 6 and ten years more than ten, 21.6% (40), Musculoskeletal disorders reported by social workers and secretaries were pain in the neck (40%), shoulders (50%), wrist and hand (50%), lower back (70%), knees (40%), ankles and swelling 50% during the last seven days. The type of tasks was quantified and with repetitive movements (45.9%), varied tasks such as copies, agendas, etc. (21.6%) and mixed 32.4% ($p < 0.05$ X2). Regarding how long these discomforts have prevented them from doing their work, 35% mentioned more than three disabilities for 3 days, 25% have received heat therapy and 100% have taken anti-inflammatories. In addition, it is observed that there is no ergonomic prevention and the risk increases 2 times more with age and work seniority (95% CI, $p < 0.05$). It can be observed that while the ten minute breaks are not respected for each working hour an 8 h/day shift will increase disorders.

Keyword: Musculosketal disorders · Prevalence · Occupational health · Discomfort

1 Introduction

According to the International Labor Organization (2013), 160 million people worldwide suffer from non-fatal diseases annually related to occupation [1]. This is in largely due to technological, social and economic changes that have contributed to the creation of new risks, or exacerbating existing ones, including musculoskeletal disorders (MSD). These

R. S. Goonetilleke et al. (Eds.): AHFE 2021, LNNS 273, pp. 78–83, 2021.
https://doi.org/10.1007/978-3-030-80713-9_11

disorders represent one of the most frequent occupational diseases, both in countries industrialized and developing [2]. It is estimated that about 30% of the Occupational morbidity corresponds to this type of pathologies, which affect the quality of life of the workers and represent an economic cost in terms of lost work days, disabilities, absenteeism, early retirement, expenses for diagnostic tests and treatments [3].

Musculoskeletal disorders (MSD) are injuries associated with locomotor apparatus. In 1997, the Centers for Disease Control and Prevention's (CDC) National Institute for Occupational Safety and Health (NIOSH) released a review of evidence for work-related MSDs. Examples of work conditions that may lead to WMSD include routine lifting of heavy objects, daily exposure to whole body vibration, routine overhead work, work with the neck in chronic flexion position, or performing repetitive forceful tasks. This report identified positive evidence for relationships between work conditions and MSDs of the neck, shoulder, elbow, hand and wrist, and back [4]. Musculoskeletal injuries have always been related to the workday and type of work. Musculoskeletal disorders (MSD) are alterations of seriousness and include symptoms such as discomfort, pain and immobility, and depending on the progress they can cause temporary sick leave and receive medical treatment. MSDs do not include disorders caused by slips, trips, falls, or similar incidents. Examples of MSDs include: Sprains, strains, and tears, Back pain, Carpal tunnel syndrome, hernia [5].

Musculoskeletal disorder affects the muscles, joints, tendons, ligaments, bones, and nerves of the human body, but it can generally be back, neck, shoulders, and upper extremities. OSHA, EU has mentioned that musculoskeletal disorders can develop as repeated trauma, or with cumulative effect in the workplace and depends on the type of company. Musculoskeletal disorders have a multifactorial aetiology. It is difficult in most cases to point out the exact cause of an individual case of disease. They are also not very commonly accepted as occupational diseases in the national compensation or reporting systems. There is little evidence of the use of standardised diagnostic criteria for MSDs across Member States of the European Union, and a range of terms and health problems have been covered in different countries to describe these disorders. This variation is reflected in the nationally reported data and makes comparisons between Member States difficult [6].

Prevalence

Musculoskeletal disorders account for nearly 70 million physician office visits in the United States annually, and an estimated 130 million total health care encounters including outpatient, hospital, and emergency room visits [7]. In 1999, nearly 1 million people took time away from work to treat and recover from work-related musculoskeletal pain or impairment of function in the low back or upper extremities [7]. The Institute in Medicine estimates the economic burden of WMSDs as measured by compensation costs, lost wages, and lost productivity, are between $45 and $54 billion annually [7]. According to Liberty Mutual, the largest workers' compensation insurance provider in the United States, overexertion injuries—lifting, pushing, pulling, holding, carrying or throwing an object—cost employers $13.4 billion every year [7].

An increasing trend at the EU level MSDs + carpal tunnel syndrome increased by 32% from 2002 to 2005 (by 39% among women). MSDs + carpal tunnel syndrome accounted for 59% of all recognised disease covered by EODS in 2005 (about 85% of all ODs among women). But all in all, the number of accepted cases of occupational disease is much smaller than the number of self-assessed work-related cases described in the previous section would suggest [8].

In the case of Mexico, 16% of economic activities correspond to the manufacturing branch, classified into ten groups: production of food, beverages, tobacco; machinery, equipment; derived from oil and coal; plastic and rubber chemical industries; metal industries, products based on non-metallic minerals; textile industries, leather; paper; wood and furniture (National Institute of Statistics and Geography [9]. Dates from the IMSS (2015), showed that during 2015 there were 3,230 cases of diseases of the osteoarticular system equivalent to 26.8% of the total risks classified as occupational disease. On the other hand, the National Rehabilitation Institute (2014) carried out a study in patients treated in its facilities due to musculoskeletal pathologies; as a result, 11.5% of the 3508 cases reported carrying out activities of operator officers, artisans of mechanical arts, various trades, plant operators, machines and assemble [10, 11].

Methods

The purpose was to identify post-work musculoskeletal alterations in social workers and secretaries of second-level hospitals. Cross-sectional study, with a sample of 185 subjects being calculated in Epidat by proportions, 20% of the population with damages at a 95% CI. The questionnaire applied was the Nordico de Kuorinka [12], this is a standardized questionnaire for the detection and analysis of musculoskeletal symptoms, applicable in the context of ergonomic or occupational health studies in order to detect the existence of initial symptoms, which have not yet constituted disease or have not yet led to consult the doctor. The questions are multiple choice and can be applied in one of two ways. One is in self-administered form, that is, it is answered by the person surveyed himself, without the presence of a respondent. The other way is to be applied by a pollster like part of an interview as was the case in this investigation, this was applied to collect information on pain, fatigue or discomfort in different areas of the body [12]. The questionnaire was applied to the evaluate the Prevalence of post-work musculoskeletal disorders in social workers and secretaries of second-level hospital in the Guanajuato region and these were selected for convenience according to their geographic location. The information was collected in a period of two weeks by researchers. With the information collected, an Excel database was created. The reliability test was carried out, obtaining a measure how percentage, and correlation test, between the items and inferential statistical analysis.

2 Results

The reliability test was carried out, obtaining a 65% correlation between the items and inferential statistical analysis. The 42.8% (150) were aged ≤35 years and 58.2 between 36 and 50 years (35), in the seniority in the position 45.9% (80) had between 6 and ten years more than ten, 21.6% (40).

Musculoskeletal disorders reported by social workers and secretaries were pain in the neck (40%), shoulders (50%), wrist and hand (50%), lower back (70%), knees (40%), Thigh (50%), ankles and swelling 50% during the last seven days (Fig. 1).

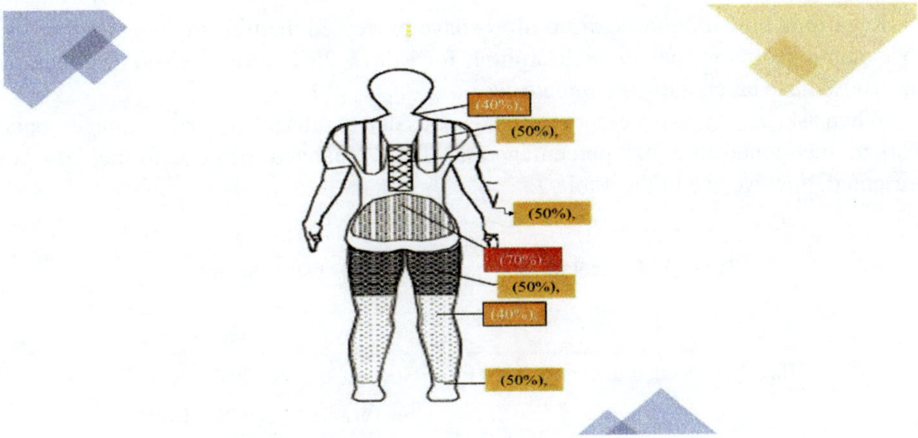

Fig. 1. Percentage of Musculoskeletal disorders.

According to the type of job, there were differences in discomfort during the same time of the working day (8 h per 5 days a week during the morning shift) as can be seen in Table 1.

Table 1. Percentage of discomfort in the last 12 months during 8 h per 5 day a week.

		Neck		Shoulders		Lower back		Ankles		Wrist and hand		Thigh		Knees	
		Yes	No	Yes	No	Yes	No	Yes	No	Yes	No	Yes	No	Yes	**No**
Do you had any discomfort in the last 12 months?	Social workers	25%	75%	20%	80%	20%	80%	20%	80%	25%	75%	25%	75%	25%	**75%**
	Secretaries	**15%**	**85%**	**30%**	**70%**	**50%**	**50%**	**30%**	**70%**	**25%**	**75%**	**25%**	**75%**	**25%**	**75%**

The type of tasks was quantified and with repetitive movements (45.9%), varied tasks such as copies, schedule, (21.6%) and mixed 32.4% ($p < 0.05$ X2) (Table 2).

Table 2. Type of task for have Musculosketal discomfort.

		Tasks	Yes
Tasks was quantified and with repetitive movements	Social workers	Copies Schedule Typing	45.9% 21.6% 100.0%
	Secretaries	Check expedient Take date, Address Other	80.0% 100.0% 70.0%

Regarding how long these discomforts have prevented them from doing their work, 35% mentioned more than three disabilities for 3 days, 25% have received heat therapy and 100% have taken anti-inflammatories.

When asked if it was necessary to change jobs due to musculoskeletal damage, social workers mentioned in a 20th percentage that if they changed jobs due to the damages presented, how we see in the Table 3.

Table 3. Percentage of person who need change the job.

		Yes	No
Have you needed to change your job?	Social workers	20%	80%
	Secretaries	**0%**	**100%**

There is increased by 20% in the prevalence of signs when performing different tasks during 8-day workdays during a week, in addition to making an effort physical mainly moving boxes, sitting for more than 6 h having an impact on the lower spine as can be seen in Table 4.

Table 4. Prevalence rate by activities during working hours (8 h/5 day).

	Yes	No	Prevalence rate	Value "p"
Performing repetitive movements	130	55	1.5	0.004
Lifting heavy objects >20 kg	133	73	1.4	0.03
Effort with the waist and back	118	67	1.1	0.048
Shoulder tension	118	67	1.3	0.05
Immobilization of legs for more than 3 h in a row	113	72	1.2	0.04
Wrists and hands, I work more than 6 h in a row	115	70	1.5	0.001

3 Conclusions

In addition, it is observed that there is no ergonomic prevention, and the risk increases 2 times more with age and work seniority (95% CI, $p < 0.05$). In this study, the presence of musculoskeletal disorders was observed in different parts of the body according to the application of the Nordic questionnaire. It is important to establish a program for the identification of risks and requirements present in all workplaces where ergonomically there are no preventive or surveillance programs specific for this type of morbidity, different articles show that forced postures and the handling of heavy loads, repetitive movements, are processes that they require studies and programs to protect the worker.

The authors recommend early diagnosis and appropriate management of the signs in early step can help people who have more than 5 years in the same place when performing different tasks during 8-h workdays during a week. Another strategy is to evaluate at least twice a year the knowledge that operating personnel have of health and safety standard. Is very important includes not only consulting with a doctor, is important put self-management education programs to help teach people who have for a lot of hours in the same position, and developing and implementing workplace controls and more when now we are more sit-down in home work by the pandemic.

References

1. Organización Internacional del Trabajo (OIT): Comunicado de prensa del 26 de abril, 2013 (2013). https://www.ilo.org/global/about-the-ilo/newsroom/news/WCMS_211645/lang--es/index. Accessed 13 Feb 2017
2. Riihimäki, H., Viikari, E.J.: Enciclopedia de la Organización Internacional del Trabajo: Sistema Musculoesquelético España. Ed. Instituto Nacional de Seguridad y Salud en el Trabajo [CD-ROM] (2014)
3. Jiménez, M.: Relación con los trastornos musculoesqueléticos en miembros superiores de una empresa cordelera. Universidad de Carabobo, Venezuela (2014). http://riuc.bc.uc.edu.ve/handle/123456789/962
4. Bernard, B.P. (ed.): U.S. Department of Health and Human Services, Centers for Disease control and Prevention, National Institute of Occupational Safety and Health. Musculoskeletal disorders and workplace factors: a critical review of epidemiologic evidence for work-related musculoskeletal disorders of the neck, upper extremity, and lower back. July 1997. DHHS (NIOSH) Publication No. 97-141 (1997). https://www.cdc.gov/niosh/docs/97-141/
5. NIOSH workers health chartbook: NIOSH Publication No. 2004-146, Washington, D.C. (2004)
6. Buckle, P., David, G.: Defining the problem. In: Magazine of the European Agency for Safety and Health at Work – 3. Preventing Work-related Musculoskeletal disorders, p. 5 (2000)
7. National Research Council and the Institute of Medicine: Musculoskeletal disorders and the workplace: low back and upper extremities. Panel on Musculoskeletal Disorders and the Workplace. Commission on Behavioral and Social Sciences and Education. Washington, DC: National Academy Press (2001). https://www.nap.edu/read/10032/chapter/1
8. Commission Recommendation (EC) 3297/2003 of 19 September 2003 concerning the European Schedule of Occupational Disease
9. Instituto Nacional de Estadística & Geografía (INEGI): Industria manufacturera (2017). http://cuentame.inegi.org.mx/economia/secundario/manufacturera/default.aspx?tema=E#uno. Accessed 11 Jan 2017
10. Instituto Mexicano del Seguro Social (IMSS): México: Coordinación de Salud en el Trabajo (2013). www.imss.gob.mx/sites/all/statics/pdf/.../06_SaludenelTrabajo.xls. Accessed 18 Feb 2017
11. Instituto Mexicano del Seguro Social (IMSS): Memoria estadística (2015). http://www.imss.gob.mx/conoce-al-imss/memoria-estadistica-2015. Accessed 27 Dec 2017

Estimation of Spine Loads During Daily Activities and Its Relationship with Musculoskeletal Disorders in Elderly Indigenous Women

Alejandra Lascano, Thalia San Antonio(✉), David Guevara, and Anita Larrea

Universidad Técnica de Ambato, Ambato, Ecuador
{am.lascano,t.sanantonio,dguevara,anitallarea}@uta.edu.ec

Abstract. This investigation aims to establish a correlation between spine deformations in indigenous elderly female of Ecuador and the magnitude and maneuver of the loads they manipulate. The women work, since infancy, in agricultural and home labors and tend to do so when carrying on their backs their siblings, children or grandchildren. Photoshoots of static body postures of daily activities were the source for body segment angles identification using 3-Dimensional Static Strength Prediction Program to calculate the biomechanical demands estimating the values of compressive and shear forces at the spine complementing the use of methods for ergonomic evaluations. The calculated lower back compression forces were compared to the NIOSH guidelines. The maximum weight of agricultural loads in the evaluated activities ranged from 50 to 100 lb. The findings should be used to modify the standard of agricultural packages reducing their capacity to protect the health of women stevedores.

Keyword: Low back compression force · Indigenous women · Daily activities · Work-related injuries

1 Introduction

Musculoskeletal disorders (MSDs) are the occupational diseases most studied for been the most frequent in the industrial environment and for producing the highest rate of absenteeism which has a direct impact on productivity [1]. Since the studied working population are self-employed indigenous women from the Ecuadorian high lands the decrease in the productivity has a direct impact in the life quality of these women and in her community [2]. These disorders known to be caused mostly by postural demands, repetitive movements, load manipulation, vibrations, poor lighting, and cold environments [1].

Do to detected prevalence of spine deformations in elderly indigenous women reports in a previous study [3, 4], the present investigation is focuses on MSDs of body mid-section which are related to heavy load manipulation, and static and dynamic muscular work both putting at risk the integrity of the spine and the pelvic area [5].

R. S. Goonetilleke et al. (Eds.): AHFE 2021, LNNS 273, pp. 84–91, 2021.
https://doi.org/10.1007/978-3-030-80713-9_12

The indigenous women of Ecuadorian highlands usually work performing a variety of agricultural and domestic labors using rudimentary tools making these chores physically demanding. Among the tasks they perform, the transport of agricultural products in sacks whose standard weight is 100 lb and agricultural tasks with their brothers, children or grandchildren in their backs stand out due to their high physical demand [6, 7].

The MREBA is an adaptation based on the REBA method that can assess the risk level of the daily activities of the indigenous women more accurately [8], MREBA due to modifications that allows to better valuate the hard tasks perform by these women especially the ones that involve heavy loads in the range of 50 to 100 lb and the important postural loads cause by the constant adoption of extreme postures (elevated trunk, arms, and wrists angles). Additionally, many of the tasks are performed with a child asymmetrically tied to their backs.

To evaluate tasks that involve loads lifting it is widely used the 3D motion capture data of the gesture involve in the task and the use of a software that allow estimating the compression force in the column to simulate the ubication and magnitude of the loads in the human body model [9, 10]. One of the great challenges when using these programs is the modeling of the external loads, as for example a child carried in the back, to find the appropriate strains in the column in each analyzed position [11].

The 3D Static Strength Prediction Program (3DSSPP) a software of the University of Michigan allows to estimate the percentage of men and women who have the strength to carry out a certain activity correlated with posture, force, and gender anthropometry data. The results include the low back compression forces, and a comparison to NIOSH guidelines [12, 13].

The goal of this work is to calculate the low back compression force using a static strength prediction software in adult's indigenous woman, performing six typical daily activities.

2 Materials and Methods

2.1 Description of the Software Used

Six daily tasks were evaluated, de video of each activity was analyzed considering duration, frequency, and deviation from the natural position of the body and in each case de photograph of the most demanding posture was model in the 3DSSPP [14]. As one aspect that forms the basis of the risk assessment is the denomination of the maximum effort in the work cycle and the maximum static duration, these parameters were also considered.

The 3DSSPP include analysis of the loads handling and the forced postures, this software is used to evaluate the biomechanical capacities and the static resistance of the labors perform by the indigenous women in relation to the physical demands of their unstructured work environment. The program allows to predict the static resistance requirements through the estimation of the compression forces in the spine when performing each activity that is analyzed, this is done by means of a 3D simulation of the posture, which includes the external loads, measurements of the body sections and the position of the joints.

To determine the most demanding posture, the duty cycle of each activity is analyzed. The duty cycle is calculated as the coefficient between the time of effort and the time of work expressed as a percentage, these times are based on the percentage of maximum voluntary contraction, which is calculated as required force divided by the average force of the population [15]. For the estimation of the percentage of women who have the strength to carry out each analyzed activity the strength of the 25th percentile population is taken, since this corresponds to the average anthropometry of indigenous women. The reported maximum static duration corresponds to the time of sustained static stress above which localized fatigue is likely to occur.

2.2 Description of the Evaluated Tasks

The six daily activities evaluated were recorder in a video, then photograph of the critical static body postures, in terms of physical demand and they were separated from the videos to be used as the source for body segment angles measurements using the 3DSSPP program. This software allows to estimate the biomechanical demands by calculating the compressive and shear forces in the lumbar spine. The body posture photograph for each activity was selected taking into consideration those positions that, a priori, suppose a greater postural load because of their duration, frequency, or their deviation from the neutral position. The six-task analyzed in are described below:

TASK 1. She performs the activity with her torso bent and her arms outstretched. Takes the garment, rotates her body to wet it in the channel on her left, then places it on a stone, that is at the level of her knees, and presses it using the weight of her body swinging forward and backward (see Fig. 1a).

TASK 2. The woman in standing position, leans her trunk forward and carries a 55-gallon container with a weight of 15 kilos to pour its contents into buckets placed on the ground. The big container is used to mix liquid and dry ingredients (see Fig. 1b).

TASK 3. Woman in a kneeling position and with her trunk bent forward uses her hands to repeatedly pull the plants, then when the tubers are on the surface, she uses her hands to remove the excess soil from them and throws them into various containers depending on their size (see Fig. 1c).

TASK 4. A standing woman bends her trunk until her arms reach a base 50 cm above the ground, on which are the clothes to be washed. Next to the base she has a water tank with which she wets and rinses the clothes using a small container. The washing process consists of pressing, with her arms and hands, the wet clothes using the weight of her body swinging forward and backward (see Fig. 1d).

TASK 5. A woman standing and leaning, carries a 4-year-old infant on her back, held by a blanket tied to her chest. She holds in her hands a tool that she raises and drops on the ground and then pulls towards her body dragging material, this movement is made constantly during approximately 10 min then she rests a few minutes and restart (see Fig. 1e).

TASK 6. The work is carried out standing and with the trunk very inclined on the edge of a water tank. It consists of rotating to take a bundle of radishes that are on the right side, immerse and shake them in the water that is below the level of her feet and rotating to deposit them on the left side. The women rest from the sharply bent position every 10 repetitions approximately (see Fig. 1f).

Fig. 1. Photographs of women performing de six daily tasks that are evaluated in this study.

The estimation of the loads to be added in the model were determined based on the knowledge of the carried weigh and the gestures of the activity. According to the 3DSSPP anthropometry database, the weight and height of each evaluated subject is estimated based on their characteristics. Figure 2 show an example of posture analysis using the 3DSSPP program.

3 Results

Analysis of a photograph of the most forced posture for each of the six daily activities perform by the indigenous women was carry on using the 3DSSPP program, an example of the report obtained for each task is show in Fig. 2.

Table 1 shows the percentage of Exertion Duty Cycle Limit (EDCL) and the Static Duration Limit (EDL) in seconds, for three joints (wrist, elbow, and shoulder) and the six analyzed activities. From the obtain results, it can be deduced that indigenous women

TASK 5. Woman tilling the land with child on her back.

LOCALIZED FATIGUE	
Exertion Duty Cycle Limit (%)*	Static Duration Limit (s) *
Wrist: 4.2 Elbow: 5 Shoulder: 18.6	Wrist: 48 Elbow: 43 Shoulder: 65

* Maximun admisible in the analyzed position

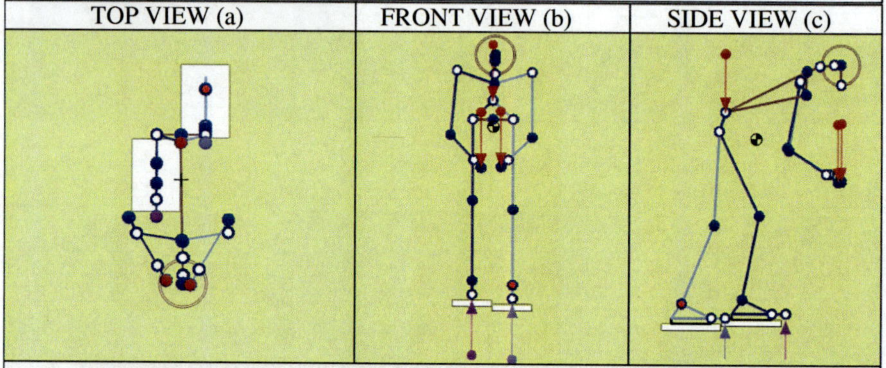

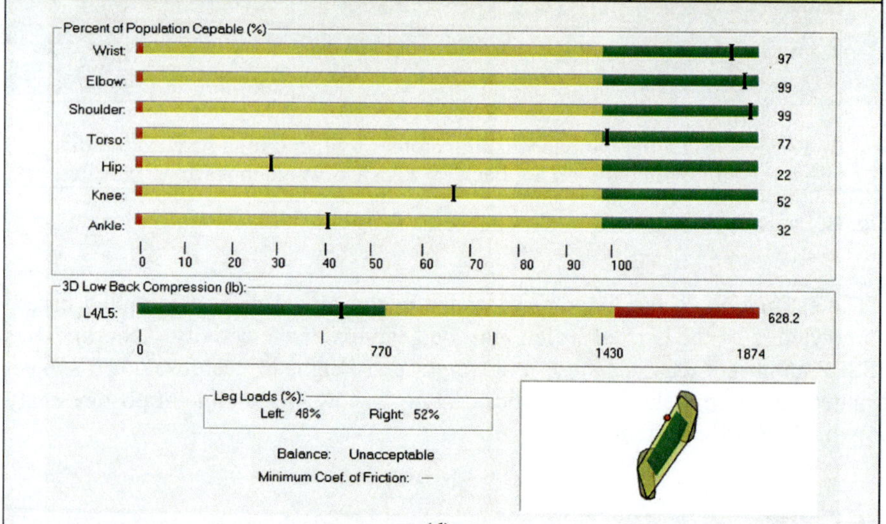

(d)

Fig. 2. Summary of results of the 3DSSPP application to task 5 evaluating: the exertion duty cycle Limit (%), the static duration cycle from three different perspectives (a), (b), (c) and the low back compression results for L4/L5 of a population with 50th percentile anthropometry (d).

with the highest EDCL in the wrist and elbow are those who perform tasks that involve repetitive movements and application of force with their hands and arms (tasks 1, 3 and 6). In the shoulder joint, the EDCL is higher in those tasks that require greater deviation from the neutral position and that involve repetitive movements of the upper extremities. The tasks that produce a higher EDL are those with the greatest static effort, however the low values of EDL were found in those tasks where is not permissible to maintain a sustained effort independently of the measure of the time intervals, this is mainly caused by repeatability and the excessive weight of the handled load.

Table 1. Exertion duty cycle limit and static duration limit for wrist, elbow, and shoulder in the six analyzed activities.

Task	Localized fatigue					
	Exertion duty cycle limit (%)			Static duration limit (s)		
	Wrist	Elbow	Shoulder	Wrist	Elbow	Shoulder
1	5.2	40.5	2.9	53	332	21
2	7.6	22.6	14.4	64	152	53
3	0	0.5	0	0	14	0
4	11	11.8	29.8	79	80	102
5	4.2	5	18.6	48	43	65
6	39.5	33	63.2	226	244	304

The compression of the lower back, specifically in L4/L5 intervertebral discs, 33.3% of the tasks are in the yellow zone (tasks 3 and 8), implying that the possibility of low back pain in indigenous women who perform these tasks is three times more likely than in the general population. On the other hand, 33.3% of the evaluated tasks are in the red zone (tasks 2 and 5), these tasks require handling and/or transportation of heavy loads, being in this zone involve the appearance of lumbar back pain is ten times more likely in the women who perform these tasks compared to the general population (see Fig. 2d).

The distribution of the loads supported by each leg; the results indicate that in 66.6% of the analyzed tasks the left leg is the most demanded. It has also been determined that the center of pressure between the body and the support surface, in the 66.6% of the tasks implies a significant loss of balance, in 16.6% this characteristic presents as a critical result, and the rest of the tasks (16.6%) the center of pressure remains within the functional stability zone.

4 Conclusions

The results of the postural load evaluation, on indigenous women musculoskeletal system, shows a high musculoskeletal injuries risk for the evaluated daily activities, with a general increase of risk level for tasks that involve handling and transportation of heavy loads.

The 3DSSPP program application for biomechanical demands of evaluated daily tasks, reveals a risk level that ranges from significant to intolerable when women are in bipedestation with trunk and neck rotations or performing repetitive movements of hands and arms with the consequent application of force. This software is an appropriate tool to estimate the maximum effort in the work cycle, the maximum static duration, and the compression of the lower back on the L4/L5 intervertebral discs.

Although the 3DSSPP is a very useful tool to estimate the biomechanical demands of work postures, it is important to highlight that a task cannot be evaluated in its entirety by a single program or method, so it is necessary to complement the evaluations with other criteria and professional judgments to perform adequately the safe design of any productive activity.

Acknowledgments. This work is being funded by (1) Dirección de Investigación y Desarrollo (DIDE), of Universidad Técnica de Ambato under the project: UTA-CONIN-2020–0298-R "*Desarrollo de un sistema de capacitación para la protección de salud física de las usuarias de chumbi mediante la creación de contenidos utilizando instrumentos pedagógicos adaptados a su cultura y entorno*" and (2) Corporación Ecuatoriana para el Desarrollo de la Investigación y la Academia (CEDIA) under the project: CEPRA XIII-2019–14, *Llankak Warmikunata Kamachikkuna (Protec the working woman).*

References

1. Arenas, L., Cantú, O.: Factores de riesgo de trastornos musculoesqueléticos crónicos laborales. Med. Int. Mex. **29**, 370–379 (2013)
2. San Antonio, T., Larrea, A., Urrutia, F., Naranjo, M., Latta, M.: Influence of the traditional use of Chumbi in lumbar weakness in women of Ecuadorian highlands. In: Duffy, V., Lightner, N. (eds.) AHFE 2017. AISC, vol. 590, pp. 229–238. Springer, Cham (2018). https://doi.org/10.1007/978-3-319-60483-1_23
3. Antonio, T., Latta, M., Naranjo, M., Guevara, D.: Postural analysis in older females from an indigenous community of Ecuador that uses a Chumbi as part of their traditional outfit. In: Kalra, J., Lightner, N.J. (eds.) AHFE 2020. AISC, vol. 1205, pp. 350–356. Springer, Cham (2020). https://doi.org/10.1007/978-3-030-50838-8_47
4. Del Sol, M., Hunter, K.: Evaluación postural de individuos mapuche de la zona costera de la IX Región de Chile. Int. J. Morphol. **22**(4), 339–342 (2004)
5. National Institute for Occupational Safety and Health: Work Practices Guide for Manual Lifting; Technical report number: 81-122, U.S. Dept. of Health and Human Services (NIOSH), Cincinnati, Ohio (1981)
6. Antonio, T.S., Larrea, A., Lascano, A., Urrutia, J.: Ergonomic evaluation of daily activities of women from a community in the Ecuadorian highlands. In: Goonetilleke, R.S., Karwowski, W. (eds.) AHFE 2018. AISC, vol. 789, pp. 102–110. Springer, Cham (2019). https://doi.org/10.1007/978-3-319-94484-5_11
7. Abdoli, M., Agababova, M., Janabi, J., Pasko, E., Damecour, C.: Evaluation and comparison of lift styles for an ideal lift among individuals with different levels of training. Appl. Ergon. **78**, 120–126 (2019)
8. Lascano, A., Antonio, T., Larrea, A., Ciaccia, M.: Ergonomic methods adaptation for risk evaluation associated to musculoskeletal disorders in elderly indigenous women of the Ecuadorian highlands. In: Karwowski, W., Goonetilleke, R.S., Xiong, S., Goossens, R.H.M., Murata, A.

(eds.) AHFE 2020. AISC, vol. 1215, pp. 266–273. Springer, Cham (2020). https://doi.org/10.1007/978-3-030-51549-2_35
9. Skotte, J.H., et al.: A dynamic 3D biomechanical evaluation of the load on the low back during different patient handling tasks. J. Biomech. **35**(10), 1357–1366 (2002)
10. Ciaccia, M., Mejía, C., Vásquez, C., San Antonio, T.: Ergonomic evaluation of agriculture-related activities performed by Ecuadorian indigenous women. In: Karwowski, W., Goonetilleke, R.S., Xiong, S., Goossens, R.H.M., Murata, A. (eds.) AHFE 2020. AISC, vol. 1215, pp. 348–353. Springer, Cham (2020). https://doi.org/10.1007/978-3-030-51549-2_46
11. Sungdo, K.: Ergonomic analysis of army backpack designs: back and shoulder stresses and their implications. Ph.D. Thesis. The University of Utah. Mechanical Engineering (2014)
12. Waters, T.R., Putz, V., Garg, A., Fine, L.J.: Revised NIOSH equation for the design and evaluation of manual lifting tasks. Ergonomics **36**(7), 749–776 (1993)
13. Mas, D.J.: Evaluación ergonómica del levantamiento de carga mediante la ecuación de NIOSH. Ergonautas, Universidad Politécnica de Valencia (2015). http://www.ergonautas.upv.es/me-todos/niosh/niosh-ayuda.php
14. Center for ergonomics University of Michigan: 3DSSPP (7.1.0) [Software] (2020). https://c4e.engin.umich.edu/tools-services/3dsspp-software/
15. Caso, A.: Técnicas de medición del trabajo. Fundación Confemental. España (2006)

Social & Occupational Ergonomics

Lateral Reaching Distances for Novice and Experienced Ladder Users

Angela Levitan(✉)

ARCCA, Inc., 2299 Second Street Pike, Penns Park, PA 18943, USA
alevitan@arcca.com

Abstract. This study intended to explore the effects of acclimation (practice) and motivation (task completion) on lateral reach distances performed by novice and experienced ladder users while working on stepladders of different heights. It was hypothesized that as the novice ladder user's experience (i.e., acclimation) working on a stepladder increased, their reach distances would increase (reflecting decreased risk perception). In addition, ladder users would reach farther when motivated to do so by a concrete task. Acclimation to the task significantly increased RD for both sets of participants, indicating that acclimation to a specific task may be as important as general experience. The introduction of a concrete task and a time component had a significant effect on reach distance. This desire to finish a task quickly may alter an individual's risk assessment of the situation, leading them to perform tasks that they deem unsafe in other circumstances.

Keywords: Falls from elevations · Ladders · Construction · Lateral reaching

1 Introduction

Falls from ladders account for a large portion of workplace injuries related to falls from elevations since ladders are widely used at different types of workplaces and falls from ladders occur at a high frequency. Falls from ladders remain a critical safety issue due to the frequency of falls and the severity of the injuries. Workers who received treatment at an emergency room due to an injury sustained after falling from a ladder were interviewed to identified factors associated with such falls. The majority of workers (51.0%) were using a stepladder at the time of the fall and 51.3% reported standing and working on the ladder when the incident occurred [1]. Guidelines for safe ladder use state that the body should remain within the rails of the ladder ("belly button" or "belt buckle" rule), yet, many falls occur while workers are moving laterally and performing extended reaches during task performance. Falls occur because the individual loses balance and falls off the ladder or because the ladder tips over, causing the individual to fall with it.

Physical manifestations of anxiety are present for many individuals standing on elevated work platforms [2]. This anxiety detrimentally affects postural control, which will likely influence lateral reach distances. Since the use of taller ladders is less common outside the workplace, it is hypothesized that novice users would initially be more careful on the taller ladders, resulting in shorter reach distances as compared to experienced

R. S. Goonetilleke et al. (Eds.): AHFE 2021, LNNS 273, pp. 95–99, 2021.
https://doi.org/10.1007/978-3-030-80713-9_13

users. Reaching guidelines are typically included in training regarding safety practices while using a ladder. However, workers still overreach with numerous factors potentially interacting to influence reaching behavior. Prior research indicates that the more familiar a user is with a product or task, the less likely they will look for, or heed, warning information [3]. That is, greater familiarity with a product or task may cause workers to perceive less of a hazard, which in turn, decreases the likelihood of their complying with associated safety precautions. Therefore, novice ladder users might be expected to take fewer risks than more experienced users and to increase their level of risk-taking after acclimatization.

In their study of falls from ladders, Lombardi et al. [1] indicated that 50% of the injured workers had <3 years of job experience and 62% had no on-site safety training. Individuals may not be aware of the repercussions associated with overreaching while working on a ladder if they do not receive proper safety training. Even with proper training, it is possible that workers will overreach due to other motivations and influences. Workers may determine that the additional risk is warranted to minimize the time required to descend the ladder, reposition it, and ascend the ladder while performing a given task. The priority given to safety as dictated by coworkers and management as compared to productivity and rapid task completion will play an important role during this decision-making process.

2 Methods

2.1 Participants

Participants in the study included 24 novice and 24 experienced male ladder users. The mean (sd) age, height and weight of the novice group was 39.7 (13.2) years, 170.9 (6.4) cm and 81.0 (12.9) kg, respectively. The mean (sd) age, height and weight of the experienced group was 47.8 (12.3) years, 173.6 (5.5) cm and 89.6 (12.1) kg, respectively. Participants had normal or corrected vision in both eyes and were free of known musculoskeletal and balance problems.

2.2 Experimental Design

This study implemented a mixed factorial design with one within subject factor (experience) and two fixed factors (ladder height and reach condition). A ladder user was defined to be a novice if they had never used a ladder as part of their employment or received training on the proper use and setup of ladders. Individuals were considered experienced users if they regularly climbed ladders as part of their employment (minimum tenure of 3 years). Two ladder heights were included in the study (6′ and 12′) and three reach conditions were evaluated for each participant: initial maximum, motivated maximum and acclimated maximum. New Type 1A fiberglass stepladders of 6′ and 12′ were used during the experiment. Extra heavy-duty industrial ladders rated for up to 300 lb (Type 1A) were chosen since they are most commonly used in commercial settings (e.g. construction sites).

2.3 Experimental Procedure

The participants were briefed on the study objectives, procedures and potential risks. In preparation for the experimental tasks, the participants changed into sportswear and low-rise trail shoes (Nike Bandolier II) provided by the research team. A full body harness attached to a belay system was donned to maintain safety of the participants throughout the experiment. A 12-camera passive motion capture system (Motion Analysis Corp., Santa Rosa CA) was used to collect motion data from a full-body marker set and the location of the belly button and ladder. The 3D trajectory data was collected at 100Hz and filtered using a zero-lag fourth-order 8 Hz low-pass Butterworth filter. Participants stood on the proximal side of the ladder with the distal side of the ladder used to stabilize the ladder. The ipsilateral side of the ladder was that closest to the target, while the contralateral side was opposite the target.

Lateral reaches towards the right side of the ladder were performed while participants stood on the third rung from the top of the ladders. For each ladder height, the order of which was randomly presented, participants initially performed a lateral reach with the instructions to "reach as far as you feel comfortable." For the second trial the participant was asked to "reach as far as you can." The initial maximum reach distance established by this trial was used to determine the lateral target locations used in the subsequent trials.

A target (key from a computer keyboard) was placed at a vertical height midway between the height of the acromion process of the right shoulder and right elbow. Participants were required to perform a concrete task by reaching and pressing the target, which required a very low level of force but some level of motor control. Participants were not allowed to lean on or brace against the target. A motivated maximum reach distance was determined using a modified method of constant stimuli procedure. The target was placed closer and farther to the ladder rail until a maximum motivated lateral reach distance was established. If the participant was able reach the target, the subsequent reach distance was presented without substantial delay. If the participants felt that they could not reach the target at the current location, they were required to climb down from the ladder and perform a short (20 s) card-sorting task. In this manner, participants were motivated to reach for the target since the instructions implied that successful reaches would result in decreased time to complete the experiment. Participants were paid a flat rate for completion of the experiment so taking longer did not result in a financial benefit. After the motivated maximum reach distance was determined, the target was removed and another unmotivated trial was performed where the participant was asked to "reach as far as you can." The total time to complete all the trials comprising each experimental condition was approximately 15 min.

Reach distance (RD) was defined as the distance from the ipsilateral ladder rail to the right wrist in the frontal plane. The distance to the end of the hand was not used to determine reach distance so the technique used to push the target would not influence the distance measured. The first trial was considered as practiced and not analyzed. The second trial at each ladder height was designated as the initial maximum reach. The maximum reach distance that resulted after the modified method of constant stimuli procedure was the motivated maximum distance. The final trial, which did not include the target, was deemed the acclimated maximum distance and used to determine the effect of practice.

3 Results

Table 1 includes the reach distances for each of the experimental conditions. RD was significantly shorter on the 12′ ladder as compared to the 6′ ladder ($p < 0.001$). The initial maximum was always the shortest reach distance with significant increases after acclimation and additional significant increases during the motivation reach condition ($p < 0.001$). There was no significant effect of experience ($p = 0.169$) or any of the interactions ($p > 0.320$).

Table 1. Reach distances.

	Ladder height	Reach condition	Reach distance (mm)
Novice	6′	Initial	914 (78)
		Motivated	1016 (58)
		Acclimated	949 (58)
	12′	Initial	853 (123)
		Motivated	957 (119)
		Acclimated	889 (114)
Experienced	6′	Initial	936 (69)
		Motivated	1040 (56)
		Acclimated	953 (62)
	12′	Initial	871 (72)
		Motivated	999 (75)
		Acclimated	915 (82)

4 Discussion

Ladders are used on a regular basis on many types of job sites. Even though falls from ladders occur relatively frequently, workers continue to disregard safety guidelines and increase fall risk by overreaching while on ladders. Workers' behaviors may be influenced by many factors, including their individual level of risk perception, peer pressure and time constraints imposed by management. Individuals that associate working on ladders with a high level of fall risk may abide by safety guidelines and be less inclined to overreach, whereas those who prioritize speed of task completion may overreach in an attempt to increase productivity.

Anxiety attributed to being at heights has been shown to detrimentally affect postural control [4, 5] so it was not surprising that users, particularly novice users, would be less willing to reach further on a taller ladder partially due to feelings of increased instability. Although RD for the experienced users was always greater than the novices, the difference was not found to be statistically significant. Acclimation to the task, based on a short timeframe, significantly increased RD for both sets of participants. Since RD increased for experienced users, it would indicate that acclimation to a specific task involving ladders may be as important as general experience. The introduction of a concrete task and a time component had a significant effect on reaching distance. Individuals generally want to succeed at completing tasks efficiently and in a timely manner. This may alter an individual's risk assessment of the situation, leading them to perform tasks that they deem unsafe in other circumstances. This desire may be intensified in newer employees who are trying to impress their coworkers and managers. It is imperative that management and coworkers clearly articulate that safety is a priority on the jobsite.

Ladder specific safety training is very important and continuous reinforcement regarding the prioritization of safety over task completion speed is paramount, especially since current findings indicate that experienced users are at the same or greater risk of falling. Further research is necessary to determine why experienced users are willing to reach beyond safety guidelines. It is possible that risk perception is reduced so management must reinforce the message that injuries caused by using ladders occur frequently and are often serious, leading to multiple days lost from work. The relationship between overreaching and ladder falls must be emphasized, possibly by recounting the history of injuries related to ladders on the jobsite.

Acknowledgments. The author would like to thank Mary F. Lesch, Jacob J. Banks and Amanda Rivard for their assistance with this investigation, which was conducted at the Liberty Mutual Research Institute for Safety.

References

1. Lombardi, D.A., Smith, G.S., Courtney, T.K., Brennan, M.J., Kim, J.Y., Perry, M.J.: Work-related falls from ladders–a follow-back study of US emergency department cases. Scand. J. Work Environ. Health **37**, 525–532 (2011)
2. Adkin, A.L., Frank, J.S., Carpenter, M.G., Peysar, G.W.: Fear of falling modifies anticipatory postural control. Exp. Brain Res. **143**(2), 160–170 (2002). https://doi.org/10.1007/s00221-001-0974-8
3. Godfrey, S.S., Laughery, K.R.: The biasing effects of product familiarity on consumers' awareness of hazard. In: Human Factors Perspectives on Warnings, pp. 58–61. The Human Factors and Ergonomics Society, Santa Monica (1993)
4. Huffman, J.L., Horslen, B.C., Carpenter, M.G., Adkin, A.L.: Does increased postural threat lead to more conscious control of posture? Gait Posture **30**, 528–532 (2009)
5. Davis, J.R., Campbell, A.D., Adkin, A.L., Carpenter, M.G.: The relationship between fear of falling and human postural control. Gait Posture **29**, 275–279 (2009)

Objective and Subjective Evaluation of Motorcycle Helmet Visors Based on ECE 22.05 Regulations

Nhât Nam Nguyên, Ellen Vanderlooven, Kevin van der Velden, Jochen Vleugels, and Regan Watts(✉)

Department of Product Development, Faculty of Design Sciences, University of Antwerp, Prinsstraat 13, 2000 Antwerp, Belgium
regan.watts@uantwerpen.be

Abstract. With the growing demand for helmet mounted displays on motorcycle helmets, it is essential not to compromise the peripheral vision of the motorcyclist. The relevant European standard of motorcycle helmet visors ECE 22-05, equivalent to the DOT standard FMVSS 218, states that the visor should permit peripheral vision horizontally through an arc of 105° from the helmet midline and vertically through an arc of 52°, which is located 7° upwards and 45° downward from the eyes. Consequently, this study compares a 3D environment to a human testing environment and an controlled testing environment using standard headforms to create an objective method to verify the correct field of view of motorcycle helmets. Firstly, the 3D environment is developed in Solidworks and provides a simulation set up that validates the necessary field of view in reference to the ECE 22-05. This environment is matched up to the testing environment with test subjects that are familiar with the use of a motorcycle helmet. The fields of view were determined by 12 indication points for 9 different motorcycle helmet models. The downwards tilt of the helmet of each test subject was individually determined. Each participant was additionally tested with a control helmet. The control helmet was also 3D scanned for use in the 3D environment. This provides a reference for the tilt of the helmet in the other two objective evaluation methods. With an average downwards angle of 21.1° down, the control helmet has a success rate of 66.7% in the testing environment with the test subjects, although a number of external factors influence the determination of the field of view of the test subjects. Lastly, a field of view was generated by means of a human headform model into which a Ricoh Theta 360 camera was inserted. This results in an objective field of view for the control helmet. The synthesized downwards angle was implemented in the 3D environment to generate a 3D render of the field of view with the 3D scan of the control helmet. The render shows a similar field of view to the field of view generated by the Ricoh Theta 360 camera, resulting in a confirmation of the validity of the 3D model. Combining these three methods of testing, guarantees an objective evaluation of motorcycle helmet visors.

Keywords: Certified motorcycle helmet · Field of view · Objective verification tool

R. S. Goonetilleke et al. (Eds.): AHFE 2021, LNNS 273, pp. 100–108, 2021.
https://doi.org/10.1007/978-3-030-80713-9_14

1 Introduction

The perception of the road conditions is critical to motorcycle riders. *A helmet limits the rider's peripheral vision, which is crucial regarding the perception of speed* [1]. The rider moves their head and eyes away from the road to monitor the dashboard. This causes a higher risk of miscalculating road events that can lead to accidents [2]. Optical see-through augmented reality (OST-AR), by means of a helmet mounted display (HMD) displays the vital information regarding the dashboard within the field of view of the motorcycle rider. However, the regulations concerning the limited field of view must be met. These regulations are conforming to the ECE 22-05 [3]. *Developers want to ensure the correct and objective use of these regulations while developing new motorcycle helmets.* However, there is not yet an objective method available that is both easy to use and accessible to all developers.

There are a number of studies regarding HMD's specifically for aviation-based helmets. While these helmets have the same requirements regarding the field of view, they do not have similar visors [4, 5]. These studies, therefore, cannot be used in the development of motorcycle helmets, more specifically, the verification of the requirements regarding the field of view of a motorcycle helmet. Although, *these studies can be used as a reference to methodology.*

The aim of this study is to create *an objective method of verification for motorcycle helmets both in simulation software and in human testing.* In order for this method to be substantiated, a number of tests were performed. *A Solidworks 3D environment* was created containing an accurate human head [3] and a test set up. This set up was replicated in a usability lab to be used during *tests with subjects that are familiar to the use of a motorcycle helmet.* These tests resulted in a number of fields of view and these were compared to the simulated field of view from the 3D environment. With the outcome of this study, we strive to generate an objective 3D environment that can be utilized by manufacturers of motorcycle helmets for the verification of the field of view of new models.

2 Materials and Methods

Three methods of simulating the field of view of motorcycle helmets were applied to a standardized helmet. The standardized helmet is a HJC RPHA Max Evo, Size L and is certified *in reference to the ECE 22.05 specifications.*

Method 1. This method is applied *in a physical usability lab with test subjects.* The test is performed on 6 different test subjects with 9 different helmet models, one being the standardized helmet. The test subjects are *frequent motorcycle users* and familiar to wearing a motorcycle helmet. In this test, the test subjects are positioned in front of *a two-piece grid.* The grid is marked in the middle (Fig. 1A). This is the focal point of the test subject. The position of the subjects is determined by two strings that are equidistant to the grid. While positioning the subject, both strings are tensioned, making the test subject stand in the correct position. This means a 1m offset perpendicular to both grids. These strings are not directly attached to the grid, because of the changes in height per test subject. By means of two rods attached to the grid (Fig. 1B), the string height can be adjusted according to the eye height of the test subject. The subject is focusing on the center mark, meanwhile a green dot that is attached to a rod is slowly being moved into the field of view. Once the dot is spotted, it is placed onto the grid resulting in a border point of the perceived field of view. With the use of 12 key points, the field of view is determined. Each test subject performs this test with two helmets, one being the standardized helmet. The field of view is captured by a Ricoh Theta 360 camera for later comparison to the other two methods.

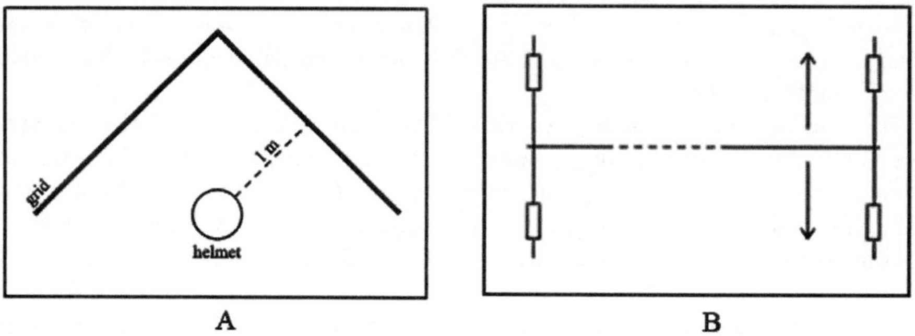

Fig. 1.

Subtest. During the subtest, that is performed in a usability lab as well, the subjects stand in a transversal position (Fig. 2). The subjects are photographed with the helmet on in a neutral position. Afterwards, the downwards angle of the helmet is processed.

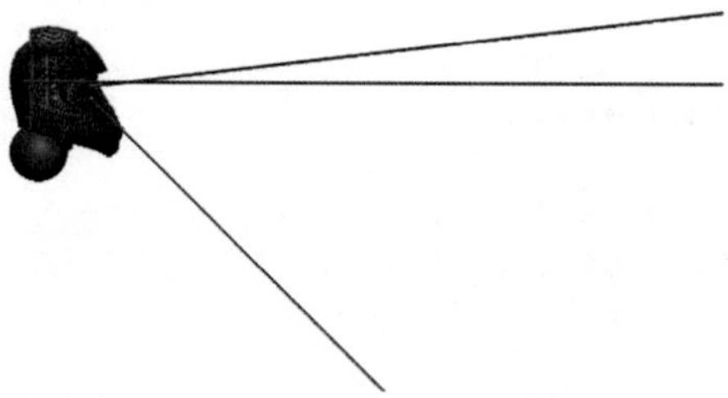

Fig. 2.

Method 2. In the second method a 3D modeling program, *Solidworks*, has been used to simulate the users view. A 3D scan of the helmet is positioned on a verified 3D head model [3]. The helmet and head are placed in *a 3D modelled environment*. In the environment the grids are simulated. This position (Fig. 3) has been recreated based on method 1. However, the tilt of the helmet affects the field of view. In order to verify the correct downwards angle of the head, a subtest (see subtest) was executed.

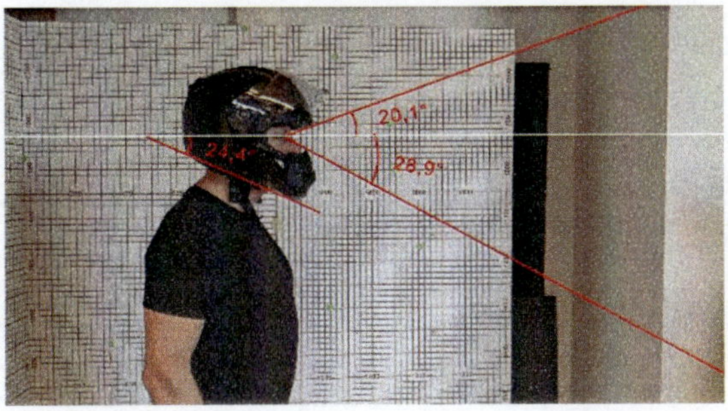

Fig. 3.

Method 3. Lastly, the field of view of the standardized helmet is determined by a Ricoh Theta 360 camera. Inside a verified human head model, the 360 camera is attached in reference to the ECE 22.05 standards (Fig. 4). The standardized helmet is placed upon the head model. Subsequently, the verified human head model is mounted on a tripod and placed in front of the two-piece grid. The position is measured according to the strings and the Ricoh Theta 360 camera captures a field of view.

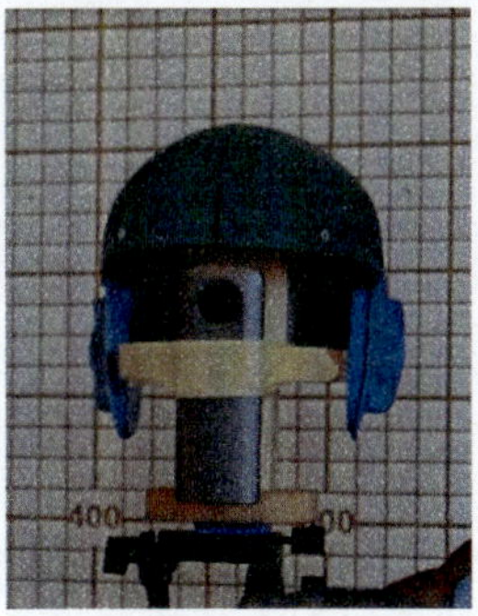

Fig. 4.

Afterwards, the fields of view are processed. These different methods of capturing the field of view are compared, in order to proof the authenticity of the verification tool in a 3D setting.

3 Results

Method 1 – Usability Lab Test: As shown in Table 1, two out of six test subjects did not pass the test regarding the minimal vertical field of view with the standardized helmet. In the test with their own helmets, again two subjects did not pass the test. However, in this test, the subjects failing the test are different subjects than in the first test using the standardized helmet. The table also shows the different angles of the helmet when the test subjects put it on. These angles, regarding the standardized helmet, are $21{,}15 \pm 4{,}11°$ (Table 2).

Table 1. Results method 1 - vertical field of view

Test subject name	Nr. Test subject	Standard helmet		Space between FoV standard helmet and FoV specification	Own Helmet		Quality mark
		Passed test?	Position (degrees)	(in decimeters)	Passed test?	Position (degrees)	
Frank	1	y	17	0	n	22,2	ECE 22.05
Ils	2	n	17	-1,5	y	17,6	FMV88 218
Kris	3	n	18,6	-1	y	18,1	ECE 22.05
Kris	4	y	23,4	1,75	y	21,2	ECE 22.05
Sylvie	5	y	26,3	1,25	y	28,3	ECE 22.05
Matthias	6	y	24,6	0,5	n	24,1	None
Kris 2	7				y	21,6	ECE 22.05
Matthias 2	8				y	30,8	ECE 22.05
Ricoh Theta 360°		y		2,5			

Table 2. Results method 1 - horizontal field of view

Test subject name	Nr. Test subject	Standard helmet		Space between FoV standard helmet and FoV specification	Own Helmet		Quality mark
		Passed test?	Position (degrees)	(in decimeters)	Passed test?	Position (degrees)	
Frank	1	n	17	0	n	22,2	ECE 22.05
Ils	2	y	17	-1,5	n	17,6	FMV88 218
Kris	3	n	18,6	-1	n	18,1	ECE 22.05
Kris	4	n	23,4	1,75	y	21,2	ECE 22.05
Sylvie	5	n	26,3	1,25	n	28,3	ECE 22.05
Matthias	6	n	24,6	0,5	n	24,1	None
Kris 2	7				y	21,6	ECE 22.05
Matthias 2	8				n	30,8	ECE 22.05
Ricoh Theta 360°		y		2,5			

In the horizontal view only one person passes the test with the standard helmet. This is an unexpected result, considering the standard helmets. For the test with their own helmet, Kris (test subject 4, 7) passes twice with his own helmets (two).

The resulting field of views were combined into one heatmap (Fig. 5) where all field of views are layered on top of each other and levelled with one eye height. The thinner more opaque line is the most saturated edge of all the field of views combined. The wider less opaque band shows the variation of the edges of the fields of view. The yellow colored center cross is the safe zone, which is the minimal field of view in reference to ECE - 22.05 standards.

Method 2 – 3D Environment: A 3D model of the standardized helmet is scanned in and brought into Solidworks 2019 (Fig. 6). The placement of the helmet strongly influences the resulting field of view [6]. Misplacement could result in a render that is not representative of the actual field of view of human test subjects. To determine the bottom angle of the helmet in this 3D environment, we refer back to the subtest. During this test, we gathered an average bottom angle of $21{,}15 \pm 4{,}11°$. We implemented these measurements into a 3D model. The render itself is captured with a 360° camera function in visualize (Solidworks).

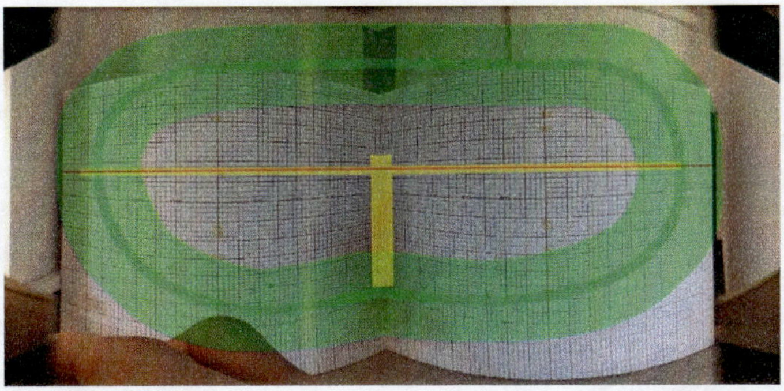

Fig. 5. Heatmap field of view – taken with Ricoh Theta 360

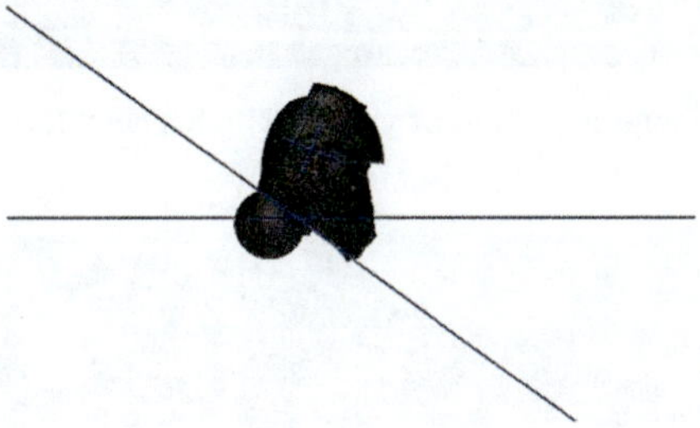

Fig. 6. Bottom angle 3D model

Method 3 – Ricoh Theta 360° Camera: This method shows the field of view of an inhuman test subject that would have a perfect spherical vision [7]. The resulting field of view, generated with the use of a Ricoh Theta 360 camera, shows a similar image (Fig. 7) to the generated heatmap of method 2. The one defining difference is the width of the image. This refers to the capabilities of the peripheral view of the test subjects, which will be discussed later in this study.

Comparison of Method 2 and 3. The heatmap of method 1 is projected onto the field of view generated by the 3D environment (Fig. 8) and the Ricoh Theta 360 image (Fig. 9). The results of both images are similar. Both the safe zones are indicated and the fields of view overlap.

4 Discussion and Conclusion

The Subjectivity of Method 1. Only one of six subjects passed the test relating to the lateral field of view. This can be accounted for in two different aspects. Firstly, these results are affected by the peripheral view [8] of the test subjects. During the test, *five out of six subjects* did not succeed in seeing the edge of the lateral field of view. When asked about it, they did not feel that the helmet was blocking their field of view, however

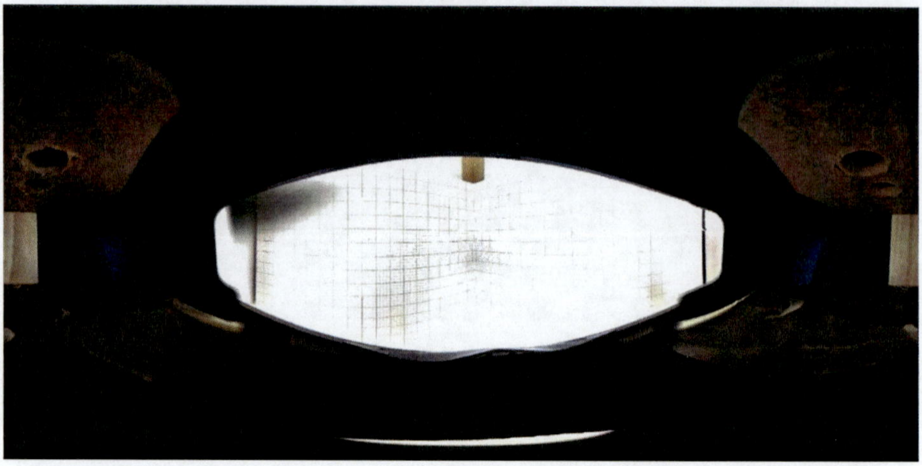

Fig. 7. Generated field of view – Ricoh Theta 360 Camera

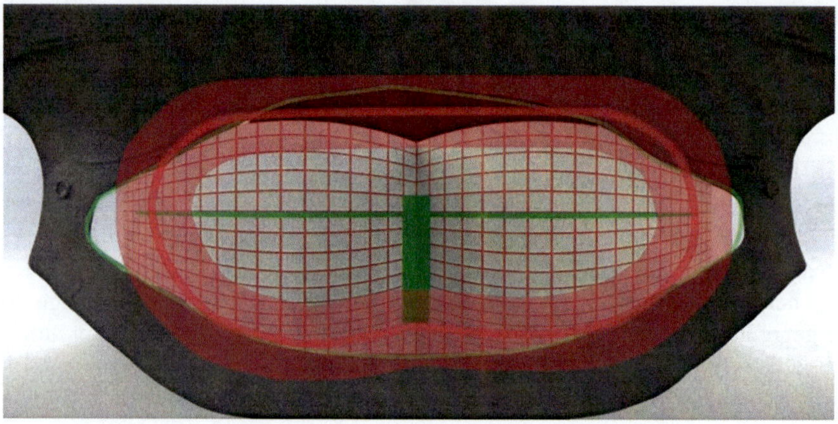

Fig. 8. 3D environment – field of view

it was their respective field of view [8] that was not as wide as the ECE -22.05 reference prescribed. During the test the subjects are asked to focus on one fixed dot at eye height. We do this to refrain the subject from moving their head during the test. But by focusing on the dot, the eyes could not move. The image of the Ricoh Theta 360° Camera visualizes the potential field of view without any peripheral hinderance except for the helmet. This image results in a wider field of view, that corresponds with the ECE - 22.05 reference. Secondly, there is a problem with perception [8]. We use people as test subjects and they each have a different manner of perceiving their surroundings [8]. Health, age, lifestyle, attention span and reflexes can influence the perception of the moving targets during the tests. Our resources are not complete enough to rule out certain anatomical and external factors that might influence the results.

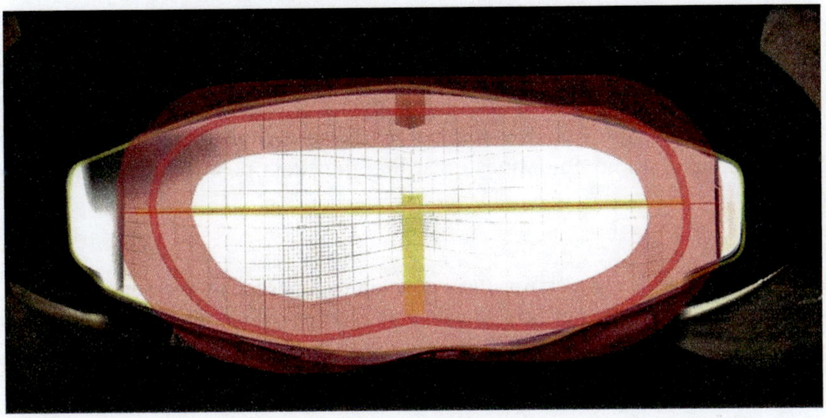

Fig. 9. 3D environment – field of view

More Test Subjects During Method 1. We tested six subjects in total. This is a relatively low number of test subjects during research. However, adding to the pool of test subjects could potentially raise more subjectivity. As mentioned before, the resources are not complete enough to filter out the external factors that might influence the method in itself. Adding more possibilities for these external factors might influence the objectivity of the paper.

Objectivity of the 3D Environment and the Ricoh Theta 360 Camera. Both the 3D environment and the camera images are based on the synthesized angle that we gathered from the first test. This angle is based off of the side images of the test subjects. We tested six subjects. The variation in the bottom angle was limited. This synthesized angle was used to both angle the helmet in the 3D environment as well as the helmet on the rigged head with the Ricoh Theta 360 Camera. This angle is dependent on the objectivity of the test subjects. We previously established that the use of more test subjects based on the field of view could potentially have a negative effect on the objectivity of this study, however it is certain that increasing the amount of side images will further create a more objective base [9] for the synthesized side angle that was used to position the helmets.

Aside from that aspect, both the Ricoh image and 3D render presented the same field of view, and created an objective method of predicting the field of view with different helmets. We can state this because the angle in itself was the same in both cases and the imagery is a match.

Conclusion – Objective Tool for the Verification of the Field of View. In this study, three different testing methods have been compared and converged in one evaluation. This way of evaluating helmets is accessible to any developer, designer or researcher for further use. The goal was to create an objective evaluation method that generates a singular definitive image that either passes or fails the helmet that is being tested, based on the field of view. For this to happen, we created a neutral testing environment and based on multiple different methods of testing, we generated an analog and a digital evaluation. The analog evaluation, using a 360 camera and a rigged head is more time consuming. The setup for testing the field of view of potential helmets will need the exact same testing situation as shown in method 3. For the digital evaluation, a simple CAD file is sufficient. The test setup in itself is already made and accessible, in order to provide an objective evaluation setup. The human testing in itself is not necessary for generating a field of view, however the side angles are vital for the positioning of the helmets, since there are no guidelines regarding the exact positioning of the helmet on the head.

Acknowledgments. This research was funded within the imec.icon project smartGLAZ, bringing academia and industry together (Flanders Innovation & Entrepreneurship Project No. HBC.2017.0634).

References

1. Cannon, M.W.: Recent advances in understanding peripheral vision. In: Proceedings of 30th Annual Meeting of the Human Factors Society, vol. 30, pp. 601–603 (1987)
2. Traffic Safety Facts. 2013 Data. Motorcycles. (Report No. DOT HS 812 148) (2015)
3. United Nations: E/ECE/324 (2012). http://www.unece.org/fileadmin/DAM/trans/main/wp29/wp29regs/r022r4e.pdf. Accessed 16 Oct 2019
4. Melzer, J.: How much is enough? The human factors of field-of-view in head-mounted displays (2017). https://doi.org/10.1117/12.2263334
5. Foote, B., Melzer, J.: A history of helmet mounted displays. 94700T (2015). https://doi.org/10.1117/12.2181337
6. Berger, R., Adams, S.: Motorcycle Helmet Design. Proc. Human Factors Soc. Annu. Meet. **28**(5), 433–437 (1984)
7. Field of view with and without motorcycle helmets. Appl. Ergon. **8**(2), 115 (1977)
8. Edwards, D., et al.: The useful field of view test: Normative data for older adults [Ebook] (2006)
9. Ladd, H., Goertz, M.: Handbook of Research in Education Finance and Policy. Routledge, New York (2015)

Attitudes of Young Generation Towards Traditional Irrigation System "Foggara" in the Southwestern Algeria: A Green Ergonomics Approach

Mohammed Mokdad[1](✉) and Bouhafs Mebarki[2]

[1] College of Arts, University of Bahrain, Zallaq, Bahrain
[2] Laboratory of Ergonomics, University of Oran 2, Bir El Djir, Algeria

Abstract. For centuries, the main source of water in the southwestern Algeria is the "Foggara" traditional irrigation system, which was preserved and well considered by the local population. However, this system is in real threat these days. This paper, aims to investigate the attitudes of the young generation of residents in the region towards the actual status of the Foggara. A sample of 159 residents of the Oasis "Ksar" of Saly (Algeria) participated in the study, with an educational level ranging from primary school to university. An attitude scale was designed for the purpose, to measure four attitude dimensions governing the relationship between the Foggara and its users. Results showed an average level of positive attitudes of the young generation of residents towards the "Foggara". There was no significant effect of educational level and age group on attitudes towards Foggara. Results were discussed from a sustainable green ergonomics point of view.

Keywords: Foggara · Sustainability · Attitudes · Participative ergonomics

1 Introduction

Sustainability and sustainable development has become an important trend for ergonomists, because it is interested in creating a more sustainable work environment, which cannot be achieved unless it tackles the daily concerns of people's livelihood. The United Nations Brundtland Commission [1] gave the following definition: "Sustainable development is development that meets the needs of the present without compromising the ability of future generations to meet their own needs."

Since Moray's [2] and Helander's [3] appeals for ergonomists to address "global environmental and social problems", ergonomists are more and more interested in sustainable development issues like, sustainable development [4], and green ergonomics [5–7].

One of the most precious natural resource on earth is water, particularly in desert environmental conditions, without its preservation green ergonomics goals cannot be achieved. These goals are the duty of local population, especially, the younger generations.

R. S. Goonetilleke et al. (Eds.): AHFE 2021, LNNS 273, pp. 109–118, 2021.
https://doi.org/10.1007/978-3-030-80713-9_15

In this perspective, the present study comes to shed light on attitudes of the younger generation towards the source of water in the southwestern Algeria (Touat, Gourara and Tidikelt regions), and their perception of the multiple threats that endanger the sustainability of their traditional irrigation system which preserved water for centuries, and guaranteed a sustainable livelihood for the local inhabitants living in Sahara desert villages called "Ksours", created around Oasis's Palmaris. Their main substantial activity is cultivation of the palmary, which are small plots of cultivable land, due to the scarcity of water in Sahara desert. Their traditional irrigation system called "Foggara" is a millenary irrigation system [8, 9], without which life cannot be in such adverse climatic conditions. It was the main reason for the reunion and settlement of inhabitants in this extremely hot and dry area [10].

The Foggara is an environmental friendly irrigation system, which preserved underground water for centuries, as water basins are a non-renewable source of water [11]. The Foggara is a very important component of the Touat, Gourara and Tidikelt regions' heritage. It occupies a very prominent position within society. With the aim of preserving this position, families of these regions educate young people to love and respect the Foggara from the earliest age.

However, changes such as the neglect of Foggaras in some villages [8, 9], as well as deep drilling for new irrigation systems that are drying up the Foggara as a whole [12, 13].

The topic of attitudes is one of the most important topics in psychology [14]. The relationship between attitudes and behavior is not direct, but rather mediated and controlled by multiple variables [15].

This paper, aims to investigate the attitudes of the younger generation towards the current status of the Foggara. Consequently, the following questions were formulated:

1. What are the attitudes of the young generation towards the Foggara?
2. Are there differences in attitudes of the young generation towards the Foggara attributed to the level of education and age groups?

2 Methodology

Method. To accomplish this research, the researchers used the descriptive approach mainly the sample survey.

Sample. A sample of 159 male residents of the Oasis "Ksar" of "Saly" (Adrar governorate, Algeria) participated in the study (age range: 10 to 40 years), with an educational level ranging from primary school to university (Table 1). Female participants were not feasible, as the Foggara work is a male business in these communities, for the simple reason is that, the Foggara work is physically demanding and highly risky for females.

Table 1. Demographic characteristics of the sample.

No	Sample demographic data			Some characteristics of the age group
	Age group	Education level	Total	
1	10–15 years	Primary school	05	They are at the school age. They can help in light tasks during Foggara maintenance works
		College & secondary school	13	
		University	00	
2	16–20 years	Primary school	07	They can do the physically demanding tasks. They are well aware of the importance of Foggara. They don't have economic responsibilities
		College & secondary school	59	
		University	01	
3	21–40 years	Primary school	19	They are generally married and have economic duties towards their families. They are well acquainted with the Foggara challenges
		College & secondary school	33	
		University	22	
Total			159	

Tools. In this research, Researchers designed and used an attitude scale to answer the above mentioned questions trying to benefit from the advantages of self-report methods in measuring attitudes according to Schwarz [16].

Attitude Scale. The scale seeks to measure attitudes towards the Foggara. It consists of (36) items centered around four dimensions (knowledge of Foggara, Foggara position in society, restoring life to Foggara and obstacles facing the Foggara's sustainability) as seen in Table 2. Each dimension includes the cognitive component, the emotional component, and the behavioral component. The scale was designed according to Likert's format. Each has five choices: Strongly agree, agree, neutral, disagree, and strongly disagree. Respondents are requested to answer all of these items without leaving any unanswered. After which, each of the dimensions is evaluated based on three levels of evaluation: low (less than 2.33), medium (2.33–3.66) and high (greater than 3.66).

Scale Validity: To ensure the validity of the scale, it was presented to six professors of Psychology at the University of Oran. Five of them (84%) agreed that the scale actually measures the attitudes towards the Foggara.

Scale Reliability: To ensure the reliability of the scale, it was applied twice to a small sample of residents of the "Saly" Ksar in governorate of Adrar, Algeria. The interval

Table 2. Scale dimensions and items.

No	Dimension	Items
1	Knowledge of Foggara	1, 2, 3, 4, 5, 6, 7, 8, 9
2	Foggara position in society	10, 11, 12, 13, 14, 15, 16, 17, 18
3	Restoring life to Foggara	19, 20, 21, 22, 23, 24, 25, 26, 27
4	Obstacles facing Foggara's sustainability	28, 29, 30, 31, 32, 33, 34, 35, 36

between them was 15 days. The reliability was found to be 0.84. The scale was considered valid and reliable. Hence, it is usable.

3 Results

3.1 What Are the Attitudes of the Young Generation Towards the Foggara?

The results obtained can be seen in Table 3.

Table 3. Results obtained.

No	Items	Mean	SD	Attitude
1	The Foggara is what unites the Ksar residents	4.60	1.03	High
16	I strive to place the Foggara in a prominent place in our community	4.26	1.22	High
11	I think the time has come to abandon the Foggara and rely on modern methods of obtaining water	4.22	1.05	High
17	I do my best to keep the Foggara safe within the community	4.18	1.23	High
34	I do not hesitate if I am asked to belong to an association for the care of Foggara	4.11	2.31	High
33	I am very concerned about the low water level in the Foggara	4.02	2.22	High
8	I regularly provide funds in support of the Foggara maintenance budget	4.01	1.45	High
2	The Foggara is the most important thing that the grandparents left for the grandchildren	4.00	1.02	High
9	I do not hesitate to punish children who mess and spoil the Foggara's water	4.00	1.01	High
32	I hate individuals who put obstacles in the way of Foggara growth	3.95	2.23	High

(*continued*)

Table 3. (*continued*)

No	Items	Mean	SD	Attitude
10	I think that the place of Foggara in Touat, Gourara and Tidikelt society is great	3.88	1.22	High
24	It is a pleasure for the government to pay for the costs of restoring the Foggara	3.88	1.20	High
35	I encourage programs aimed at restoring and maintaining the Foggara	3.88	2.42	High
25	I participate in volunteer work aimed at saving the Foggara from collapse	3.74	1.20	High
23	I hate all the forms of marginalization that led to the destruction of Foggara in my region	3.66	1.01	Medium
19	I think that holding a ceremony honoring the winners of the best Foggara competition is significant	3.56	1.22	Medium
22	I abhor all the fraudulent methods that corrupt Foggara restoration projects in my region	3.54	1.22	Medium
26	I am ready to quarrel with everyone who throws dirt beside or in the Foggara wells	3.47	1.15	Medium
21	It would be useful to allocate a financial envelope to restore the Foggara in my region	3.45	1.45	Medium
18	I educate my sons to love the Foggara since their age	3.33	1.23	Medium
36	I attend the national and international forums that are organized around Foggara development	3.22	2.23	Medium
3	The Foggara is obsolete, and it is necessary to search for alternatives to it	3.20	1.23	Medium
12	Modern irrigation systems threaten the Foggara system	3.20	1.23	Medium
31	I like to attend meetings that aim to find solutions to the difficulties that hinder Foggara development	3.18	2.01	Medium
5	I hate everyone who attacks the Foggara	3.10	1.55	Medium
6	The Foggara is part of my cultural being	3.08	1.26	Medium
13	I love everyone who contributes to the stabilization of the Foggara in the region or in the society	3.02	1.32	Medium
28	I imagine that the hot climate is a major obstacle to the continuation of the Foggara in the region	2.99	1.20	Medium
20	Interest in the Foggara is at the core of caring for the national heritage in general	2.98	1.56	Medium
14	I prefer Foggara watering to other watering methods	2.89	1.55	Medium

(*continued*)

Table 3. (*continued*)

No	Items	Mean	SD	Attitude
29	I see that the lack of specialized labor is one of a major obstacle to developing Foggara	2.89	1.55	Medium
4	The Foggara system embodies the principle of social cooperation and solidarity between partners	2.80	1.00	Medium
27	I am always present in the social cooperation system, "Twizeh", which seeks to promote Foggara	2.59	1.16	Medium
7	I participate in volunteer work for the maintenance of the Foggara	2.57	1.56	Medium
30	I think that the inability to use technology in Foggara maintenance contributed to its neglect	2.35	1.30	Medium
15	The Foggara provided all the best for the community	1.89	1.26	Low

As is evident from Table 2, the mean values of the statements are arranged in descending order. They ranged between 4.60 (with a standard deviation of 1.03) for the item (The Foggara is what unites the Ksar residents), and 2.12 (with a standard deviation of 1.26) for the item (The Foggara provided all the best for the community). In light of what has been mentioned above, the general average of these results reached 3.44 (with SD of 0.62).

What does this result represent? If we go back to the criterion that was referred to previously (scale correction), we will find that this result reflects an average attitude level because it is in the range (2.33 and 3.66). However, it must be noted that it is very close to the great end of this field (3.66). This makes us say that the overall level of youth attitudes is close to good.

With regard to the four scale dimensions, the obtained results can be seen in Table 4.

Table 4. Scale dimensions and items.

Dimension	Mean	SD
Knowledge of Foggara	3.48	0.68
Foggara position in society	3.82	0.78
Restoring life to Foggara	3.29	0.40
Obstacles facing Foggara's sustainability	3.39	0.61

It can be seen that the position of Foggara dimension has got the highest mean (3.82, SD: 0.78). Then the dimension Knowledge of Foggara in the second position (Mean 3.48, SD: 0.68). The Dimension Obstacles facing the Foggara in the third position (Mean: 3.39,

SD: 0.61). Finally, in the fourth position the dimension restoring life to Foggara with a mean of 3.29 and SD of 0.40.

3.2 Are There Differences in Attitudes of Inhabitants (the Young Generation) Towards the Foggara Attributed to the Level of Education and age Groups?

To answer this question, the means and standard deviations of the attitude scale items according to the age and educational level variables were calculated.

1. First, are there differences between educational level groups in their attitudes towards the Foggara? Table 5 shows the results.

Table 5. Descriptive results.

Educational level	Mean	SD	N
Primary school	3.9677	1.11007	31
College & secondary school	4.0286	1.09595	105
University	3.9565	1.14726	23
Total	4.0063	1.09946	159

Table 6. ANOVA results.

	Sum of squares	df	Mean square	F	Sig.
Between groups	.155	2	.078	.063	.939
Within groups	190.839	156	1.223		
Total	190.994	158			

Results in Tables 5 and 6 show that there is no significant effect of educational level on attitudes towards Foggara at the $p < .05$ level for the three educational levels [$F(2,156) = .063$, $p = .939$].

2. Second, are there differences between age groups in their attitudes towards the Foggara?

Results show that there is no significant effect of age group on attitudes towards Foggara at the $p < .05$ level for the three age groups [$F(2,156) = .458$, $p = .634$].

Table 7. Descriptive results.

Age group (Years)	Mean	SD	N
10–16	3.77	1.21	18
17–21	4.01	1.08	67
21–40	4.05	1.09	74
Total	3.94	1.12	159

Table 8. ANOVA results.

	Sum of squares	df	Mean square	F	Sig.
Between groups	1.114	2	.557	.458	.634
Within groups	189.880	156	1.217		
Total	190.994	158			

4 Discussion

Results of the present study have shown that, the general level of the sample's attitudes towards Foggara are in the middle range (2.33–3.66), but it's actually close to the good. This may mean that the attitudes of the young generation of residents towards the Foggara system of irrigation is not up to the challenges facing not only this irrigation system, but the whole eco-system of the region [10–12]. The obvious and urgent action in this case is to boost the positive attitudes aspects of the young generation in order to change their behavior. Thatcher [6], stipulates that "Perhaps the most important role for green ergonomics would be in facilitating larger, and systemic behavior change". This can be done through different ways, one of which is through training interventions to facilitate appropriate behavior change for the conservation, preservation and restoration of the natural capital. Education can also play an important role to change behavior of young generation. Notables and elders of the Ksar can sit with the young generation and explain everything to them related to the change. This is an effective method for change [17, 18]. Other researchers [19] propose a global attitudes and behavior change strategies to support a global sustainability transition, which is out of the scope of the present study.

Changing behavior of the young generation, at micro level, will not be a hard exercise, as showed in (Table 4) the four attitude dimensions towards the Foggara are well rated by respondents of the present study, which indicates that the Foggara has still, a "high position" within the young generation community, they know it very well, they are acquainted with the obstacles facing it, and they are quiet aware of the difficulties of restoring life to Foggara. These difficulties can be overcome through the introduction of some technical equipment which facilitates the maintenance work operations that are physically demanding and risky at the same time [20].

As shown in (Tables 5, 6, 7 and 8) there is no significant effect of educational level and age group on attitudes towards Foggara. which can be explained by the fact that young generation in the "Ksours" community of southwest Algerian regions are brought up with love and respect values for the Foggara, which is translated into positive attitudes towards it, as revealed in the above mentioned results, no matter the age or the educational level of individuals.

The present study, advocates for a clear strategic work among the younger generation of inhabitants of the southwest Algerian regions, to preserve and promote the Foggara irrigation system. The young generation inhabitants should be directly involved in developing and implementing such strategy.

It is time for "green ergonomics" to intervene, through a "participative ergonomics approach", in order to preserve Foggara irrigation system, and brings attention of the stakeholders of Oases in the southwestern Algeria, to gather their efforts around a sustainable living environment. The direct and most important stakeholders in this process are the new generation of inhabitants, who inherited such a millenary irrigation system, which preserved men and nature in a complete harmony for centuries, but this harmony is in real jeopardy these times.

5 Conclusion

This research was conducted to study the attitudes of the young generation of the habitants of the Ksours in the Southwestern Algeria, towards Foggara as a system that provides water to the Ksar residents for drinking, watering farms and meeting other life needs. The results of the research showed that the attitudes of the Ksours population were generally of a medium level, which means that to preserve Foggara, attitudes towards Foggara should be boosted among young generation. It was expected that these attitudes would be of a high level, considering that the Foggara occupies a prominent place among the inhabitants of the Ksours, which it obtained from the importance of water in life, especially in areas with extreme heat in summer.

Although, local efforts to preserve Foggara exist, here and there, in some Ksour localities, but actions are not coordinated, nor well studied. A sustainable development strategy has to be a national one, incorporating Foggara preservation aspects as components in the ecosystem as a whole.

Acknowledgments. Thanks and gratitude go to all who participated in the research paper including the sample population as well as researchers who helped facilitate the data collection and distribution.

References

1. United Nations General Assembly: Report of the world commission on environment and development: Our common future. United Nations General Assembly, Development and International Co-operation: Environment, Oslo, Norway (1987)

2. Moray, N.: Ergonomics and the global problems of the twenty-first century. Ergonomics **38**(8), 1691–1707 (1995)
3. Helander, M.G.: Forty years of IEA: some reflections on the evolution of ergonomics. Ergonomics **40**(10), 952–961 (1997)
4. Steimle, U., Zink, K.J.: Sustainable development and human factors. In: Karwowski, W. (ed.) International Encyclopedia of Ergonomics and Human Factors, vol. 2, pp. 2355–2360. Taylor & Francis, London (2006)
5. Hanson, M.: Green ergonomics: embracing the challenges of climate change. Ergon. **480**, 12–13 (2010)
6. Thatcher, A.: Green ergonomics: definition and scope. Ergonomics **56**(3), 389–398 (2013)
7. Sharan, D.: How green is ergonomics in India? In: Bagnara, S., Tartaglia, R., Albolino, S., Alexander, T., Fujita, Y. (eds.) Proceedings of the 20th Congress of the International Ergonomics Association (IEA 2018): Volume VIII: Ergonomics and Human Factors in Manufacturing, Agriculture, Building and Construction, Sustainable Development and Mining, pp. 1009–1010. Springer, Cham (2019). https://doi.org/10.1007/978-3-319-96068-5_112
8. Saadi, M.A., Remini, B.: Water in the Ksours: what genius? Case of Timimoun and Tiout (Algeria). J. Fund. App. Sci. **12**(2), 525–537 (2020)
9. Hadidi, A., Yaichi, M., Saba, D., Habi, M.: An overview of the traditional irrigation system (Foggaras) in Taouat Region. In: AIP Conference Proceedings, vol. 2123, no. 1, pp. 020008-1–020008-8. AIP Publishing LLC, USA (2019)
10. Biara, R.W., Alkama, D.: Future of an oasis to the detriment of a hydraulic monument "Case of Timimoun." IMPACT: Int. J. Res. Eng. Technol. **2**(7), 113–122 (2014)
11. Mohamed, B., Remini, B.: Water wells' exploitation and its impact on the drying up of Foggaras. Appl. Water Sci. **7**(1), 349–359 (2014). https://doi.org/10.1007/s13201-014-0250-2
12. Remini, B., Achour, B.: The water supply of oasis by Albian Foggara: an irrigation system in degradation. LARHYSS J. **26**, 167–181 (2016)
13. Benaricha, B., Khaldi, A., Elouissi, A.: Physico-chemical characterization of irrigation water Foggaras of Aougrout (Southwest of Algeria). J. Appl. Environ. Biol. Sci. **7**(1), 19–29 (2017)
14. Allport, G.W.: Attitudes. In: Murchison, C. (ed.) A Handbook of Social Psychology, vol. 2, pp. 798–884. Clark University Press, Worcester (1935)
15. Ajzen, I., Fishbein, M.: Attitude-behavior relations: a theoretical analysis and review of empirical research. Psychol. Bull. **84**(5), 888–918 (1977)
16. Schwarz, N.: Attitude measurement. Attit. Attit. Change **3**, 41–60 (2008)
17. Recardo, R.J.: Overcoming resistance to change. Natl. Prod. Rev. **14**, 5 (1995)
18. Şerban, A., Aviana, A.E., Popescu, D.M.: Resistance to change and ways of reducing resistance in educational organizations. Ovidius Univ. Ann. Ser. Econ. Sci. **20**(1), 785–788 (2020)
19. Leiserowitz, A.A., Kates, R.W., Parris, T.M.: Do Global Attitudes and Behaviors Support Sustainable Development? Environ.: Sci. Pol. Sustain. Dev. **47**(9), 22–38 (2005)
20. Mokdad, M., Mebarki, B., Mokdad, I., Bouabdallah, L.: Ergonomics of date palm irrigation work: Algerian Foggara as an example. In: Karwowski, W., Goonetilleke, R.S., Xiong, S., Goossens, R.H.M., Murata, A. (eds.) AHFE 2020. AISC, vol. 1215, pp. 282–288. Springer, Cham (2020). https://doi.org/10.1007/978-3-030-51549-2_37

Check-App Voice®: A Tool to Self-evaluate Dysphonia in Speaking Voice Among Teachers

Maria Patrizia Orlando[1], Fabio Lo Castro[2(✉)], Maurizio Diano[3], Raffaele Palomba[4], Raffaele Mariconte[4], Martina Amodeo[1], and Claudia Giliberti[4]

[1] Dipartimento Organi di Senso, Università Sapienza Rome, Rome, Italy
[2] CNR-INM, Sect. of Acoustics and Sensors O.M. Corbino, Rome, Italy
fabio.locastro@cnr.it
[3] Direzione Regionale Calabria, Unità Operativa Territoriale, INAIL, Catanzaro, Italy
[4] Dipartimento Innovazioni Tecnologiche e Sicurezza degli Impianti, Prodotti ed Insediamenti Antropici, INAIL, Rome, Italy

Abstract. During the SARS-Cov-2 pandemic, a tailor-made app (Check-app voice®) has been developed to submit Voice Handicap Index (VHI) questionnaire for self-assessment of dysphonia to a sample of 219 teachers. Check-app voice® provides VHI calculation and collects answers for post-processing analysis. Results have been statistically analysed taking age, gender, school categories (nursery and primary, middle, high school, and university), length of service, anamnestic data (illnesses, smoke, alcohol habits) into account. The sample of female teachers (the most numerous) has been analysed in detail. The contribution of three domains, functional, physical, and emotional to the overall VHI, has been discussed per type of school and age group. The results show that this tool represents an easy, quick to use prevention method for promptly identifying the groups most at risk of the onset of vocal disorders.

Keywords: App · Voice impairments · Dysphonia · VHI · Teachers

1 Introduction

Dysphonia is a voice disorder characterized by a qualitative and/or quantitative alteration of spoken voice, which follows a structural and/or functional modification of one or more organs involved in its production. Vocal impairments occur to varying degrees, from hoarseness, weakness, breathiness or harshness, to pain, tension, or effort when talking [1]. American Academy of Otolaryngology, defines dysphonia as any "deviation in the vocal quality, pitch, loudness, and vocal effort that affect communication or produces a negative impact on the voice-related quality of life" [2]. Many factors may predispose the development of voice disorders: individual factors, as the presence of pathologies (nasal respiratory disorders, allergic rhinopathies, gastric disorders); particular lifestyles as voice over-using for job or recreational activities, smoking, etc.; harmful vocal habits (screaming, singing, etc.); professional factors (voice-demanding

R. S. Goonetilleke et al. (Eds.): AHFE 2021, LNNS 273, pp. 119–127, 2021.
https://doi.org/10.1007/978-3-030-80713-9_16

occupations); environmental factors (noisy background, dusty environments, thermal changes, etc.) [3].

Only few studies have investigated voice disorder proportion in general population. Voice problems upset 1 in 13 adults annually and affect patients of all ages and sex, but have an increased prevalence in women and elderly [4].

In modern economies, for about a third of the workforce, voice is the primary working tool. Telesales, sport coaches, singers, actors, call-center employees, and teachers, can be considered as professional voice users, particularly at risk for developing the so-called Work Related Voice Disorders [5, 6].

During the last decade, voice disorders among teachers have increased and several researches were conducted to evaluate the prevalence and etiology of these dysfunctions [6]. About one-third of the teacher workforce experiences voice problems [7], occurring 32 times more than in other professions [8]. Teaching needs a moderate voice quality, although the vocal load is heavy, involving extra factors such as continuous background noise, long and loud speaking, poor acoustic environments, poor air quality, lack of voice amplifiers, lack of opportunity for vocal rest [5, 6]. Another important risk factor is the lack of professional voice education, although the results of properly designing courses for the correct use of the voice are encouraging, showing improved vocal health and increased vocal awareness [7]. The most common symptoms have been found to be tired or strained voice, felt tired when speaking, temporary voice loss, and sore or painful throat, while the most common organic conditions, acute laryngitis, vocal nodules, polyps, and edema [6]. Therefore, it is essential to detect and manage any voice issues at early stages to enhance treatment outcomes and avoid progression to more serious complications that could affect the quality of life [9].

The assessment of a patient with a voice problem is multidimensional and usually includes the use of clinician instruments (fibro-laryngoscope, endoscopy, aerodynamic and acoustic analysis tests, etc.). The subjective voice evaluation using self-perceived questionnaires has been proven to effectively record the patients' experience of their voice disorder [10, 11]. Studies have reported that the validated Voice Handicap Index (VHI) [12] questionnaire represents an important tool for specialists and every person to check preliminarily how voice impairments can affect social, emotional, and professional comfort [13].

The aim of this study is to evaluate, in a sample of teachers, the self-perceived vocal state using a tailor-made app (Check-app voice®) which implemented the validated VHI questionnaire for the self-assessment of dysphonia.

2 Methods

Check-app voice® was built in MIT App Inventor [14] and developed for Android® mobile phones. It was employed for remote administration of the VHI questionnaire, in a sample of 219 teachers of nursery and primary, middle, high school, and university. VHI is a 30-item questionnaire, divided into three domains: functional, physical, and emotional (Table 1). The functional subscale investigates the consequences of a voice disturbance on daily activities; the physical subscale is related to the perception of dysphonia in terms of physical symptoms; the emotional subscale measures the effects on the emotional life of a voice problem [11]. Participants choose the answer to every item in each subscale, between five, scored: 0 *never*, 1 *almost never*, 2 *sometimes*, 3 *almost always*, 4 *always*. The total score ranging from 0 to 120 and is divided into four grades of severity. Score 0 corresponds to *no disturbance*, between 1–40 to a *slight disturbance*, between 41 and 80 to a *moderate disturbance*, and over 81 to 120 a *severe disturbance* [12, 15]. The VHI typically takes less than 5 min to complete and is usually done without assistance from health care providers.

Table 1. Voice Handicap Index (VHI) questions in the version validated by Jacobson.

VHI questions			
#	Functional	Physical	Emotional
1	My voice makes it difficult for people to hear me	I run out of air when I talk	I am tense when talking to others because of my voice
2	People have difficulty understanding me in a noisy room	The sound of my voice varies throughout the day	People seem irritated with my voice
3	My family has difficulty hearing me when I call them throughout the house	People ask, "What's wrong with your voice?"	I find other people don't understand my voice problem
4	I use the phone less often than I would like	My voice sounds creaky and dry	My voice problem upsets me
5	I tend to avoid groups of people because of my voice	I feel as though I have to strain to produce voice	I am less outgoing because of my voice problem
6	I speak with friends, neighbors, or relatives less often because of my voice	The clarity of my voice is unpredictable	My voice makes me feels handicapped
7	People ask me to repeat myself when speaking face-to-face	I try to change my voice to sound different	I feel annoyed when people ask me to repeat
8	My voice difficulties restrict my personal and social life	I use a great deal of effort to speak	I feel embarrassed when people ask me to repeat
9	I feel left out of conversations because of my voice	My voice is worse in the evening	My voice makes me feel incompetent
10	My voice problem causes me to lose income	My voice "gives out" on me in the middle of speaking	I am ashamed of my voice problem

After starting Check-app voice®, a preliminary screen reminds the rights of personal data processing. Providing consent, the button to continue is enabled. An initial screen collects patient's anamnestic data regarding age, gender, profession (teacher of nursery and primary, middle, high school, university), length of service, previous illnesses (none, respiratory, intestinal, gastric, auditory and other), smoke and alcohol consumption. This information is mandatory and a warning appears if an incongruence is checked. Then, three vertically scrolling screens are displayed, referring to the three VHI domains, each reporting ten questions.

Table 2. Warning messages in the Check-app voice ® final screen.

Score range	Perceived dysphonia	Icon	Warning messages
0	No disturbance		No related issues
1-40	Slight disturbance		It is advisable to consult your family doctor, if the recommended therapy does not have any effect, consult the ENT specialist or speech therapist
41-80	Moderate disturbance		It is advisable to consult in a short time the ENT specialist or speech therapist
81-120	Severe disturbance		It is necessary to consult immediately the ENT specialist or speech therapist

The answers are selectable via pop-ups. At the end of the questionnaire, four traffic light icon warning messages are shown, depending on the score achieved (green: slight, yellow: moderate, red: severe perceived dysphonia) and useful additional information, regarding vocal hygiene rules together with advice to a pharmacological or speech therapy, are provided (Table 2). Tapping the button "send form" all data are updated and collected in a GoogleSheet® for post-processing analysis.

The sample was clustered by school categories, age groups (up to 35 years old, from 36 to 45, from 46 to 55, 56 and more), and length of service (up to 10 years on the job, between 11 and 20, between 21 and 30 and over 30). The entire sample was statistically analysed according to the four levels of perceived dysphonia and the contribution of the three domains to the total score was verified by gender, school categories, age groups, and length of service. VHI values per gender were statistically analysed by presence/absence of pathologies, smoking habits, and alcohol consumption (t-test $p < 0.01$).

3 Results and Discussion

The sample of 219 teachers, 181 females and 38 males, was distributed into the following school categories: 14.1% nursery and primary, 19.3% middle, 52.5% high school teachers, and 14.1% university. Among females, the median age was 51.2 years (SD = 9.3 years) with the extreme value of 28 and 67 years. Among males, the median age was 50.6 years (SD = 10.2 years) with extreme values of 26 and 71 years. The prevailing age group was between 46 and 55 years for both females (37.6%) and males (36.8%).

The high school was prevalent among both females (49.7%) and males (66%). Females were mostly in the range 21 - 30 years on job (34.2%), while males had fewer years of service (between 11 and 20 years, 33.3%). Pathologies were absent in 72% of the sample, without gender difference. In particular, 70.2% of females and 81.6% of males declared no pathologies. Among females, 6.6% had gastric pathologies, 3.3% intestinal, 2.8% respiratory pathologies, 17.1% other pathologies. About smoke and alcohol consumption, the percentages were respectively 16% and 1.1%. Among males, 13.2% reported smoking and 18.4% alcohol.

Table 3. Distributions of VHI levels among female teachers in the school categories.

VHI	University teachers (%)	Nursery and primary school teachers (%)	Middle school teachers (%)	High school teachers (%)	Total Female (%)
0	4.2	–	2.8	3.3	2.8
1–40	95.8	90.3	75.0	85.6	85.6
41–80	–	9.7	22.2	11.1	11.6
81–120	–	–	–	–	–

The VHI data, divided into the four severity ranges of perceived dysphonia, analysed by age group, length of service, and gender, showed that, in almost all cases, the total VHI was included in the range 1–40 (slight dysphonia) for all age groups, regardless years on the job. In Table 3, the distribution of VHI levels among female teachers in the four school categories was reported. In particular, 85.6% of females and 97.4% of males showed slight dysphonia, moderate for 11.6% of females and 4% of males, no disturbance for only 2.8% of females (no results of VHI = 0 among males). No severe dysphonia was found in the sample, both among males and females. For slight dysphonia, the most relevant percentages were obtained for university (95.8%) and nursery and primary female teachers (90.3%). The latter had never scored no perceived disturbance. Among the other categories, the percentages of no disturbance remained at very low values (4.2% was the maximum value for university teachers). The highest value for moderate perceived dysphonia was found among middle school teachers (22.2%). Even in the male sample, the most representative condition was related to a slight dysphonia, mainly in the high school (96% of the sample).

In Fig. 1 the total VHI for the female sample, pooled by age and years on the job was shown. The greater statistical variability was referred to the age groups up to 35 years (1st quartile = 1.0, 3rd quartile = 15.5) and between 46 and 55 years (1st quartile = 7.0, 3rd quartile = 21.5, the highest statistical VHI value, Fig. 1, left). Similarly, depending on the length of service (Fig. 1, right), an important statistical variability could be noted up to 10 years (1st quartile = 5, 3rd quartile = 19.8), and from 11 to 20 years (1st quartile = 8, 3rd quartile = 20.5) while the lower one for teachers over 30 years on the job (1st quartile = 4, 3rd quartile = 10). An adaptation to the vocal discomfort, increasing the length of service and ageing, could explain this result.

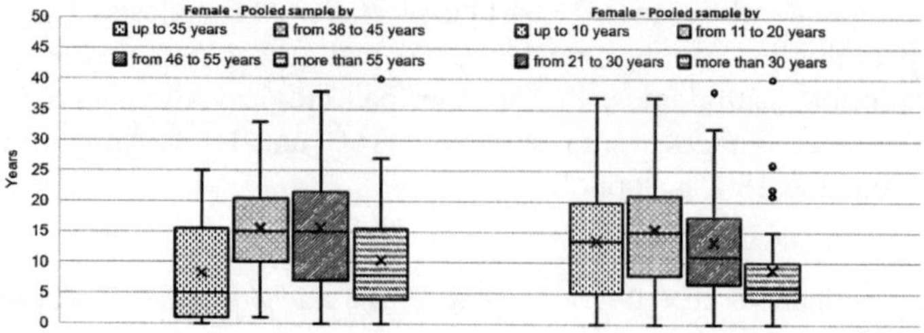

Fig. 1. Total VHI for female sample pooled by age (left) and years on the job (right).

In Fig. 2, the contribution of three domains (functional, physical, and emotional) to the overall VHI for the entire sample was presented. In terms of detected and statistical values (1st and 3rd quartile, mean and median), the most significant contribution was provided by the physical aspect, while the least contribution to the emotional one, both among female and male.

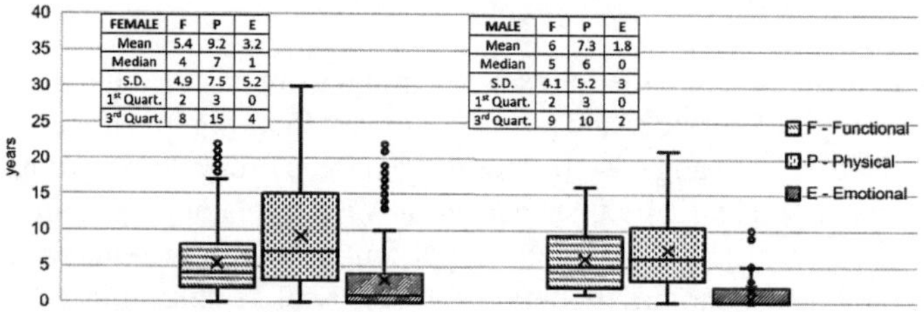

Fig. 2. Contributions of three domains, functional (F), physical (P), and emotional (E), to the overall VHI for the entire sample.

In the female group of slight dysphonia, answers provided by teachers with few years of service (up to 10 years) were compared with those by teachers over 30 years. The difference was statistically significant ($p < 0.01$) for each domain (Fig. 3), mainly for high school teachers.

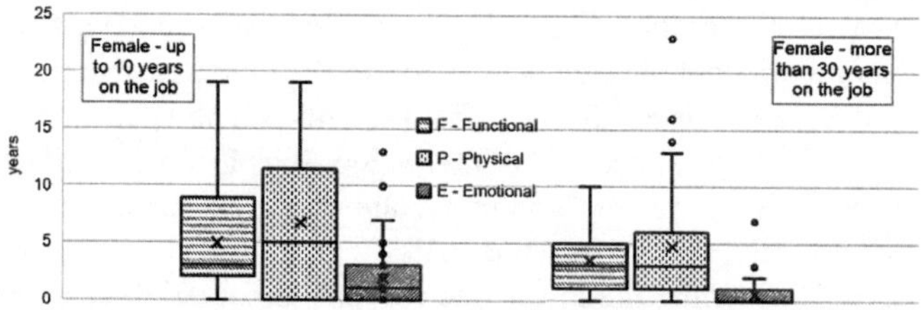

Fig. 3. VHI for the functional, physical, and emotional components in the female sample of slight dysphonia, in case of few years and many years on the job.

In both cases, the emotional impact was residual and the most relevant contribution to total VHI was observed for the physical domain (up to 10 years: 1st quartile =

0, 3rd quartile = 11.5; over 30 years: 1st quartile = 1, 3rd quartile = 6). Increasing years of service, the perception of vocal discomfort among female teachers with slight dysphonia decreases drastically. VHI data were analysed by gender, presence/absence of pathologies, and smoking habits (Table 4). Only 4.1% declared to consume alcohol (77.8% were males). Among alcohol-consuming males, 28.6% were also smokers.

Table 4. VHI Statistics for the entire sample per gender for smoke habits and absence/presence of disease.

VHI	Female				Male			
	Smoke	No smoke	Disease	No disease	Smoke	No smoke	Disease	No disease
Mean	14.1	18.5	18.9	17.3	13.6	15.2	17.0	14.7
SD	15.1	15.7	16.5	15.3	5.3	10.9	7,2	10,8
1stquart	3.0	7.0	6.0	7.0	11.0	6.0	10.5	5.5
3rdquart	20.0	25.0	28.8	21.0	17.0	22.0	22.0	21.0

The results for smoke showed that the perception of vocal dysfunctions, in terms of VHI average and maximum values, was more accentuated among no-smokers, for both genders, without statistical significance. Then, the total VHI could be influenced by a healthy lifestyle that represents a fundamental rule of vocal hygiene. Lower dispersion of VHI values was found among males, presumably associated with the small sample size. Among females, data dispersion was more significant, with standard deviation values close to the average ones. For the sample that declared diseases, the VHI average value was higher for both genders, without statistical significance. It would appear that total VHI is influenced by a greater sensitivity to voice issues for teachers suffering from pathologies.

The frequency of each answer (*never*, *almost never*, *sometimes*, *almost always*, *always*) in the three domains (functional, physical, and emotional) was analysed for the largest sample (females), for the type of school and age group. The emotional perception was usually zero, for all school levels and age groups. For the functional domain, the results showed that voice problems did not interfere in family, social, and work interactions, for all school types and age groups. In this domain, a congruence of the results, on question 2: *People have difficulty understanding me in a noisy room* for all age groups and school levels was recorded, in particular among youngest middle schools teachers and nursery and primary school ones. For the physical domain, a medium-level vocal perceived effort had the greatest contribution, especially among middle school teachers, up to 45 years of age (Fig. 4).

For the other educational levels, the perception was always lower, with an important congruence of answers for questions 2 and 9 (2: *The sound of my voice varies throughout the day*; 9: *My voice is worse in the evening*). These questions are related to each other and highlight a greater vocal commitment at school levels (nursery and primary school and high school) where the involvement of pupils in the didactic activity requires a

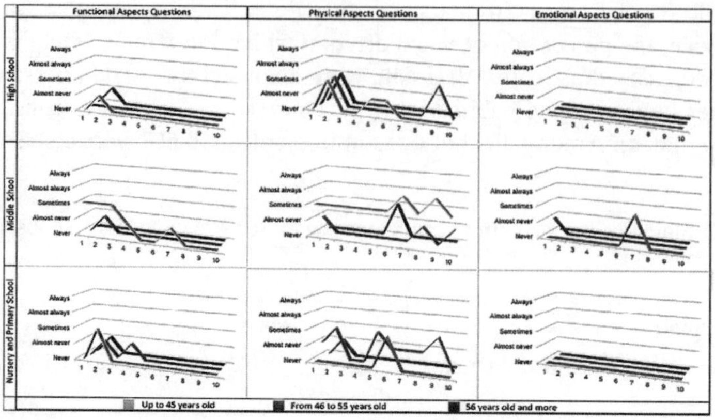

Fig. 4. The highest frequencies of answers in each set of questions are ordered by teaching school level and type of questions. The results are collected by age of teachers, grouped into three classes: up to 45 years old, from 46 to 55 years old, and 56 years old and more.

prolonged effort. These teachers are more prone to perceive a variability of their voice during the day and especially at the evening [16].

These results agree with the risk profile for voice disorder among female teachers, which includes age between 40 and 59 years, having 16 or more years of education, vocal fatigue more than male teachers, and prevalent towards the end of the working day [3, 5, 9].

4 Conclusion

This study shows that Check-app voice®, administered remotely during the pandemic period, to check the perceived voice status, represents an easy, quick to use prevention tool for promptly identifying workers most at risk of the onset of vocal disorders. Data collected and analysed, exploring the effect of variables on the prevalence of VHI among teachers, showed that, according to the literature, female teachers above 30 years of services were more prone to perceive slight voice disorders in the long-term, varying acoustic abilities throughout the day. Implementation of effective preventive actions (noise control, remediation interventions, etc.) together with the design of personalized vocal ergonomics education paths (speech courses, vocal and postural hygiene, support of speech language pathologists, etc.) could avoid progression towards complications that could affect teachers' quality of life.

References

1. Rossi, R.P., Kortte, J.H., Palmer, J.B.: Speech and Language Disorders Essentials of Physical Medicine and Rehabilitation, 4th Ed. (2020)
2. Behlau, M., Madazio, G., Oliveira, G.: Functional dysphonia: strategies to improve patient outcomes. Patient Relat. Outcome Meas. **1**(6), 243–253 (2015)

3. Ravall, S., Simberg, S.: Voice disorders and voice knowledge in choir singers. J. Voice **34**(1), 157 (1997)
4. Roy, N., Merrill, R.M., Thibeault, S., et al.: Prevalence of voice disorders in teachers and the general population. J. Speech Lang. Hear. Res. **47**, 281–293 (2004)
5. Boltezar, L., Šereg, B.M.: Voice disorders in occupations with vocal load in Slovenia. Zdrav Var **53**(4), 304–310 (2014)
6. Greve, K., Bryn, E., Simberg, S.: Voice disorders and impact of voice handicap in Norwegian student teachers. J.Voice **33**(4), 445–452 (2019)
7. Ohlsson, A.C., Andersson, E.M., Södersten, M., Simberg, S., Barregård, L.: Prevalence of voice symptoms and risk factors in teacher students. J Voice. **26**(5), 629–634 (2012)
8. NIDCD.Statistics on voice, speech and language. http://www.nidcd.nih.gov
9. Alarouj, H., Althekerallah, J.M., AlAli, H., Ebrahim, M.A., Ebrahim, M.A.K.: Comparative study utilizing the voice handicap Index-10 (VHI-10) in teachers and the general population of Kuwait. Voice, S0892–1997(20)30178–8 (2020)
10. Ugulino, A.C., Oliveira, G., Behlau, M.: Perceived dysphonia by the clinician's and patient's viewpoint. J. Soc. Bras. Fonoaudiol. **24**(2), 113–118 (2012)
11. Fussi, F., Giovanna, B.: Self-administered questionnaires for the assessment of voice disorders in normal and professional users. https://www.francofussi.com/wp-content/uploads/CSHI-MSHI.pdf
12. Jacobson, B., et al.: The voice handicap index (VHI): development and validation. Am. J. Speech-Lang. Pathol. **6**(3), 66–70 (1997)
13. Tafiadis, D., et al.: Checking for voice disorders without clinical intervention: the Greek and global VHI thresholds for voice disordered patients. Sci. Rep. **9**, 9366 (2019)
14. MIT App Inventor. https://appinventor.mit.edu/
15. Ricci Maccarini, A., Limarzi, M., Pieri, F., Stacchini, M., Lucchini, E., Magnani, M.: La refertazione e l'interpretazione dei tracciati e dei questionari nello studio della disfonia, Quaderni di Aggiornamento AOOI, Ed. Pacini, Pisa (2008)
16. Spadone, M., Garzaro, G., Bergamaschi, E.: Fattori di rischio occupazionale nell'insorgenza di disturbi vocali negli insegnanti, Prevenzione in Corso n.3 (2018)

Analysis of Head Size Related to the Design of Eye and Face Protection Products

Linghua Ran(✉), Xin Zhang, Taijie Liu, Chaoyi Zhao, and He Zhao

China National Institute of Standardization, No.4. Zhichun Road, Haidian District, Beijng, China
ranlh@cnis.ac.cn

Abstract. This study presents two-dimensional distribution tables for Chinese men and women aged 18–60, including the two-dimensional distribution table from ear to pupil distance and pupil distance, and the two-dimensional distribution table of face width and face length, which can be used for the design of eye and face protection articles. The two-dimensional distribution tables can be merged reasonably according to the actual needs, calculate the coverage rate and set the model.

Keywords: Head size · Eye and face protection · Products design

1 The Function of Eye and Face Protection Devices

Eye and face protection device is used to prevent flying debris, dust, gas, droplets, chemicals, hot solids, thermal radiation, optical radiation and other harmful substances from damaging eyes or face. With the increase of occupational safety and health supervision in China, people's awareness of safety and health in production and life has been improved, and the users of eye and face protection devices have higher requirements for safety, comfort and beauty.

Eye and face protective equipment mainly includes goggles and protective mask. There are many kinds of goggles. According to the different protection purposes, they are divided into Anti foreign body impact glasses, anti chemical liquid splash glasses, anti dust and various harmful gas glasses, anti arc radiation glasses, anti laser glasses, anti microwave glasses and anti X-ray glasses. The protective mask is used to shield the facial products and protect the face and neck from flying metal debris, harmful gas splashing, molten metal and high temperature solvent droplets. The protective masks are divided into Anti strike mask, anti radiation mask, anti chemical liquid splash mask, anti smoke and gas mask and heat insulation mask according to their uses.

2 Application of Head and Face Data in the Design of Eye and Face Protection Device

For the eye and face protection device, in order to achieve good protection effect, the coincidence of the optical center of the lens and the horizontal line of sight of the pupil

R. S. Goonetilleke et al. (Eds.): AHFE 2021, LNNS 273, pp. 128–133, 2021.
https://doi.org/10.1007/978-3-030-80713-9_17

is a necessary condition to ensure that the wearer will not suffer from visual impairment. At the same time, the protection area of the lens is closely related to the position, size, length and other dimensions of the eyes (for example, the eye and face protection device should at least cover the soft tissue in the orbit). Therefore, the standards of various countries have made corresponding provisions on the size or protection range of eye protection devices. For people from different countries or regions, the head shape and face shape are quite different. Even if the eye and face protection products that have passed the European and American standards are allowed to be worn by Chinese people, if the products can not fit the face shape of Chinese people, there is still a risk of eye injury. For example, the average exophthalmos of Chinese is 12 mm to 14 mm, while the side protection length (E = 10 mm) specified in ISO / FDIS 16321 [1] (eye and face protection part 2: Additional requirements for protective articles used in welding and related processes) is 2 mm to 4 mm shorter than the average exophthalmos of Chinese. If the eye protection device is designed, manufactured and tested according to the standard, the eye protection device can not effectively protect the eyes of Chinese people.

3 Current Standards About Eye and Face Protection Equipment

Eye and face protection equipment is the last line of defense to protect the eyes and face safety, and its product quality is very important.

In the American Standard ANSI/ISEA Z87.1-2010 "occupational eye and face protective equipment" [2], Canadian standardCSA Z94.3-2007 "eye and face protective equipment" [3], and BS EN 168:2001 "non optical test method for personal eye and face protection" [4], the head mold is used for the detection of eye and face protective equipment. In the newly released ISO 18526-4:2020 [5] "eye and face protection test methods Part 4: headform", six headforms with three sizes, large, medium and small, are required for the detection of eye and face protection appliances, and the key size values of eye parts are given. These headforms were developed by the national occupational safety and Health Council of the United States.

Up to now, there are 8 standards of eye and face protective equipment in China. Among them, there are three standards related to detection methods: GB 14866-2006 technical requirements for personal eye protectors, GB/T 32166.2-2015 [6] personal protective equipment eye and face protection occupational eye and face protectors Part 2: measurement methods, and GB 30863-2014 [7] personal protective equipment eye and face protection laser protective glasses. Among them, GB 14866-2006 stipulates that the head mould used for personal eye protectors should meet the requirements of adult men's head and face dimensions in GB/T 2428-1998 adult head and face dimensions [8]; GB/T 32166.2-2015 and GB 30863-2014 both require that the head mould should meet the requirements of GB/T 23461-2009 adult men's head three dimensional dimensions [9].

At present, there are two problems in the national standards of eye and face protective equipment: (1) GB /T 2428-1998 "adult head and face size" is based on the data obtained from the first national adult body size survey in 1988. Through the analysis and statistical calculation of small sample human head and face size data, the national standard is established. The data age is relatively long. (2) GB/T 23461-2009 is a national standard formed on the basis of the research on the three-dimensional dimensions of the head shape of men's army. The measured samples are only adult men aged 16–36 years old, which can not fully cover the actual user group of eye and face protection products. (3) These two standards do not cover some key reference points and positions for eye protection areas such as orbit. If the eye protection device is designed, manufactured and tested according to the existing data and standards, it can not effectively protect the eyes of Chinese people.

4 Data Resources

In this paper, the head size related to the design of eye and face protection products in Chinese population were analyzed. The data used in this paper are collected from 2014 to 2018 by China National Institute of Standardization from 32 survey points in 6 natural areas across the country, including nearly 17000 head and face 3D data, covering people aged 18–75.

5 Data Analysis

For eye protection products, the two-dimensional distribution table is set according to the pupillary distance of adult men (18–60 years old) and adult women (18–60 years old) and the two-dimensional distribution coverage of the attachment point on the ear to the eyeball. For facial protective equipment, the two-dimensional distribution table is set according to the two-dimensional distribution coverage of head width and face length of adult men (18–60 years old) and adult women (18–60 years old).

Table 1 and Table 2 show the two-dimensional distribution of pupil distance and supraauricular attachment point to eyeball in men and women, respectively. The pupillary distance was graded by 10 mm, and the distance from the attachment point to the pupil was graded by 5 mm. In use, it can merge reasonably according to the actual needs, calculate the coverage and set the model.

It can be seen from Table 1 and Table 2 that when the distance between men's ears and pupils is 85-88mm and 58-60mm, the proportion is 6.2%. It shows that the distance between men's ears and pupils is 85–88 mm, and the proportion of men with pupil distance between 58–60 mm is 6.2% in the whole country. At this time, the population coverage rate is the highest, which indicates that the eye and face protective mask produced by this size can meet the normal wearing of more people, It has practical production value. For women, the coverage rate is the highest when the distance from ear to pupil is 82-85mm and the distance from pupil is 56-58mm; the two-dimensional distribution can be reasonably merged according to the actual needs, and the coverage rate after merging can be calculated to set the corresponding model.

Table 3 and Table 4 show the two-dimensional distributions of face width and morphological face length for men and women, respectively. The surface width is graded by 10 mm, and the surface length is graded by 5 mm. In use, it can merge reasonably according to the actual needs, calculate the coverage and set the model.

Table 1. The two-dimensional distribution of pupillary distance and auricular attachment point to pupil for Male

Pupillary distance (mm)	Distance from auricular attachment point to pupil (mm)								
	<76	76–79	79–82	82–85	85–88	88–91	91–94	94–97	>97
<54	0.2%	0.4%	0.8%	0.3%	0.3%	0.5%	0.5%	0.2%	0.1%
54–56	0.2%	0.5%	1.5%	1.8%	2.0%	1.3%	0.8%	0.3%	0.1%
56–58	0.3%	1.1%	2.3%	3.7%	4.5%	3.4%	1.7%	0.6%	0.2%
58–60	0.1%	0.6%	3.0%	4.8%	6.2%	4.1%	3.0%	1.3%	0.4%
60–62	0.1%	0.6%	1.5%	3.3%	4.5%	5.0%	2.8%	1.4%	0.5%
62–64	0.0%	0.1%	0.8%	2.1%	3.0%	2.8%	2.7%	0.9%	0.9%
64–66	0.0%	–	0.4%	0.7%	1.4%	2.0%	1.7%	0.6%	0.6%
66–68	0.0%	–	0.2%	0.2%	0.5%	0.6%	0.8%	0.4%	0.3%
>68	0.0%	–	0.2%	0.1%	0.0%	0.4%	0.1%	0.3%	0.1%

Table 2. The two-dimensional distribution of pupillary distance and auricular attachment point to pupil for Male

Pupillary distance (mm)	Distance from auricular attachment point to pupil (mm)								
	<73	73–76	76–79	79–82	82–85	85–88	88–91	91–94	>94
<52	0.4%	0.7%	0.9%	1.0%	1.1%	0.4%	0.1%	–	0.1%
52–54	0.2%	1.0%	2.4%	2.2%	2.3%	1.1%	0.7%	0.1%	–
54–56	0.5%	1.5%	3.6%	4.2%	5.0%	2.3%	1.6%	0.1%	0.1%
56–58	0.5%	1.3%	3.5%	5.1%	5.7%	4.3%	1.9%	0.4%	0.2%
58–60	0.4%	1.1%	3.6%	4.7%	5.4%	3.9%	2.6%	1.0%	0.1%
60–62	–	0.3%	1.3%	2.5%	3.1%	3.2%	1.4%	0.8%	0.3%
62–64	0.1%	0.2%	0.5%	0.8%	1.1%	1.3%	0.7%	0.5%	–
>64	–	–	0.2%	0.5%	0.4%	0.8%	0.6%	0.2%	–

Table 3. Two dimensional distribution of facial width and facial length for Male

Facial length (mm)	Facial width (mm)						
	<133	133–136	136–139	139–142	142–145	145–148	>148
<100	–	–	–	–	–	0.1%	0.1%
100–105	–	–	–	0.1%	0.3%	0.5%	0.5%
105–110	–	0.1%	0.3%	1.0%	2.3%	2.3%	1.8%
110–115	–	0.1%	0.7%	2.4%	5.3%	6.0%	4.7%
115–120	0.1%	0.2%	1.3%	4.3%	7.9%	8.1%	6.4%
120–125	0.1%	0.1%	1.3%	3.7%	6.9%	6.5%	4.9%
125–130	–	0.2%	0.8%	2.1%	3.6%	2.6%	2.3%
130–135	–	0.1%	0.4%	0.9%	1.5%	0.9%	0.6%
>135	0.2%	0.3%	0.4%	0.6%	0.7%	0.4%	0.4%

Table 4. Two dimensional distribution of facial width and facial length Female

Facial length (mm)	Facial width (mm)						
	<130	130–133	133–136	136–139	139–142	142–145	>148
<100	0.2%	0.9%	1.6%	1.4%	0.8%	0.1%	0.1%
100–105	0.4%	2.0%	5.3%	5.5%	3.4%	0.9%	0.3%
105–110	0.6%	2.6%	7.3%	9.9%	6.6%	2.5%	0.8%
110–115	0.4%	1.5%	4.9%	8.4%	7.2%	3.1%	1.2%
115–120	0.1%	0.6%	2.4%	4.0%	4.0%	2.0%	1.0%
120–125	–	0.1%	0.4%	1.0%	1.2%	0.8%	0.4%
125–130	–	–	0.2%	0.2%	0.2%	0.3%	0.1%
130–135	–	–	–	0.1%	0.1%	0.1%	0.1%
>135	–	–	–	0.1%	0.1%	0.1%	–

6 Conclusion

Ergonomics is not only to provide an accurate and scientific description of the head and face, but also to simplify the size system through scientific and reasonable design, so as to reduce the model and specification of products on the premise of improving the comfort rate. Based on the human head and face size data related to eye and face protection design, this paper presents a two-dimensional distribution table for Chinese men and women aged 18–60, including the two-dimensional distribution table from ear to pupil distance and pupil distance, and the two-dimensional distribution table of face width and face length, which can be used for the design of eye and face protection articles.

Acknowledgments. This work is supported by the National Key R&D Program of China (2017YFF0206602) and China National Institute of Standardization through the "special funds for the basic R&D undertakings by welfare research institutions" (project number 522018Y-5984).

References

1. ISO/FDIS 16321: Eye and face protection part 2: Additional requirements for protective articles used in welding and related processes
2. ANSI/ISEA Z87.1-2010: Occupational eye and face protective equipment (2010)
3. CSA Z94.3-2007: Eye and face protective equipment (2007)
4. BS EN 168:2001: Non optical test method for personal eye and face protection (2001)
5. ISO 18526-4:2020: Eye and face protection test methods Part 4: headform (2020)
6. GB/T 32166.2-2015: Personal protective equipment eye and face protection occupational eye and face protectors Part 2: measurement methods (2015)
7. GB 30863-2014: Personal protective equipment eye and face protection laser protective glasses (2014)
8. GB/T 2428-1998: Adult head and face dimensions (1998)
9. GB/T 23461-2009: Adult men's head three dimensional dimensions (2009)

Differences in the Perception of the Quality of Work Life According to Gender in Health Care Workers

Raquel González-Baltazar(✉), Silvia G. León-Cortés, Mónica I. Contreras-Estrada, Liliana Hidalgo-González, and Brenda J. Hidalgo-González

Health Sciences University Center, University of Guadalajara, Guadalajara, Jalisco, México

Abstract. The aim of this study was to identify gender differences in the evaluation of the Quality of Work Life (QWL) in health professionals. The study was descriptive and was conducted with doctors and nurses of both genders in an institution in the health sector in Guadalajara, Mexico. Included 322 subjects evaluated with the QWL with CVT-GOHISALO validated instrument. This allowed us to measure the satisfaction of the QWL in seven dimensions for each gender in order to make comparisons. 322 subjects were interviewed, of whom 32% (102) were male and 68% (220) female. Males predominated in the doctors with a 65.3% (98), while in the nursing industry 97.7% (168) were women. The differences in the perception of workers according to their gender were present and so we have that the percentage of people who reported being satisfied with each of the dimensions of the QWL was higher in four of the seven dimensions for men, while women only had higher satisfaction than men in the dimensions: job satisfaction 82% to 75%, personal development achieved 84% against 75% and free time management 86% to 75%. The dimensions in which women expressed greater satisfaction than men, show us that they find their work activity more satisfactory and identify greater achievements in terms of their personal development, in contrast to men. The dimension most affected for both sexes was job satisfaction, which constitutes a serious problem for stating that at least a fifth of our workers, men and women, are dissatisfied with the work they do. The differences in the evaluation of the QWL in health professionals, revealed the dissatisfaction of women by the conditions of inequality of opportunity. Women in the study showed with statistically significant, they are more satisfied than men on personal development that they achieved through their work and managing their free time.

Keyword: Gender · Quality of Work Life · Health care workers

1 Introduction

The QWL is a relatively recent construction, which, along with other meanings, such as Health-Related Quality of Life (HRQoL), has its origin in the study of Quality of Life (QoL), which went from being studied in general, to relate to health and later with work [1].

R. S. Goonetilleke et al. (Eds.): AHFE 2021, LNNS 273, pp. 134–148, 2021.
https://doi.org/10.1007/978-3-030-80713-9_18

With the accelerated economic, political, social and environmental changes, which affect our way of living and the way we relate to our fellow human beings, we can realize the increasing need to improve our QoL as individuals and as society.

The QoL constitutes a relevant value for human beings, due to the aspects that determine it; it is associated with objective conditions found in the environment of people and simultaneously with subjective conditions that emanate from the expectations and perceptions of the subjects, at the same time that it is affected by the gender relations that occur in different contexts [2].

This is a social construction that reflects differences in perception about the state of physical, mental, social and spiritual well-being of people; depending in turn largely on their values and beliefs, their cultural context and personal history. Simultaneously, different perspectives of disciplinary approach are identified, such as the philosophical, economic, medical, humanistic, ethical and functionalist, among others, the philosophical perspective being the most used, each of them relates QoL with different variables [3].

Among the variables that can significantly reflect differences in the perception of QoL in general and its impact on health, is gender; this is one of the variables with which we can express significant differences in the perception of QWL and its impact on the health of workers. The gender has been studied by several authors and in different contexts; however, the focus of their relationship with work has been poorly addressed up to now and sometimes outside the approach of researchers.

The QWL also includes satisfaction with work that is done, the achievements through work and personal fulfillment among other aspects. Their level of satisfaction in the dimensions that compose has a great importance in the health of workers, the quality of work done, productivity and business efficiency [1].

International organizations in charge of establishing health policies, including labor and its relationship with the QoL, such as the World Health Organization (WHO) and the Pan-American Health Organization (PAHO), have recognized in their international conferences since 1994 in Cairo and 1995 in Beijing, the need to integrate the gender perspective in their agreements and commitments to correct inequities that violate the right to health of men and women. They also recommend to include the gender perspective in the teaching-learning processes and in the information related to the health care of the population [4].

To understand the differences in the results in the measurement of QWL in health professionals that are presented, it is important to clarify that the gender perspective implies recognizing that beyond a sexual difference, in which societies have traditionally structured their lives and their culture, are the attributions, ideas and social representations that are built based on the anatomical differences that characterize men and women; differences that mark the destiny of people [5].

The gender approach or perspective, provides us with the necessary elements to identify and analyze the conditions of inequality that women still maintain in these times, inequalities that are reproduced from generation to generation. In this chapter, attention is focused on the health system due to the persistence of practices that legitimize and institutionalize these inequalities within the social spaces where they are trained and worked.

In general, the commitment to develop actions aimed at the care and preservation of the well-being of those around us, mainly towards the family, has traditionally been assigned to women as an activity of their gender. The care offered by women has historically been a role assumed by themselves, in a common way in all cultures and societies, which has allowed growth and social development both in number and well-being.

Even PAHO recognizes that distinctions are made in the work of health caregivers based on gender and that these are consistently associated with a lower appreciation of the work that women do, since both in the formal and informal sectors they are concentrated in occupations with lower remuneration, prestige and decision-making power, reaching gratuity when this work is transferred to informality, especially in the care of other members of the community and the family, without appearing in the national accounts. This body also mentions that there is an interdependence between formal and informal health care work, because when policies are defined that affect the incorporation of personnel into formal health services, informal health services are also affected, which they fall predominantly on women [6].

The activities included in caring for other people are related to the QoL of both the people to whom such care is granted, as well as the people who provide it, since they involve maintaining an adequate diet in the person being cared for, a temperature comfortable, good personal and home hygiene, avoid dangers and accidents, while taking care of aspects related to a harmonious and healthy life, good emotional and affective development, social relationships, work activity and entertainment. Forgetting that providing care to other people and based on the transfer that may exist between the caregiver and the person or persons to whom care is provided, can significantly affect the QoL of caregivers.

The foregoing makes it seem more unfair that the contributions made in most cases remain outside of social recognition and narrative; few women stand out in history books, in politics, in social, academic and cultural recognition, thereby contributing to the high appreciation of the role that men play in social development and the invisibility of women as a pillar essential for the maintenance of life and human development [7].

Trying to give fair value to female care, in no way does it intend to detract from the care that is provided by men, since at present the tasks both inside and outside the home, in education, care and support of the family as well as the role of caregivers is being exercised in a greater proportion by both genders, but without yet reaching the ideal proportion of equity. The care that mostly women and increasingly men provide in a constant, silent and committed way, acquires relevance when facing phenomena such as the aging of the population, high incidences of chronic and disabling diseases, changes in the structure, dynamics and family values, greater incorporation of women into work and loss of purchasing power. Therefore, it is constituted in a well-being strategy that it is necessary to promote and maintain through due recognition and as a factor of success for current societies, of great relevance for our cultural context since it contributes to the family and social economy.

In 2007, Astrid Perafán [8], presented the differences in the perception of QoL between men and women in a locality in Colombia where they specifically related it to three variables: work, personal development and material assets. In this study, for men, work is intrinsically related to socioeconomic and work conditions, reaching goals, having a good work environment and being able to carry out social activities; on the other hand, the most important thing for women is the personal development provided by work, integration into their job, being able to carry out other types of activities and finally achieve their goals, reflecting the recent incorporation of women into the field labor. Regarding personal development, while men integrate the family and social with their personal development, women privilege family relationships over social ones and try to reconcile them with their personal development. Regarding material goods, men consider them as essential, possibly associated with power and social prestige, while women consider as fundamental the aspects that revolve around the improvement of home conditions, relating the family and collective interests of the individual female gender.

In a study published in 2001, gender differences in the perception of QoL are measured, finding that women have a worse evaluation of sports, health and recreation services, which is directly related to the responsibility assigned to them women in caring for their children's health. In access to recreation, women also give lower qualifications than men, since the amount of free time available is significantly less, since they have the responsibility of family care, many times they also develop informal work and when they have free time the alternatives are few and not compatible with the other activities they carry out. In general, although multiple indicators are presented for the evaluation of QoL, women have a poorer perception than men and this study concludes that this is due to the assigned roles and the limitations that these roles confer on women [2].

Once what the gender perspective means in relation to health care has been reviewed and taking QoL as a starting point, the object of study is retaken, which is to establish the differences in the evaluation of the QWL, according to the gender.

Inequalities based on gender in the workplace can be found in working conditions, role assignments, task demands and salaries, among others. In a study carried out with respect to economic remuneration in two department stores in Chile in 2009, it is mentioned that, despite having a higher proportion of women in both stores, in one of them men are paid an average 41% more, while the other store pays them 36% more. The above taking into account that the working hours are similar; In addition to this, other important gaps against women are identified, especially in sales commissions, income par excellence, which theoretically correspond to the individual effort of the seller and also in the payment of bonuses [9].

On the other hand, the impact on health and QoL of caregivers was studied, finding that with regard to the assignment of the caregiver role in families, this assignment is not homogeneous and that the typical profile of the main caregiver is that of a woman, a housewife, with a direct kinship relationship with the person she cares for. The most relevant variables in the caregiver's role are gender, coexistence and kinship, with an important difference in the burden between men and women, which is manifested in the fact that informal care is mostly female [10].

The incorporation of an increasing number of women in productive activities, in an increasing and constant way, occurs mainly in the service sector, especially in those occupations that imply their extension of the role of caregivers, particularly in health. Jobs seen as "suitable for women" are nurses, nutritionists, social workers and dentists among others. Comparatively, in the medical occupation men are mainly located on the basis of the dominant role and decision-maker regarding the health of patients. The division of work carried out in the home by gender is transferred to the workplace, the latter becoming an extension of their domestic role. Professions that reveal femininity are exercised, under a dependent social role and culturally acceptable "for women" [11].

This is explained by the cultural training of women, where gender education makes girls and young women feminine and constitutes a job training that qualifies them for certain jobs, which are an extension of domestic work and care and care that women give to boys and men. Highly valued occupational characteristics such as thoroughness and submission are also identified in women [5].

Health care implies the possession and practice of values that are not required in other professions, such as the sense of responsibility, commitment and availability to the needs of the sick, working many times under inappropriate conditions such as absence of rest periods or feeding schedules, emotional exhaustion caused by the transfer between patients and the service provider, in addition to the institutional demands of productivity and optimization of resources. Therefore, these professions generate high degrees of stress that as a consequence can be accompanied by varying degrees of dissatisfaction [5].

Varying according to historical moments, different sexed professional identities (male or female) have been built for health professions and activities, these constructions have been based on the selection of the members of the professional group based on their sex and gender. incorporation of gender values into health practices. The progressive increase during the twentieth century and the beginning of the twenty-first, of women in the medical profession, has been permeated by internal discriminatory processes and the creation of new professional spaces, where men try to delimit the activities of women giving as the result is their concentration in some specialties, without facilities to access decision-making or representation positions, without respect for their preferences in the use of time and with little access to research work; while women participate in the search for spaces professionals that give them greater identity [12].

In the report of the 116th meeting of the WHO Executive Board, held on 2005, it is spelled out "the correction of gender inequalities among health workers is essential for public health systems to be effective. Globally, although women make up more than half of the conventional health care workforce, they receive lower wages and have no decision-making power. Furthermore, as part of unpaid health workers in the informal sector, girls and women also bear a disproportionate burden of care in households and the community, often to the detriment of their own health and well-being" [13].

Directing attention to the participation and status of women employed in the health services in Mexico, it is important to mention that the professions in this area are linked to historical, social, political, cultural, economic processes and scientific and technological progress of the country, which determines the variation in the data presented regarding the medical and nursing professions, where according to reports from the Mexican Foundation for Health (FUNSALUD), the medical profession that was traditionally dominated by the male gender, shows the gradual incorporation of women into the medical profession, with a gradual increase in the average graduation from 1970, going from 19% in that year to 50% in 1999 and to 63% in 2014 [14].

In both the data from the female doctors and those from the female nurses, women consider housework as a second activity, a situation that did not occur in men despite being questioned about it. With regard to nursing as it is historically known; it is a field totally dominated by women because in the records used for the FUNSALUD study, they represent 95%, also showing low education, since the technical level predominates in 64% over the bachelor's degree, which has 35%, reaching only 1% of nurses with postgraduate level. Of graduate technical nurses, only 60% work compared to 70% of those with bachelor and postgraduate degrees; in this study, it was concluded that despite the feminization of health services, participation continues to be unequal, since intervention and income continue with very marked differences. This, increases the probability that women end up dedicating themselves to different activities of those in which they were trained and give up their spaces for participation, which surely leads to different levels of dissatisfaction, affecting the QoL of health professionals [15].

Despite the fact that it has become very common to talk about the gender perspective in regard to health professions, it has not been possible to transcend the university education of students in these professions, most likely due to the fact that the study plans have not been updated and reoriented with this perspective, which is reflected in their training as health educators, with more equitable and humane practices, that make differences by gender and that empower female users and female providers of health services, which in the majority are women, in the exercise of their rights [4].

Social acceptance for female doctors in primary care is very high among the female population, with better communication, more open and focused on the psychosocial than when doctors are male. This is reflected with greater emphasis when the problems that affect patients are of a gynecological nature and oriented towards detections, since the observation of serious health problems such as cervical cancer and breast cancer, which are detectable in very early stages, in our population they continue to present a high incidence and are a cause of death because they did not want to see male doctors.

In a study carried out in 2004, in Granada, Spain, the gender of the patients and that of the health professionals were not shown to be related to the perception of quality, since it has not been shown that the health results are better if medical care is received or provided by people of the same gender. However, their findings showed that the sex of the service provider modulates a very important aspect, which is communication, since female doctors grant longer consultations, provide more information and are more interested in the emotional and social aspects related to problems of health of their patients, compared to male doctors. It concludes that the information available on these variables and medical practice is still very poor, and that in order to understand a concept as complex as the perception of quality, gender must be taken into account as a determinant of individual and social identity, since it conditions the interaction of the patient and health professional relationship [16].

Health services in Mexican institutions, both public and private, as in most countries in the world, have to be provided continuously every day of the year, except in primary care units where they work from Monday to Friday. Having to adapt work activity to institutional needs, with shifts that in many cases include night shifts, alters the family and social life of those who provide these services, which can even have repercussions on their own health, since working conditions in health units, including family medicine units and hospitals, carry the risk of affecting the perception of satisfaction with the work activity carried out, as well as with other personal needs within and outside of work.

Work is the activity of greatest importance for the human being, since he or she, individually or collectively, makes a series of contributions such as effort, time, aptitudes and skills, among others, pending certain rewards that help to satisfy his needs, whether economic, material, psychological or social, among others, so it is essential to know the factors that influence the QWL of workers. Work activity is the main contributor to raising the QoL of people and when the individual has the possibility of doing a job that corresponds to their ability and vocation and identifies elements for their personal growth, the level of their QoL will be longer [17].

The QWL has objective and subjective components, so for its evaluation, the way in which people live the daily life of their work environment must be considered, taking into account their working conditions, both physical and contractual and remuneration, as well as their social relationships both between colleagues and between the worker and the organization. The attitudes and values of the workers and the perception of satisfaction or dissatisfaction derived from this set of factors are also very important. The QWL presents methodological difficulties for its evaluation, mainly based on the fact that within the multiple criteria found to evaluate it, there are few studies that present a concrete proposal to do so. The theoretical and methodological models found have been applied in other areas and in other places, without necessarily having the same application in the local context [18].

The results of this research allowed to account for the gender differences regarding the perception of QWL and contribute to its knowledge. When carried out with medical and nursing personnel with workers of both genders in the two branches, the different perceptions between those who adopt the role of caregivers and those who have the role of decision-makers regarding the health of the patients could be identified [19].

The evaluation of the QWL was carried out based on the definition on which the elaboration of the CVT-GOHISALO instrument was supported: QWL is "a multidimensional concept that is integrated when the worker, through employment and under his own perception, sees the following personal needs covered; institutional work support, reliability and integration to the job and satisfaction for it, identifying the well-being achieved through their work activity and the personal development achieved, as well as the management of their free time".

This concept of QWL identifies seven dimensions, which can be evaluated through the CVT-GOHISALO instrument; Institutional work support, Job reliability, Integration to the job, Satisfaction with the work, Well-being obtained through the job, Worker's personal development and Free time [19].

2 Aim

The aim of this study was to identify gender differences in the evaluation of the QWL in health professionals.

3 Methodology

A descriptive and analytical study was conducted with doctors and nurses of both genders in a health sector institution in Guadalajara Metropolitan Area, Mexico.

The work universe was made by 1866 active workers of a health institution in its three levels of care, doctors and nurses, which included all types of administrative units and offices, in all work shifts. A stratified probabilistic sample was calculated, where six strata were identified; one from doctors and the other from nurses in each of the three levels of care (see Tables 1 and 2).

The proportions in the population universe were calculated for each stratum and the same proportions were applied to the sample strata; according to the calculation, with a formula for finite populations, the sample was 351 and only those randomly selected were included, with at least six months seniority, the CVT-GOHISALO questionnaire was applied to all of them, all those persons commissioned or those who were searched twice in their workplace and were not present were excluded, applying substitution criteria; the questionnaire was applied to the next person from the health unit staff with the same category (doctor or nurse).

Each person selected was located in the health unit and in the corresponding shift, where they worked, visiting a total of 98 health units in all shifts; 88 of first level, 6 of second level and 4 of third level.

The analysis was descriptive and inferential and identified the differences in satisfaction with the QWL in each of the seven dimensions of the instrument for each gender and profession.

Table 1. Study universe

Levels of care	Doctors	%	Nurses	%	Total staff
First level	481	25.8	333	17.9	814
Second level	122	6.5	247	13.2	369
Third level	215	11.5	468	25.1	683
Totals	818	43.8	1048	56.2	1866

Table 2. Stratified sample for the study

Levels of care	Doctors		Nurses		Total staff surveyed
	%	Surveyed	%	Surveyed	
First level	25.8	91	17.9	63	154
Second level	6.5	23	13.2	46	69
Third level	11.5	40	25.1	88	128
Totals	43.8	154	56.2	197	351

4 Results

322 subjects, 150 doctors and 172 nurses were surveyed, of which 32% (102) were male and 68% (220) female. The male sex predominated in physicians with 65.3% (98), while in the nursing branch 97.7% (168) were women (see Table 3).

Table 3. Distribution by profession and gender in study population

Profession	Male	%	Female	%	Total
Doctors	98	65.3	52	34.7	150
Nurses	4	2.3	168	97.7	172
Total	102	32	220	68	322

Source: Direct

The distribution by profession was equitable and directly proportional to the universe, since 150 male and female doctors were included, representing 46.6% of the total number of people surveyed, and 172 male and female nurses, who constituted 53.4% of the sample (see Table 4).

Workers' perception of their QWL in each of the dimensions that make up the instrument varied according to their gender and also according to their occupation as a doctor or nurse (see Tables 5 and 6).

The following table, shows that the percentage of people who reported being satisfied with each of the QWL dimensions was higher in four of the seven for men, Institutional work support, Job reliability, Integration to the job and Well-being obtained through the job, finding the greatest difference in the latter, with 88.2% satisfaction among men versus 80% among women.

Women only had greater satisfaction than men in the dimensions of Satisfaction with the work with 81.8% against 74.5% of men, in that of Worker's Personal Development, where the percentage difference was 8.1% in favor of women and regarding the free time administration, the percentages were even more significant, because while 86.4% of the women were satisfied, only 75.5% of the men were satisfied.

However, it should be noted that the percentage of people dissatisfied with their QWL in general ranges from 10 to 25% for men, while women showed dissatisfaction

Table 4. Distribution by profession, sex and level of care of the study population

Level of care	Medical professional					Nursing professional				
	Male	%	Female	%	Total	Male	%	Female	%	Total
First level	59	38.3	33	21.4	92	1	0.7	61	39.6	62
Second level	9	22.5	9	22.5	18	1	2.5	21	52.5	22
Third level	30	23.4	10	7.8	40	2	1.6	86	67.2	88
Total	98	30.4	52	16.2	150	4	1.2	168	52.2	172

Source: Direct

in 13 to 20%. The dimension most affected for men was Satisfaction with the work and for women Well-being obtained through the job (see Table 5).

Table 5. QWL, satisfaction by dimension, according to gender, in the health personnel participating in the study

Gender	D1	D2	D3	D4	D5	D6	D7
Men	86.3	90.2	88.2	74.5	88.2	75.5	75.5
Female	81.8	86.8	87.2	81.8	80	83.6	86.4

Source: Direct, D1 Institutional work support, D2 Job reliability, D3 Integration to the job, D4 Satisfaction with the work, D5 Well-being obtained through the job, D6 Worker's personal development, D7 Free time administration

Table 6 shows the percentages of satisfaction and dissatisfaction of both medical and nursing personnel in each of the seven dimensions of the instrument. They highlight the highest satisfaction for male doctors in the Dimension of Job reliability, while in male nurses the highest satisfaction was found in the dimensions of Well-being obtained through the job, Worker's personal development and Free time administration.

For female doctors, the dimension with the highest satisfaction was that of Integration to the Job, while for female nurses the dimension with the best qualification was that of Job reliability.

Regarding dissatisfaction, male physicians have the worst score for the dimension of Satisfaction with the work, while male nurses score the same, with 25% dissatisfaction, for the dimensions of Institutional work support, Job reliability, Integration to the job and Satisfaction with the work.

For female physicians, the dimension with the highest dissatisfaction is the Worker's personal development, while for female nurses the dimension of Well-being obtained through the job is the one with the highest dissatisfaction.

It is highlighted that by combining the variables of profession and gender, female doctors and nurses surpass male nurses in satisfaction in the first four dimensions and that male nurses reach 100% satisfaction in the last three dimensions (see Table 6).

Table 6. Percentage of satisfaction and dissatisfaction with the QWL by dimension, profession and gender in the study population

Profession and gender	D1		D2		D3		D4		D5		D6		D7	
	Si	No	Si	No	Si	No	Si	No	Si	No	Si	No	Si	No
Male doctors	86.7	13.3	90.1	9.9	88.8	11.2	74.5	25.5	87.8	12.2	74.5	25.5	74.5	25.5
Male nurses	75	25	75	25	75	25	75	25	100	0	100	0	100	0
Female doctors	84.6	15.4	76.9	23.1	94.2	5.8	76.9	23.1	84.6	15.4	73.1	26.9	84.6	15.4
Female nurses	81	19	89.9	10.1	85.1	14.9	83.3	16.7	78.6	21.4	86.9	13.1	86.9	13.1

Source: Direct, D1 Institutional work support, D2 Job reliability, D3 Integration to the job, D4 Satisfaction with the work, D5 Well-being obtained through the job, D6 Worker's personal development, D7 Free time administration

The next graph shows the differentiation by dimensions and gender in terms of dissatisfaction with QWL, adding the profession variable to the analysis. It shows that female nurses exceed male nursing dissatisfaction in three dimensions: Well-being obtained through the job, Worker's personal development and Free time administration. Male nurses have greater dissatisfaction than women in four dimensions: Institutional work support, Job reliability, Integration to the job and Satisfaction with the work.

Regarding the medical branch, medical women are more dissatisfied than men in four dimensions: Institutional work support, Job reliability, Well-being obtained through the job and Worker's personal development. While men are dissatisfied in the dimensions: Integration to the job, Satisfaction with the work and Free time administration.

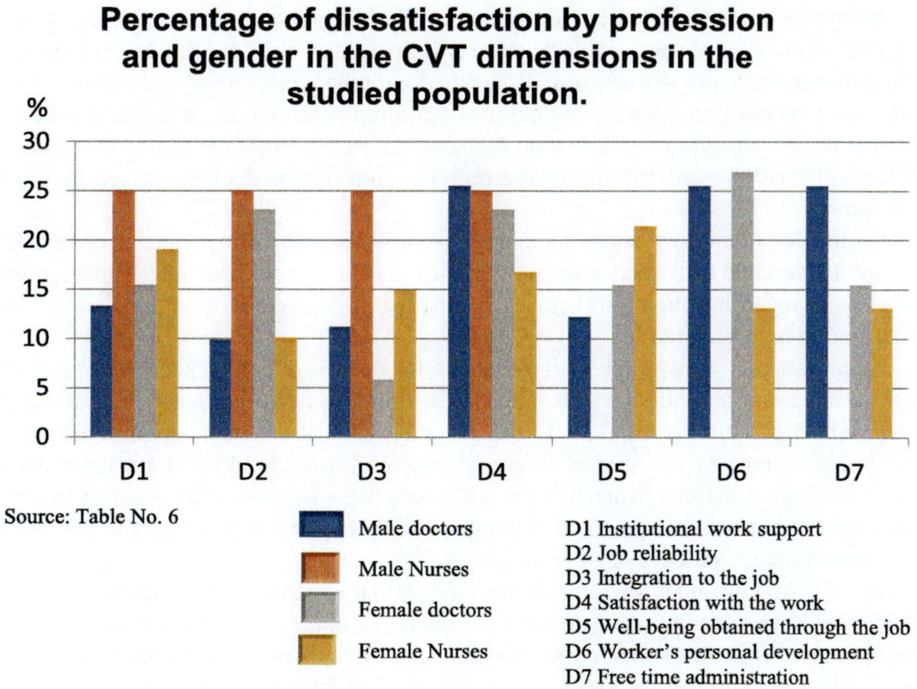

Source: Table No. 6

5 Discussion

To analyze the differentiation by gender and occupation of the results; It is necessary to mention that of the 322 subjects surveyed, 32% (102) were male and 68% (220) female, which is consistent with the reviewed publications, which speak of a strong feminization of health services. The distribution by profession was equitable and directly proportional to the universe, since 150 doctors were included, representing 46.6% of the total number of people interviewed, and 172 male and female nurses, who constituted 53.4% of the sample.

Due to the predominance of the male gender in doctors (65.3%) and that of the female gender in the nursing branch (97.7%), the androgenism already referenced in the medical profession is confirmed, where the role of decision-maker remains with men and activities related to care and service, is placed with women.

The distribution presented with a greater burden for the female gender in nursing and for the male gender in the medical profession, was maintained when analyzing by levels of care. For only at the second level is an equitable distribution for both genders observed among the medical professional, highlighting that at the third level of care, where the degree of specialization is greater, the presence of women in this study is maintained, while than that of men increases. On the other hand, the distribution by gender in nursing is almost entirely female, and this feminization was found uniformly in the three levels of care.

In both genders, Job satisfaction is one of the first two most affected dimensions, which constitutes a serious problem, since it shows that a quarter of male workers and a fifth of female health workers, are dissatisfied with the work activity carried out. This dimension allows us to identify the degree of commitment that the person has with the mission of the institution and with their own objectives, highlighting the positive aspects of the work and the satisfaction for the remuneration obtained when performing their functions [19].

On the other hand, when the percentage of satisfaction by dimensions is compared, adding the variable of doctor or nurse to that of gender, we observe that the perception of workers regarding their QWL in each dimension presented variations that can be attributed to both variables.

According to results (graph 1), it is evident that doctors of both genders make a much lower evaluation than nursing personnel in the dimension Worker's personal development, which perhaps corresponds to the expectations that are made about their work. In this dimension, the dissatisfied worker expresses insecurity about the achievements achieved, feels vulnerable when dealing with users or co-workers, may consider that the work has diminished their physical or emotional capacity and they are pessimistic faced with institutional challenges and goals [19].

By directing the analysis to the dimensions with the greatest dissatisfaction for both genders, differentiated by the profession of doctor or nurse, there are not many coincidences, since in female gender, nurses are more dissatisfied in the dimensions: Well-being obtained through the job, Worker's personal development and Free time administration, while female doctors are most dissatisfied in the dimensions: Institutional work support, Job reliability, Well-being obtained through the job and Worker's personal development. On the other hand, male nurses show greater dissatisfaction in the dimensions: Institutional work support, Job reliability, Integration to the job and Satisfaction with the work, while male physicians show greater dissatisfaction in the dimensions: Integration to the job, Satisfaction with the work and Free time administration.

In general terms, the highest percentage of dissatisfaction for both genders and professions is found in Satisfaction with the work dimension, and it stands out that the medical branch in both genders shows a very high dissatisfaction in Worker's personal development.

6 Conclusions

In conclusion, there must be a match between the capabilities of the worker and the content of their work for their QWL to be satisfactory, but this condition is not sufficient as long as discrimination in the workplace on grounds of race, color, religion, political opinion, national or social origin, age, sexual orientation and very importantly gender, among others are not eliminated.

The gender perspective complements the concepts of QoL and QWL, based on the identification of differences in gender relations, needs, roles, access, decision-making capacity and rights over goods and services that they affect the perception and satisfaction of human beings [2].

Since the United Nations assumed gender equality as one of its lines of action, the International Labor Organization conventions include the principles of non-discrimination and equal opportunities and treatment in the workplace for women. Consequently, Western governments have incorporated it as a policy in all their official documents [20].

Differences in the evaluation of QWL in health professionals reveal dissatisfaction on the part of women with the conditions of unequal opportunities, despite the increasing number of women present in the medical profession and their greater participation in positions of responsibility and decision-making. These inequalities are the result of gender segregation that places them in caregiver roles and distances them from positions of responsibility and authority, both professional and scientific [12].

References

1. González-Baltazar, R.: "Calidad de vida en el trabajo" Elaboración y validación de un instrumento en prestadores de servicios de salud. Tesis doctoral para obtener el grado de Doctora en Ciencias de la salud en el Trabajo, junio 2007. Universidad de Guadalajara (2007)
2. Fadda, G., Jirón, P.: Calidad de vida y género en sectores populares urbanos. Un estudio de caso en Santiago de Chile. Síntesis final y conclusiones. Boletín del Instituto de la Vivienda, Universidad de Chile, vol. 16, Núm. 042, ISSN 0716-5668, Santiago de Chile, pp. 105–138 (2001)
3. Pichardo Arlette. Bibliografía sobre calidad de vida y temas conexos; en (2005). http://www.cinpe.una.ac.cr/investigacion/proyectos/calidad-de-vida/bibliografia/Bibliograf%EDa.pdf
4. Rosete, M.G., Arenas, G., Santillán, M., Torres, M.A.: Análisis desde la perspectiva de género de los planes de estudios de las carreras de: Cirujano Dentista, enfermería y médico cirujano de la Universidad Nacional Autónoma de México. Memoria Académica del Primer Congreso Latinoamericano de Ciencias de la Educación (2010)
5. Lamas, M.: La perspectiva de género. La tarea, Revista de educación y cultura de la Sección 47 del S.N.T.E. No. 8 (1996). http://www.latarea.com.mx/index.htm
6. Panamerican Health Organization and World Health Organization. Towards gender equity in health sector reform policies. 18th Session of the Subcommittee on Women, Health and Development, item 5 of the agenda (1999)
7. Bover, A.: Cuidadores informales de salud del ámbito domiciliario: percepciones y estrategias de cuidado ligadas al género y a la generación. Tesis doctoral. Universitat de les Illes Balears, Departament de Ciències de L'educació (2004)
8. Perafán, A.: Calidad de vida y género, un estudio exploratorio en la ciudad de Popayán, Colombia, en Estudios de Antropología Biológica, Vol. XIII, Ed. Instituto de Investigaciones Antropológicas, Universidad Nacional Autónoma de México, ISSN 1405-5066, pp. 515–532 (2007)
9. Gálvez, T., Henríquez, H., Morales, G., Uribe-Echevarría, V., Yáñez, S.: Remuneración de vendedores y vendedoras en grandes tiendas. Temas laborales nueva época, No. 24. División de Estudios de la Dirección del Trabajo, Gobierno de Chile (2009)
10. García, M.M., Mateo, I., Maroto, G.: El impacto de cuidar en la salud y la calidad de vida de las mujeres. Gaceta Sanitaria No. 18 (Suplemento 2) pp. 83–92 (2004)
11. Enciso V. Género y Trabajo: La Enfermería, La Ventana, Núm. 6 (1997)
12. Ortiz-Gómez, T., Birriel-Salcedo, J., Ortega, R.: Género, profesiones sanitarias y salud pública. Gaceta Sanitaria [online], ISSN 0213-9111, vol. 18, suppl.1, pp. 189–194 (2004)

13. World Health Organization. Gender, women and health: incorporating a gender perspective into the overall framework of WHO policies and programs, 116th Meeting of the Executive Council, item 4.4 of the agenda, report of the Secretariat (2005)
14. Fernández-Ortega, M., Ortiz-Montalvo, A., Ponce-Rosas, E., Fajardo-Ortiz, G., Mazón-Ramírez, J.: Caracterización de alumnos de la carrera de Medicina. Investigación en Educación Médica **5**(19), 148–154 (2016)
15. Ramírez, M.A., Méndez-Carniado, O., Nigenda, G., Vargas, M.M.: Recursos humanos en los Servicios de Salud: Una perspectiva de género. Fundación Mexicana para la salud, Centro de Análisis Social y Económico (2002)
16. Delgado, A., López-Fernández, L.A.: Práctica profesional y género en atención primaria. Gaceta Sanitaria No. 18 (Suplemento 1) pp. 112–117 (2004)
17. Herrera, R., Cassals, M.: Algunos factores influyentes en la calidad de vida laboral de enfermería. Revista Cubana de Enfermería 2005 No. 21 (2005)
18. Espinoza, M.: Calidad de Vida en el Trabajo: Reflexiones en Torno a la Inseguridad y el Malestar Social. Temas laborales No. 18 (2001)
19. González, R., Hidalgo, G., Salazar, J.G., Preciado, M.L.: Instrumento para medir la Calidad de Vida en el Trabajo CVT-GOHISALO, Manual para su aplicación e interpretación. Ediciones de la Noche, Guadalajara, México, ISBN 978-970-764-866-1 (2009)
20. Godinho, D., Cappellin, P.: La perspectiva de género en la dimensión socio-laboral de los acuerdos de asociación, Análisis y propuestas Ed. Friedrich-Ebert-Stiftung, Uruguay (2009)

Open Learning: The New Normal of Design Education

Rosa Retuerto Luna(✉)

CIAUD, Lisbon School of Architecture, Universidade de Lisboa, R. Sá Nogueira, 1349-063 Lisbon, Portugal
rosa.luna@edu.ulisboa.pt

Abstract. Digital acceleration has uncovered the need to understand how design education is developing today, and how to take advantage of the possibilities of interaction provided by Open Educational Resources (OER) such as Massive Open Online Courses (MOOC) in the new globalized and increasingly technological educational scenario. Allowing exploration, experimentation, research, and creation of new educational models for learning that can adapt to different environments and make it possible for transmission of knowledge to continue even in times of crisis.

Keywords: Design education · MOOC · Interaction · OER

1 Introduction

Global pandemic forced the education sector to change the way it was teaching and visualized a great need for massive digital learning tools such as OER (Open Educational Resources) and the development of MOOCs, which provide a collaborative and open learning space where people can interact through different formats such as videos, e-books, recorded lectures, animations, among others.

Words like 'distance learning', 'remote learning' or 'virtual learning', became important in the life of the whole world in the wake of Covid-19. With educational institutions closed due to the health crisis, the only thing left to do was to rely on distance learning that would allow students to continue bringing knowledge amid a pandemic that for the moment only showed an uncertain future.

This work sought to identify the problems within the discipline of design in higher education and the use of open educational resources for learning in this area. Norman [1, p. 6] had already identified that "Today's designers are poorly trained to meet the today's demands: We need a new form of design education, one with more rigor, more science, and more attention to the social and behavioral sciences, to modern technology, and to business".

We performed semi-structured interviews using open-ended questions, conducted with some design teachers from several higher education institutions in Portugal, to know different realities about higher education in design, to perceive the main teaching/learning difficulties before and during Covid-19, as well as the teachers' experience

R. S. Goonetilleke et al. (Eds.): AHFE 2021, LNNS 273, pp. 149–156, 2021.
https://doi.org/10.1007/978-3-030-80713-9_19

with open educational resources and their perspective on the needs of this discipline in the current context.

Some of the responses obtained were that this pandemic made it possible to take better advantage of new technologies for knowledge transmission and interaction with students. Although at the beginning there were some cases of adaptation, which caused an overload of work for both teachers and students, the academic semester was completed.

We observed that it is necessary to provide spaces for discussion to discover current concepts, tools, and platforms that allow us to take advantage of the possibilities of instant interaction provided by OER such as MOOCs. Also that ensuring that this practice helps in improving a good educational experience, promotes the development of collaborative work and facilitates the acquisition of new digital skills.

2 OER: Open Educational Resources

Within institutions of higher education, with all that is causing the pandemic, it is being allowed to experiment with educational technologies that challenge the curriculum and the way teachers –who are often unprepared for these digital spaces–, teach.

OER, free-access education, or open education, is a clear example of this and several definitions can be found. In this study we will use "The concept of open education which is closely linked to the British university The Open University, which was founded in 1969 with the intention of being open to all people, places, methods and ideas". [2, p. 8].

This notion has been transformed over time to consider OERs "more than with distance education (…) with online teaching and digital resources that are offered openly". [2, p. 8] and which easy access has been one of its particularities, not involving many requirements and being able to join from anywhere in the world.

Now more than ever, due to the health crisis that we are experiencing, it is urgent that HEIs [Higher Education Institutions] invest in the development of digital content, such as the implementation of OERs, such as MOOCs in their different typologies, at a distance or blended- learning [3] Such developments would contribute to democratize education and continue to provide knowledge, this time, with a global reach.

3 MOOCs in Higher Education and Design

Since its creation, in 2008, MOOC has evolved and has grown exponentially due coronavirus. In its early years, Dave Cormier, an educational researcher, defined MOOCs as open, participatory courses that promote lifelong learning. They are also considered as "Classes taught through technological platforms that enable the teaching-learning process to thousands of students" [4, p. 4].

MOOCs within higher education are moving towards different actions, one of them is their contribution for to scientific knowledge, in which MOOCs function as an instrument for "(IDI) Research Data Exchange where universities that enter can share information and data for research purposes, big data of learning information" [5]. In other words, educational institutions can identify how students are learning to then monitor and evaluate learning, measure results, to discover new tools or products for online education.

MOOCs for higher education in design show a range of possibilities, both in research for the promotion of collaborative work and interaction, as well as to train professionals for a more current and global market. Students should be trained for modern learning, providing them with the necessary instruments to face the challenges of society that allow them to open the way for the development of their professions and their environment [6].

4 Methodological Procedures

This work was developed with an exploratory research model, making use of interviews to allow greater flexibility in data collection.

Explorative research aims to be able to develop and modify ideas with the intention of manifesting more detailed problems and hypotheses for future studies. It is considered the first phase of a more extensive investigation because it allows a rough initial look at a specific topic [7].

For this study it was decided to interview professors of Portuguese nationality from different higher education institutions in Portugal, with the intention of knowing this context and its different realities on design education.

Semi-structured interviews were carried out that were based on previously designed scripts, as well as informed consent forms, which were sent by email to the selected teachers who have had the experience of teaching in a design master's degree. The scripts served to structure the topics to be addressed, facilitate the interview to maintain the original plan [8] and take advantage of the reflective potential offered by the virtual environment [9].

Due to the state of emergency caused by Covid-19, all interviews were conducted virtually / synchronously (in real time) through Google meet and Colibri Zoom, University of Lisbon institutional platform and lasted from 60 to 90 min. It began by giving a brief explanation to the participants of the research setting, their context and the purpose of the interview.

It is important to mention that in Portugal, confinement began on March 18, 2020, the following months corresponding to the 2nd semester, classes had to be at a distance and making use of digital tools that allowed classrooms not to stop. However, in the 1st semester during the months of September to December 2020, we opted for semi-face-to-face classes and/or B-learning.

The sample used is non-probabilistic and intentional, where 7 professors were interviewed, 3 professors from the Lisbon School of Architecture, University of Lisbon-FAUL, a professor from IADE - Faculty of Design, Technology and Communication, European University, a professor from the University of Algarve, a professor from the Tomar Polytechnic Institute - IPT, and a professor from the Beja Polytechnic Institute - IPBeja (4 women and 3 men).

The data analysis method used was Thematic Analysis (AT). One of the most widely used qualitative methods in different research areas for data analysis, which allows to recognize themes and patterns in a group or set of data [10]. Within the TA approach, Reflective and Interpretive Thematic Research was used. With the intention of exploring concepts and data in depth and going beyond descriptive analysis [10].

The following Table 1 shows the profile of the interviews and due to a data protection issue for this research, the name of the teachers is kept anonymous, considering only the name of the institutions they represent.

Table 1. Profile of the interviewees during the semesters 2020 – I y II

Interview	Institution	Type of teacher	Courses with more hours taught in 2020 - I and/or II
Professor 1	Faculty of Architecture of the University of Lisbon	Impartial time	Product Design and Sustainability
Professor 2		Full time	Fashion design
Professor 3			Communication Design
Professor 4	IADE-Faculty of Design, Technology and Communication, European University	Full time	Digital animation
Professor 5	University of Algarve	Full time	Communication Design
Professor 6	Polytechnic Institute of Tomar	Full time	Editorial Design and Infographics
Professor 7	Polytechnic Institute of Beija	Full time	Graphic Design and Communication Design

5 Analysis and Results

The following set of questions was constructed to obtain perception of professors about most common problems regarding design knowledge in graduate students such as master's degrees; the use of digital tools such as OER, specifically MOOCs; and how they have been facing the challenges caused by Covid-19 and distance learning concerning design training:

Q1. In your experience as a teacher, what do you think of student's design knowledge when completing their master's in design and whether they are prepared for the challenges that design faces today?
Q2. What are the most common problems that your design students have about the topics covered in a class?
Q3. Before Covid-19 and the suspension of face-to-face classes, what kind of educational resources did you use in your classes?
Q4. Did it include OER (Open Educational Resources) like MOOCs?
Q5. What are the challenges that Covid-19 and distance education brought to higher education institutions concerning design teaching in Portugal?
Q6. Do you know if your design master's students have knowledge of OER about learning like MOOCs?

Q7. Did you have an opportunity to be a student or teacher of a MOOC course?
Q8. If MOOCs can be used as a resource to help level the knowledge of students who will attend master's courses in design, what do you think will be the main implementation difficulties?

After conducting the interviews, we proceeded with interpretations that allowed identification of some essential exploration factors, related to the connection between design knowledge of postgraduate students, the use of OER and MOOC as educational tools, as well as the skills that teachers had to develop during Covid-19 and distance learning.

The main ideas and lessons learned from the conversations with the interviewees are presented below:

Interviewees mentioned that students in many cases enter with little experience to study a postgraduate degree in design, specifically a master's degree. They have the foundations but not the maturity to provide solutions with design thinking according to the current problem. They also have poorly developed critical thinking and suffer when confronted with ideas.

Regarding how do design teachers have been coping with their classes amid a pandemic, it was observed that they had to use their skills to take on this new challenge and gradually learn digital tools that in some cases they were using for the first time.

Several professors claimed to have used some e-learning resources such as digital globes, on-line delivery of tests, use of web sites, e-mail, videos, and dropbox to manage students' work. The use of slides was maintained, as well as the theoretical and practical classes, where the first hours were destined for theory and the following hours for the development of practical exercises.

They also affirmed that in this new non-face-to-face modality it is necessary to change the way of thinking, how content is produced, how information is passed on, how knowledge is transmitted, that is, everything must be rethought.

This online context gave rise to several difficulties in the teaching of classes, among them the professors said they had experienced fatigue from being several hours in front of the computer, and they realized that in the face-to-face classes it was much easier to answer student's questions than in virtual mode.

Another problem identified because of the distancing for teaching design was the development of practical work in polytechnics and universities workshops that had to be interrupted. Spaces that are generally used for drawing sketches, manual work and the construction of models, had to abandoned and professors had complications when monitoring non-face-to-face learning. Added to this was the enormous task of ensuring that students were motivated throughout the semester and could complete their design projects.

Regarding MOOCs, only one of the teachers claimed to have had experience developing a MOOC course, the other answers were that they did not know about this resource or in any case, they had only heard about this tool but never used it. However, some commented that as a digital educational resource they used the Moodle tool to manage their classes.

They commented that students take online classrooms, use distance learning platforms, such as Moodle, Zoom, and Teams, Domestika being the educational website most used by design students, but they are not sure if they know about OER or MOOC.

However, they considered that it would be good for MOOCs to become a common and widespread practice. Despite difficulties that the construction of courses like these may entail, among them: budget, time for their development, choice of content, teachers willing to create courses, methodology, and learning needs of students with OER.

Additionally, they believed that these courses, due to their complexity of organization, should take into account the allocation of credits that are worth the effort and can motivate students to enroll in a MOOC (Tables 2 and 3).

Table 2. Outputs of interviews with design higher education professors

Factors	Outputs
Master's degree students and their knowledge of design to face the challenges of today's design	They have the bases: 3 They do not have sufficient professional experience: 3 They have more notions of technology: 1 It depends on the curriculum of the institution and the quality of the teachers to be more prepared: 2
Most common problems of design classes today	Lack of critical thinking There is a lack of balance between the theoretical and the practical Difficulties in providing solutions to current problems More teamwork needs to be promoted Lack of independence to carry out design projects
Most used educational resources before Covid-19	Presentations (slides): 7 Work in the ateliers: 3 Face-to-face tutorials: 3 Use of Moodle: 2
Use of OER or MOOC	Does not use OER or MOOC: 6 If you use OER and MOOC for classes: 1
Challenges of COVID-19 and distance learning	Social distancing Not being able to use design ateliers Not being able to do manual work Difficulties for the construction of models
Students and their knowledge of OER	They have little knowledge: 4 They do not know if students use OER: 3

(*continued*)

Table 2. (*continued*)

Factors	Outputs
Experience as a student or teacher of a MOOC	As a teacher: 1 As a student: 0
Difficulties of implement a MOOC	Time Interest of universities and facilitate obtaining credits Construction and implementation Complex courses to organize Find teachers available for these courses Motivation of students by MOOC courses

Table 3. Difficulties and educational resources used in the distance education of Design

Difficulties distance education	Most used educational resources during Covid-19
Inattention from students	Digital presentations (Slides), Zoom and Google Meet
Lack of literature	Websites and books
Body and visual fatigue	Web, Zoom, Moodle, Times, Dropbox and Padlet
Little time for tutorials	Zoom and Google Meet groups
Little time to resolve doubts	Zoom (Video conferences)
Greater adaptation	Zoom and the virtual classroom Moodle
To transmit design content	Zoom and the virtual classroom Moodle
Teamwork	Zoom Groups and WhatsApp Groups

6 Conclusions and Future Work

This work made it possible to detect that the current scientific knowledge on OER and MOOC in Portugal is not very developed or disseminated and that is why many teachers stated that they did not know or have not used this tool before.

The task for higher education institutions will be that the percentage of ignorance of MOOCs decreases and evaluate what actions can be taken for their dissemination.

The advancement and innovation of MOOCs will occur as more institutions are willing to take on the challenge of developing this model in their teaching centers, therefore it is necessary to invest in professionals who are ready to carry out research and monitoring work regarding this topic is very current but has not yet achieved the expansion required to discover and exploit the benefits it can generate for higher education organizations.

Regarding students, MOOCs are a good option to motivate them in self-learning and networking.

It will be essential to develop more research on OER, mainly MOOCs in Europe, specifically in Portugal, to disseminate the results of these to measure how beneficial it can be to adopt this teaching modality within universities or other higher education institutions.

For this reason, this exploratory study that is part of a Ph.D. will investigate further the open learning needs, in disciplines such as design.

Acknowledgements. This work was financially supported by the Centro de Investigação em Arquitetura, Urbanismo e Design - CIAUD (Center for Research in Architecture, Urbanism and Design) of the Faculdade de Arquitetura da Universidade de Lisboa (Faculty of Architecture of the University of Lisbon).

References

1. Norman, D.: Why design education must change (2010). http://www.core77.com/posts/17993/why-design-education-must-change-17993
2. Castrillo, M.D., Martín, E., Vásquez, E.: Guía práctica para el diseño y tutorización de MOOC. 1ª Edición. España (2018)
3. João, M., Osório, A., Valente, L.: Atas da IX Conferencia Internacional de TIC na Educação: Meio século de TIC na Educação, Half a century of ICT in Education. Universidad do Minho, Braga. Portugal (2015)
4. Observatorio de Innovación Educativa.: Reporte Edutrends: MOOC. Tecnológico de Monterrey (2014). https://observatorio.tec.mx/edutrendsmooc
5. Agarwal, A.: Aprendizaje activo en: E-learning, MOOCs y Micro credenciales. Observatorio de Innovación Educativa, Tecnológico de Monterrey (2020). https://observatorio.tec.mx/entrevistas/anant-agarwal-aprendizaje-activo
6. Frascara, J.: Enseñando Diseño. Buenos Aires. Infinito (2018)
7. Gil, A.C.: Métodos e Técnicas de Pesquisa Social, 6ª ed. Atlas, São Paulo (2008)
8. Flick, U.: Métodos Qualitativos na Investigação Científica. Monitor, Lisboa (2005)
9. Murray, C.D., Sixsmith, J.: Email: a qualitative research medium for interviewing? Int. J. Soc. Res. Methodol.: Theory Pract. **1**, 102–121 (1998)
10. Braun, V., Clarke, V.: Successful Qualitative Research A Practical Guide for Beginners. SAGE Publications Inc., London (2013)

Ergonomic Risk Assessment of Sea Fisherman Part IV: Tunisian Chapter

Alessio Silvetti[1(✉)], Elio Munafò[2], Ari Fiorelli[1], Lorenzo Fiori[1], Antonella Tatarelli[1], Alberto Ranavolo[1], and Francesco Draicchio[1]

[1] DIMEILA, INAIL Research Area, Rome, Monte Porzio Catone, Italy
{al.silvetti,lo.fiori,a.tatarelli,a.ranavolo, f.draicchio}@inail.it
[2] ITAL UIL, Rome, Italy

Abstract. Following our experiences in Italy in Mazara del Vallo and Chioggia Marinas, we had another one in Tunisia in Sfax Marina. Aim of this paper is a biomechanical risk assessment of fishermen working on a vessel smaller than those previously analyzed in Italy and in a different socio-cultural context. The task that involved most of the crew's working time was the sorting. We assessed this task by ISO Standard 11226 and 3DSSPP software. The maximum recommended exertion duty cycle, listed as a percentage of job time from the 14 analyzed frames with 3DSSPP, was between 2.3 and 14%. This data confirms that workers' postures during sorting task were critical. In these task workers flexed the trunk over 90° for approximately 10 min for every catch. Both maximum exposure time (1 min) and maximum acceptable trunk flexion (60°) threshold, defined by ISO 11226, were exceeded. Moreover, it should be considered that ISO 11226 does not consider boat instability that significantly impacts biomechanical risks. The maximum posture holding time before the occurence of fatigue, according to Potvin's mathematical model, is lower than the time held by workers. Trunk maximal values were between 35s and 125s and the lower limb joints values rarely exceeded two minutes of maximum exposure. Estimated values of the percentage of Maximum Voluntary Contraction (%MVC) were between 35% and 61% for the trunk and between 31.4% and 39.2% for the neck. The lower limbs %MVC values were remarkable for the hips (between 28% and 68%), knees (between 31% and 88%) and ankles (between 27% and 94%). This new experience confirmed medium-high risk levels during the sorting process. Furthermore given the reduced size of the boat, the crew couldn't improve their working conditions with low-cost solutions, as in Italy. For operational problems, manual handling of the fish crates wasn't assessed although the placement of the freezer on the deck of the boat, instead of the bottom floor as in Mazara del Vallo, is certely less stressful.

Keywords: Biomechanical overload · MSDs · Fatigue · 3DSSPP · Awkward posture

1 Introduction

Biomechanical risk in the fisheries has been widely investigated in our previous studies in Mazara del Vallo and Chioggia. Our activities focused on the analysis of two different

R. S. Goonetilleke et al. (Eds.): AHFE 2021, LNNS 273, pp. 157–167, 2021.
https://doi.org/10.1007/978-3-030-80713-9_20

kinds of fishing (driftnet and rake trawl with depressor) and with different work organization [1–3]. Except for slight differences in the drop and the set sail of the net, our results showed similar critical issues. The most hazardous tasks were the sorting and the manual handling of crates. The sorting involved the crew for almost all time. Our previous research showed that workers, based on their experience, developed customized and low-cost solutions to reduce the biomechanical load. Other epidemiological surveys have been recently added to the literature reported in our previous papers, that highlights how the MSDs in fisheries workers is still a common and widespread problem [4–6]. These reports, mainly conducted in Northern Europe, further confirm a high presence of MSDs in the fishery sector. It also emerges that the design of ships still doesn't include technological improvements to reduce biomechanical risk.

Besides our two previous boardings, we did a new experience in Sfax, Tunisia. The fishing technique was the same as in Mazara del Vallo (trawling) but with a smaller boat length (30 m vs 35 m). The crew were younger and without MSDs.

The aim of this paper is a biomechanical risk assessment aboard a ship in a socio-cultural scenario other than the ones analyzed. Furthermore, a discussion between the activities on board the vessels of Mazara del Vallo and Sfax will be carried out to identify analogies and differences.

2 Material and Methods

During boarding, we did video recordings of the working cycle, consisting of drop and in the set sail of the net in the sea and sorting, cleaning, and storing the catch in a refrigerator. The work cycle lasted three hours, two hours of work, and one of break. From video recordings we sampled some frames afterward analyzed with 3DSSPP software (v 7.1.3) [7, 8]. The software provides an estimation of various biomechanical parameters: compressive and shear forces at L4/L5 and L5/S1 levels, strength percent capable (SPC), an esteem of the percentage of Maximum Voluntary Contraction in major muscles of the trunk, upper and lower limbs (%MVC), the maximum recommended exertion duty cycle listed as a percentage of job time (MAX%DC) and the maximum static (continuous) allowed exertion time (MaxT). All parameters other than SPC are referred to 50th percentile of the working population.

The thresholds of the orthogonal compression forces are those proposed by Jager [9]. These are based on studies earlier than the NIOSH [10] ones used in the 3DSSPP software. Unlike NIOSH, where a single threshold value of 3400N is recommended, the thresholds proposed by Jager are gender- and age-specific (Table 1). The shear force threshold used as a reference is 700N according to Gallagher's review [11].

The thresholds of Max%DC and MaxT parameters are those proposed by ACGIH [12] based on Potvin's research on muscle fatigue for upper limb [13] and trunk [14].

All proposed thresholds refer to healthy people working on stable ground.

Besides the 3DSSPP software, the trunk posture in the sorting task was assessed with ISO 11226 standard [15]. This old standard, and unsupported by recent physiological data, remains an international benchmark for static postures risk assessment. Figure 1, shows a diagram that relates trunk flexion (x-axis) and maximum exposure time (y-axis). According to the diagram in Fig. 1 the maximum acceptable trunk flexion is 60° and can be sustained up to one minute continuously.

Table 1. Gender-specific age-related reference values for maximum lumbar compressive forces according to the Revised Dortmund Recommendations [9]

Age	Female	Male
20 years	4.1 kN	5.4 kN
30 years	3.8 kN	5.0 kN
40 years	3.1 kN	4.0 kN
50 years	2.4 kN	3.1 kN
≥60 years	1.8 kN	2.2 kN

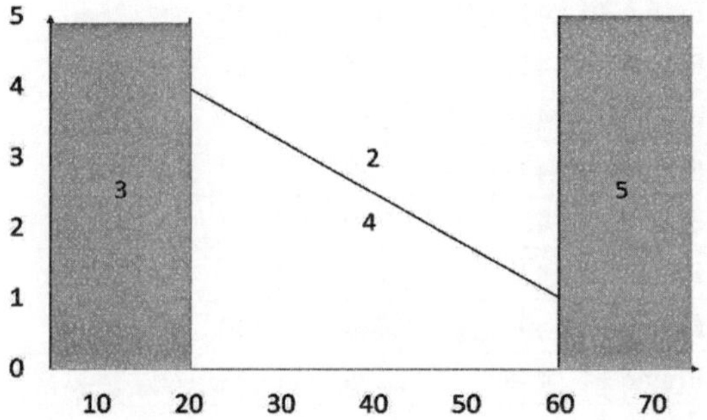

Fig. 1. Picture shows the thresholds for trunk flexion. In the x-axis is shown trunk flexion (degrees), in the y-axis maximum exposure time (minutes). 3 identifies the acceptable zone, 4 identifies the acceptable area using trunk support, 2 identifies the unrecommended area, 5 identifies the unacceptable area. Modified from ISO standard 11226.

3 Results

3DSSPP

Results of the 14 frames analyzed on the sorting task are summarized in Tables 2 and 3. Table 2 shows the results of orthogonal (L4/L5 or; L5/S1 or) and shear forces (L4/L5 sh; L5/S1 sh) at two levels. Table 3 reports the most relevant results for SPC, %MVC, MAX%DC and MaxT.

Figures 2 and 3 (frames 1 and 13) show selected reconstructions from 3DSSPP for the sorting task.

Shear forces at L4/L5 level were between 262N and 491N; 10 frames out of 14 exceeded 400N. Shear forces at L5/S1 level were between 360N and 442N; 8 frames out of 14 exceeded 400N. Orthogonal forces ranged from 1739N to 2913N at L4/L5 level and from 1786N to 2947N at L5/S1 level.

Upper limbs had no noticeable values of %MVC. The most affected joints were neck, trunk, and lower limb ones (hip, knee, and ankle). Neck flexion was noticeable for 8 of the 14 analyzed frames. 3DSSPP data were between 20% and 23%. Trunk flexion was relevant on all frames and the %MVC was from 35% to 61%. The left hip showed remarkable values of %MVC in all frames ranging from 33% to 68%; the

Table 2. For each of the 14 investigated frames, the table shows the results of shear and orthogonal forces at L4/L5 and L5/S1 levels during the sorting task.

	L4/L5 sh	L4/L5 or	L5/S1 sh	L5/S1 or
Frame 1	423	2656	439	2947
Frame 2	418	2305	442	2597
Frame 3	362	1922	379	2167
Frame 4	372	2464	399	2446
Frame 5	262	2308	360	2366
Frame 6	384	2372	409	2544
Frame 7	412	2321	425	2540
Frame 8	417	2335	427	2547
Frame 9	429	2236	425	2292
Frame 10	428	1739	377	1786
Frame 11	428	1739	377	1786
Frame 12	428	1775	377	1825
Frame 13	491	2913	440	2699
Frame 14	448	2235	438	2443

Table 3. The table shows, for each of the 14 frames, the values of %MVC, Max%DC, SPC, and MaxT of the sorting task. Only the most relevant values have been reported in the table.

	%MVC	Max%DC	SPC	MaxT (s)
Frame1	23% neck flex; 47% tru flex; 44% hip R/L; 41% knee L; 36% knee R; 30% ank L; 29% ank R	31.4% neck flex 6.1% tru flex	Hip 92%	64s tru flex; 57s hip L; 55s hip R;51s knee L; 65s knee R
Frame2	23% neck flex; 40% tru flex; 38% hip L; 41% hip R; 44% knee L; 38% knee R;53% ank L; 55% ank R	32.5% neck flex 9.8% tru flex	Hip 93% Ankle 91%	92s tru flex; 75s hip L; 63s hip R; 45s knee L; 60s knee R; 63s ank L;59s ank R
Frame3	37% tru flex; 33% hip L; 36% hip R; 38% knee L; 44% knee R;46% ank L; 49% ank R	11.9% tru flex	Hip 95% Ankle 94%	108s tru flex; 96s hip L; 85s hip R; 59s knee L; 46s knee R; 84s ank L; 75s ank R
Frame4	20% neck flex; 37% tru flex; 66% hip L; 67% knee L; 52% ank L	39% neck flex 11.6% tru flex	Hip 80% Knee 87% Ankle 93%	106s tru flex; 25s hip L; 21s knee L; 67s ank L;

(continued)

Table 3. (*continued*)

	%MVC	Max%DC	SPC	MaxT (s)
Frame5	36% tru flex; 56% hip L; 57% knee L; 28% knee R; 53% ank L; 31% ank R	12.7% tru flex	Hip 86% Knee 93% Ankle 92%	115s tru flex; 34s hip L; 32s knee L; 64s ank L;
Frame6	20% neck flex; 37% tru flex; 55% hip L; 59% knee L; 53% ank L; 33% ank R	39.2% neck flex 12.1% tru flex	Hip 87% Knee 92% Ankle 92%	110s tru flex; 36s hip L; 26s knee L; 63s ank L;
Frame7	36% tru flex; 36% hip L; 37% hip R; 23% knee L; 24% knee R; 27% ank L; 28% ank R	12.5% tru flex	Hip 94%	113s tru flex; 85s hip L; 77s hip R; 152s knee L; 141s knee R;
Frame8	20% neck flex; 38% tru flex; 34% hip L; 40% hip R; 24% knee L; 32% knee R; 24% ank L; 27% ank R	39% neck flex 10.8% tru flex	Hip 93%	99s tru flex; 93s hip L; 67s hip R; 141s knee L; 85s knee R;
Frame9	21% neck flex; 40% tru flex; 41% hip L; 36% hip R; 65% knee L; 48% knee R; 71% ank L; 61% ank R	36.4% neck flex 9.3% tru flex	Hip 93% Knee 88% Ankle 81%	88s tru flex; 65s hip L; 82s hip R;22s knee L; 39s knee R; 35s ank L; 48s ank R
Frame10	35% tru flex; 36% hip L; 28% hip R; 40% knee L; 30% knee R; 42% ank L 30% ank R	14% tru flex	Hip 92%	90s tru flex; 90s hip L; 55s hip R; 44s knee R; 156s ank R
Frame11	35% tru flex; 36% hip L; 28% hip R; 40% knee L; 30% knee R; 42% ank L 30% ank R	14% tru flex	Hip 94%	125s tru flex; 81s hip L; 136s hip R; 54s knee L; 93s knee R; 104s ank L;
Frame12	36% tru flex; 37% hip L; 29% hip R; 41% knee L; 31% knee R; 43% ank L; 31% ank R	13.2% tru flex	Hip 94%	119s tru flex; 79s hip L; 131s hip R; 51s knee L; 88s knee R; 98s ank L; 193s ank R

(*continued*)

Table 3. (*continued*)

	%MVC	Max%DC	SPC	MaxT (s)
Frame13	21% neck flex; 47% tru flex; 24% tru rot; 68% hip L; 88% knee L; 76% ank L; 27% ank R	36.5% neck flex 5.8% tru flex 29.5% tru rot	Hip 79% Knee 66% Ankle 76%	62s tru flex; 23s hip L; 13s knee L; 32s ank L;
Frame14	21% neck flex; 61% tru flex; 51% hip L; 36% hip R; 88% knee L; 48% knee R; 94% ank L; 68% ank R	37.7% neck flex 2.3% tru flex	Torso 89% Hip 89% Knee 66% Ankle 58%	35s tru flex; 41s hip L; 85s hip R; 13s knee L; 47s knee R; 19s ank L;40s ank R

Fig. 2. Left side: picture shows 3DSSPP reconstruction (frame 1) involving a worker while sorting. Trunk flexion exceeds 90°

right hip showed important values in 10 frames and their values were between 28% and 44%. The right knee showed values between 24% and 48% of MVC on 11 frames; the left knee showed values between 23% and 88% of MVC on all frames. Right ankle presented notable values of %MVC in 13 frames ranging between 27% and 68%; left ankle presented notable values on all frames, they ranged from 24% to 94%.

MaxT is strictly related to %MVC values. All the reconstructions showed the maximum time of 1200s for the upper limbs. Otherwise lower limbs and trunk, showed noticeable values. Trunk presented values ranging from 34s to 125s. As for the lower

Fig. 3. Right side: picture shows 3DSSPP reconstruction (frame 13) involving a worker while sorting. Trunk flexion exceeds 90°, the trunk is twisted and the neck is extended.

limbs, the hips ranged from 23s to 96s on the left and from 25s to 136s on the right; the knees ranged from 13s to 152s on the left and from 39s to 141s on the right; the ankles ranged from 19s to 98s on the left and from 40s to 193s on the right.

Trunk flexion showed the lowest values for Max%DC; they were between 2.3% and 14% DC. The only other joint that evidenced low Max%DC values was the neck; its values ranged from 31.4% to 39.2%. No significant values were found for upper limbs.

The lower limbs (hip, knee, and ankle) reported the lowest SPC values. Their values were between 79% and 94% for the hip, 66% and 93% for the knee, and 58% and 94% for the ankle. The trunk had a single remarkable value in frame 14 (89%).

ISO 11226

The crews, during all the sorting task, flexed their trunk beyond 90° most of the cases for more of 10 min continuously. As described previously, according to ISO 11226 thresholds, trunk flexion over than 60° is unacceptable.

Images show the sorting task in the Tunisian ship (Fig. 2, 3, and 4) and in the Italian ship (Fig. 5 and 6). The reduced length of the Tunisian ship made it hard for the crew to use a table to sort fish standing up (Fig. 5) or seating (Fig. 6). Figure 6 shows, moreover, crew members in Mazara using their backs for mutual support.

Manual Material Handling

Dues to organizational problems it was not possible to assess the task of handling crates. Anyway the placement of the cooling room at the stern level (Fig. 7) it's more advantageous within respect Mazara placementadvantages contrary to Mazara where it was located at the lower level (Fig. 8 and 9).

Fig. 4. (Left) Image shows the Tunisian crew while sorting.

Fig. 5. (Center) Image show the crew of Mazara del Vallo while sorting.

Fig. 6. (Right) The boat's dimensions allow the placement of a table (Fig. 5) and a raised floor at the stern to sit and support each other.

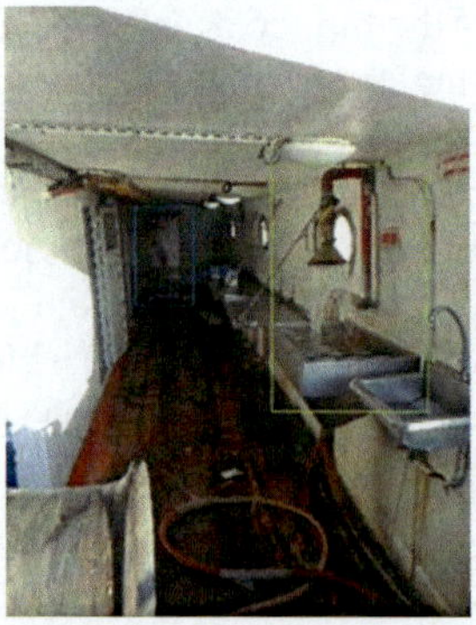

Fig. 7. (Left) image shows the corridor of the Tunisian ship. A blue rectangle marks the placement of the cooling room. The green rectangle shows the washing area.

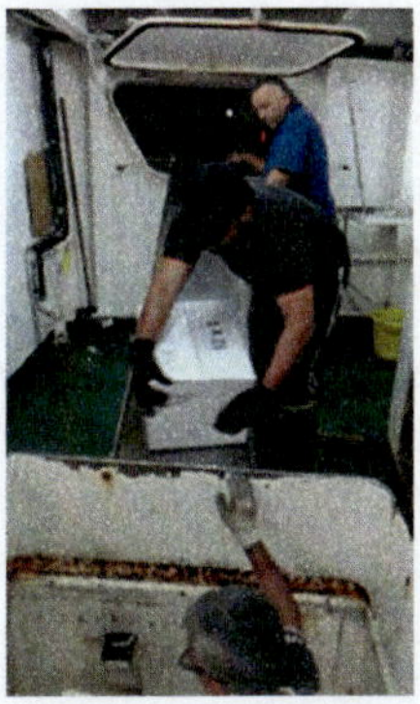

Fig. 8. (Center) Image show the location of the cooling room on the Italian ship.

Fig. 9. (Right) The cold room is located on the lower level of the ship.

4 Discussion

Our new onboard experience in Tunisia confirmed our previous findings.

The most demanding task which took up a considerable part of the crew's work cycle was sorting. Given the smaller dimensions of the ship, compared to the one in Mazara, it was not possible to execute the sorting task while sitting or standing on a table.

Workers' posture while sorting, according to ISO 11226, is considered unacceptable. Data from 3DSSPP software of the maximum percentage of the working cycle (Max%DC), based on Potvin's researches, are coherent with our previous data and also confirm the issues highlighted with ISO 11226. 3DSSPP data were very low, mainly for trunk flexion, ranging from 2.3% to 14% of DC. Even neck flexion showed maximum holding time that was exceeded.

As described before the mathematical model of Potvin has been validated for the trunk and upper limbs, not for lower limbs. Concerning the lower limbs, it is important to consider that all joints involved (hip, knee, and ankle) showed high values of %MVC that may cause musculoskeletal disorders.

Shear forces at L4/L5 and L5/S1 levels were always lower than the threshold of 700N proposed by Gallagher. For both joints examined, the highest orthogonal force value found was 2947N. According to Jager's table (Table 1), this value appeared to be safe for men up to 60 years without pathologies. Compression and shear forces are below the suggested thresholds, however, it has to be noted that they are the result of the worker's posture only.

Lastly, it has to be highlighted that all the considered parameters of 3DSSPP software are based on mathematical models that consider the worker's balance. Continuous boat oscillations due to the waves increase the biomechanical risk. Oscillations are very hard to quantify, but they are remarkable, from a biomechanical load viewpoint, also under rough sea conditions as those occurred while boarding.

As already noted during our previous research, activities onboard may be optimized through an ergonomic ship design. It would be feasible, even in the Tunisian ship, to use automated equipment, like that shown in Fig. 10, that automatically washes and conveys the catch, into a covered area where workers can work without adopting awkward postures and being safe from adverse weather conditions.

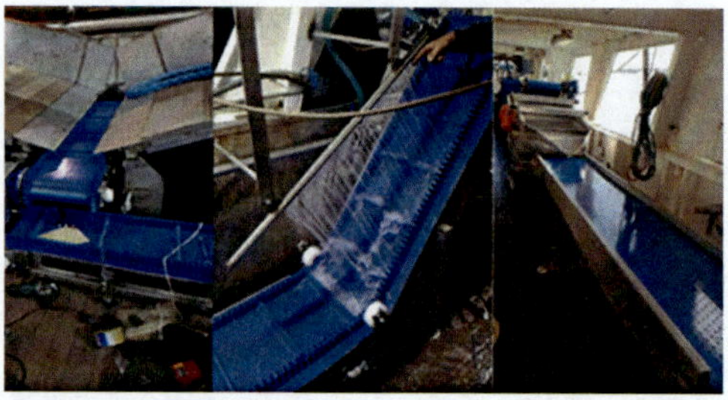

Fig. 10. Image shows equipment that, conveys the catch to an automatic washing system than to an area where workers don't assume awkward posture.

Automatic equipment as one shown in Fig. 10 are still expensive anyway. To reduce the biomechanical workload in fisheries, these technologies could be partially financed through public funds aimed to prevent fisheries workers' health.

Aknowledgements. We would like to thank UILA Pesca for promoting this study and [1, 2, 3].

References

1. Silvetti, A., Munafò, E., Ranavolo, A., Iavicoli, S., Draicchio, F.: Ergonomic risk assessment of sea fishermen Part I: manual material handling. In: Goossens, R.H.M. (ed.) Advances in Social & Occupational Ergonomics. AISC, vol. 487, pp. 325–332. Springer, Cham (2017). https://doi.org/10.1007/978-3-319-41688-5_29
2. Silvetti, A., Munafò, E., Ranavolo, A., Iavicoli, S., Draicchio, F.: Ergonomic risk assessment of sea fishermen Part II: upper limb repetitive movements. In: Goossens, R.H.M. (ed.) Advances in Social & Occupational Ergonomics. AISC, vol. 487, pp. 333–340. Springer, Cham (2017). https://doi.org/10.1007/978-3-319-41688-5_30
3. Silvetti, A., et al.: Ergonomic risk assessment of sea fisherman Part III: manual handling and static posture. In: Goossens, R.H.M., Murata, A. (eds.) AHFE 2019. AISC, vol. 970, pp. 379–392. Springer, Cham (2020). https://doi.org/10.1007/978-3-030-20145-6_38

4. Sønvisen, S.A., Thorvaldsen, T., Holmen, I.M., Øren, A.: Work environment and health in the fishing fleet: results from a survey amongst Norwegian fishers. Int. Marit. Health **68**(4), 203–210 (2017)
5. Sandsund, M., et al.: Musculoskeletal symptoms among workers in the commercial fishing fleet of Norway. Int. Marit. Health **70**(2), 100–106 (2019)
6. Øren, A., et al.: Sickness absence and hospitalization among workers on board norwegian fishing vessels. J. Agromed. **24**(4), 357–363 (2019). https://doi.org/10.1080/1059924X.2019.1640150
7. Chaffin, D.B., Erig, M.: Three-dimensional biomechanical static strength prediction model sensitivity to postural and anthropometric inaccuracies. IIE Trans. **23**(3), 215–227 (1991)
8. Chaffin, D.B.: Biomechanical Modeling for Simulation of 3D Static Human Exertions Computer Applications in Ergonomics, Occupational Safety, and Health. Elsevier Publishers, Amsterdam (1992)
9. Jäger, M.: Extended compilation of autopsy-material measurements on lumbar ultimate compressive strength for deriving reference values in ergonomic work design: the revised Dortmund recommendations. EXCLI J. **17**, 362–385 (2018)
10. Waters, T.R., et al.: Revised NIOSH equation for the design and evaluation of manual lifting tasks. Ergonomics **36**(7), 749–776 (1993)
11. Gallagher, S., Marras, W.S.: Tolerance of the lumbar spine to shear: a review and recommended exposure limits. Clin. Biomech. **27**(10), 973–978 (2012)
12. ACGIH. TLVs and BEIs. 2015. 202-204 ISBN 978-1-607260-77-6
13. Potvin, J.: Predicting maximum acceptable efforts for repetitive tasks: an equation-based on duty cycle. Hum. Factors **54**(2), 175–188 (2012)
14. Potvin, J.: An equation to predict maximum acceptable loads for repetitive tasks based on duty cycle: evaluation with lifting and lowering tasks. Work **41**(Supplement 1), 397–400 (2012)
15. ISO 11226:2000/Cor 1:2006 Ergonomics - Evaluation of static working postures

Vocal Disability Index in Teachers from Ecuador

Rommel Silva[1(✉)], Oswaldo Jara[2], Esteban Carrera[3], Pablo Davila[2], and José Luis Saá[1]

[1] Universidad de Especialidades Espíritu Santo, Samborndón, Ecuador
[2] Universidad Internacional Sek, Quito, Ecuador
[3] Instituto Superior Tecnológico Edward Deming, Quito, Ecuador

Abstract. Introduction: One of the main characteristics of human beings is the voice because it is essential to be able to communicate. In the workplace we can say that it is one of the main tools that teachers have to teach their classes, but the frequent use of the voice in different tones can imply a great effort or demand, and the vocal quality can be affected by different factors environmental and personal habits. Teachers have requirements in their work activity, such as adapting to the demands of teaching, developing more dynamic classes, and in current conditions the high interactivity demanded by virtual classes generates a perception of high vocal effort. **Objective:** To assess the level of vocal impact on university teachers from different universities, colleges and schools in Ecuador through the self-perception of vocal disadvantage. **Material and Methods:** It is a descriptive and cross-sectional study, which included 106 male and female teachers. The Voice Handicap Index questionnaire was applied. This questionnaire allows evaluating the impact of vocal quality on teachers' working lives. In addition, a survey was conducted on variables such as sex, age, average number of classes and hours taught per week, and smoking habits. **Results:** The average of the Voice Handicap Index of the total sample corresponded to a moderate handicap. 17% of teachers report that the clarity of their voice is unpredictable; 13.2% have smoking habits; 3.8% perceive their voice as hoarse and dry; 2.8% that other people do not understand their voice problem; 1.9% of those surveyed state that they try to change their voice to make it sound different; 1.9% feel that they must strain to speak; and 0.9% mentioned hearing their voice with difficulty. **Conclusions:** Teachers are one of the professional groups in which there is significant vocal involvement. The lack of vocal training that teachers have in the actions of their professional lives was evidenced.

Keywords: Dysphonia · Vocal quality · Voice handicap index

1 Introduction

The voice according to the dictionary of the Spanish Royal Academy of the Language, is the sound caused by the vibrations of the vocal cords, the same ones that are responsible for producing the sound of the voice emitted by teachers when they teach classes, because these They are the components of the speech apparatus that makes it possible to produce the voice both when speaking and when singing [1]. The World Health Organization

R. S. Goonetilleke et al. (Eds.): AHFE 2021, LNNS 273, pp. 168–173, 2021.
https://doi.org/10.1007/978-3-030-80713-9_21

(WHO) [2] defines disability as "a restriction or lack of capacity manifested in the performance of daily tasks".

The voice is a multidimensional phenomenon and essential for human communication. For this reason, a vocal alteration, even slight, has a great psychosocial impact on the life of an individual [3].

All jobs have risks inherent to their profession, which is why it is mandatory that they know the risks and be advised on preventive measures. However, the inevitable first step is to determine the level of risk. In the case of teachers, they are known to be prone to voice disorders of different tones, at some point in their career [4].

The incapacity of the voice is the impossibility for a person to produce a loud sound or speak aloud with some difficulty. This disadvantage is defined as "a social, economic or environmental disadvantage, an advantage resulting from an impairment or disability" [5].

Likewise, dysphonia is considered as a disorder of oral communication, where the voice is unable to fulfill its basic role in the verbal and emotional transmission of messages and limits the effectiveness of this transmission and in the case of women, it is a disturbance that can bring harm to the individual, and is represented 14 by any change, or difficulty of production, in the natural emission of the voice, making it unpleasant and socially unacceptable [6].

Likewise, dysphonia is the precursor of vocal damage or vocal pathology. It is a multifactorial event where the person who notices this type of condition reports greater effort, discomfort when speaking, reduced voice characteristics and improvement in vocal symptoms with a period of vocal rest. Vocal fatigue is a common malady among voice professionals, such as teachers, as they have substantially higher vocal demands. Studies indicate that around 18% to 33% of professional voice users experience symptoms of vocal fatigue. Which may increase as the day progresses and may decrease after a period of voice rest [7].

Establishing the severity of a vocal disorder is difficult, techniques have varied in subjectively establishing the severity of the voice disorder, which includes perceptual judgments (e.g., the quality of the voice) to much more objective measures of the particularities of the voice. voice (physiological measure of voice in comparison). Although these methods can produce valuable reviews, they do not provide a clear idea why patients with similar voice disorders perceive different levels of disability [5].

In this study, the Voice Handicap Index (VHI) was used, which is a questionnaire developed by Jacobson et al. In order to quantify the impact perceived by a person affected by a vocal disorder in conditions of the vocal function itself, in the physical capacity related to it and in the shocks caused by dysphonia. The VHI measures the influence of voice problems on a patient's quality of life [7].

In a study conducted on teachers in Kuwait in 2020, it was possible to observe the prevalence of voice disorders among teachers, it was observed that vocal dysfunction among teachers was very significant compared to other professions. Among the factors that contribute to this anomaly are; vocalizing out loud for long periods of time, to back-to-back classes with little rest in between, poor air quality, and talking about background noise [8].

The points of interest and the variables that deserve to be taken into account to explain a possible vocal pathology The multifactorial etiopathology, [4] that can trigger an occupational disease in teachers, are; the number of hours of teaching per day, type of subject, number of students per classroom, the use or not of amplification equipment, years of work, the type of training that is instructed (elementary, preparatory, upper secondary), habits toxic (liquor, cigarette), among others [9]. A study conducted in China reveals that the younger the students, the higher the risk teachers have for developing a voice disorder [10].

The main objective of the following research is to know the vocal perception of teachers. Evaluate the degree of vocal impact in university teachers from different universities, colleges and schools in Ecuador through the self-perception of vocal disadvantage through the VHI-30 test.

2 Methodology

The HIV questionnaire developed by (Jacobson et al., 1997) was used and consists of 30 items, where the teacher's perception of laryngeal discomfort or the characteristics of vocal production functional subscale is evaluated, where the effect of the disorder is evaluated of the teacher's voice in their activities and emotional subscale, where the affective response of the patient to the vocal disorder can be observed [11].

The VHI-30 is divided into 3 groups (Physical, Functional and Emotional) of 10 questions, in turn the questions in each group have a rating of 0 to 4 being (0 = Never, 1 = Almost never 3 = Almost always and 4 = Always), each scale has a maximum score of 40 points. When tabulating the responses, the degrees or levels of disability are established between 0–30 points as mild, between 31–60 points as moderate, between 61–90 points as severe and between 91 to 120 points as severe [5].

It has subsequently been shown that these subscales are not separate measures of vocal impairment and that they lack validity as such [7].

This is a descriptive and analytical cross-sectional study. This study had the participation of n = 106 teachers randomly, carried out between June and August 2020, which included most of the provinces of Ecuador, framed in all areas from basic education to higher education, as well as they were considered teachers who taught classes online.

There were no exclusion criteria, and the only inclusion criterion was that the respondent was a teacher of some type of education level in the country.

The study also surveyed the participants on socio-demographic issues, such as age, gender, and certain personal habits.

3 Data Collection

The study population were teachers from all over the country, through a cross-sectional study. For this activity, a questionnaire was carried out where socio-demographic questions were included and questions from the translated version of the original VHI were also included. Respondents answered the questionnaire through a link created in Google Forms, a tool that was used to prepare the questionnaire, the way of sending was by mail and using the Social Network What's APP and subsequently sent collectively to various teachers in the country, of which we obtained 106 respondents, who answered all the questions.

4 Results

The vocal disability assessment test was applied to 106 teachers from different educational institutions (basic, middle, and higher) in various cities of Ecuador, they are 56 male (52.8%) and 50 females (47.2%). 65.1% of people are between 31 and 50 years old. 74.5% of teachers teach between 8 and 30 average hours of classes per week. 55.4% have between 5 and 10 years as a teacher and 13.2% have smoking habits, of which 28.5% smoke daily. In graph No. 1 we can see that the average value of the total index of vocal disability was 12.8, which corresponds to a mild vocal disability. 94 employees (88.7%) obtained a score considered mild and 12 people a moderate score (11.3%) (Fig. 1).

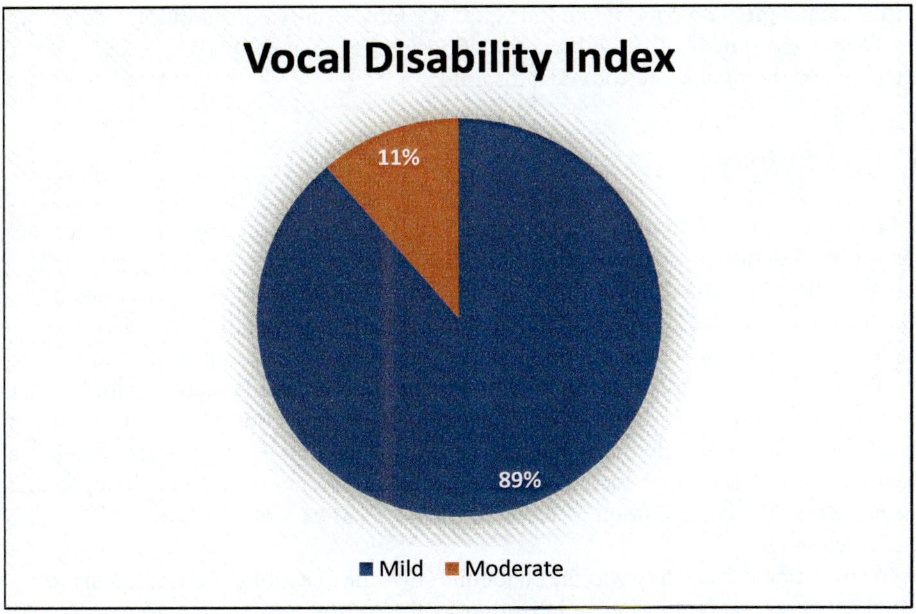

Fig. 1. Handicap index VHI-30

To process the results of the questions and obtain the results, the SPSS V.23.0 tool was used. Table No. 1 details statistical variables of the total score, for each scale that constitutes the index (Table 1).

4.7% of people mentioned that their voice almost always worsens in the afternoon, 3.8% of teachers almost always feel upset when other people ask them to repeat what was said, and 2.8% almost always perceive their voice as broken and dry. 16% of employees report that the quality of their voice is unpredictable. 25.5% of people report that sometimes other people do not understand them in noisy places, and 22.6% report that on certain occasions their voice sounds different throughout the day. Among teachers with

Table 1. Statistical variables of the total score.

	Total index	Functional scale	Physical scale	Emotional scale
Mean	12.8	5	5.3	2.5
Median	8	4	4	0.5
Typical deviation	12.3	3.9	5.2	4.1
Minimum	0	0	0	0
Maximum	58	15	24	19
Confidence interval 95%	12.73–12.87	4.98–5.02	5.27–5.33	2.48–2.52

a smoking habit, 78.6% obtained a mild vocal disability index, and 21.45 a moderate index.

Of the people with a moderate index, 66.6% have a daily consumption frequency.

There is no statistically significant relationship between people's age, sex, years of teaching and the total score for vocal disability.

5 Conclusions

In Ecuador, within the field of occupational health and safety, the risk faced by teachers has not been identified or given importance due to the vocal effort made in teaching their classes. The educational institutions within their health surveillance programs do not take the mentioned factor into account and do not carry out specific clinical evaluations that could identify any pathology in the larynx and vocal cords at an early stage.

The clinical assessment of this risk factor should be implemented within the occupational health programs in all educational centers of the country, as well as teachers should be trained on the elements that intervene in the use of the voice, and the conditions that could affect its quality such as environmental situations, and habitus inappropriate foods such as the consumption of alcohol, tobacco and excess seasonings, coffee, and foods rich in fat.

A limitation in our study was the pandemic since the questionnaire was made through social networks, without prior socialization, of the instructions reasons for carrying out this study.

Even though in our study the level of vocal disability is slight, an assessment should be implemented at the national level in order to determine the general condition of teachers in Ecuador.

6 Discussion

At present and with the new demands for online classes due to the pandemic, in our country and throughout the world, teachers must maintain excellent vocal health, since their main challenge is communicating with their students. Taking into account that the teacher's performance in the educational process is fundamental, and his voice is an indispensable work "tool" and is used during educational classes, exposing himself to the risk of vocal overload or overexertion that can impair the ability to perform professional performance. Routine or abuse in the use of the voice by teachers can lead to wear and tear or injury to the speech apparatus. Due to these particularities in the use of the voice by teachers, it should be considered "professional speech therapy, otorhinolaryngology and speech therapy for a timely prevention of occupational diseases in teachers" [12].

References

1. Begonya, T.: Anatomía Funcional de La Voz (2008)
2. Organization Mundial de la Salud: Clasificación internacional de deficiencias, discapacidades y minusvalías. Ginebra: Mundo Organización de la salud (1980)
3. Zubeldía, J., Elias, L., Bentorino, E., Wilder, F.: Relación entre VHI (Voice Handicap Index) y valoración objetiva de la perturbación vocal. Rev. Faso (1), 24–30 (2019)
4. Menon, U.K., Raj, M., Antony, L., Soman, S., Bhaskaran, R.: Prevalence of voice disorders in school teachers in a district in South India. J. Voice **35**(1), 1–8 (2021). https://doi.org/10.1016/j.jvoice.2019.07.005
5. Jacobson, B.H., et al.: The voice handicap index (VHI): development and validation. Am. J. Speech Lang. Pathol. **6**(3), 66–69 (1997). https://doi.org/10.1044/1058-0360.0603.66
6. En, L., et al.: Percepción de la discapacidad vocal de los pacientes con disfonía funcional del programa de terapia de lenguaje en el hospital guillermo almenara (2019)
7. Núñez-Batalla, F., Corte-Santos, P., Señaris-González, B., Llorente-Pendás, J.L., Górriz-Gil, C., Suárez-Nieto, C.: Adaptación y validación del índice de incapacidad vocal (VHI-30) y su versión abreviada (VHI-10) al español. Acta Otorrinolaringol. Espanola **58**(9), 386–392 (2007). https://doi.org/10.1016/S0001-6519(07)74954-3
8. Alarouj, H., Althekerallah, J.M., AlAli, H., Ebrahim, M.A., Ebrahim, M.A.K.: A comparative study utilizing the voice handicap index-10 (VHI-10) in teachers and the general population of Kuwait. J. Voice **10**, 1–10 (2020). https://doi.org/10.1016/j.jvoice.2020.05.006
9. Magdalena, U., Revollo-zúñiga, F., Hernández-blanco, J.: Prevalence of voice disorders in teachers Universidad del. **17**(2), 2017–2018 (2020)
10. Tao, Y., Lee, C.T.C., Hu, Y.J., Liu, Q.: Relevant work factors associated with voice disorders in early childhood teachers: a comparison between kindergarten and elementary school teachers in Yancheng, China. Int. J. Environ. Res. Public Health **17**(9), 1–16 (2020). https://doi.org/10.3390/ijerph17093081
11. Matas, A.: Diseño del formato de escalas tipo Likert: un estado de la cuestión. Rev. Electrón. Invest Educ. **20**(1), 38–47 (2018)
12. Domínguez, A., López, A., Núñez, S., Portela, I., Elia, V.: Perturbación de la Voz en Docentes. 13 (2020)

Occupational Health Management in Informal Work: A Theoretical Analysis of the Field

Luis Betancourt-Sánchez(✉)

Corporación Universitaria Minuto de Dios, Bogotá, Colombia
luis.betancourt@uniminuto.edu

Abstract. Informal employment has become the income generator for millions of human beings around the world. It is estimated that at least 2 billion people earn a living from work in informal conditions. However, given the characteristics of informal employment, there are no occupational health and safety management systems that allow accounting for the conditions under which these activities are carried out. Worse still, the health effects of informal work are completely unknown. Worldwide, it is estimated that up to 85% of workers in the world lack access to occupational health and safety services. For this reason, the identification of hazards, risk assessment and the analysis of their effects constitute a problematic situation, especially for workers who lack insurance coverage for accidents and occupational diseases. From a theoretical point of view, the health-work relationship in the informal sector is seen from multiple angles, which implies a diversified analysis of its realities and characteristics. In this sense, focusing attention only on working conditions or health conditions implies ignoring the variety of views that underlie the informal economy.

Keywords: Occupational health · Informal workers · Health inequalities · Occupational exposure · Social security

1 Introduction

The informality is one of the great challenges in terms of safety and health worldwide. Work, as one of the health determinants with the highest incidence, has positive or negative effects on the life of the human being, hence the employment and material conditions, as well as individual and non-work factors play an important role in the health-disease process of millions of workers.

Despite the constant adoption of strategies and programs for the protection of health at work, it is estimated that 85% of workers worldwide lack access to basic occupational health services that mitigate the impact of exposures to hazards. and demands emanating from forms of organization and working and employment conditions that are harmful to the human collective [1].

Deep effects of macroeconomic measures aggressively promoted in Latin America and the Caribbean have stimulated changes in the productive structures and labor markets that have brought with them new dynamics for the region such as: technological

R. S. Goonetilleke et al. (Eds.): AHFE 2021, LNNS 273, pp. 174–178, 2021.
https://doi.org/10.1007/978-3-030-80713-9_22

changes and in the distribution of work with an important foundation in the organization with mixtures between Taylorism-Toyotaism and a low technological turnover, a fact that promotes productivity centered on the distribution of labor and not the impulse of technology; second, labor flexibility that significantly breaks with the organization of workers and generates new forms of hiring as well as a displacement of the workforce to the informal economy; third, the dynamics of the labor markets, which generate important reflections on informal work, domestic work, female work and the general precariousness of employment [2].

However, the problems generated around informal work cannot be seen only as an effect of economic policies but must be analyzed as a set of factors that trigger conditions of employment, work and health with peculiar characteristics in different parts of the world.

Thus, the problem of informality and health at work must be approached as a matter of public policy, based on economic models, relations between workers and employers, deregulation around employment conditions, and the dynamics of employment. access to social security, the disadvantages and competitive advantages between formal and informal jobs and production, the absence of universal safety and health programs at work, and finally the personal living and employment conditions that generate job insecurity and poverty for individuals.

For Benach, Santana, Solar & Muntaner Quinlan [3] one of the most important factors for social inequality in the world is the conditions of employment and work. Although, it considers that the working conditions (tasks, physical conditions, of the organization) have received a great deal of attention and their implications on human health are recognized, this has not happened with the employment conditions that were named the way a worker is linked to work from different perspectives (full-time employment, unemployment, precarious employment, informal employment, child labor, slavery, etc.).

In turn, Vélez, Escobar and Pico [4] argue that relationships in the informal sector are basically mediated by casual employment, kinship, and social and personal relationships in which the characteristic contracts of formal companies do not intervene. Likewise, they determine that characteristics related to informality include the lack of legal protection that generates vulnerability to both workers and informal entrepreneurs; job instability and low and irregular income; competitive disadvantages compared to those in formal economies; insecurity emanating from the absence of benefits from macroeconomic policies, the absence of organization and association, among other factors.

Vélez, Escobar and Pico [4] point out that it is as a result of the reforms derived from the Washington Consensus that Latin America is determined to promote processes that deregulate rules that promote market economies, the expansion of the private sector, financial and commercial liberalization which in turn induces a reduction in the role of the state in the dynamics of work. They emphasize that, due to this, economic interests prevailed over social ones and equity, producing consequences in the labor markets that tended towards flexibility, joblessness and labor informality.

One of the effects of work in informality (which in turn is the cause of a circular dynamic of exclusion) is the absence of social protection, a defining characteristic of the notion of informality and a critical aspect that generates social exclusion. The overwhelming growth of informality also supposes the loss of forms of protection typical of formal workers, such as contribution to retirement, access to recreation, health, culture and sport with greater opportunity. In this case, informal workers, due to their low-income margin, must give priority to elements of basic needs such as food, housing, private health care, etc., generating a large-scale exclusion mechanism for millions of workers [5].

2 Workers' Health and Informality

2.1 Informality in the World

The ILO [6] estimates that about 60% of the world's employed population is informal. In turn, it argues that informality exists in all countries of the world regardless of the level of socioeconomic development, although with greater prevalence in developing and emerging nations.

This organization estimates that around the world close to two billion male and female workers live in the informal economy and are deprived of a job in decent conditions, also showing that informality is not in most cases a choice but a consequence. the absence of opportunities in formal economic settings and the absence of other forms of subsistence.

The ILO statistical report on men and women in informality [6] that takes into account at least 100 nations shows a mixed picture in terms of levels of informal employment in a regional perspective, indicating, for example, that for Africa the informality can reach 85.5% of the total workforce on the continent, being the region of the world with the highest rate. It also indicates that the Asia-Pacific region has 68.2% of its workforce in informality, a percentage very similar to that of the Arab states with informalities that are around 68.6%.

In the Americas region, there is evidence of a lower percentage of informality (40%), however, with significant asymmetries between the countries of North and South America and Central America. Finally, there is the region of Europe and Central Asia with the lowest percentage of informal labor force with 25.1%.

As mentioned, informality is mainly prevalent in developing and emerging nations, which if they were taken into account only for the calculation of informality would increase it until reaching 70% of the working population. While if informality were measured only in developed countries, it would be less than 19%.

Additionally, the composition of informal employment is diverse, since it can occur in the informal economy sector, in the formal sector or particularly in domestic work. Thus, the ILO [6] estimates that of the 61.2% of people who have an informal job, 51.9% have it in the informal sector of the economy, while 6.7% have it in the formal sector of the economy and 2.5% in domestic work.

An important element of informality is the occupational position that the worker occupies within the employment scenario. Globally, the highest percentage of informal workers does so under the category of self-employed, a position that is especially relevant in developing and emerging countries where up to 87% of self-employed workers are informal, while in countries developed this figure becomes 69%.

Another important characteristic of informality is that it is mainly high in the young population and in the elderly. The ILO statistical study estimates that 3 out of every 4 young and old are informal workers worldwide, figures that are especially representative of developing and emerging countries. In the emerging and developing countries of the Americas, the level of informality in young people under 25 years of age reaches up to 62% and in the elderly, it reaches figures of 80%.

2.2 Social Security and Informal Work

The informal sector is defined as the set of informal economic activities, which, because they do not pay taxes, do not enter the social and labor security regime. Unlike informal employment, which is the activity as such that a person performs independently outside of labor law, without social or labor security guarantees, which causes a high rate of accidents and occupational diseases, which are not covered in a job security system created specifically for the informal sector. Furthermore, such accidents and illnesses are not reported as such. Therefore, informal employment is one of the great problems and great challenges that all governments have in terms of job security in the world [7].

According to the International Labor Organization 5 out of 10 people can be vulnerable workers, so that about half of the workers do not have health coverage. In addition, every day 6,300 people die from injuries caused by occupational accidents or diseases, there are 2.34 million accidents per year and 60 million DALYs (disability-adjusted life years). The figures may be higher, because only 15% of workers worldwide would be covered by occupational health services [7].

In this sense, employment relationships exist at two levels: macro and micro. At the macro level they are determined by the social institutions and social relations that respond to the place of countries and activities in a global division of production and their historical determinants. Power relations between relevant organized actors (trade unions, corporate groups, non-governmental organizations, etc.) have a broad influence on the structure of the labor market and the welfare state. Macro-level power is the force behind interrelated socio-economic transformations, such as the reduction of social security, the loss of public sector jobs, the growth of precarious employment - all of these weaken regulatory protections, worldwide. Going to the micro level, the multidimensional concept of job insecurity enables the specific job to be rated with reference to the ideal type of full-time or standard job. At this point, the differences in power linked to social positions categorized as gender, race/ethnicity, social class or immigrant status are included [8].

Therefore, adequate, timely and rigorous monitoring of the working conditions of those who carry out their activities in the informal sector is essential, since it involves directing actions for the correct identification of hazards and requirements and the assessment of their risks in order to mitigate accidents, morbidity and hidden mortality in informal jobs. Likewise, it makes it possible to contribute significantly to decent work policies, ensuring the protection of workers' rights and the promotion of health in all workplaces in line with the sustainable development objectives that seek to eradicate poverty (Goal 1) and promoting decent work and economic growth (Goal 8).

References

1. Luna, J.: Es posible la salud ocupacional para todos? Una estrategia de entornos de trabajo saludable en atención primaria de salud. In: Narcea (ed.) Salud Laboral: aspectos sociales de la prevención: libro de actas del congreso "Managua 2012", pp. 39–51, Madrid (2013)
2. De la Garza, E.: Teoría sociológica del mercado de trabajo. In: de la Garza, E. (ed.) Primera Ed. Tratado latinoamericano de sociología del trabajo. México, pp. 511–539 (2000). http://www.flacsoandes.edu.ec/libros/digital/51932.pdf
3. Benach, J., Muntaner, C., Solar, O., Santana, V., Quinlan, M.: Empleo, trabajo y desigualdades en salud: una visión global. Primera. Icaria Editores, Madrid (2010)
4. Vélez, C., Escobar, M., Pico, M.: Determinantes sociales de la salud y el trabajo informal. Rev Costarric Salud Pública. **22**(2), 156–162 (2013)
5. Ministerio de la Protección Social. Plan Nacional de Salud Ocupacional 2008–2012. Imprenta Nacional de Colombia. 76 p. (2008). https://www.minsalud.gov.co/DocumentosyPublicaciones/Plannacionaldesaludocupacional.pdf
6. OIT-ILO. Women and Men in Informal Economy: A statistical picture. Geneva, 164 p. (2018)
7. Carretero, J.L., Cueva, O., Martínez, B., Martínez, M.V., Lobato, J.R.: The informal economy: an occupational health issue. Arch. Prev. Riesgos. Labor **20**(1), 30–32 (2017). http://www.ncbi.nlm.nih.gov/pubmed/28079324
8. Muntaner, C., et al.: Unemployment, informal work, precarious employment, child labor, slavery, and health inequalities: pathways and mechanisms. Int. J. Heal. Serv. **40**(2), 281–295 (2010). http://journals.sagepub.com/doi/10.2190/HS.40.2.h

Didactic Adaptation with ICT's Preliminary Educational Proposal

Luis Serpa-Andrade(✉), Erika Pinos-Velez, and Freddy Rivera-Calle

GI-IATa, Cátedra UNESCO Tecnologías de apoyo para la Incl. Educativa, Universidad Politécnica Salesiana, Cuenca, Ecuador
{lserpa,frivera}@ups.edu.ec

Abstract. First, a collection of information from the educational system of Ecuador corresponding to basic general education was carried out, from this document the information on the general and specific objectives of the subjects from first to sixth in which "Caso Mile.." is found is derived, with which we aim to a single subject that is Language and Literature that according to the ministerial report works in 5 blocks that are detailed in the graph, all this at the macro level of curricular concretion, going to the meso curricular level, there is already the annual curricular planning with its evaluation criteria and evaluation indicator going down to the micro level where you can already see the planning of the didactic unit and within this it is necessary that through a IDCA (Individual Diagnosis of Curricular Adaptation) adaptation planning can be considered. Now the IDCA is enriched with the information obtained from the reports presented annually by "Caso Mile..", so we have the most relevant things explained in the table from the second year to the fifth and it is studying the sixth. Then the proposal of didactic adaptation with ICTs is launched, where reviewing the ministerial objectives the minimum achievements of scope, an initial evaluation must be carried out using the resources that currently have to cope with the teaching of "Mile.. Case" and with these results propose the use of Ramps and adaptations if necessary of hardware and software to strengthen your skills. The problem that was seen when reviewing the history is that the reports have little or no curricular adaptations, the maximum is to do fewer repetitions and achieve the minimum achievements in the subjects, the potential we have in this case is that we work at the therapy level At the pedagogical level, there is speech therapy, occupational therapy and physical therapy that can be fed to exercise the adaptation of specialized hardware and software.

Keywords: Didactic adaptation · Educational system · ICT · IDCA

1 Introducción

In the task of introducing children with special educational needs to learning, some digital ramps and aid programs have been chosen, but the same curriculum must be analyzed to achieve the minimum expected results of each grade, it is more difficult to assimilate this fact when a starting point is not established with an evaluation, for which

R. S. Goonetilleke et al. (Eds.): AHFE 2021, LNNS 273, pp. 179–185, 2021.
https://doi.org/10.1007/978-3-030-80713-9_23

it is decided to enter the DIAC tips that allow us to obtain an individualized diagnosis history in general terms and thus be able to propose a series of tasks associated or not with ICT devices [1].

The digital ramps allow to achieve proposed goals based on a curriculum through individualized and socialized strategies with the diagnosis and intervention group of children with special educational needs, so there are many assistants or systems based on artificial intelligence that acquire expertise through therapists and can be used as helpful tools [2].

However, we are going to note that the curricula adopted by the ministry of education do contemplate inclusive education, only that this treatment can be given when you have the tools and assistance personnel prepared in these specific areas, education is placed as a right that has all the people so it is necessary to feed this axis with more tools [3].

An analysis of the curricular adaptation plans is then carried out based on the diagnosis of a specific case study to propose a system that allows both the institution, the student, the therapists, and parents to achieve the objectives and goals set. In the next section, we will review some works that present proposals for curricular adaptation and the context in which they are developed, then the specific table to be studied will be analyzed based on preliminary information and in a final section the proposal will be presented and future work will be considered for this case study.

2 Related Work

Some works are presented in which curricular adaptations are established in various areas such as literacy, based on the decree of curricular adaptation of the Ministry of Education, adaptations for prevention of behavior disorders in a specific area such as Physical Education, The approach that is carried out is with the sole objective of achieving significant learning from students with special educational needs, without losing the line and order, in the first place is the well-being of the student with SEN, the study environment must be adapted and adapted with the interest and participation of the student's environment and above all make teachers aware that each case is unique, do not generalize adaptations and expect similar results, rather each case study is unique and each adaptation is individualized with plans that include times, efforts and achievable objectives [4–8].

The curricular adaptation establishes study scenarios that contemplate the use of software and hardware with the capacity to adapt students with special educational needs, being a support in the development of the proposed activities, then we can talk about work with children with autism, with hearing difficulties with motor difficulties with cognitive difficulties, programs that enhance some variables allowing to overcome the barriers that are established by their condition [9, 10], but in other related works, the aptitudes of children with special educational needs are established to enhance these skills and achieve learning outcomes by enhancing their strengths [11–14].

The strategies or methods used to achieve the study results are even more essential than the tools or ramps used since it starts from a systematized study of the entry profile, aptitudes are declared and a progress analysis is carried out in the group of students.

intervention [15]. There are several ICT's applications to support children with disabilities in the learning process. These projects looks for improve the life quality of children with disabilities [16].

3 Case of Study Proposal

It is based on the concept of using a single proposal for a special case study, we present the DIAC in the following Table 1.

Reviewing the previous picture of a girl diagnosed with Spastic Quadriplegia Motor Disability, that is, she has difficulty in her upper and lower limbs, she remains lying down, she uses technical aids to sit her down. She needs visual support when she has to give a speech or present work. He follows the fourth year school program of general basic education according to his age, in terms of abilities we can highlight, his level of intelligence and development of different skills both verbal and visual, which has allowed him to follow the school program, the Evaluations of each subject are carried out orally, or the teacher reads the question and the girl answers.

To be able to contrast a form of learning without graphology, we incorporated the Tobii Eye Tracker 4c eye tracking device developed by the Tobii company, technology that allows controlling a computer with sight through projectors, sensors, and algorithms to calculate the position of the eyes., the device is designed to improve the experience in video games, we take advantage of the potential and compatible tools to achieve the access of a person with disabilities to the independent use of the computer (Fig. 1).

The incorporation and adaptation process is carried out over 3 months, in one-hour sessions with the advice and guidance of speech therapists who are familiar with the girl and know her skills. The eye tracking device is used in conjunction with Windows 10's accessibility tools, which provide support tools such as the keyboard and mouse to be used together. The final objective is to achieve independence in the use of a computer as a basic tool for education, to achieve this, it begins using Tobbi's tools through the process of calibration and eye-tracking. However, we focus on the acquisition of writing through Windows' digital keyboards, then we continue until the writing is achieved in word processors.

The use of the eye tracking device is carried out in 10 sessions of 40 min, the final objective is 3 instructions with the use of the eye tracking device which are:

- Correct control of the Computer through the sense of vision.
- Access to entertainment content.
- Access to educational content and study material.

The first setpoint is the one with the greatest difficulty, once the following setpoints are passed, the following setpoints can be more easily fulfilled. To achieve correct control of the Computer using sight, short-term objectives are used starting from the sweep in the device's calibration process mobile to the use of Windows' accessibility tools. The objectives are exceeded based on the criteria of time and the analysis of the experts in speech therapy shown in this knowledge diagram.

Table 1. Individual diagnosis for curricular adaptation "Mile case.."

Case study data	Age: 11 years, 7 months Diagnosis: Cerebral Motor Disability. Spastic Quadriplegia Educational Level: Sixth of basic with adaptations in access. Assists in a group of 14 students. Good interaction with his peers and teacher Characteristics: Non-marching girl, uses a wheelchair, dependent on activities of daily living, does not access physical writing, communicates orally, with short phrases due to phono-respiratory incoordination, deficiency in visual capture and monitoring Follow the pedagogical programming of the degree, with oral evaluations Emotional situations interfere with the quality of speech Observations: The girl also receives individualized attention in therapies: four sessions of physical therapy, 1 of occupational therapy, 1 of language		
History of pedagogical progress	The girl enters her inclusive educational process at the age of 5 years in initial 2 of 4 years, from that moment until the sixth of basic, she has been advancing in her learning with the direct help of the teacher and the family		
Work areas	Initial Competences	Work proposal	Correlation with Hardware and software
Occupational therapy	"Mile case.." has 3 s of visual fixation which is deficient, it has visual tracking, but it does not dissociate the eyes from the head, its contractions are very strong, which prevents being relaxed and in a good functional posture	Perform fixation exercises, visual monitoring, passive mobility of upper limbs to achieve relaxation and thus a better functional posture that allows us to carry out the activities planned for access to the computer Discovering the computer: keyboard, mouse, touch screen… Reading: letters, words, reading practice, writing text…	GCompris Sebran

(*continued*)

Table 1. (*continued*)

Language therapy	"Mile case..", does not have a good respiratory capacity due to its strong muscular contraction, which translates into dysfluent speech and a limited grammatical structure	Re-educate the muscles: build up their residual function to the maximum and replace or compensate for motor weaknesses to strengthen damaged function, use an eye-tracking tool for reading	Sebran Print Tobbi eye tracker

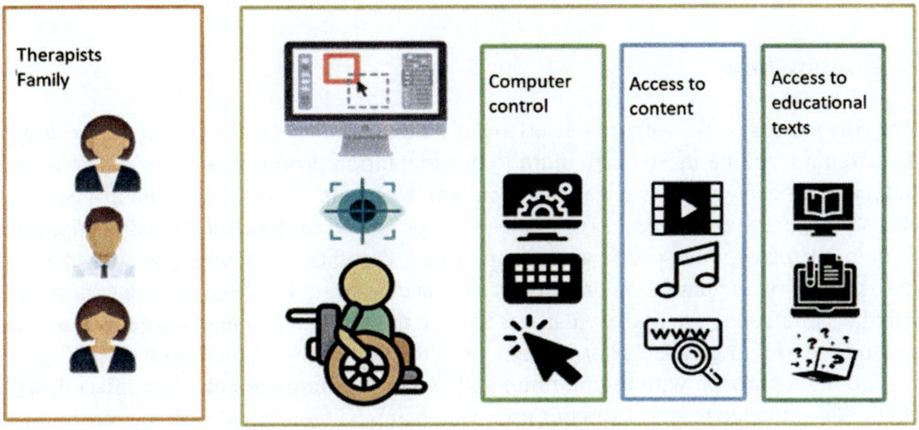

Fig. 1. Intervention process through eye-tracking

1. The calibration process is not passed in the first session, it is attributed to stress and the presence of more professionals, whether they are therapists or teachers. For the second session, access is restricted to two people, completing it in an average of fourteen minutes. In the fourth session, this objective is considered exceeded since the girl develops her skills and performs the calibration process in approximately two minutes.
2. From the fourth session, the use of the virtual keyboard is established, restricting autocompletion. In this objective, you are asked to write the name you like the most, which comprises six letters, in the fourth session, the objective is not achieved, this is attributed to the motor problems it presents and the ignorance of the commonly used QWERTY keyboard layout. To overcome these inconveniences, recognition and sweeping tasks are carried out at home and in the classroom.
3. The process of writing words is overcome in the sixth work session by managing to write words in an average of two to three minutes, jointly using the mouse tools to access programs and scroll to read documents.

4. In the seventh work session, the dictation of sentences begins, which must be written using the virtual keyboard and the autocompletion of words, this activity generates distractions, the girl selects a letter and searches the autocomplete without succeeding in the writing of sentences, at the same time freedom in Internet access is given where the girl accesses music videos of her liking in approximately five minutes.
5. In the eighth and ninth work sessions, the child is expected to write complete sentences without success due to the distractions of autocompletion and the desire to access entertainment content on the Internet. However, this desire is used as a stimulus to gain access to educational content and sentence writing.
6. In the tenth work session, the girl begins to use the text complement auto efficiently and writes complete sentences on a word processor. As for entertainment content, she has a greater disposition and ease of access, similar to educational content. This is attributed to the girl's tastes and motivations.

4 Conclusions

When the related works were reviewed in the state of art, they internalize in the hierarchy of curricula from the meso curriculum to the adaptation proposal, we have taken a case study, the same one that has its diagnosis in search of a curricular adaptation in The areas where it is being reinforced to reach the skills proposed in the educational curriculum, we help ourselves with speech and occupational therapies, through some software and with the help of an adaptation or hardware for access to it, it is described an experiment with the same Tobbi eye tracker that was used in the two therapy areas to be able to read the study texts and present an improvement in the autonomy of the "Mile Case..".

The first approach with digital ramp and associated hardware has been placed, with which, it is intended to start from this approach using ICT, enriched with the information from the DIAC and the academic record, in addition to abstracting all the information from the profile. and propose a system based on competencies and coupled with the academic curriculum.

We have material to propose a comprehensive study based on the good progress results seen when using the ramps and software for curricular adaptations.

References

1. Tomalá, A.M.V., Muñoz, K.A.V., Electrónicos, C.: La educación especial en la realidad ecuatoriana del siglo XXI. In: Memorias del quinto Congreso Internacional de Ciencias Pedagógicas de Ecuador: Aprendizaje en la sociedad del conocimiento: modelos, experiencias y propuestos: Instituto Superior Tecnológico Bolivariano, pp. 637–646 (2019)
2. Pinos-Chuya, B., Uguña-Uguña, V., Pérez-Muñóz, A., Suquilanda-Cuesta, P., Robles-Bykbaev, V., González-Arias. K.: A robotic assistant based on fuzzy logic and kinesthetic stimuli for gross motor rehabilitation of children with disabilities. (E33), 213–224 (2020)
3. Llumipanta Viscaino, R.A.: La educación en derechos humanos en la educación general básica preparatoria: un análisis del currículum en acción. Universidad Andina Simón Bolívar, Quito (2020)
4. Aldás M.R.L., Sandoval, S.E.A.: Percepciones de los docentes hacia las adaptaciones curriculares para estudiantes con necesidades educativas especiales. **5**(8), 1056–1078 (2020)

5. Sánchez Herrera, J.G.: Adaptaciones curriculares para la lectoescritura. Latacunga: Universidad Técnica de Cotopaxi: UTC, Ecuador (2020)
6. García, Y.P., Carracedo, J.E.G.: Adaptaciones curriculares para la prevención de trastornos de la conducta desde la Educación Física escolar (Original). Rev. Científica Olimpia **17**, 1232–1244 (2020)
7. Porras Sevillano, S.S., Rodríguez López, N.M.: Las adaptaciones curriculares que utilizaron los docentes en los procesos de aprendizaje con estudiantes que poseen necesidades educativas especiales. Universidad de Guayaquil-Facultad de Ciencias Psicológicas (2020)
8. Pomasqui Chuquín, V.E.: Adaptaciones curriculares en necesidades educativas especiales asociadas a una discapacidad en educación básica de la unidad educativa "Daniel Reyes" de San Antonio de Ibarra, período académico 2018–2019 (2020)
9. Berrezueta-Guzmán, J., Coronel-Maldonado, F., Serpa-Andrade, L., Robles-Bykbaev, V.: A didactic transmitter to support the communication and learning process of children with cerebral palsy. In: MATEC Web of Conferences: EDP Sciences, p. 05004 (2016)
10. Berrezueta-Guzman, J., Serpa-Andrade, L.: Embedded system as a third version of a didactic transmitter of needs that provides a way of communication to children with cerebral palsy of the spastic type. In: 2018 IEEE International Systems Engineering Symposium (ISSE), pp. 1–4. IEEE (2018)
11. Cabanillas-Tello, A., Cabanillas-Carbonell, M.: Application software analysis for children with autism spectrum disorder: a review of the scientific literature from 2005–2020. In: 2020 International Conference on e-Health and Bioengineering (EHB), pp. 1–4. IEEE (2020)
12. Hersh, M., Leporini, B., Buzzi, M.: ICT to support inclusive education. In: Miesenberger, K., Manduchi, R., Covarrubias Rodriguez, M., Peňáz, P. (eds.) ICCHP 2020. LNCS, vol. 12377, pp. 123–128. Springer, Cham (2020). https://doi.org/10.1007/978-3-030-58805-2_15
13. Zambrano, C.I.P.: Intervención pedagógica para la inclusión escolar de estudiantes con necesidades educativas especiales. Ciencia Educ.-Rev. Científica **1**(9), 52–67 (2020)
14. Berrezueta-Guzmán, J., Serpa-Andrade, L., Robles-Bykbaev, V., Pinos-Velez, E.: Digital trainer for the development of the fine motor ability in children with cerebral palsy. In: MATEC Web of Conferences: EDP Sciences, p. 20006 (2016)
15. Alban Delgado, W.S., Muñoz Bravo, C.I.: Análisis de las estrategias psicoeducativas en niños con necesidades educativas especiales asociadas o no a discapacidad del CDI "Manitos Traviesas. Universidad de Guayaquil-Facultad de Ciencias Psicológicas (2020)
16. Berrezueta-Guzman, J., Robles-Bykbaev, V., Serpa-Andrade, L.: How is the quality of life of patients with cerebral palsy improved? Qualitative and quantitative evaluation of a communication and learning assistance system based on ICTs. In: Freund, L.E., Cellary, W. (eds.) AHFE 2017. AISC, vol. 601, pp. 73–81. Springer, Cham (2018). https://doi.org/10.1007/978-3-319-60486-2_7

Challenges for an Observatory of the 2030 Goals, SDG and Social Economy, in Northern Mexico

Rodolfo Martinez-Gutierrez[1](✉), Maria Marcela Solis-Quinteros[2], Carlos Sanchez-Hurtado[1], and Carmen Esther Carey-Raygoza[1]

[1] Tecnológico Nacional de México/IT de Tijuana, Calzada Del Tecnológico S/N, Fraccionamiento Tomas Aquino, 22414 Tijuana, Baja California, Mexico
{rodolfo.martinez,carlos.sanchez,carmen.carey}@tectijuana.edu.mx
[2] Universidad Autonoma de Baja California Campus Tijuana, Calzada Universidad S/N, Fraccionamiento Otay Universidad, 22414 Tijuana, Baja California, Mexico
marcela.solis@uabc.edu.mx

Abstract. The objective of this research is to identify the best practices in promoting the Sustainable Social Economy that are aligned with the objectives of the national development plan to reduce the social inequality gaps that currently exist in the city of Tijuana, Baja California, as well as the instrumentalization of an observatory of the social economy that is capable of identifying the objectives for the sustainable development of the 2030 agenda; Therefore, it is essential to know precisely the needs of the business exercise, of the representatives of the social, solidarity and inclusive economy that is capable of generating development opportunities. It is also essential to analyze the wills to work as a team in a coordinated way between government, academia, social sector, representatives of civil society, as well as the citizens of Tijuana, Baja California, under a systemic development approach, according to the SDG Agenda. According to INEGI (2020) the city of Tijuana has 57 thousand SMEs that require the implementation of projects to support them and promote their development and growth; It also has more than 130 outstanding positive aspects, some worldwide and others at the country level. It is a quantitative type of research with a descriptive scope through documentary analysis to support the theoretical framework, as well as in-depth interviews with experts who have the best practices in the field of social economy; and the application of a questionnaire to a representative sample of SMEs to determine if there is a relationship between the critical factors that have an impact on the degree of implementation of the sustainable social economy. The results will provide necessary information that may be useful for a program of actions aimed at the formation of Nodes for the Promotion of the Social and Solidarity Economy (NODESS) in Baja California.

Keywords: Social economy · Ciriec Mexico North · Sustainable development

1 Introduction

As part of the applied research and sectoral linkage activities from the academy, the Tecnologico Nacional de México Campus Tijuana [1] holds the vice-presidency of CIRIEC

R. S. Goonetilleke et al. (Eds.): AHFE 2021, LNNS 273, pp. 186–191, 2021.
https://doi.org/10.1007/978-3-030-80713-9_24

Mexico for the North Region, the Honorary Presidency of the Commission for Social Economy of the Economic Development Ecosystem of the City Council of the City of Tijuana, Baja California [2, 3]. Likewise, it participates in the metropolitan activities in the Municipalities of Tijuana and Playas de Rosarito of the NODESS-BC (Nodes for the Promotion of Social and Solidarity Economy in the State of Baja California), the above through the Project of an Academic Body of nominated: Competitiveness and innovation for sustainable regional development, con-formed in the Department of Economic-Administrative Sciences that integrates the postgraduate Master's in Administration and the Doctorate in Administration Project [4], with three lines of research; 1. Systemic, sectoral and global competitiveness, 2. Models of innovation, ict and organizational management and 3. Capacities of sectoral chains and clusters. Research activities towards economics from the social academy include the plan to analyze the design and development of a proposal for an Observatory of the evolution and behavior of the UN Sustainable Development Goals 2030 [5].

2 Framework

According to a resolution approved by the UN General Assembly on September 25, 2015, the focus was on actions in favor of people, the planet and prosperity. Including the strengthening of universal peace and freedom. Having the recognition of poverty eradication, one of the great challenges of the world and indispensable for sustainable development. The 17 Sustainable Development Goals (SDGs) and 169 targets are the basis of the Universal Agenda 2030 [6]. The Goals and targets combine the three dimensions of sustainable development: economic, social and environmental. The SDGs are listed below: 1. End poverty in all its forms worldwide, 2. End hunger, achieve food security and improved nutrition, and promote sustainable agriculture, 3. Ensure healthy lives and promote well-being for all at all ages, 4. Ensure inclusive, equitable and quality education and promote lifelong learning opportunities for all, 5. Achieve gender equality and empower to all women and girls, 6. Guarantee the availability of water and its sustainable management and sanitation for all, 7. Guarantee access to affordable, safe, sustainable and modern energy for all, 8. Promote growth sustained, inclusive and sustainable economy, full and productive employment and decent work for all, 9. Industry, innovation and infrastructure, 10. Reduce inequality within and between countries, 11. Make cities and human settlements more inclusive, safe, resilient and sustainable, 12. Ensure sustainable consumption and production patterns, 13. Take urgent action to combat climate change and its effects, 14. Conserve and sustainably use the oceans, seas and marine resources for sustainable development, 15. Sustainably manage forests, combat desertification, halt and reverse land degradation and halt the loss of biodiversity, 16. Promote just, peaceful and inclusive societies, 17. Revitalize the World Alliance for Sustainable Development. The statistical data of the platform of the Government of Mexico for the Agenda of Sustainable Goals are presented in Table 1, which indicates the gap of the objectives and goals for each State of the Northern Region of Mexico. CIRIEC Mexico, International Center for Research and Information on the Public, Social and Cooperative Economy is an international scientific non-governmental organization, whose objectives are the collection of information, scientific research and the publication of works on the sectors economic activities and activities oriented towards services of general interest such as: the actions of public authorities in the economic field, public utilities, public and mixed companies at national, regional and municipal levels, the social economy, participation participation of workers, cooperatives, non-profit organizations and social enterprises [7].

Table 1. SDG indicators by state in the Northern Region of Mexico

SDG	Baja California	Sonora	Chihuahua	Nuevo Leon	Tamaulipas
1.1	2.2	1.6	5.6	2.3	0.9
2.1	14.1	21.5	17.7	18.3	12.2
3.1	28.3	39.9	39.1	42.7	35.3
4.1	100.4	92	97.1	94.4	98.6
5.2	NA	NA	NA	NA	NA
6	NA	NA	NA	NA	NA
7	NA	NA	NA	NA	NA
8.5	2.6	4.3	3	4.5	3.5
9.1	98.7	98.7	88.8	99.3	99.2
10	NA	NA	NA	NA	NA
11	12.36	9.39	6.49	5.4	7.97
12	NA	NA	NA	NA	NA
13	NA	NA	NA	NA	NA
14	NA	NA	NA	NA	NA
15	NA	NA	NA	NA	NA
16.1	78.9	45.6	78	9.6	17.8
17.8	80.78	79.08	2.98	69.94	78.48

Source: INEGI, Government of Mexico (2021) NA: Information not available

With the support of the National Institute of Social Economy (INAES), it is a concentrated administrative body of the economy secretariat) of Mexico, which has technical, operational and managerial autonomy, and its objective is to implement, as part of the Policy National Social Development of Mexico, public policies for the promotion and development of the social sector of the economy, in order to strengthen and consolidate the Sector as one of the pillars of economic and social development of the country, through participation, training, research, dissemination and support for productive projects in the Sector. INAES supports the Development and training of NODESS with the strategy of systematic actions of participation in Society, such as the involvement of the higher education sector (Universities and Technological Institutes), the cooperative sector and the social economy through civil associations. And INAES as a promoter of the formation of the social economy in the country, with international collaboration activities [8].

2.1 Model and Methodology

The applied research and sector linkage work has generated the analysis of research projects from the academy such as Tapia & Martinez with its Hackathon-Edu: Hackathon-Edu: A Global Competitiveness Perspective Model [9], Bonilla which points out the importance of the impacts of clusters In the Baja California region [10], likewise, the methodology for the design and development of the Model and Methodology of Dictionaries of Sectorial Competencies [11–13], required for the development of this institutional project, is the Methodology of the Fifth Systemic Helix [14–16] that will allow, through the involvement of all the participants in the sector strategy (Government, Education, Companies, Associations-Clusters and Consultants-Society), define the Sectoral Development Agenda (SDA); Based on them, the priority topics of each Sector can be defined to develop the Sector Skills Dictionaries (SCD) and, consequently, the Professional Skills Dictionaries (PCD) mentioned above for Level 6 and the Research Skills Dictionary (DRC) for Levels 7 and 8. The process of construction of the Model for the Elaboration of Dictionaries of Sectorial Competencies (DSC) and Dictionary of Professional Competencies (DPC), begins with the identification of the strategic sector for which the Dictionaries Sector Competences (DSC) [17–19]. It will be developed from the occupational analysis of the competencies of the strategic sector. Subsequently, a priority categorization of these positions must be generated to develop the analysis of occupational competencies; These positions are identified and validated within the framework of the Diagnosis of the Sectorial Agenda that each cluster develops and, with this, the roadmap for the development of the Dictionary of Sectorial Competencies (DCS) can be developed. This article is part of a collection of dissemination products of the Academic Body Academic Body called: Competitiveness and innovation for sustainable regional development, made up in the Department of Economic-Administrative Sciences that integrates the postgraduate Master's Degree in Administration and the Project Doctorate in Administration from the Tecnologico Nacional de México Campus Tijuana (TecNM Tijuana).

3 Conclusions

The challenges of applied research from the academy to develop a proposal that contributes to the analysis of the evolution and traceability of the variables of the objectives and goals of sustainable development in the northern region of Mexico implies a collaborative effort of the different sectors of the Society, but with policies of interinstitutional linkage and collaboration it is possible to advance towards the empowerment of the different sectors of the social economy.

Academic work will always be an effort that can cement good practices of collaboration and unity, hence, organizations such as CIRIEC International in this case CIRIEC Mexico in the Northern Region, has formed the unit of a team of specialists, researchers, and representatives of the sectors of the social economy, civil associations, as well as the great contribution of the facilities of the local government sector, for these sectorial studies the intervention and collaboration of the authorities of the Ministries with their programs and initiatives in their plans of Municipal and State Development and in Mexico the National Development Plan as a policy of articulation from the academy, government and civil society. An example of the research generated as a product in the applied research process is the case of the social economy developed in the Municipality of Playas de Rosarito [20, 21], through the analysis of the supply chain and logistics to join the efforts of cooperatives that produce Piñatas, handcrafted products of artistic figures of wood and metal for local sales as tourist souvenirs in Baja California and even export to the southern area of the State of California in the United States of America.

References

1. Tecnologico Nacional de México Campus Tijuana (TecNM). https://www.tijuana.tecnm.mx/
2. Comisión de Impulso a la Economía Social de Tijuana. https://www.tijuana.tecnm.mx/instalan-comision-de-impulso-a-la-economia-social/
3. Ecosistema de Desarrollo Economico de Tijuana. https://www.tijuana.tecnm.mx/%EF%BB%BFitt-colabora-en-comision-de-economia-social-del-ecosistema-para-el-desarrollo-economico-de-tijuana/
4. Foros de Consulta Plan Municipal de Tijuana. https://www.tijuana.tecnm.mx/maestria-en-administracion-participa-en-los-foros-de-consulta-para-plan-municipal-de-desarrollo-2020-2021/
5. INEGI Sistema de Información de los Objetivos de Desarrollo Sostenible, México
6. ONU Objetivos de Desarrollo Sostenible. http://agenda2030.mx/#/home. https://www.un.org/sustainabledevelopment/es/objetivos-de-desarrollo-sostenible/
7. United Nations Sustainable Development Summit 2015. https://www.un.org/en/development/desa/population/migration/generalassembly/docs/globalcompact/A_RES_70_1_E.pdf
8. CIRIEC Mexico. http://www.ciriec.uliege.be/en/our-network/national-sections/mexico/
9. INAES Nodos de Impulso a la Economía Social y Solidaria (NODESS). https://www.gob.mx/inaes/documentos/nodos-de-impulso-a-la-economia-social-y-solidaria-nodess
10. Tapia-González, G., Martínez-Gutiérrez, R., Tapia-González, F.: Hackathon-Edu: a global competitiveness perspective. In: Russo, D., Ahram, T., Karwowski, W., Di Bucchianico, G., Taiar, R. (eds.) IHSI 2021. AISC, vol. 1322, pp. 294–300. Springer, Cham (2021). https://doi.org/10.1007/978-3-030-68017-6_44
11. Bonilla-Hernández, A.L.: Accountability in assertive decisions making. In: Kantola, J.I., Nazir, S., Salminen, V. (eds.) AHFE 2020. AISC, vol. 1209, pp. 143–148. Springer, Cham (2020). https://doi.org/10.1007/978-3-030-50791-6_18
12. Martínez Gutiérrez, R.: Quinta Hélice Sistémica (QHS), un método para evaluar la competitividad internacional del sector electrónico en Baja California, México. Invest. Adm. (2012). ISSN 1870-6614 . (Julio-Diciembre): [Fecha de consulta: 20 de agosto de 2018] Disponible. http://www.redalyc.org/articulo.oa?id=456045338003
13. Martínez Gutiérrez, R.: Análisis de productividad y trazabilidad académica caso:Quinta Hélice Sistémica 2010–2020 [Reseña]. Adm. Lo Público, 8va. Compilación (8), 62–81 (2020). ISBN 978-9968-932-36-3 CICAP - Universidad de Costa Rica, San José
14. Martinez-Gutierrez, R.: Methodology of dictionaries of sector competences (DCS), to design standards of professional competences, research and labor. In: Nazir, S., Ahram, T., Karwowski, W. (eds.) AHFE 2020. AISC, vol. 1211, pp. 323–329. Springer, Cham (2020). https://doi.org/10.1007/978-3-030-50896-8_46
15. Martinez-Gutierrez, R., Correa, M.E.S.: Dictionary of competencies for sustainable development in the municipalities of Northern México. In: Kantola, J.I., Nazir, S., Salminen, V. (eds.) AHFE 2020. AISC, vol. 1209, pp. 312–317. Springer, Cham (2020). https://doi.org/10.1007/978-3-030-50791-6_40

16. Martínez Gutiérrez, R.: Quinta Hélice Sistémica (qhs), un modelo para el desarrollo de políticas públicas. Rev. Cooper. Desarrollo **20**(101), 28–51 (2012)
17. Martinez-Gutierrez, R.: Methodology of the fifth systemic helix for the development of public sector policies. J. Competit. Stud. **22**(3–4), 147–168 (2014)
18. Solis-Quinteros, M.M., Avila-Lopez, L.A., Zayas-Márquez, C., Carrillo-Gutierrez, T., Arredondo-Soto, K.C.: Analysis of the technological capability of linking SMEs in the electronic sector to integrate into the maquiladora industry electronic sector in Tijuana, Baja California, Mexico. In: Karwowski, W., Trzcielinski, S., Mrugalska, B. (eds.) AHFE 2019. AISC, vol. 971, pp. 376–387. Springer, Cham (2020). https://doi.org/10.1007/978-3-030-20494-5_35
19. Solís-Quinteros, M.M., Ávila-López, L.A.: Effective design of service supply chains in México. In: García-Alcaraz, J.L., Sánchez-Ramírez, C., Avelar-Sosa, L., Alor-Hernández, G. (eds.) Techniques, Tools and Methodologies Applied to Global Supply Chain Ecosystems. ISRL, vol. 166, pp. 237–258. Springer, Cham (2020). https://doi.org/10.1007/978-3-030-26488-8_11
20. Solis-Quinteros, M.M., Avila-Lopez, L.A., Carrillo-Gutierrez, T., Arredondo-Soto, K.C.: Technological capability of the small and medium-sized enterprises manufacturing sector. In: Kantola, J., Barath, T., Nazir, S. (eds.) Advances in Human Factors, Business Management and Leadership. AHFE 2017. Advances in Intelligent Systems and Computing, vol 594. Springer, Cham (2018). https://doi.org/10.1007/978-3-319-60372-8_19
21. Municipio de Playas de Rosarito Baja California. http://www.rosarito.gob.mx/VIII/comunicados/2020/convoca-gobierno-de-playas-de-rosarito-a-programa-de-apoyo-economico-para-micro-y-pequenas-empresas/
22. Plan Estrategico Metropolitano 2012 – 2034 Tijuana – Tecate – Rosarito. https://implan.tijuana.gob.mx/pdf/PEM.pdf

Computer Science Engineers their Profile and Competencies for Generations X, Y and Z

Carlos Hurtado-Sanchez(✉), Rodolfo Martinez-Gutierrez, Carmen Carey, and Artemio Lara-Chavez

Tecnológico Nacional de México/IT de Tijuana, Calzada Del Tecnológico S/N, Fraccionamiento Tomas Aquino, 22414 Tijuana, Baja California, Mexico
{carlos.hurtado,rodolfo.martinez,carmen.carey,alara}@tectijuana.edu.mx

Abstract. The 2017 report of the Organization for Economic Cooperation and Development (OECD) on competencies in Mexico indicates the opportunity to rethink actions aimed at strengthening competitiveness, development, and social progress, considering the impact of capabilities. This research is a series of methodological analyzes from the methodological perspective of the Systematic Fifth Helix (QHS) [1] to determine the level of evolution and the relevant needs of educational programs in the area of economic-administrative sciences, in this research on Engineers in Computer Science from the Tecnológico Nacional de México Campus Tijuana, considering the systematic approach of the QHS with representatives of the business sector, government, education, associations, consultants from the sectors under study and thereby assessing the perspectives of generational competitiveness and educational gaps in professional training. In this research, the generations of graduates X, Y, and Z, are taken as a sample of the universe. Considering representatives of the different sectors of employability such as the industrial sector, commerce, services, government, or the entrepreneur. With these research results, we propose the work situational analysis studies of each study program to formulate labor initiatives of labor, professional, and research competencies. Likewise, design and development of specialty strategies according to the sectorial feedback of generational competitiveness.

Keywords: Job competencies · Professional competencies · Research competencies

1 Introduction

This work is Computer Science Engineers' competencies review considering the contextual needs of generations X, Y, and Z. According to the research carried out by the OECD [2], in turn, an analysis of their Graduate profiles of the best universities in the world, national ranking with the National Technology of Mexico campus Tijuana [3] as a frame of reference for an institutional development project in the National Technology of Mexico Campus Tijuana for the formation of an Academic Body that has lines of research-oriented to the analysis and innovation of the needs of the study programs and

R. S. Goonetilleke et al. (Eds.): AHFE 2021, LNNS 273, pp. 192–196, 2021.
https://doi.org/10.1007/978-3-030-80713-9_25

their relevance for an effective process of professional integration and thus contribute to the sectoral development of their area of specialty.

2 Framework

The term generation [4] refers to an age group that shares throughout its history a set of formative experiences that distinguish them from its predecessors. Currently, in higher education institutions it is possible to see four generations coexisting: generation X (from 1965 to 1980), generation Y (from 1980 to 1994), and generation Z (from 1995 to 2010). Each generation responds, according to researchers on the subject, to different attitudes and expectations about work and their career.

Generation X individuals, according to this model, are nomadic archetypes. They have been independent since childhood. They have little reason to employ or adopt the visions of the postwar generation since they have rarely used their practical reality. Generation Y, offspring that make up the relief staff in the workplace, in today's generation, these are school children, the older ones are finishing postgraduate studies. They have grown up with planned lives.

Generation Z [5] is the first generation considered digital natives, that is, they were born immersed in digital culture. Its characteristics include the following: Experts in understanding technology; Multitasking; Socially open from technologies; Quickness and impatience; Interactive; and Resilient. Table 1 shows the main characteristics of each generation regarding technology.

Table 1. Main characteristics of each generation

Generation	Born	Technology	Digital competence
X	1965–1980	PC	Early digital adopters
Y	1981–1994	Smartphone	Digital natives
Z	1995–2010	AR/VR	Digital innates

The OECD highlighted that the main tool for conducting higher education was evaluation, accreditation of programs and institutions, information, relationships on formal and explicit agreements between the government and universities, as well as the distribution of financial resources. It also proposed the creation of a national accreditation system for university study programs that would not cross the official knowledge of a state body such as the Secretary of Public Education.

In this line of reconstruction, the OECD requested a group of researchers of different theoretical stature the Project Definition and selection of competencies: Theoretical and Conceptual Foundations, whose main objective was to offer a resource for the process of defining, selecting and measuring the competencies that an individual needs to lead a successful and responsible life and for societies to face the challenges of the present and the future.

2.1 Model and methodology

According to QS World University Rankings [6] places the Massachusetts Institute of Technology (MIT) in 2020 in the first position worldwide in computer science engineering, this institute was Founded in 1975 and has 777 students enrolled, among its most outstanding graduate students, is Sal Khan creator of Khan Academy, Drew Houston and Arash Ferdowsi creators of dropbox.

At MIT [7] the objectives for a Computer Science Engineer are in the study program and include engineering ethos, leadership, versatility, and engagement. Computing Commission Student Outcomes are focused on analyzing complex computer problems, design, implement and evaluate computing-based solutions, communicate effectively, recognize professional responsibilities, teamwork, and apply computer theory to solve computing-based solutions.

According to QS World University Rankings [6], it places the National Autonomous University of Mexico (UNAM) currently is at the first position in computer science in Mexico, it was founded in 1977 and has an enrollment of 530 students In the science faculty of the UNAM the student obtains the following knowledge: analysis, design, and implementation of computer applications, networks, and computer science. Mathematical foundations for the model and automation of scientific applications and coordination of work teams for computer applications development.

The UNAM [8] is a compound of 254 institutions around the Mexican Republic, being the institution of higher education with the most territorial presence in the country. In these institutions, TecNM serves a school population of more than 600 thousand undergraduate and graduate students throughout the national territory, including Mexico City, serving 41% of engineering students in the country. The computer science career was founded in 1983, taught on 189 campuses, and the Tijuana campus currently has an enrollment of 700 students.

TecNM [9] computer systems engineering has the general objective: To train leading professionals with strategic vision and a broad ethical sense; able to design, develop, implement and manage computer technology to provide innovative solutions for the benefit of society; in a global, multidisciplinary, and sustainable context. In his graduation profile, the student obtains knowledge to: implement computer applications, design, develop, evaluate and apply interfaces, networks, databases, software, computational models to solve problems through the use of mathematical tools, coordinates, participates in multidisciplinary teams, and detects areas of opportunity using a business vision to create information and communication technology projects. Tables 2, shows the skills develop by institutes.

Table 2. Skills developed by institutes

Skills to develop	MIT	UNAM	TecNM
Team work	Yes	Yes	Yes
Troubles hooting	Yes	Yes	Yes
Computational modeling	Yes	Yes	Yes
Computational theory	Yes	Yes	No
Digital competences development	Yes	Yes	Yes

3 Conclusions

After analyzing the profile and characteristics of generations X, Y, and Z, as well as their development in the industry, it is observed that in the case of the TecNM campus in Tijuana, refresher courses are required for the management of information technology in education. It is important to use these technological tools since current educational practice should not be exclusive to students in the area of computer science [10]. Graduate profile evaluations were carried out in the most prestigious universities in the world and the country at a technological level in the area of computer science, it is observed that the TecNM campus Tijuana and its study program in the area of computer science are relevant and complies with the labor competencies requested by the industry at a local, national and international level, however, it is necessary to strengthen competencies regarding the theory of computing, scientific and applied research. The challenges of applied research from the academy to develop a proposal that contributes to the analysis of the evolution and traceability of the variables of the objectives and goals of sustainable development in the northern region of Mexico implies a collaborative effort of the different sectors of Society, but with policies of inter-institutional collaboration and linkage, it is possible to advance towards the empowerment of the different sectors of the social economy.

References

1. Martínez Gutiérrez, R.: Quinta Hélice Sistémica (QHS), un método para evaluar la competitividad internacional del sector electrónico en Baja California, México. Invest. Adm. (2012). ISSN 1870-6614. http://www.redalyc.org/articulo.oa?id=456045338003. Accessed 20 Aug 2018
2. Organización para la Cooperación y el Desarrollo Económicos (OCDE): Consultado el 15 de Agosto (2017). https://www.oecd.org/eco/surveys/mexico-2017-OECD-Estudios-economicos-de-la-ocde-vision-general.pdf
3. Tecnologico Nacional de México Campus Tijuana (TecNM). https://www.tijuana.tecnm.mx/
4. Gilburg, D.: Brecha de Liderazgo (2007). www.sun.com/emrkt/boardroom/newsletter/latam/0407expertinsight.html-29k. Accessed 17 Feb 2009
5. Schroer, W.: Defining, Managing, and Marketing to Ge- nerations X, Y, and Z. The Portal **10**, 9 (2008). http://goo.gl/Fc40dB. Accessed 15 Feb 2015
6. QS World University Rankings https://www.topuniversities.com/qs-world-university-rankings

7. Massachusetts Institute of Technology (MIT). https://www.eecs.mit.edu/academics-admissions/program-objectives
8. Universidad Nacional Autonoma de Mexico (UNAM). http://oferta.unam.mx/ciencias-de-la-computacion.html
9. Tecnológico Nacional de Mexico (TecNM). https://www.tecnm.mx
10. Hurtado, C., Licea, G., Garcia-Valdez, M.: Integrating learning styles in an adaptive hypermedia system with adaptive resources. In: Sanchez, M.A., Aguilar, L., Castañón-Puga, M., Rodríguez-Díaz, A. (eds.) Computer Science and Engineering—Theory and Applications. SSDC, vol. 143, pp. 49–67. Springer, Cham (2018). https://doi.org/10.1007/978-3-319-74060-7_3

Postgraduate Administration Education: Profiles and Skills Contribution to the Knowledge Society

Rodolfo Martinez-Gutierrez[1](✉), Eduardo Ahumada-Tello[2], Ramon Galvan-Sanchez[2], Carlos Hurtado-Sanchez[1], and Beatriz Chavez-Ceja[1]

[1] Tecnológico Nacional de México/IT de Tijuana, Calzada Del Tecnológico S/N, Fraccionamiento Tomas Aquino, 22414 Tijuana, Baja California, Mexico
{rodolfo.martinez,carlos.sanchez, beatriz.chavez}@tectijuana.edu.mx

[2] Universidad Autonoma de Baja California Campus Tijuana, Calzada Universidad S/N, Fraccionamiento Otay Universidad, 22414 Tijuana, Baja California, Mexico
{eduardo.tello,ramon.galvan}@uabc.edu.mx

Abstract. Higher education and especially the postgraduate Master's and Doctorate programs in northern Mexico, particularly in the border region of the Tijuana Baja California area, have become highly competitive at the international level and demanding qualified personnel, this research focuses on a comparative research through a series of methodological analyzes from the methodological perspective of the Fifth Systematic Helix (QHS) to determine the level of evolution and relevance of educational programs in the area of economic-administrative sciences of the Master's Degree in Administration and Doctorate in Administration in the two institutions with the greatest tradition and roots in the training of professionals and scientists in the region. The National Technological Institute of Mexico Tijuana Campus and the Autonomous University of Baja California. From the perspective of the systematic approach of the QHS Methodology, with representatives of the business sector, government, education, associations, consultants from the sector under study and with this, evaluate the perspectives of generational competitiveness and the educational gaps in professional training and research. In this research, the generations of graduates are taken as a sample of the universe, considering representatives of the different sectors of employability such as the industrial sector, commerce, services, government or the entrepreneur. With the results of the research, studies of the impact on sectoral development and the contribution to the society of knowledge, innovation and competitiveness of the region at an international level are proposed.

Keywords: MBA · PhD in administration · Postgraduate in administration

1 Introduction

One of the institutional challenges of the Tecnologico Nacional de Mexico Campus Tijuana within the framework of its 50th Anniversary [1] and more than 25 years of

R. S. Goonetilleke et al. (Eds.): AHFE 2021, LNNS 273, pp. 197–202, 2021.
https://doi.org/10.1007/978-3-030-80713-9_26

offering a post-graduate program of Master in Administration [2–6] at the regional level in the northwestern area of Mexico is the opening of a Doctorate in Administration program that covers the needs of the strategic economic sectors of the Cali-Baja region [7], since a postgraduate course with links and research strengthens the competence of the region and raises the level of progress and Development of the Society. The main elements for the analysis of a quality and relevance study program profile with its lines of research will be to consider the best international University that offers a Doctorate in Administration, as well as the frame of reference of the best University in Mexico and the Baja California region.

2 Framework

UNESCO, within the framework of Scholarships Without Borders [8], for countries with challenges to strengthen science and technology considers that doctoral programs should have as specific objectives to cover the following aspects mentioned in Table 1.

Table 1. Scholarships without borders for PhD

Objective	Purpose
1	Improve access to educational and training opportunities in science and technology for talented young women, from countries lagging behind in science and technology, who have university degrees
2	Increase scientific productivity and the creativity of women scientists in countries lagging behind in science and technology
3	Empowering a new generation of talented women to take on a leadership role in science and technology
4	Encourage women scientists to contribute to the sustainable development of their countries of origin
5	Enable women scientists from the South to collaborate and network at the regional and international level
6	Combat HIV/AIDS, malaria and other diseases

Considering the Institutional Development Plan of the Tecnologico Nacional de México Campus Tijuana, the Project for the analysis and relevance of the feasibility study of the opening of a Doctorate in Administration is oriented to the challenges of sectorial and country competitiveness. The Cali-Baja region is faced with the growing need for talent management and highly qualified specialized personnel, coupled with the challenges of the Free Trade Agreement between Mexico the United States of America and Canada (T-MEC) [9], making it necessary to review the supply and educational offer capacity aligned to the global challenges of competitiveness and Development of the Society.

According to the recommendations of the OECD report [10] on competencies in Mexico, he points out that the current model of the Mexico Competency System has the opportunity to rethink actions aimed at strengthening the model. Therefore, the proposal in this article is based on the challenges that the OECD points out in order to strengthen competitiveness, development and social progress, considering the impact of education at the eight levels indicated by ISCED-UNESCO [12] which makes it necessary to reflect on the current progress and results on the use of the Labor Competency Standards (LCS) in the sectors of society, where it is imperative to build a model and appropriate methodology for Professional Competencies and Research, making clear the differentiation of the eight levels and their requirements for training, alignment, evaluation and eventual certification of job skills, professional skills and applied research skills. To identify the gaps between the regulations and inputs required for sectoral strategic planning processes in terms of instruments called Sector Competence Dictionaries (SCD), they must respond to each regional sector, city or State vocation.

The analysis of the feasibility study for the opening of the Doctorate in Administration or in its case Doctorate in Sciences in Administration is based on the analysis of 3 reference institutions see Table 2. Firstly, the National Autonomous University of Mexico (UNAM) that in its Faculty of Accounting and Administration has the Doctorate of Science in Administration program that aims to train doctors who are capable of carrying out original research, with theoretical and methodological rigor, that contributes to the expansion of knowledge of the Sciences of the Administration, as well as empowered to perform professionally, with high levels of quality, both in teaching and in solving organizational problems [13]. In the second case of the National Polytechnic Institute (IPN), the analysis of the structure of the Study Plan and its characteristics is carried out, the General objective of the Doctorate in Administration Sciences is to train researchers in the field of administration sciences with a focus towards solving problems of public, educational and business organizations at the local, national and global levels. It considers three specific objectives: 1. To train researchers in the field of administration with international standards in an academic environment that fosters basic and applied research, 2. To focus research on the great local, national and world problems and 3. Generate links with the educational, public, business sectors, non-governmental organizations and communities to study and contribute to transforming their reality [14].

The third educational institution in the frame of reference is the Autonomous University of Baja California (UABC) which considers the objective of the Doctorate in Administrative Sciences The objective of the Program is the training of scientists in Administrative Sciences capable of developing in various professional fields - them to contribute to the approach and resolution of problems in a context of a knowledge-based society. Likewise, it is ensuring that knowledge and innovation are valued in society and in the academic, government and business sectors. Establish strategies to promote the knowledge society where innovation is the advantage to compete and improve the quality of life. Implement the criteria with the objective of aligning institutional efforts with international requirements through the best practices provided by the National Council of Science and Technology (CONACYT). According to the feasibility study of the state of the art and the characteristics of the variables that intervene in the design and development of a Project proposal that meets and aligns with the needs of a quality postgraduate program including the requisites sites of a National Quality Postgraduate Program with accreditation before CONACYT this first analysis reflected in this article reflects the challenge of the roadmap for the opening of a doctoral program [15].

Table 2. Research lines for a doctorate of science in administration in Mexico

Characteristics	UNAM	IPN	UABC
Lines of investigation	Finance, Tax, Human Resources, Marketing, International Business, Knowledge Management, Informatics, Operations, Theory of Administration and Organization	Organizational management, Competitiveness and technological innovation, Organizacion-ales Strategies	Administration and Development of organizations, Studies to promote Competitiveness, Regional development and innovation systems
Quantity	9	3	3

2.1 Model and Methodology

Considering the institutional guidelines of the Tecnologico Nacional de México and the requirements of CONACYT itself, sectoral analysis methodologies will be included to strengthen the study of relevance and feasibility, as is the case of Tapia & Martinez with its Hackathon-Edu: Hackathon-Edu: A Global Competitiveness Perspective Model [16], Bonilla which points out the importance of the impacts of clusters In the Baja California region [17], likewise, the methodology for the design and development of the Model and Methodology of Dictionaries of Sectorial Competencies [18–20], required for the development of this institutional project, is the Methodology of the Fifth Systemic Helix [21–23] that will allow, through the involvement of all the participants in the sector strategy (Government, Education, Companies, Associations-Clusters and Consultants-Society), define the Sectoral Development Agenda (SDA); Based on them, the priority topics of each Sector can be defined to develop the Sector Skills Dictionaries (SCD) and, consequently, the roadmap for the development of the Dictionary of Sectorial Competencies (DCS) can be developed. This article is part of a collection of dissemination products of the Academic Body Academic Body called: Competitiveness and innovation for sustainable regional development, made up in the Department of Economic-Administrative Sciences that integrates the postgraduate Master's Degree in Administration and the Project Doctorate in Administration from the Tecnologico Nacional de México Campus Tijuana (TecNM Tijuana).

3 Conclusions

The challenges of opening a postgraduate program in a federal public institution, not only lies in the design and development of the feasibility study and the regulatory requirements of the study program, lines of research, a great challenge is the to form a Basic Academic Nucleus (NAB) which must cover various requirements not only of academic and professional training, aspects of positions must be covered, which implies the contractual situation of full-time professors, researchers with registration in the National System of researchers SNI of CONACYT and National Recognition of the Desirable Profile (PRODEP) [24] the above generates institutional challenges, since it is a budgetary aspect at the national level.

References

1. Tecnologico Nacional de México Campus Tijuana (TecNM). https://www.tijuana.tecnm.mx/
2. Maestría en Administración del Tecnologico Nacional de Mexico Campus Tijuana. https://www.tijuana.tecnm.mx/maestria-en-administracion-celebra-25-aniversario-de-fortalecer-competitividad-del-talento-humano-en-baja-california-y-mexico/
3. Tecnologico Nacional de Mexico Campus Tijuana en Feria de Competitividad. https://www.tijuana.tecnm.mx/el-tecnologico-nacional-de-mexico-campus-tijuana-de-nueva-forma-fue-invitado-especial-en-el-panel-de-competitividad-del-sector-academico-en-baja-california/
4. Comisión de Impulso a la Economía Social de Tijuana. https://www.tijuana.tecnm.mx/instalan-comision-de-impulso-a-la-economia-social/
5. Ecosistema de Desarrollo Economico de Tijuana. https://www.tijuana.tecnm.mx/%EF%BB%BFitt-colabora-en-comision-de-economia-social-del-ecosistema-para-el-desarrollo-economico-de-tijuana/
6. Foros de Consulta Plan Municipal de Tijuana. https://www.tijuana.tecnm.mx/maestria-en-administracion-participa-en-los-foros-de-consulta-para-plan-municipal-de-desarrollo-2020-2021/
7. Cali-Baja, una mega region llena de oportunidades. http://www.startupstijuana.com/es/noticias/170/CaliBaja-una-mega-region-llena-de-oportunidades
8. Becas Sin Fronteras para Doctorado UNESCO-OWSD. https://becas-sin-fronteras.com/beca/becas-de-doctorado-unesco-owsd-a1520/
9. México entra al T-MEC sin condiciones para competir en innovación y tecnología. http://www.anuies.mx/noticias_ies/mxico-entra-al-t-mec-sin-condiciones-para-competir-eninnovacin-y
10. Organización para la Cooperación y el Desarrollo Económicos (OCDE) (2017).https://www.oecd.org/eco/surveys/mexico-2017-OECD-Estudios-economicos-de-la-ocde-vision-general.pdf. Accessed 15 Aug 2017
11. Proyecto Tuning América Latina: Reporte Proyecto Tuning, 2011–2013 Innovación Educativa y social (2018). http://www.tuningal.org/
12. UNAM Programa de Doctorado en Ciencias Administrativas. http://posgrado.fca.unam.mx/doctorado.php
13. IPN Programa de Doctorado en Ciencias Administrativas. https://www.sepi.escasto.ipn.mx/oferta-educativa/dca/programa-academico/mision-vision-y-objetivo-general.html
14. UABC Doctorado en Ciencias Administrativas. http://www.uabc-fca.com/Doctorado/objetivo.html
15. CONACYT PNPC Marco Normativo. https://www.conacyt.gob.mx/index.php/el-conacyt/convocatorias-y-resultados-conacyt/convocatorias-pnpc/marcos-de-referencia-pnpc
16. Tapia-González, G., Martínez-Gutiérrez, R., Tapia-González, F.: Hackathon-Edu: a global competitiveness perspective. In: Russo, D., Ahram, T., Karwowski, W., Di Bucchianico, G., Taiar, R. (eds.) IHSI 2021. AISC, vol. 1322, pp. 294–300. Springer, Cham (2021). https://doi.org/10.1007/978-3-030-68017-6_44

17. Bonilla-Hernández, A.L.: Accountability in assertive decisions making. In: Kantola, J.I., Nazir, S., Salminen, V. (eds.) AHFE 2020. AISC, vol. 1209, pp. 143–148. Springer, Cham (2020). https://doi.org/10.1007/978-3-030-50791-6_18
18. Martínez Gutiérrez, R.: Quinta Hélice Sistémica (QHS), un método para evaluar la competitividad internacional del sector electrónico en Baja California, México. Invest. Adm. (2012). ISSN 1870-6614. http://www.redalyc.org/articulo.oa?id=456045338003. Accessed 20 Aug 2018
19. Martínez Gutiérrez, R.: Análisis de productividad y trazabilidad académica caso:Quinta Hélice Sistémica 2010–2020 [Reseña]. Adm. Lo Público, 8va. Compilación, (8), 62–81 (2020). ISBN 978-9968-932-36-3 CICAP - Universidad de Costa Rica, San José
20. Martinez-Gutierrez, R.: Methodology of dictionaries of sector competences (DCS), to design standards of professional competences, research and labor. In: Nazir, S., Ahram, T., Karwowski, W. (eds.) AHFE 2020. AISC, vol. 1211, pp. 323–329. Springer, Cham (2020). https://doi.org/10.1007/978-3-030-50896-8_46
21. Martinez-Gutierrez, R., Correa, M.E.S.: Dictionary of competencies for sustainable development in the municipalities of Northern México. In: Kantola, J.I., Nazir, S., Salminen, V. (eds.) AHFE 2020. AISC, vol. 1209, pp. 312–317. Springer, Cham (2020). https://doi.org/10.1007/978-3-030-50791-6_40
22. Martínez Gutiérrez, R.: Quinta Hélice Sistémica (QHS), un modelo para el desarrollo de políticas públicas. Rev. Cooper. Desarrollo **20**(101), 28–51 (2012)
23. Martinez-Gutierrez, R.: Methodology of the fifth systemic helix for the development of public sector policies. J. Competit. Stud. **22**(3–4), 147–168 (2014)
24. Perfil Deseable PRODEP. http://promep.sep.gob.mx/

Holistic Approach in Safety Management During the Pandemic

Holistic Approach in Risk Reduction Processes of the Machinery Equipment

Hana Pačaiová(✉), Michaela Balážiková, Marianna Tomašková, Katarína Firmentová, Katarína Chomová, Lukáš Ďuriš, Peter Darvaši, Lukáš Salaj, and Ján Kán

Faculty of Mechanical Engineering, Technical University of Kosice, Letna 1/9, Kosice, Slovakia
{hana.pacaiova,michaela.balazikova,marianna.tomaskova}@tuke.sk,
{katarina.firmentova,katarina.chomova,lukas.duris,peter.darvasi,
lukas.salaj,jan.kan}@student.tuke.sk

Abstract. Safety of machinery is an area, which, at the first glance, appears as solved and sufficiently specified by regulations and harmonized standards in the EU. However, the reality is that, according to statistics, the machinery, including lifters, conveyors, and similar equipment, is a source of as much as 25% of all serious work injuries annually. Machinery condition, method of performed activities, environment in which the machine is located, are basic factors projected into hazardous event origination. This contribution, based on the status analysis protective measures of more than 100 machineries, evaluates the level of measures implemented on the equipment throughout its life cycle. Within the research, the methodology was proposed, evaluating the efficiency of protective measures on three levels: on the machinery itself; in relation to machines existing within the given operation, as well as complex level of machinery safety in an organization.

Keywords: Machinery safety · Human factors · Risk reduction

1 Introduction

Occupational health and safety during machinery use is a key area in OHS management in industrial plants [1]. The condition of machinery changes during its life cycle, so do employees and the conditions of the operation, as well. Despite numerous legislative requirements and supervision over their observance (e.g.: EC declaration of CE mark conformity), number of accidents when operating the machinery reaches up to 25% of all serious occupational injuries per year [2].

Permanently sustainable development of an organization requires holistic approach to evaluation of both goals as well as activities of corporation, including the machinery risk assessment. Monitoring and inspection are critical elements in the management of processes. Therefore, it is inevitable to identify key performance indicators (KPI) [3], which are relevant to the analyzed processes. It is possible to carry out the monitoring of the processes on various management levels, as well as from different points of view: on the managerial, operational, logistical level; from the viewpoint of finances, safety or maintenance provision [4, 5]. Holistic approach means a general view on the system.

R. S. Goonetilleke et al. (Eds.): AHFE 2021, LNNS 273, pp. 205–212, 2021.
https://doi.org/10.1007/978-3-030-80713-9_27

System properties, of the machinery in this case, depend on the properties of its parts and its functioning in the given area.

Occupational Health and Safety currently represents one of the most significant areas of social politics of the EU and developed countries of the world.

Holistic approach to dealing with OHS issue starts as early as the machinery is designed and is concluded with a feedback of the safety management during its operation.

In 1989, legal framework determining mutual relation between machinery equipment design and its safe usage in practice was created in Europe. These were two basic regulations 89/391/EEC, known as the "OHS Directive" [6] and the Directive 89/392/EEC – known as the "Machinery Directive". Requirements for construction of "safe machinery equipment" were developed and the Machinery Directive is currently known as the European Directive 2006/42/ES [7].

Condition of machinery, method of activities performed, environment in which the machine is located are basic factors projecting into creation of hazardous events. Although the unified EU market, based on more than 31 years of defined rules, expects that the approach in risk reduction process as early as by the machinery design shall be observed (or developed). However, mainly within the organizations having out-of-date machinery equipment, changes of machinery equipment are realized with insufficient consideration of the basic requirements. The next issue is education and training of technical workers [8] who have no knowledge of principles and requirements for safe designing.

Risk assessment realized as early as in the stage of machine design is an essential prerequisite for safe machinery construction, so that the residual risks were on the lowest possible level. Although modernization of old machinery provides higher effectivity and enables significant increase of productivity by means of new automated functions, but only when using the right principles of safe designing [9]. Currently, machinery safety requirements are more and more connected with programmable electronic control systems (SRP/CS). Risk estimation is a significant constituent part of risk analysis in the process of machinery development because categorization and allocation of safety requirements is based on this [10]. It is important to know how the risk estimation is carried out, because incorrect selection of the safety integrity level (SIL) or performance level (PL) may lead to multiple costs on a part of management system connected with safety [11].

2 Machinery Risk Assessment Principle

The principle of risk assessment lies in the succession of individual steps mentioned in the algorithm shown in Fig. 1.

Risk analysis begins with classification of system into elements. The aim is to describe and identify the source of unwanted event as clearly as possible, i.e., what represents threat in relation to a human; description of hazard situation – what may happen during the given activity and how. Next comes the estimation of probability and consequence, which requires selecting the appropriate methodology to match the parameters of probability and consequence (e.g.: risk matrix, risk graph). This methodology may be in a form of qualitative, semi-quantitative or quantitative approaches [12]. The next important step of this logical process is expression of the probability relation and consequence by a combination of their values, so-called risk estimation. Risk evaluation is a step that comes only after risk analysis and it basically compares the estimated risk extent (related to the identified hazard) with the "set" parameters of its acceptance (or tolerability).

Taking measures is related to such risk values, which exceed or reach an inacceptable level. It may happen that even though the estimated risk level ranges within the acceptability zone, a manager (designer, producer) will decide to take certain measures in order to reduce it. This approach is marked as ALARP (as Low as Reasonable Practicable) – represents risk reduction to the most reasonable level, i.e., effective and efficient [13]. This entire process is marked as risk reduction or also as risk control [14] and requires its re-assessment, in order to verify whether the proposed and implemented measures are really effective [15, 16].

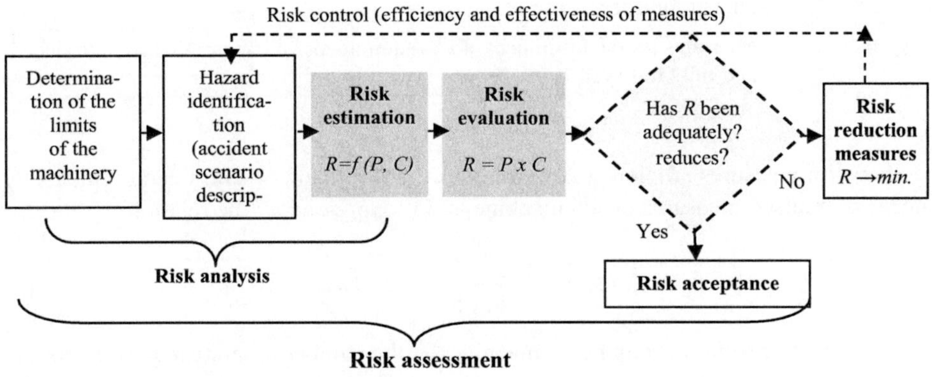

Fig. 1. Modified simplified procedure for machinery risk assessment/risk control (inspired by ISO 12100).

3 Analysis and Evaluation of the Machinery Protective Measures' Efficiency

Within the running research, experts from safety field (2 from practice and 3 from university) cooperated on the creation of the CLMS (Comprehensive Level of Machine Safety) methodology, the aim of which was, based on risk assessment results of operated machines (for the activities of operators and maintenance according to the own methodology of an organization), with regard to the phase of their operation, to analyze the status and efficiency of current measures on particular equipment, or the total efficiency level of taken protective/safety measures applied on the equipment in operations [17, 18].

Main parameters of the model were the following assumptions: there are n operation machines at each production facility; the number of safety requirements (SR) for machinery safety is m. Then the status of ith requirement ($i = 1, 2, 3, \ldots, m$) is assessed on each kth machine ($k = 1, 2, \ldots, n$) by means of implemented suitable safety measures.

3.1 Proposal of Methodology

Current status of safety requirements on an assessed machine is expressed by means of the so-called coefficient of current measure status $w_{k,i}$, for which holds $w_{k,i} \in \{0, 1, 2\}$. The $w_{k,i}$ coefficient represents a categorial variable, which can reach three possible values (see Table 1).

Table 1. Coefficient of current measure status $w_{k,i}$.

Coefficient $w_{k,i}$	Explanation
$w_{k,i} = 0$	Measures for the fulfillment of ith requirement on k^{th} machine are not introduced
$w_{k,i} = 1$	Measures for the fulfillment of ith requirement on k^{th} machine are introduced but not followed
$w_{k,i} = 2$	Measures for the fulfillment of ith requirement on k^{th} machine are introduced and fully followed

Level of measures efficiency Δ_k expresses the fulfilment of safety requirements by means of realised measures on k^{th} machine and is expressed by the relation:

$$\Delta_k = \frac{\sum_{i=1}^{m} w_{k,i}}{S_{MAX}} \times 100\%. \tag{1}$$

Where n represents the number of machines, m the number of safety requirements and $w_{k,i}$ is the coefficient of current measures status of kth machine by i^{th} safety requirement.

S_{MAX} variable represents the coefficient of maximum reached efficiency of all measures on a given machine and is determined by the relation:

$$S_{MAX} = w_{max} m. \tag{2}$$

Where m is the number of safety requirements (SR) and w_{max} is the maximum value of measures evaluation, in our case $w_{max} = 2$.

Total efficiency level of measures Δ in the given operation is expressed by the relation:

$$\Delta = \frac{\sum_{k=1}^{n} \Delta_k}{n} \times 100\%. \tag{3}$$

Where Δ_k is the level of protective measures efficiency on machines in the case of kth machine and n is the total number of machines located in the given operation.

Total level of measures efficiency of machinery protective devices Δ in the given operation takes on values from the interval $\langle 0, 100 \rangle$.

For a complex safety level of a given operation conditioned by the status of introduced measures on machinery, the following evaluation levels were suggested – see Table 2:

Table 2. Levels of measures efficiency of an operation Δ.

Coefficient Δ	Level of measures
$0\% < \Delta < 29\%$	Low
$30\% < \Delta < 49\%$	Negligible
$50\% < \Delta < 69\%$	Middle
$70\% < \Delta < 89\%$	High
$90\% < \Delta$	Very high

If a production organization has p operations, then the complex level of safety measures efficiency $\bar{\bar{\Delta}}$ would be possible to be expressed by the relation:

$$\bar{\bar{\Delta}} = \frac{\sum_{j=1}^{p} \Delta_j}{p} \times 100\%, \; j = 1, 2, \ldots, p. \tag{4}$$

Where p is the number of operations (in our case 3) and Δ_j is the level of measures efficiency in the case of jth operation, for which holds the relation:

$$\Delta_j = \frac{\sum_{k=1}^{n} \Delta_{k,j}}{n} \times 100\%, \; k = 1, 2, \ldots, nj = 1, 2, \ldots, p. \tag{5}$$

Where $\Delta_{(k,j)}$ is the level of measures efficiency in the case of kth machine in the jth operation and n is the total number of machines within the given operation.

3.2 Results of Applied Methodology

The aim of suggested and applied methodology in order to assess the safety level of machines in a phase of their use was to inspect the status of the machinery that was operated in a given organization from 1 to more than 30 years. Within the particular operations (3 operations for the production of plastic components, marked I, II, III), there were new as well as older machines, such as automatic or semi-automatic assembly workstations, with one or maximum two control places (loading of components, checking and unloading of ready products). For the research purposes, a questionnaire, stemming from the requirements from the Directive on machinery, which consisted of 19 safety requirements, was created.

The experimental research was carried out in several parts:

- risk assessment of each machine based on unified methodology in accordance with ISO 12100 (risk matrix – own methodology of an organization),
- status evaluation of already established (current) safety measures for each machine according to determined safety requirements (SR1–SR19) by means of the coefficient of current measures status,
- evaluation of the efficiency level of such measures (with regard to the outcomes from risk assessment) for each machine and for each operation,

- evaluation of complex efficiency level of introduced protection/safety measures for the whole organization. evaluation of complex efficiency level of introduced protection/safety measures for the whole organization.

Each ith safety requirement of the assessed j^{th} machine was assigned the coefficient of current measures status $w_{k,i}$. The assigned value (0, 1 or 2) was the result of a consensus of 5 reviewers with the aim to decrease the uncertainty rate at subjective deciding (mainly when assessing the safety status of older machines).

From the results of the evaluation of the status of measures of implemented in operation I it is obvious that safety requirements SR1, SR4, SR5, SR11, SR14, SR16 and SR19 are fulfilled on all machines. Requirements SR 17 and SR 18 are not fulfilled on none of the machines, i.e. the requirements for the application of devices for hazardous energy isolation and application of LOTO (Lockout, Tagout) means for machinery [18] were not fulfilled.

The analysis of evaluation shows that in operation I, there is no machinery that would meet all safety requirements. Each machine met on average only 52% of requirements, where measures were implemented and followed. For almost 39% of requirements, safety measures were only introduced but not followed.

For each kth machinery equipment, requirement fulfilment efficiency Δ_k is determined according to the relation (1). Total level of measures efficiency Δ in operation I (marked Δ_I) is determined according to the relation (3). The average value of the total efficiency of current measures in the operation I is $\Delta_I = 69.66\%$.

Three criteria were analyzed and evaluated in a similar way (Δ_k, Δ) in two other operations: operation II (20 machines) and operation III (23 machines). The basic figures of total efficiency of current measures Δ in individual operations are shown in Table 3.

Table 3. Figures of total efficiency of measures Δ in operations [%].

Operation	Number	Average	max	min	Rv	s	IS – 95%
I	17	69.66	78.95	63.16	15.79	4.57	(67.24, 72.02)
II	20	75.66	78.95	47.37	31.58	6.77	(72.49, 78.83)
III	23	53.89	81.58	52.63	28.95	6.03	(51.28, 56.50)
Δ	60	65.61	81.58	47.37	34.21	11.25	(48.66, 68.52)

The analysis shows that the total evaluation of current applied safety measures in operation I reaches the value of 69.66% of total efficiency rate, which represents the *middle efficiency level* of current safety measures.

Operation II reaches the value of 75.66% of total efficiency rate, which represents *high efficiency level* of measures. On each machine of the operation II, there were on average only 62% such requirements, where measures were introduced and followed. For almost 27% of the requirements, measures were only introduced but not followed.

Operation III reaches only 53.89% of total efficiency level, which means almost lower borderline of the middle level of measures efficiency. On each machine of the operation

III were on average only 18% of such requirements, where the measures were introduced and followed. For almost 71% of the requirements, measures were only introduced but not thoroughly followed.

To evaluate the status of machinery safety in the entire organization, a complex efficiency level of introduced measures $\bar{\bar{\Delta}}$ (4) was determined, according to safety requirements SR1–SR19. Based on the results of applied CLMS methodology of introduced safety measures on machines and by means of the evaluation of total efficiency level in the entire organization, it is possible to state that the value of complex efficiency level $\bar{\bar{\Delta}}$ is on the value of 65.61%, which means middle efficiency level.

4 Conclusions

The proposed and in practice verified CMLS methodology has proved that even despite the fact that the machinery safety is managed already at its design by a producer, and throughout its operation by an operator, regular evaluation of efficiency of introduced measures shall point at its insufficiencies related to the entire organization [19, 20]. By analysis of status of implemented safety measures of operated machinery, which, already during the process of their procurement, must meet current requirements of EU directives, weaknesses of management of their safety at the operator itself were "uncovered".

Development of protective devices implemented on machinery equipment (so-called integrated safety) is very fast and they are normally applied in complex module solutions already (e.g., robots). However, with older machines, responsibility for the level of protective measures is on the shoulders of the operator [21, 22].

It is possible to extend the CLMS methodology, which is also a subject of further research consisting of more detailed examination of implemented types of protective devices in relation to the machinery equipment age and level of its maintenance.

Acknowledgments. This contribution is within the project KEGA No. 015TUKE-4/2019 Audit management using software application according to standard ISO 9001:2015 and APVV No. 19-0367 Framework of the Integrated Process Safety Management Approach for the Intelligent Enterprise.

References

1. Çakit, E., Olak, A.J., Karwowski, W., Marek, T., Hejduk, I., Taiar, R.: Assessing safety at work using an adaptive neuro-fuzzy inference system (ANFIS) approach aided by partial least squares structural equation modeling (PLS-SEM). Int. J. Ind. Ergon. **76**, 102925 (2020)
2. Lombardi, M., Fargnoli, M., Parise, G.: Risk profiling from the European statistics on accidents at work (ESAW) accidents' databases: a case study in construction sites. J. Environ. Res. Public Health **16**(23), 4748 (2019)
3. Namesanska, J., Nagyova, A., Markulik, S., Pacaiova, H.: Proposal of KPI methodology and Structure for Industrial Companies. In: Conference Asia-Pacific Management and Engineering, Asia-Pacific Management and Engineering Conference APME, Shanghai, pp. 362–370 (2014)

4. Maletic, D., Maletic, M., Al-Najjar, B., Gomiscek, B.: An analysis of physical asset management core practices and their influence on operational performance. Sustainability **12**(21), 9097 (2020)
5. Zhang, P., Qin, G., Wang, Y.: Optimal maintenance decision method for urban gas pipe lines based on as low as reasonably practicable principle. Sustainability **11**, 153 (2019)
6. Council Directive. Directive 89/391/EEC of the Council Directive of 12 June 1989 on the Introduction of Measures to Encourage Improvements in the Safety and Health of Workers at Work OHSI. Directive 89/391/EEC; Council Directive, Brussels, Belgium (1989)
7. European Parliament and of the Council. Directive 2006/42/EC of the European Parliament and of the Council of 17 May 2006 on Machinery, and Amending Directive 95/16/EC (Recast); European Parliament and of the Council, Brussels, Belgium (2006)
8. Turekova, I., Depesova, J., Bagalova, T.: Machinery risk analysis application in the system of employee training. Appl. Mech. Mater. **Vols**, 635–637 (2014)
9. Galvez, N., Marsot, J., Martin, P., Siadat, A., Etienne, A.: A new approach to hazard identification during the design process by analysing energy transfers. Saf. Sci. **95**, 1–14 (2017)
10. Cordero, C.A., Muñoz Sanz, J.L., Wiña, P.L.: Measurement of machinery safety level in the European market: a real case based on market surveillance data. Saf. Sci. **47**, 1351–1358 (2009)
11. Gabbar, H.A., Koraz, Y.: Risk assessment of micro energy grid protection layers. Energies **10**(8), 1176 (2017)
12. Glatz, J., Gorzas, M., Hovanec, M.: Oil tank fire modelling for the purposes of emergency. In: Production Management and Engineering Sciences, pp. 73–77. CRC Pre-ss/Balkema, Leiden (2016)
13. Jones-Lee, M., Aven, T.: ALARP-what does it really mean? Reliab. Eng. Syst. Saf. **96**(8), 877–882 (2011)
14. Oravec, M., Kulikov, A., Jurc, D., Kulikova, O., Vahovsky, J., Abraham, M.: Analysis of the effect of non-ionizing radiation in the area of magnetism under the limit of 1Hz generated by means of transport from the perspective of neurobehavioral effects. In: SAMI 2019: IEEE 17th World Symposium on Applied Machine Intelligence and Informatics, pp. 89–94. Institute of Electrical and Electronics Engineers, Danvers (2019)
15. Standard ISO 13949-1. Safety of Machinery-Safety-Related Parts of Control Systems – Part 1: General Principles for Design; ISO, Geneva, Switzerland (2006)
16. Comparea, M., Zioa, E., Moronie, E., Portinarif, G., Zaninif, T.: Development of a methodology for systematic analysis of risk reduction by protective measures in tyre production machinery. Saf. Sci. **110**, 13–28 (2018)
17. Jolecyn, S., Chinniah, Y., Ouali, M.S.: Contribution of dynamic experience feedback to the quantitative estimation of risks for preventing accidents: a proposed methodology for machinery safety. Saf. Sci. **88**, 64–75 (2016)
18. Pacaiova, H., et al.: Methodology for complex efficiency evaluation of machinery safety measures in a production organization. Appl. Sci. **11**, 453 (2021)
19. Markulik, S., Kamenicky, L.: How to transform the requirements into the management system? In: SGEM 2015, pp. 689–693. STEF92 Technology, Sofia (2015)
20. Turisova, R., Mihok, J., Kadarova, J.: Verification of the risk assessment model through an expert judgment. Qual. Innov. Prosperity. **16**(1), 37–48 (2012)
21. Lumnitzer, E., Andrejiova, M., Yehorova, A.: Analysis of the dependence of the apparent sound reduction index on excitation noise parameters. Appl. Sci. **10**, 8557 (2020)
22. Stejskal, T., Svetlík, J., Dovica, M., Demeč, P., Krâľ, J.: Measurement of static stiffness after motion on a three-axis CNC milling table. Appl. Sci. **8**, 15 (2018)

Digital Technologies for Monitoring the Vital Functions of Employees with Diseases Accompanied by Seizures with Loss of Balance

Juraj Glatz(✉), Milan Oravec, Zuzana Kotianova, Michal Gorzas, Jan Hijj, and Ivan Habala

Technical University of Kosice, Letna 9, 04200 Kosice-Sever, Slovak Republic
{Juraj.Glatz,Milan.Oravec,Zuzana.Kotianova, Michal.Gorzas}@tuke.sk, JHijj@sk.uss.com, Ivan.habala@zkw.sk

Abstract. The presented paper points to the use of available devices of wearable electronics for people with epilepsy to increase safety in the work process. The essence of the device is the detection of a change in the acceleration of the wrist movement of the person being monitored and, in the case of the development of an epileptic seizure, the identification of a repeated change in the acceleration of the wrist movement of the person being monitored. Subsequently, after evaluation in the analytical element, this information is sent to people in the vicinity or to a third party to assist the person being monitored. The proposed technical solution in the form of a digital device is usable not only for epileptic seizures but also for other diseases with the potential for loss of stability and subsequent injury.

Keywords: Loss of stability · Seizure · Epilepsy · Safety

1 Introduction

Digital environment devices enable monitoring of selected vital human functions and their use in creating new functionalities for human protection in the work process. Current trends are based on digitization, quality production and measurement [1].

Today, smart devices are equipped with sensors that are not used to the extent possible. In humans, it is possible to identify not only their external behavior with smart devices, e.g., change of movement, speed, etc. but also their selected vital functions.

Loss of consciousness, caused by various causes, creates human hazards at the human - machine and human - environment interface [2]. Loss of consciousness can also be caused by an epileptic seizure, as well as other diagnoses, e.g., stroke, but also more common diagnoses such as heat shock, hypoglycemia, dehydration, but also drunkenness, etc. With the loss of consciousness, each person's movement changes. A change in the movement of machine elements can be similarly identified, e.g., by changing the local magnetic field [3]. These changes can be measured, quantified, evaluated and measures developed [4]. It is important in chemical plants (human - surroundings, environment) [5, 6] but also in production plants (human - device) [7, 8].

R. S. Goonetilleke et al. (Eds.): AHFE 2021, LNNS 273, pp. 213–219, 2021.
https://doi.org/10.1007/978-3-030-80713-9_28

In today's digital environment, it is necessary that data be secured by encryption [9] during its transmission and processing. By combining surveillance systems and staff training, it is also possible to identify hazards from loss of consciousness [10–12].

It is also possible to use principles from other areas, the principles of which can also be used in monitoring [13]. Simple algorithms and the use of simple means are required.

If the person (machine operator) does not meet the process requirement at the man - machine [14, 15] man - environment interfaces, in the digital environment it is possible to create a blocking of such behavior of man or machines [16]. Each person is unique, and such are the changes in movement characteristics during the attack. To monitor the properties of objects, including machines, it is appropriate to use a combination of existing parameters from the digital environment [17].

Epilepsy is one of the diagnoses, the external phenomenon of which is a change in a person's movement behavior. Similarly, such changes can be observed in other diagnoses. The identification of these conditions at the human - machine interface is crucial for minimizing injuries caused by both a fall and a fall into the active space of a machine or into the object with the potential for human damage. In the event of a fall or rapid changes in body movement, there is a change in the acceleration of the movement of the limbs of the person affected by the attack, including the wrist. The measurement of the acceleration of the wrist movement during a fall and an epileptic seizure can be identified with sufficient accuracy by an acceleration sensor built into smart bracelets, watches, mobile phones, smart devices, etc.

2 Measurement

2.1 Measuring Chain

The change in the loss of stability, the cyclic change in the movements of the wrist during an epileptic seizure, was performed by measuring the wrist of the affected person with a smart device. The measuring chain consisted of a smart device (acceleration measurement), an analytical element (data evaluation), a communication element (WiFi). The elements communicated via protocols in the Android environment. In the analytical element, a comparison of changes in the measured data from the acceleration sensor in the smart device is performed. With loss of stability and a fall, the motion vector changes abruptly. In the case of an epileptic seizure, the acceleration vector over time has a cyclically repeating trajectory. After evaluation in the analytical element, this information was sent to the environment, for comparison with the information about the movement of the machine.

The proposed device was created alternatively from smart devices with an element for measuring the acceleration of wrist movement, an analytical element, and a communication element. The individual elements of the device can form one unit, or they can be created from digital space devices, Table 1.

All elements of the device can be applied to devices such as smart watches with the possibility of programming, or if they have enough free memory for an existing program.

Table 1. Examples of combining existing devices with the required functionalities to create the desired functionality.

Existing device	Functionality of device elements			
	Acceleration measurement	Analytical element	Communication element	Interface with protocol
Smart bracelet	x	–	x	Bluetooth, WiFi
Smart watches	x	–	x	
Mobile phone	x	x	x	
Device with IoT element	x	x	x	

2.2 Used Measuring Technique

A smart watch with built-in acceleration sensing with the Android operating system was used to verify the required functionality. The device was attached to the wrist and changes in stability loss and seizures were simulated. The orientation of the coordinate smart device with the acceleration sensor was oriented so that the x-axis was in the direction of the person's forward movement, the y-axis to the ground and the z-axis perpendicular to both axes. The analytical element was the excel spreadsheet in the Android tablet operating system environment. In this environment, the change of measured values was analyzed. When the acceleration of the wrist movement, which characterizes a fall and an epileptic seizure, changes, information was sent for subsequent processing. Table 2 shows the parameters of the sensor located in the smart device.

Table 2. Possibilities of measurement of selected quantities by smart devices

Parameter	Description	Sensor	Note
Time	Sensing frequency	Sensor bmi 160, Range to 250 Hz	Part of a smart device
ax	Acceleration in the x direction	Range to100 Hz, smart device element, the linear accelerometer is derived from the G meter	The linear accelerometer measures the acceleration in the appropriate direction
ay	Acceleration in the y direction		
az	Acceleration in the z direction		

3 Discussion

Figure 1 shows the measured values of the acceleration of the wrist movement ax, ay, az for the interval:

- Within three seconds, represented by normal walking, where the dominant movement is in the x, y plane, which is caused by fluctuating walking movement and hand movement.
- The interval between the third and sixth seconds, represented by stopping before an epileptic seizure, when a seizure occurs, stiffening. In this interval, the acceleration of the limb is almost zero in all directions.
- The interval between the sixth and eleventh seconds, representing the fall of a person to the ground. The person fell by twisting around the axis of the body and landed on his back. A slight forward tilt in this fall can be observed for the acceleration ax, the subsequent twisting to the left, the positive acceleration az and the change in position with respect to the change in wrist height represented by the change ay.
- The interval between the eleventh to the twenty-fifth second, representing an epileptic seizure. The affected person was lying on the back of the hand next to the body and there was a constant path of the hands, which can be seen in the change of acceleration ax and az.
- Interval after the twenty-fifth second, representing release after an epileptic seizure.

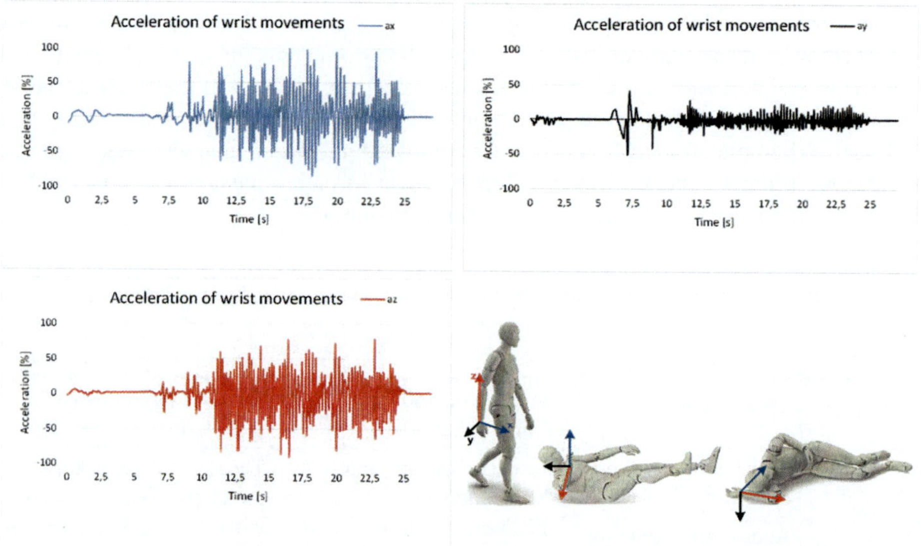

Fig. 1. Measured accelerations in the direction of the x, y, z axes for selected changes of motions

The y-axis of Fig. 1 is a percentage representation of wrist acceleration to compare the change between walking and seizure. The difference between normal walking and an epileptic seizure is 10 to 20 times depending on the direction of acceleration ax, ay, az. Figure 2 shows the procedure of using the measured movement changes at the human - machine and human - environment interface.

To ensure adequate safety function, as part of the overall functional safety of the machine controlled by its safety electronic system, it is necessary to specify the parameters and the value of the risk in more detail. It is necessary to proceed from the basic

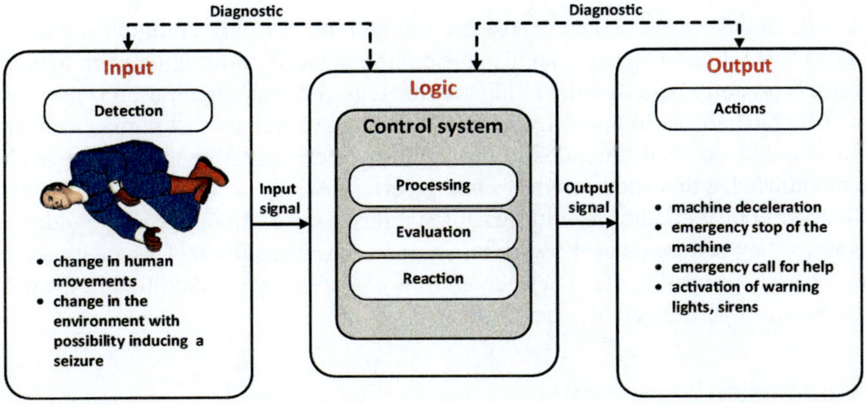

Fig. 2. The procedure of using the measured movement changes at the interface man - machine and man - environment.

factors of reliability, fault detection, architectural and system integrity. This principle is set out in EN ISO 13849-1. Most systems can be divided into three basic functions; input, logic solution and output - activation. The categories of safety control functions represent the circuit diagrams of the individual parts of these systems, the structures of which ensure the achievement of the required performance of the safety functions.

From the measured data for the monitored physical activities in case of loss of balance, it is clear that the measurement of acceleration allows the monitoring of a person's movement. The distance between the individual phases of movement, tangents are significant and characteristic for a specific movement, both rectilinear and rotational. Other physical activities caused by loss of consciousness and fall were also measured, where they were monitored as direction, rotation and speed of fall. They had a similar informative value of the observed causality as the above-mentioned representation in Fig. 1.

Knowledge of the operation of the machine, e.g., speed, feed, etc. by means of measured signals representing the specific activity of the machine and the knowledge of vital functions of a person by means of measured signals of his movement functions gives the possibility to minimize the hazard at the human - machine interface.

Advantages of digitization of processes at the human - machine and human - environment interface:

- the possibility of notifying people in the vicinity or third-party by sending information (SMS, recorded call) about potential collapse and the possibility of injury and seizures,
- in case of a fall injury and in case of loss of consciousness, sending information about this state to people in the vicinity or a third party to assist the disabled person,
- new, required functionalities to increase human safety can be created by a combination of existing smart devices commonly used and machine controls via IIoT in operating premises.

Today, digital space makes it possible to combine not only common information from identified motion signals, but also the combination of information from metadata. If a filter is properly built in order to identify the required functionalities, it is possible to create suitable barriers to minimize hazards at any interface, not just human - machine. Monitoring human vital functions in the digital world is possible today, even without human knowledge that such activity is being performed.

Recognition of conditions with loss of stability is also important in the older generation, as they can be caused by trivial causes such as insufficient water intake, late administration of insulin, etc. These conditions can not only be identified in the digital space, but third parties can also be notified.

4 Conclusion

Digital space gives the possibility to identify all movements performed by humans but also by machines and objects. These options are a prerequisite for creating new tools for minimizing hazards at the human - machine and human - environment interface. Unsuitable combinations of movements in the work environment can be minimized, as they can not only be identified in real time but also controlled.

Monitoring of human vital functions in the digital world is possible today, even without human knowledge that such activity is being performed. At present, humans have transferred a part of their identity to digital space, which can be used for his protection in the work process but also misused in everyday life. Knowing the physical and mental health through knowing the vital functions of each individual opens up many new issues that did not need to be addressed in the past.

Recognition of conditions with loss of stability is also important in the older generation, as they can be caused by trivial causes such as lack of water intake, miss an insulin dose, etc. These conditions in the digital space can not only be identified but third parties can be notified of them.

This paper was prepared thanks to the support of project 013TUKE-4/2020 Transfer of knowledge of research of means using augmented reality into the educational process in the field of safety of technical systems.

References

1. Balážiková, M., Sinay, J., Dulebová, M., Markulik, Š, Kotianová, Z.: Measurement of low-frequency noise during CNC machining and its assessment. Measurement **119**, 190–195 (2018). ISSN 0263-2241
2. Pačaiová, H., Nagyová, A., Markulik, Š.: BBS program (behavior based safety) - way to raising the level of health and safety in practice, 2012. - 1 electronic optical disc (CD-ROM). In: AHFE 2012 4th International Conference on Applied Human Factors and Ergonomics: 21–25. July 2012, San Francisco, pp. 2097–2104 (2012). ISBN 978-0-9796435-5-2
3. Pačaiová, H., Oravec, M., Šmelko, M., et al.: Extra low frequency magnetic fields of welding machines and personal safety. J. Electr. Eng.-Elektrotech. Casopis **69**(6), 493–494 (2018)
4. Balážiková, M., Tomašková, M., Dulebová, M.: Assessment of ergomonic risk for firefighters. In: 2016. In: Production Management and Engineering Sciences, pp. 3–6. CRC Press/Balkema, Leiden (2016). ISBN 978-1-138-02856–2

5. Pačaiová, H., Kotianová, Z., Brestovič, T.: Modelling ammonia pipeline leakage for the proposed pipeline change. 2016. In: Production Management and Engineering Sciences. - Leiden : CRC Press/Balkema, pp. 509–515 (2016). ISBN 978-1-138-02856-2
6. Blist'an, P., Pačaiová, H.: Modelling environmental influence on the pipelines integrity. In: Conference: 11th International Multidisciplinary Scientific GeoConference, Albena, Bulgaria, 20–25 June 2011 (2011)
7. Nagyová, A., Balážiková, M., Markulik, Š.,Sinay, J., Pačaiová, H.: Implementation proposal of OH&S management system according to the standard ISODIS 45001. In: Arezes, P. (ed.) Advances in Intelligent Systems and Computing. AHFE 2017, vol. 604, pp. 472–485. Springer, Cham (2018). ISBN 978-3-319-60525-8
8. Markulik, Š., Kamenický, L.: How to transform the requirements into the management system? 2015. In: SGEM 2015, pp. 689–693. STEF92 Technology, Sofia (2015). ISBN 978-619-7105-40-7
9. Ovseník, Ľ., Turán, J., Huszaník, T., Oravec, J., Kováč, O., Oravec, M.: Image encryption algorithm with plaintext related chaining. In: Computing and Informatics: Computers and Artificial Intelligence, vol. 38, no. 3, pp. 647–678. Institute of Informatics, Bratislava (2019). ISSN 1335-9150
10. Veľas, A., Kutaj, M., Ďurovec, M.: Influence of changing the parameters of the camera system on video-based motion detection. In: ICCST 2017: 51st international carnahan conference on security technology: 23–26 October 2017, Madrid, Spain. IEEE, New York (2017). ISBN 978-1-5386-1585-0. - [5]. (International carnahan conference on security technology proceedings. - ISSN 1071-6572)
11. Boroš, M., Kutaj, M., Mariš, L., Veľas, A.: Development of security at the local level through practical students training. In: INTED 2018, 05–07 March 2018, Spain/INTED 2018: Proceedings from 12th International Technology, Education and Development Conference, Valencia, 5th–7th March 2018, pp. 725–729. IATED Academy, Valencia (2018). ISBN 978-84-697-9480-7. ISSN 2340-1079
12. Boroš, M., Jasenčák, Ľ, Kutaj, M., Veľas, A.: The design of a testing device for passive infrared motion detectors. In: Brebbia, C.A., Garzia, F., Lombardi, M. (eds.) SAFE VII, 06–08 September 2017, Italy, Safety and Security Engineering VII, pp. 457–463. WIT Press, Southampton (2018). ISBN 978-1-78466-211-0. ISBN 978-1-78466-212-7
13. Veľas, A., Svetlík, J.: Design of equipment to improve safety on roads. In: Transport Means 2016: Proceedings of the 20th International Scientific Conference, 5–7 October 2016, Juodkrante, Lithuania, pp. 614–617. Kaunas University of Technology, Kaunas (2016). ISSN 1822-296X
14. Turisová, R., Sinay, J., Pačaiová, H., Kotianová, Z., Glatz, J.: Application of the EFQM model to assess the readiness and sustainability of the implementation of I4.0 in Slovakian companies. Sustainability. **12**(14), 1–14 (2020). ISSN 2071-1050
15. Pačaiová, H., Sinay, J., Turisová, R., et al.: Measuring the qualitative factors on copper wire surface. Measurement **109**, 359–365 (2017)
16. Balážiková, M., Andrejiová, M., Wysoczańská, B.: Methodology for risk assessment of noise effect at workplace. MM Sci. J. 3426–3430 (2019). ISSN 1803-1269
17. Sinay, J., Pačaiová, H., Oravec, M.: Present state of machinery safety assessment. In: Annals of DAAAM for & Proceedings of the International DAAAM Symposium (2005)

Production Process Optimization by Reducing Downtime and Minimization of Costs

Stefan Markulik, Renata Turisova, Anna Nagyova(✉), Tomas Vilinsky, Robert Kozel, and Katarina Vaskovicova

Faculty of Mechanical Engineering, Technical University of Kosice, Letna 9, 042 00 Kosice, Slovakia
{stefan.markulik,renata.turisova,anna.nagyova}@tuke.sk, {tomas.vilinsky,robert.kozel, katarina.vaskovicova}@student.tuke.sk

Abstract. This contribution is a case study, in which a requirement for an increase in production from customer's part occurred. From the alternatives offered, the way of production line optimization was chosen, the adjustment of which made it possible to response to this reality. Initial analysis of the production line confirmed a significant number of downtimes, which were identified almost in every part of the production line. By using the improvement tool – Ishikawa diagram (Fishbone diagram) – particular categories of reasons causing these downtimes were identified. Application of '5Why?' method was the next step, which helped to determine root causes of these downtimes. Then, p-diagram was implemented to reduce the downtimes which helped to minimize costs.

Keywords: Quality · Production · Optimization · Improvement

1 Introduction

In an effort to maintain the competitiveness, a lot of organizations are forced to accept more and more demanding requirements from customers, even when the given requirements approach borderline values of the capacity of production facilities dealing with the relevant order. If the organization, in such situation, wants to maintain a high quality of production, they have to generally respect the increased costs related to the production often carried out on the very borders of production equipment capacity.

One of the possible solutions is organization's orientation to sustainability [1] or process improvement [2] by means of sophisticated methods (Six Sigma etc.) as well as simple tools for quality improvement [3]. The implementation of basic quality improvement tools has many times proved legitimate in practice. Simple and easily comprehensible principles, on which the mentioned basic tools are built, when implemented correctly, have brought improvements in practice in a form of cost reduction, quality improvement, or the elimination of the negative impact of production on the environment [3, 4] or at safety at work [5, 6, 15]. Within this, it is common to use also the p-diagram statistic tool [7], which may, in an efficient way, help reduce downtimes. In this case study, it is

R. S. Goonetilleke et al. (Eds.): AHFE 2021, LNNS 273, pp. 220–227, 2021.
https://doi.org/10.1007/978-3-030-80713-9_29

the demonstration of using the statistic tool in the process of packing on a packing and stamping line (PSL).

Process of Packing and Stamping

The inputs into packing and stamping process using the combined packing and stamping line (PSL) are letters, envelopes, leaflets, postcards or various gifts (keychains, pens, etc.) which are produced by the company itself at their own cost or bought from suppliers. As the company competes for customers' favor, it usually respects their very specific requirements, as well [12]. This subsequently leads to borderline setting of PSL input parameters, hence to an increased likelihood of mistake occurrence. It is sometimes not even possible to pack 100% components on the PSL. These are mainly non-homogenous objects of various shapes, which head for envelopes. With most orders, however, the packing machine is fully able to meet the requirements for packing. PSL is equipped by six usable sections, a section for empty envelopes and the output section. The machine works on the principle of under-pressure suction discs. Empty envelopes are first stamped, i.e. relevant stamps are printed on the selected parts of the envelope. By means of the electromagnet and tappet system, the machine transmits the signal to particular sections, where under-pressure is generated – a particular component is taken out of the feeder and placed into the inserted conveyor, which transports the components from sections 1–6 into the collecting point. There, the particular components are, in the exactly determined order, taken by a feed arm, which inserts the components into a prepared envelope that is led in the output conveyor and is, by means of one bottom and three top suction discs slightly opened in order for a trouble-free inserting of components into the corresponding envelope. Then, the machine moves the filled and open envelope one step further, where the envelope-moistening mechanism applies a fine film of distilled water on the adhesive part of the envelope. With the following step, the machine automatically closes and seals the envelope. In the packing process, it is possible to, according to customers' requirements, select the number of used sections from 1 to the maximum of 6. The variability of packing possibilities is high, not to mention that each feeder in a section can be modified in two axis – thus change the dimension of the capacity feeder with regard to the letter width and the dimension with regard to the letter length. PSL is equipped by a number of control mechanisms, the task of which is to, in case of a problem or anomaly occurrence on the packing line, ensure the line running to stop. There is a high occurrence of such situations in practice that, during pulling-in a letter by a suction disc, when the paper is greasy and glued together, two or three letters are not separated from each other, the suction disc pulls them downwards and separation rails separate them from the remaining part of letters which are in the capacity feeder. As it is not desirable for an envelope to contain several letters of the same kind, a control mechanism is installed which warns about the error in question and concurrently stops the line. Specifically, when the incorrect thickness is detected (number of pulled-in letters), PSL is automatically blocked, at the same time a red diode flashes up on the arm where the problem has occurred. Another quite frequently used control mechanism on the packing machine is empty envelope presence detection on the conveyor strip. In order to correctly pack the components into an envelope, there is a position switch in the so-called collecting point – place where components from sections 1–6 are inserted into the envelope. If the switch detects a discrepancy, the electric circuit turns on and the line is subsequently stopped. It happens during the last step that the handle on the conveyor strip does not supply the filled envelope to the capacity feeder completely. The most frequent cause of this error is the incorrect arrestment of the distribution chain, on which the handles are placed. If such a situation occurs, the machine is equipped with a position switch that works in the same way as the one in the collecting point. PSL is usually operated by one or two employees. The need for the other operator was argued as the effect of various factors, such as employee's experience, set speed of packing, number of required attributes in one envelope or an ability of an employee to quickly and effectively eliminate the error on the machine [8]. During the full use of shears – 6 pieces and the highest recommended speed, it was a common practice for two operators working on the packing machine.

2 Implementation of Quality Tools in Practice

In order to find out the weakest points in the packing process with regard to a frequent occurrence of downtimes, aspect which may potentially lead to the occurrence of a downtime on the line, were identified by a form of brainstorming. A diagram of causes and consequences – Ishikawa diagram was used (Fig. 1).

This diagram presents possible causes leading to downtimes, classified into five categories according to the source causing the occurrence of downtime. The results of brainstorming have revealed some previously unidentified potential causes of downtime occurrence. For example, the generation of electrostatic induction during pulling out a letter from the feeder, when by the letter friction against the iron edge, static electric charge is induced on the bottom side of the letter, causing a slight floating of the letter during the transport to the collection point and subsequent complications during packing. These phenomena are in the technical practice identifiable [9] and it is possible to minimize their effect also by means of suitable maintenance [10, 11].

The causes of frequent downtime occurrence were analyzed also by means of '5 Why?' method. By repeated answers to suitable structured question 'Why?', some potential causes of downtime occurrence were identified. An example of this implemented method is shown in (Table 1).

By means of root cause analysis using the 5Why? method, we have come to a possible solution to one of the whole range of problems causing frequent downtimes on the line.

In order to decrease PSL downtime rate, it was necessary to define selected factors that significantly affect the downtime rate. For this purpose, FMEA method was used [13, 14]. The height of one component feeder is 25 cm. If we fill the feeder to the top, one feeder is completely emptied within 8.5 min with a thicker component (850 pcs) up to 12.5 min with a thinner component (1250 pcs). There was an unwritten rule for PSL, with full use of sections and one-operator service, to set the speed for 5250 pieces per hour. The argument for the decrease in speed was that the operator in control had to, apart from the duty to take care of feeders, take pieces from the previous process and stock other components, also stock their packed envelopes for the next process – binding. If such an order is processed where it is necessary to use all feeders, the worker has no time to carry out these tasks without occasional downtime occurrence. In Table 2, there is a record of a 200 000-piece order course. The order was planned in such a way that

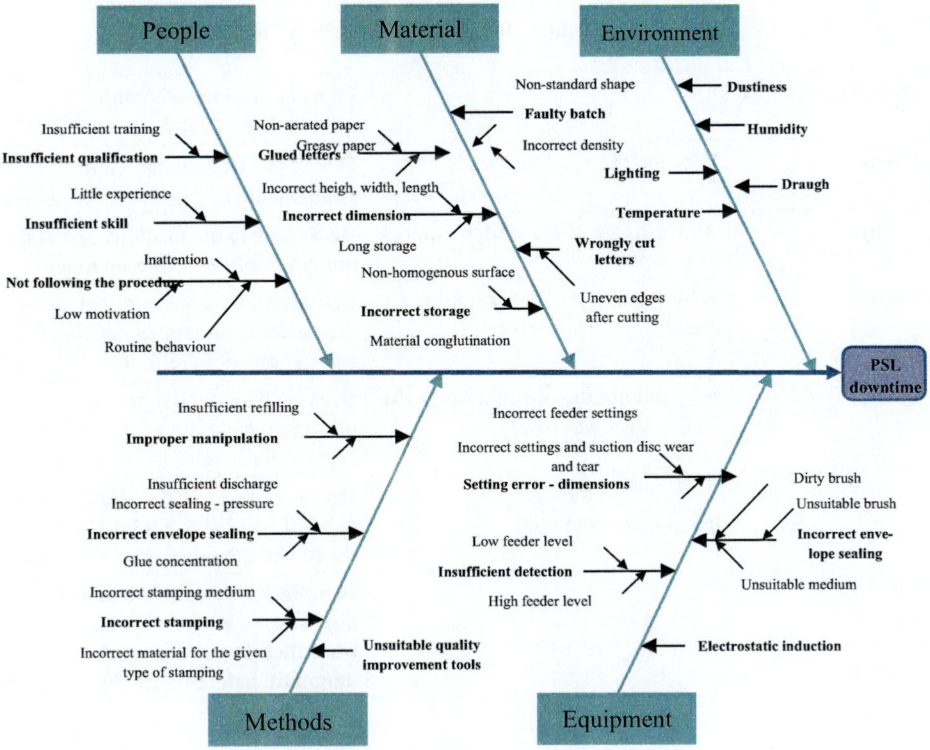

Fig. 1. Ishikawa diagram – PSL downtime

at the beginning of the shift, everything was set and it was possible to start packing. In the second column, we can see the code marking of the order. The first number means quantity, second means the number of components in one envelope, the letter L means that the envelopes are sealed directly on the packing line. The last entry is the time after which it comes to the planned shipment of order. Hourly target – the sixth column, was determined based on several factors, primarily it was logistic planning, operator's skill assessment, etc. For this order, the set target was to pack 5250 envelopes per hour.

It is apparent from the table that the target was followed only for 2 days. Further, we can see the numbers of downtimes within feeders and the total downtime duration in minutes during a day. As we can see, the hourly target of 5250 pieces was achieved twice in five days. By wrongly set hourly target as well as many downtimes, it was necessary to work overtime, i.e.: to come to work also on Saturday. Yet the overall evaluation was generally positive. Despite this, the order was successfully finished on time. This is a bad approach to the improvement, which we can encounter in many organizations. The improvers focus on one cause, mainly the one that causes an error the most frequently (generally based on Paret analysis) and based on the reduction of its effect, they evaluate the improvement degree often by means of a static method. Such an approach may be highly effective if the conditions, settings or process environment do not significantly change. Should there be frequent changes of setting, weight, shapes or

Table 1. Root cause analysis using the 5Why? method

Problem definition		Packing machine stopping – downtime occurrence
1 Why?	Why did the machine stop?	One of the feeders ran out of components
2 Why?	Why did one of the feeders run out of components?	Because packing machine operator did not refill the feeder on time
3 Why?	Why was not the feeder refilled on time?	Because the operator did not know the feeder level was so low that it was necessary to refill it
4 Why?	Why did not the operator know the feeder level was low?	Because they did not get the information about the necessity to preferentially refill the components
5 Why?	Why did not they get the information about it?	Because during feeder operation on one PSL side, it is not possible to see feeder the other side of the line
Solution		**Installation of suitably placed mirrors** **Identification of the local magnetic field change** [9]

Table 2. Downtime record on PSL caused by non-refilling of feeders.

Downtime monitoring														
Shift	Order	Target	Total per shift	Hrs.	Target (hrs)	Average (hrs)	Downtime occurrence in feeder (Z)						Sum	Down-time (min)
							Z_1	Z_2	Z_3	Z_4	Z_5	Z_6		
Shift 1	200 000 6-L 7 days	200 000	38 918	7,5	5250	5189	5	2	2	1	0	1	11	8,5
Shift 2		161 082	39 473	7,5		5263	4	0	3	2	0	2	11	7,4
Shift 3		121 609	39 045	7,5		5206	2	1	1	1	1	1	7	7,6
Shift 4		82 564	39 263	7,5		5235	2	2	1	0	0	1	6	6,8
Shift 5		43 301	39 645	7,5		5286	2	2	2	2	0	0	8	5,2
Over-time		3 656	3 656	0,75		4875	1	1	0	1	0	0	3	1,9
Total	52,5 h	0	200 000	38,25	5250	5229	16	8	9	7	1	5	46	37,4

quality of entries, as it is in the process analyzed by us, an error being dominant today may be minor tomorrow or vice versa. Therefore, it is better to use dynamic quality improvement tools in such conditions. It is also totally inappropriate to determine target values (hourly target), the determination of which is generally performed based on some habitual rules. Lower target values than PSL capacity declared by the manufacturer then cause the downtime values declared in records do not look so redoubtable. Therefore, in the company we analyzed, it was inevitable to radically change the approach to process quality improvement. Further, we will describe the change of approach to downtime minimization.

2.1 Downtime Minimization

A part of the change in attitude to downtimes was also the implementation of a regulatory diagram. The regulatory p-diagram describes the change of proportion of inconsistent units within the researched sample (subgroups) depending on time. It is the most frequently used attribute diagram. The standard deviation of the selective division of faulty product fraction changes inversely with the subgroup size. Selection size change naturally changes this value. As the selection size increases, the standard deviation decreases and vice versa. Since the regulatory limits and zone borders are calculated based on this value, when the standard deviation changes, regulatory limits and zone borders change, as well. There are two basic ways of using p-diagrams. The first is p-diagram with pre-set limits and the other one is p-diagram with calculated regulatory borders, which we used to analyze 20 work shifts, which, with the exception of public holidays, ran consecutively. All cases were extensive orders using at least five document sources. Historical data were processed, having selected only those shifts when only one operator was working. Unlike in the previous evaluation, we have used, as the size of logical subgroup, the hourly value of 6000 envelopes, i.e.: the maximum PSL capacity recommended by the manufacturer. Individual diagram points are arithmetic average of hourly values of really processed outputs per one 7.5-h shift. As we can wee in the Fig. 2a, with such setting, the p-diagram shows average hourly downtimes on the level exceeding 13%.

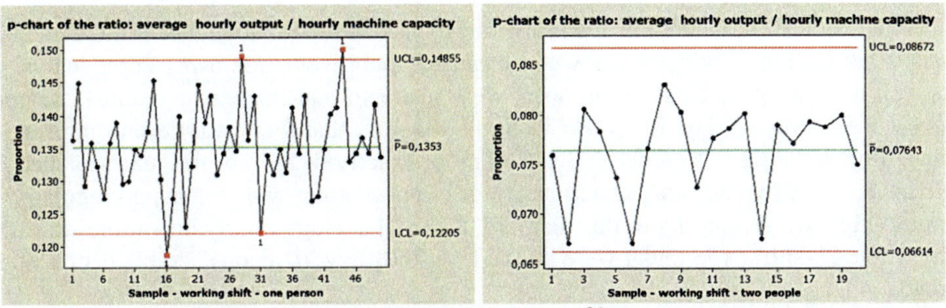

a) b)

Fig. 2. p-diagram: a) downtimes on PSL, b) with pre-set regulatory limits.

The diagram also shows an apparently high variability and exceeding both top and bottom regulatory limits. Significantly worse results with this output than with previous evaluations are caused by the fact that no other target values than the maximum value proposed by the manufacturer are considered. Further, we conducted a following experiment that stemmed from the fact that, within 20 consecutive shifts (except for public holidays and the day when PSL line was adjusted), i.e. when the maximum performance was required, we researched the downtimes of the line in question, yet, with two-member operation. One operator was on the control side of the line, while the other operated the feeders on the other side. The p-diagram of downtime rate made from the results of 20 shifts from such an experiment is in Fig. 2b. It is evident from the diagram that the average hourly downtime rate decreased to as little as to 7%, i.e., by almost a half. Within the mentioned experiment, all parallel operations carried out at the same time were recorded and an image of worktime use was made. The results of this analysis were formulated into 10 basic rules, which we cannot present (as it is the intellectual property of the company). Then, the experiment continued again with full 20 shifts, however, with only one-member operation. P-diagram with pre-set regulatory borders was used (Fig. 3a) for regulation. Pre-setting of regulatory limits was done based on the values acquired on PSL with two-member operation.

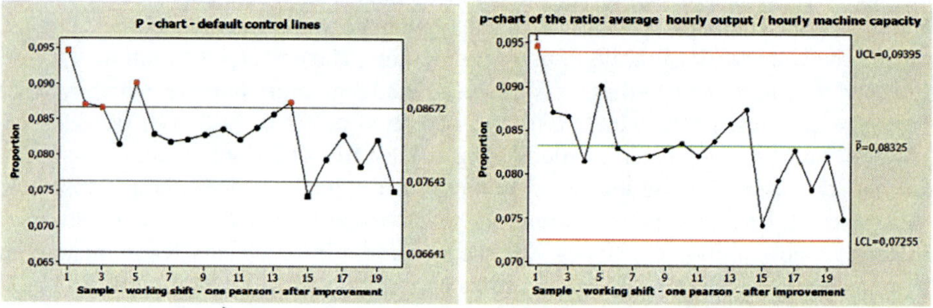

a) b)

Fig. 3. p-diagram: a) with pre-set regulatory limits b) with calculated regulatory limits.

After each shift, where the downtime rate exceeded the top tolerance limit, an analysis, stratification of particular downtime causes, their evaluation using Paret analysis, as well as preventive measure proposal were made so that the existing status did not repeat the following day. It is clear from the diagram that there was certain improvement after some time. In order to evaluate more exactly what improvement was made, we have made, for the same downtime values, a p-diagram with calculated regulatory limits (Fig. 3b). According to this diagram, the total average hourly downtimes during the whole 20-shift experiment were on the level of 8%, with a considerable decreasing trend evident from the diagram.

3 Conclusion

On a specific example of a PSL operation, we have demonstrated the effectivity of using dynamic downtime monitoring by means of p-diagram. It is apparent from the result of the experiment that, with some processes, especially with processes where entry parameters, machinery or environment setting may significantly change, it is more efficient to use dynamic quality improvement tools. These, unlike static tools, consider the changes of particular parameters in time.

Acknowledgement. This contribution is the result of the project implementation APVV-19-0367 "Integrated Process Safety Management Approach Framework for the Smart Enterprise" and project KEGA 019TUKE-4/2020 "Application-oriented education in the area of ISO 9001:2015 requirements implementation".

References

1. Turisova, R., et al.: Application of the EFQM model to assess the readiness and sustainability of the implementation of I4. 0 in Slovakian companies. Sustainability **12**(14), 5591 (2020)
2. Zgodavova, K., Slimak, I.: Advanced improvement of quality. In: Annals of DAAAM & Proceedings, pp. 1551–1553 (2008)
3. Bosák, M., et al.: Testing the effects of a thermal power plant combustion by-product stabilizer on water impermeability. Pol. J. Environ. Stud. **29**(2), 1087–1094 (2020)
4. Madzík, P., et al.: Effects of the energy and mining industry on management of national competitiveness. Acta Montanistica Slovaca **21**(1), 67–75 (2016)
5. Pačaiová, H., et al.: Extra low frequency magnetic fields of welding machines and personal safety. J. Electr. Eng. **69**(6), 493–496 (2018)
6. Sinay, J., Pacaiova, H., Oravec, M.: Present state of machinery safety assessment. In: Annals of DAAAM & Proceedings, pp. 347–349 (2005)
7. Sousa, S., Nuno, R., Eusébio, N.: Application of SPC and quality tools for process improvement. Proc. Manuf. **11**, 1215–1222 (2017)
8. Davahli, M.R., et al.: Identification and prediction of human behavior through mining of unstructured textual data. Symmetry **12**(11), 1902 (2020)
9. Oravec, M., et al.: Magnetic field image–source of information for action causality description. In: 2019 IEEE 17th World Symposium on Applied Machine Intelligence and Informatics (SAMI). IEEE (2019)
10. Pačaiová, H., Ižaríková, G.: Base principles and practices for implementation of total productive maintenance in automotive industry. Qual. Innov. Prosperity **23**(1), 45–59 (2019)
11. Stodola, P., Stodola, J.: Model of predictive maintenance of machines and equipment. Appl. Sci. **10**(1), 213 (2020)
12. Šmelko, M., et al.: Low frequency magnetic fields and safety. Acta Physica Polonica A **137**(5) (2020)
13. Turisová, R., Tkáč, M.: Error prevention at creation of concept of product under development. In: Production Management and Engineering Sciences: Proceedings of the International Conference on Engineering Science and Production Management (ESPM 2015), Tatranské Matliare, High Tatras Mountains, Slovak Republic, 16th–17th April 2015, vol. 10. CRC Press (2015)
14. Nguyen, T.L., Shu, M.H., Hsu, B.M.: Extended FMEA for sustainable manufacturing: an empirical study in the non-woven fabrics industry. Sustainability **8**(9), 939 (2016)
15. Turekova, I., Bagalova, T., Osvaldova, L.: Makovicka: analysis of safety and health awareness among primary students. In: ICERI 2016: 9th Annual International Conference of Education, Research and Innovation (iCERi), Seville, Spain, 14–16 November 2016, BookSeries: ICERI Proceedings, pp. 432–442 (2016). ISBN 978-84-617-5895-1/ISSN 2340-1095

Consideration for Experimental Verification of the Effectiveness and Safety of Exoskeletons

Daniela Onofrejová, Michaela Balážiková(✉), and Michal Hovanec

Technical University of Kosice, Letna 1/9, 042 00 Kosice-Sever, Slovakia
{daniela.onofrejova,michaela.balazikova,michal.hovanec}@tuke.sk

Abstract. Human in the work environment interacts with various means of work and technical equipment, so it is important that the work environment is adapted to eliminate negative effects on health and productivity at work.

A new trend in the world is to introduce solutions in the form of exoskeletons to the given workplaces within the framework of prevention and a healthy workplace. These are mostly mechanized devices, some with electrically driven parts. These devices work according to a biomechanical principle, distribute a workload and transfer it to different body parts with a help of the exoskeleton's support structure. In industrial practice, employers still have little experience with exoskeletons. A certain shortcoming is the lack of methodology for ergonomic analysis and optimization of workplaces with human operation, due to new elements of support tools and equipment at the time of digitization, increasing production efficiency and ensuring production quality and occupational safety.

Keywords: Human factors · Musculoskeletal disorders · Exoskeletons · Physical load

1 Introduction

Human factor in production system must have certain qualitative and quantitative characteristics based on physical, physiological, and psychological aspects. Human needs to be adapted and ready for a given job in terms of psycho - physiological and qualification. The workplace should allow maximum productivity with minimum effort. The main idea is to adapt work and working conditions to a person, improve working conditions without endangering health, make the working environment more pleasant and increase the efficiency of work activities. Basic factors that are needed in a complex work system to improve are:

- Human - adaptability and readiness for a given job in terms of psycho – physiologi-cal and qualification.
- Work process - taking into account the possibilities of man, ensuring the optimal and economic use of his abilities.
- Workplace - achieving maximum productivity with minimal effort.

R. S. Goonetilleke et al. (Eds.): AHFE 2021, LNNS 273, pp. 228–235, 2021.
https://doi.org/10.1007/978-3-030-80713-9_30

- Work environment - elimination of disturbing factors that adversely affect human health, performance and safety.
- Work equipment - machines, tools, implements should be designed and placed in accordance with anthropometric conditions and psycho - physiological capabilities of man.

Human in the work process has its abilities and limitations, therefore it is necessary to adapt the workplace to his possibilities, abilities and knowledge. Human performance can be increased by applying ergonomic knowledge to the design of the work environment, and thus reduce its load. This brings an increase in economic profit and a reduction in the cost of repairing the damage caused by the high workload of workers, incorrect organization of the working environment, etc. [1].

The National and the European Agency for Safety and Health at Work unanimously note that damage to the musculoskeletal system (MSS) is becoming an increasing problem and needs to be prevented, with a thorough ergonomic risk assessment in the interests of safety and health at work. Injuries at the workplace also increase with the increase in the average age of the workforce, as well as with the insufficient training of new workers and the use of personal protective equipment (PPE) at work [2, 5].

A person's performance is determined by physical constitution, motor skills, muscular strength, function of sensory organs, mental ability. In addition, performance is also affected by gender, age and working conditions. The adaptation of a person to working conditions is influenced by several factors: the type and content of the work activity, i.e. the nature of work tasks and operations, the technology used, the machines, the equipment, the materials; risk associated with the performance of work - harmful working environment, risky machines; work and rest regime - system of breaks during change, forced work pace, unevenly distributed workload; social climate in the workplace - interpersonal relationships, level of social support. The boundary of the adaptation mechanism is formed by a state in which the parameters of the working environment exceed the limits endangering human health, motivation and work productivity [3].

With the development of automated workplaces, the workload of operators is reduced, which does not apply to all activities, especially assembly and handling [4]. Up to 60% of all reported health problems are diseases of the musculoskeletal system. As many as 45.0% of workers report back pain from a proportion of reported health problems, 44% of workers report upper limb pain, 30% of lower limb pain, and 60% of workers have one or more problems [5].

A new ergonomic prospective solution as personal protective equipment within protection and prevention of health at work projects the wearable exoskeletons. These are mostly mechanized devices, some with electrically driven parts. These devices are intended to change the way a person exerts forces against other objects, or the way a person receives or experiences externally applied forces. In industrial practice, employers still have little experience with wearable exoskeletons, and for wider deployment in industry, targeted research is needed to give companies arguments for the effective use of wearable exoskeletons according to the type of load and activity in the interest of safety and health at work.

2 Consequences of Physical Activity and Other Health Risks of Workers

In the European Union, around a third of all employees have to work with workloads daily and spend most of their working time there. Hand handling weights mean lifting, holding, carrying, storing or pulling heavy objects, either individually or by several employees at the same time. Manual handling of the weight can cause: damage to health caused by constant, gradual deterioration of the skeleton and muscles (e.g. lower back pain) and serious damage (e.g. bone fractures; bruises or injuries caused by accidents).

The differentiation of the prevalence of Musculoskeletal disorders (MSDs) by gender, age and level of education underlines that there is a need for diversity-sensitive approaches/ risk assessments to better prevent and manage MSDs.

2.1 Exoskeletons – Features, Function

There are several classifications of exoskeletons. According to the principle of action, they are divided into active and passive exoskeletons. Active devices use an external energy source, while mechanical passive exoskeletons rely on kinetic energy and human strength [6]. Exoskeletons can also be classified according to the focus of human limb support on: upper limb exoskeletons (including or without hands), lower limbs, and whole body exoskeletons. The difference between active and passive exoskeletons based on their application is that robotic/active devices replace human labor, while mechanical exoskeletons are based on knowledge of robotics and biomechatronics in order to expand human opportunities in various movement activities [6, 7].

Exoskeletons are located on the user's body and act as amplifiers that support, enhance or restore human performance. The opposite would be a mechanical prosthesis, such as a robotic arm or leg, to replace the original part of the body. They can be made of hard materials, such as metals or carbon fibers, or they can be made entirely of soft and elastic parts. Can be powered and equipped with sensors and controls or can be completely passive. May be movable or fixed/suspended (usually for rehabilitation or tele-surgery), may cover the whole body, only the upper or lower limbs or even a certain segment of the body, such as the ankle.

Fig. 1. Wearable upper body exoskeleton suitable for overhead work (left) and for lower body suitable for work in various working positions (right)

Placing tools on a high shelf or changing production parts may not be difficult, while performing one of these tasks thousands times a day or about 1 million times a year significantly increases the load on individual body segments. This rate significantly increases the likelihood of fatigue or body injury. A new wearable upper body exoskeleton helps reducing the likelihood of injury (Fig. 1 left).

The "Chair" exoskeleton (Fig. 1 on the right) allows employees to sit at short assembly intervals. Body weight is transferred to the floor using two adjustable elements. It allows work that is light for the joints and back, and improves posture. The intention is preventive protection against health damage to the back and knees. Health effects have not yet been recorded, neither positive nor negative, on the other hand, mostly positive feedback from employees is known. Therefore, it is necessary to carry out scientific studies to verify the reduction of workload when using "chairs" and other types of exoskeletons, as well as to define criteria for occupational health.

2.2 Basic Requirements for Exoskeletons in an Industrial Environment

The exoskeleton should be anthropomorphic and ergonomic, not only in shape but also in function. The interaction of the exoskeleton with the user involves three main modules: senses, decision and performance [8]. The exoskeleton must be able to interact with the human body, a complex kinematic structure that contains multiple degrees of freedom of movement. Exoskeletons must have a large number of active joints, each with a wide range of motion, to be able to monitor and assist in movements in a large workspace. When using exoskeletons, new potential health risks may arise due to the redistribution of the load to other areas of the body. The use of exoskeletons also affects motor control, joint stability and altered kinematics [9]. One reason may be, for example, the level of discomfort associated with wearing an exoskeleton [10]. Other concerns regarding passive devices are related to the potentially increased activity of the leg or arm muscles. The team of authors [11] address the need for improved management strategies that allow smooth movements at normal to fast rates. In creating such a management strategy, there are demands for human-machine information exchange, real-time motion planning and security control.

3 Risk Factors at Workplace Ergonomics

It is obvious that work environment factors can significantly affect a human behavior and health. This is also the case of ergonomic risk factors, which considerably contribute to human well-being and health. In any work system, the employee is thus exposed to a load, whether one-time or continuous, overload or, reasonable load. The relationship, causes, and consequences of a load are illustrated in a flow chart in Fig. 2.

According to the available literature dealing with ergonomic issues, the risk factors affecting the employee during work are as follows: extreme and unnatural positions of the joints, insufficient rest, frequency of repetition of the same movements, static load on the muscles at work, temperature, heat generally increases physical activity at work, cold reduces tactile ability, individual factors (pregnancy), diabetes, vitamin B6 deficiency, contraceptive use, endocrine disorders, etc., mechanical hazards (contact with edges and pressure on soft tissues), noise and vibration (damage nerves, blood vessels and the musculoskeletal system [12]), electromagnetic field [13], lighting, force load, other factors - task and shifts work, lack of freedom for the initiative of the worker, social aspects of work organization, etc.

Many of the above-mentioned factors lead by their action, either one or cumulatively, to unpleasant occupational diseases [14].

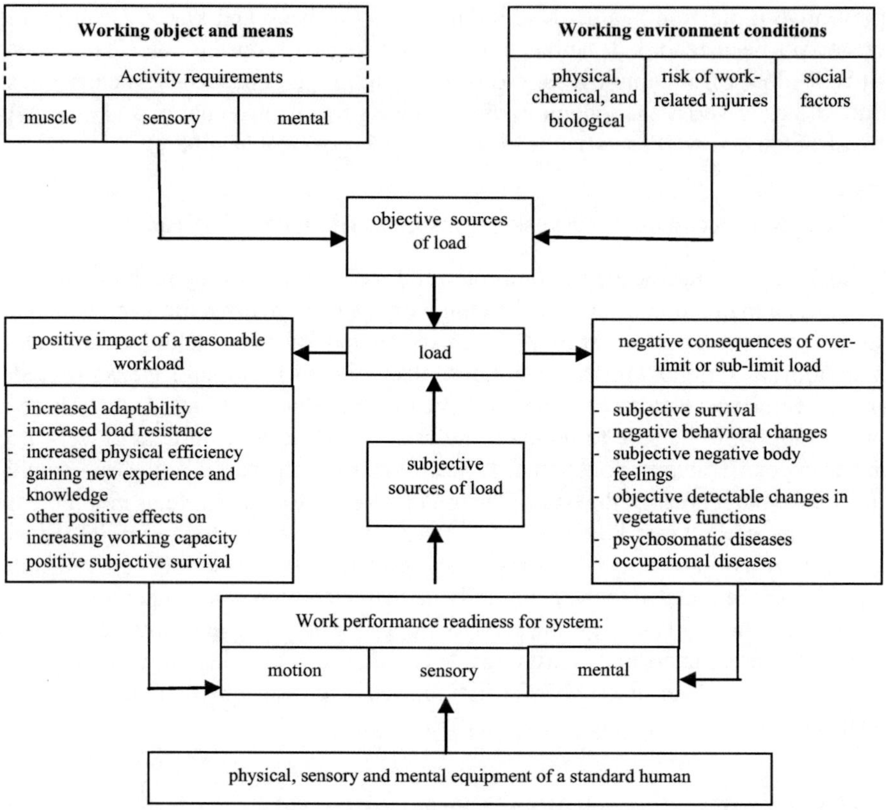

Fig. 2. Scheme of relationships, causes and consequences of workload in the work system

4 Application of the Ergonomic Evaluation Method in the Workplace

Physical activities in occupations such as manipulation, static and dynamic positions, unexpected movements and repetitive postures are among the most important risk factors for work-related musculoskeletal disorders (WMSD) [15]. There are many ergonomic methods for evaluating WMSDs that can be used in a variety of cases, but these evaluations and analyzes cannot be beneficial in themselves and should be supported by useful empirical suggestions for improving working conditions. The method for evaluation and analysis of the state of work of the upper body were the RULA method (quick evaluation of the upper limb) was applied in the working position in Fig. 3 (left). The RULA method is divided into three parts with eleven steps. Section 1 analyzes the positions of the hands and wrists by adding points related to the selected situation. Part 2 analyzes the posture of the neck, upper body and leg, but without the effects of the exoskeleton. In Sect. 3, the results from Sect. 1 and Sect. 2 are combined into a holistic result. Positive changes are observed due to the support from the armrest (part 1) and the assumption of the compensating force (part 2). The result is based on a table that uses the input parameters calculated in parts 1 and 2.

In Fig. 3, there is the output of the RULA method without the use of upper limb and back support, i.e. without an exoskeleton, in the Fig. 4 can be seen the output of the RULA method using the support of the upper limbs, i.e. the work activity performed by means of the exoskeleton.

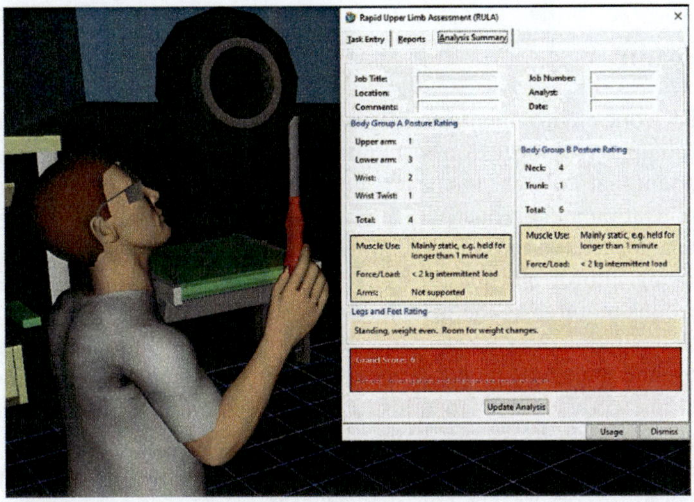

Fig. 3. RULA analysis of the working posture without exoskeleton

The ergonomic risk in the case of a job position without the use of an exoskeleton was assessed as unacceptable, there is a high probability of damage to the worker's upper limbs. When using the exoskeleton in the same working position, the risk was reduced to acceptable, there was a change in the support of the upper limbs and spine. The resulting assessment of ergonomic risks by simulation of the working position shows that exoskeletons can be an important personal protective equipment in order to minimize damage to the musculoskeletal system. Further research will be done to assess the risks of whether the use of an exoskeleton could lead to other risks in the work environment.

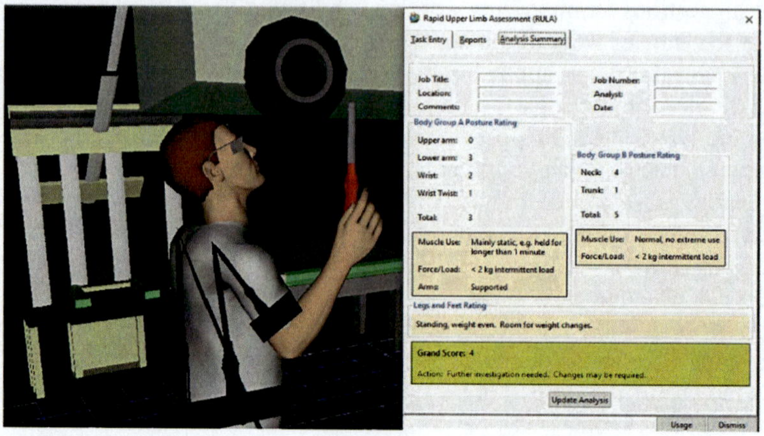

Fig. 4. RULA analysis of the working posture with exoskeleton applied

5 Conclusion

People in the work environment interact with various work equipment and technical equipment (computers, production machines, handling and transport equipment, etc.), so it is important that the work environment is adapted to eliminate the negative effects on health, motivation and productivity at work.

The use of exoskeletons has the potential to benefit workers by augmenting a person's strength, making him or her able to exert more force than normal, and reducing the strength exertions required of the person to perform a specific task.

The implementation of exoskeletons could be one of the solutions to increase human capacity and at the same time to reduce the large number of damage to the musculoskeletal system during work activities in industry. Safety is essential for both commercial and legal acceptance of industrial exoskeletons [16].

The obtained results showed that it should be safe and quite effective to apply industrial exoskeletons to workers who had to do physical work while performing work tasks. The applied procedures can significantly improve approaches to examining the functional state of the worker and the results obtained will significantly contribute to the development of the regulatory and technological basis for promising personal protective equipment [17] used to protect the musculoskeletal system within the existing system of safety standards [18]. Behavioral safety is a popular approach to increasing occupational safety [19].

References

1. Onofrejova, D., Panda, A., Agalarev, M.: Exoskeletons in workplace ergonomics. In: Studia i Materiały, vol. 40, no. 6, pp. 26–29, Warsaw University, Warsaw (2020)
2. National Centre of Health Information. Health Yearbook of Slovak Republic, Bratislava (2019). ISBN 978-80-89292-71-4
3. Onofrejova, D., Sebo, J.: Ergonomics in industry. Technical University of Kosice, Kosice (2020)

4. Turisova, R., Sinay, J., Pacaiova, H., Kotianova, Z., Glatz, J.: Application of the EFQM model to assess the readiness and sustainability of the implementation of I4.0 in Slovakian companies. Sustainability **12**(14), 1–14 (2020)
5. de Kok, J., et al.: Work-related musculoskeletal disorders: prevalence, costs and demographics in the EUs. European Risk Observatory Report. European Agency for Safety and Health at Work, Luxembourg (2019)
6. Ruiz-Olaya A.F., Lopez-Delis, A., da Rocha, A.F.: Chapter eight - upper and low extremity exoskeletons. In: Handbook of Biomechatronics, p. 283–317. Academic Press (2019)
7. Pons, J.L.: Wearable Robots: Biomechatronic Exokeletons. Willey, England (2008)
8. Knaepen, K., Beyl, P., Duerinck, S., Hagman, F., Lefeber, D., Meeusen, R.: Human-robot interaction: kinematics and muscle activity inside a powered compliant knee exoskeleton. IEEE Trans. Neuralal Syst. Rehabil. Eng. **22**(6), 1128–1137 (2014)
9. INRS. Exosquelettes au travail: impact sur la santé et la sécurité des opérateurs – état des connaissances. Institut National de Recherche et de Sécurité, Paris (2018)
10. Abdoli-E, M., Agnew, M.J., Stevenson, J.M.: An on-body personal lift augmentation device (PLAD) reduces EMG amplitude of erector spinae during lifting tasks. Clin. Biomech. **21**(5), 456–465 (2006)
11. Yang, C.J., Zhang, J.F., Chen, Y., Dong, Y.M., Zhang, Y.: A review of exoskeleton-type systems and their key technologies. Proc. Inst. Mech. Eng. Part C: J. Mech. Eng. Sci. **222**(8), 1599–1612 (2008)
12. Balazikova, M., Andrejiova, M., Wysoczanska, B.: Methodology for risk assessment of noise effects at workplace. MM Sci. J. (2019)
13. Pacaiova, H., Oravec, M., Smelko, M.: Extra low frequency magnetic fields of welding machines and personal safety. J. Electr. Eng. **69**(6), 493–496 (2018)
14. Sinay, J., Pacaiova, H., Oravec, M.: Present state of machinery safety assessment. In: Proceedings of the 16th International DAAAM Symposium, pp. 347–349 (2005)
15. Turisova, R., Sinay, J.: Ergonomics versus product attractiveness. Theor. Issues Ergon. Sci. **18**(1), 1–13 (2017)
16. Masood, J., Mateos, L.A., Ortiz, J., Toxiri, S., O'Sullivan, L., Caldwell, D.: Active safety functions for industrial lower body exoskeletons: concept and assessment. In: Gonzalez-Vargas, J., Ibáñez, J., Contreras-Vidal, J.L., van der Kooij, H., Pons, J.L. (eds.) Wearable Robotics: Challenges and Trends. BB, vol. 16, pp. 299–303. Springer, Cham (2017). https://doi.org/10.1007/978-3-319-46532-6_49
17. Jobbagy, B., Simsik, D., Karchnak, J., Onofrejova, D.: Robotic arm with artificial muscles in rehabilitation. Proc. Eng. Modell. Mech. Mechatron. Syst. MMaMS **96**, 195–202 (2014)
18. Erdelyi, S, Balazikova M., Hovanec, M.: Designing a work environment using an ergonomic application model of the process of lathing two-wheel assemblies. In. 14th International Multidisciplinary Scientific GeoConference SGEM 2014, Book 2, Albena, Bulgaria, vol.1, pp. 105–112 (2014)
19. Pacaiova, H., Nagyova, A., Markulik, S.: BBS (Behavior based safety) - way to raising the level of health and safety in practice. In: AHFE 2012 4th International Conference on Applied Human Factors and Ergonomics, San Francisco, pp. 2097–2104 (2012)

Magnetometry for Security Applications

Milan Oravec[1(✉)], Frantisek Hesko[2], Zoltan Szőke[2], Miroslav Smelko[2], and Tomas Gazda[1]

[1] Faculty of Mechanical Engineering, Technical University of Kosice, Letna 9, 042 00 Kosice, Slovakia
milan.oravec@tuke.sk, tomas.gazda@student.tuke.sk

[2] Faculty of Aeronautics, Technical University of Kosice, Rampová 7, 041 21 Kosice, Slovakia
{frantisek.hesko,zoltan.szoke,miroslav.smelko}@tuke.sk

Abstract. Magnetic field is an information medium for security applications as well as other physical fields. From the security point of view, the magnetic field is the second safest physical field just after the gravitation field, because it is hard to manipulate the magnetic field in wide area without notice and it is impossible to shield it completely. Magnetic field variations can be generated by several different sources - by the movement of the object in the Earth's magnetic field, by the current flow between the device and the power lines, by the changes in the residual magnetization caused by the temperature change of the object and others. The presented article describes examples of the possibilities when securing the objects by magnetic sensors, the experimental part of the article presents a set of basic experiments to demonstrate the possibility of using magnetometry for security applications.

Keywords: Security · Safety · Protection · Magnetometry

1 Introduction

Magnetic field is an information medium. Its fluctuations may be generated by a moving object in the Earth's magnetic field or by electrical currents running in appliances or to them through wiring. Magnetic fields can be measured around many different sources. The appearance of magnetic fields of devices depends on the quality of materials used, their permeability, geometry, or defects [1, 2]. With a sufficiently sensitive measurement device – magnetometer – it is possible to pick up changes in the states or movements of machines at production facilities and also physical values, like noise for example [3–6]. It is also possible to transmit image information with appropriate encryption [7].

The security of areas and objects is very important nowadays. Anyone can acquire different devices and aids to abuse them for illegal activities in the security related fields. Development of security system solutions for areas of interest or buildings is necessary [8–10], in some cases it leads to the ideas about dualism – in an operational mode as a monitoring system for operational safety and in a guard mode as a security (sub)system protecting against unauthorized intrusion. However, also in these cases it is important to pay attention to increase the reliability and innovation of these systems, for example by an efficient EFQM method [11].

R. S. Goonetilleke et al. (Eds.): AHFE 2021, LNNS 273, pp. 236–243, 2021.
https://doi.org/10.1007/978-3-030-80713-9_31

2 Spatial and Object Security Systems

Every security system consists of several basic types of protection: operational protection, physical protection, and technical protection. Together, these three types of protection form the integrated security system, the most important of which is the technical protection. The technical protection has priority in preemptive and detection functions; therefore, it is necessary to combine it with physical protection. The basis for active technical protection is reliable and suitably installed electrical/electronic security systems. In terms of spatial focus, technical protection can be divided into perimeter, envelope, spatial and object protection.

Perimeter protection elements protect and signal the disturbance of the external parts of large buildings or building complexes on a separate plot. Unlike sensors used inside a building, sensors used around the perimeter of a building usually have a greater range (tens to hundreds of meters). According to the principle used to detect the violation of the protected area, we can divide electrical circuit protection into several groups. These are passive infrared, microwave and dual detectors, infrared and microwave barriers. This also includes systems for detecting the crossing of a fence and movement in the perimeter. The envelope protection of buildings is used to control the disturbance or attempts to disturb the envelope of the protected building. Detectors for envelope protection mainly check construction openings in the envelope of buildings - windows, doors, ventilation and air conditioning shafts, etc. Spatial protection is a very good alternative or complement to envelope protection, based on active and passive detectors [12–14]. The elements of the spatial protection are intended to protect the interiors of the objects by indicating the movement or presence of an intruder during guarding. These are mainly key points such as stairs, halls, connecting corridors, special rooms, etc.

3 Technical Security by Magnetic Field

A specific area of application of magnetic sensors are security systems for perimeter protection and spatial protection. The simplest magnetism-based solutions use magnetic switches and magnetic proximity sensors, which are commercially available in various sizes, casings and designs. Another way of simple security is the so-called magnetic tag. Its task is to prevent unauthorized removal of the marked object from the protected area. The idea of using a vector magnetometer in the role of a security system is addressed in several development workplaces around the world.

The system with a sensitive vector magnetometer is also able to detect shocks and peaks in the magnetic field arising from the activity in the distribution network, turning on/off electrical appliances. Thanks to digital data processing, it is possible to set the system so that it is insensitive to the frequency of 50 Hz and higher, while changes in the low-frequency field are sensed as disturbance. The system works by defining a certain limit value of the magnetic field (threshold deviation from the reference value). The advantage of this solution is the possibility of placing the permanent magnet at a greater distance from the magnetometer than just a few centimeters like with magnetic switches. Another possible use of the vector magnetometer in security is protection against a car entering the protection zone. These principles served as inspiration in testing the application of the multichannel magnetometer VEMA-041 [15, 16]. The magnetometer is a vector magnetic analyzer with the frequency band from DC to 250 Hz (1000 Hz sampling). The sensitivity of the magnetometer is approximately 2 nT/LSB (the specific value depends on the probe used) in the range of measured fields $\pm 70{,}000\ \mu T$. Its four probes allow different geometric configurations, and their calibration is performed using neural networks to achieve optimal results [17, 18]. The latest experimental version of this magnetometer [19] improves its metrological properties.

When using a vector magnetometer in a security system, it is necessary to consider the existing magnetic field background of the secured area. Items of daily consumption and equipment complete the character of the magnetic image of the room. Fields generated by the electrical distribution network at frequencies of 50/60 Hz and their harmonic multiples, which arise due to the use of reactance loads and uneven loading of phase conductors, are also a significant signal of the magnetic background. Due to the influences of all local fields of objects in space, the measured magnetic field represents a dynamic image. Long-term measurements of the 24-h magnetic field evolution, performed in the area of interest prior to the system placement, help tune signal processing algorithms to minimize false alarms. The simplest method of detection is to process the deviations ΔB arising as the difference of the nth sample of the measured averaged anomalous field B_a and the background field B_b according to:

$$\Delta B = \bar{B}_{a(n)} - \bar{B}_{b(n)} = \frac{\bar{B}_{a(n-1)} \cdot k - \bar{B}_{a(n-k)} + \bar{B}_{a(n)}}{k} - \frac{\bar{B}_{b(n-1)} \cdot l - \bar{B}_{a(n-l)} + \bar{B}_{a(n)}}{l} \tag{1}$$

where k is the width of the averaging window for the anomalous field and l is the width of the averaging window for the background field (l should be greater than 30,000, i.e. an interval of 30 s for optimal results). As another source of information about what is happening in the area, it is possible to use the variance or standard deviation of signal in the basic processing, which corresponds to the effective value of changing signals in the area of interest. The intruder may use power tools to generate alternating fields, which can also be identified by a magnetometer. Different types of digital filters can be used for improved signal processing. To verify the basic assumptions of the functionality of the magnetometer in the detection of anomalous fields in the role of the security subsystem, several experiments were performed with a distribution of the magnetometer sensors into the corners of a square with sides of 2 m in the test room, as seen on Fig. 1. The tested sensors in one case were oriented axially in the same direction (parallel arrangement), and in the orientation to the center of the square (X arrangement). The basic experiments consisted of a passing car in the vicinity, opening a door marked with a permanent magnet, moving a chair, manipulating an object marked with a magnetic tag, turning on the lights in the room (Fig. 2, Fig. 3, Fig. 4 and Fig. 5).

The route of the car with a speed of approximately 15 km.h^{-1} led at a distance of 5 m from the edge of the square created by the layout of the magnetometer sensors. As it can be seen from the graphs, the system detected changes when objects were moved or changed inside and outside the room. The natural properties of the construction materials of the objects (chair, PC cabinet, etc.) or their marking with a permanent magnet (wooden door, rare object, etc.) were used.

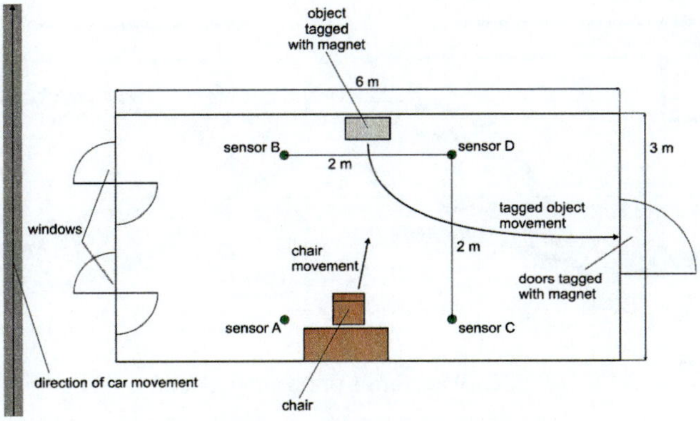

Fig. 1. Experimental room

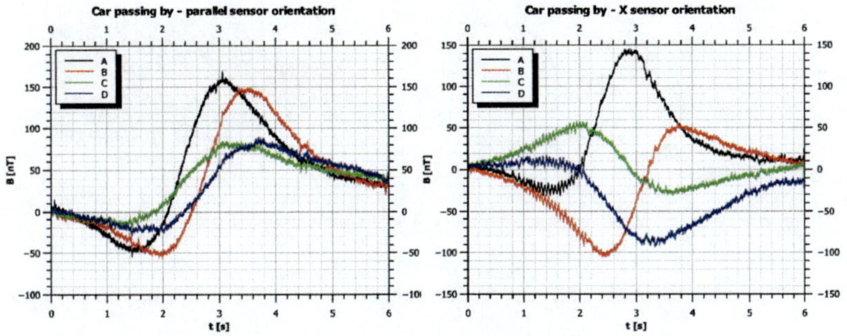

Fig. 2. Waveforms of a car passing by

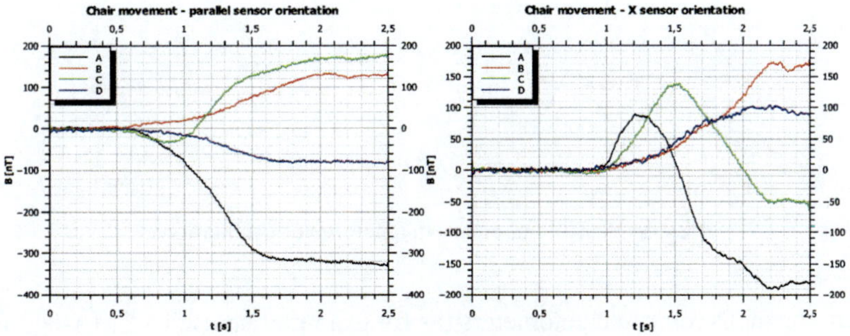

Fig. 3. Waveforms of a chair moving

Plastic windows can also be easily marked with a small neodymium magnet. Upon activation of the system, changes in signal scattering may indicate human activity in the area of interest or its protected environment. Switching electrical appliances on/off will cause the load on the electrical grid to change. This change is manifested by the change in the amplitude of the alternating magnetic fields.

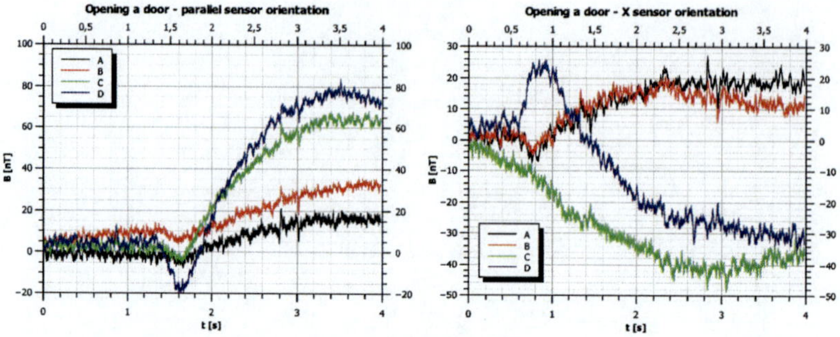

Fig. 4. Waveforms of opening a door

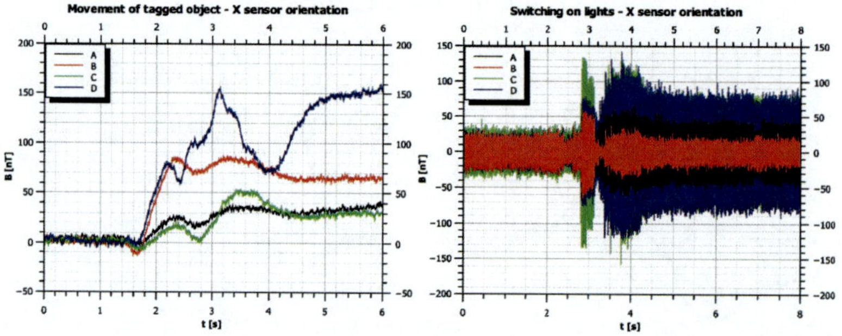

Fig. 5. Waveforms of the movement of a tagged object and switching on the lights

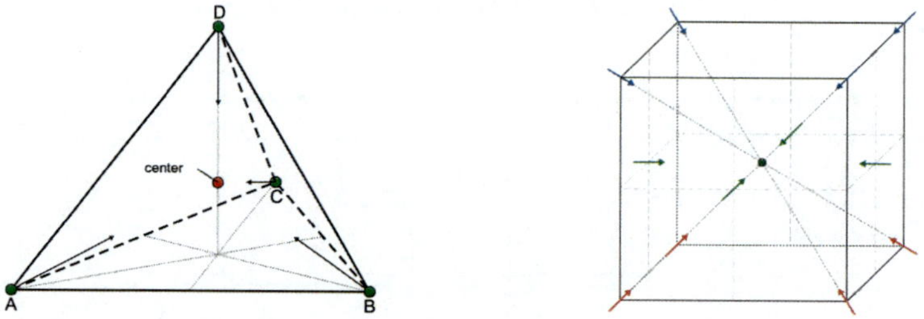

Fig. 6. Possible complex magnetometer orientations

Multi-channel vector magnetometers (as for example the used VEMA-041 magnetometer) can be effectively used as part of a security system. Experiments have shown that a magnetometer with equivalent parameters to the VEMA-041 should be able to secure a space of at least 6 m x 6 m x 3 m. The arrangement of the sensors in X, with respect to the directional properties of this arrangement, has proven to be more suitable when using four channels. For a more complex security system, it is possible to use several 4-channel magnetometers connected to one superior system (Fig. 6).

The magnetometer as a security device can also work outdoors, where it is able to create a protective perimeter using several measurement channels, which can be used to guard the entry of vehicles and low-flying objects into the guarded area. This concept is based on the detection of objects with electromagnetic fields in the frequency range of 3 Hz to 30 kHz. For example, small unmanned aerial vehicles (UAVs) can also be detected in this band. The concept of the exterior detection system with respect to the results considers the coverage of the band of 3 Hz to 300 Hz by a relaxation or microwire magnetometer [20, 21], 300 Hz to 30 kHz by means of an induction magnetometer [22]. By combining these two systems, it is possible to implement a detection subsystem for ground and air intruders based on the measurement of magnetic fields [23–25].

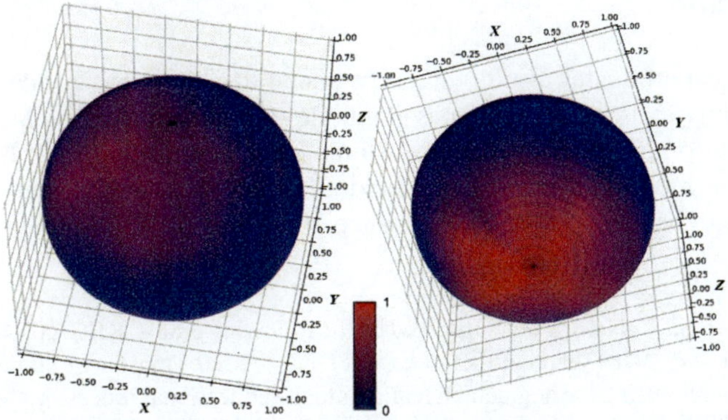

Fig. 7. Pseudo-color visualization on the surface of the unit sphere

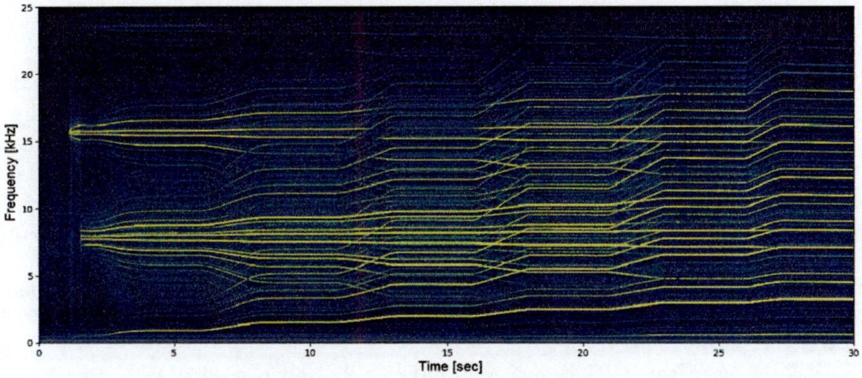

Fig. 8. Spectrum of BLDC motor in the speed range from 40% to 100%

Figure 7 shows the normalized spatial directional characteristics obtained from the measurements of a small quadcopter at a distance of about 1 m with an induction magnetometer on a special non-magnetic device (tangential orientation of the sensor). For its creation, the largest spectral line among the lines of the corresponding motors was being searched in the 8 kHz region. Based on the results, it can be stated that in the lower hemisphere (in the z-axis) the measured UAV produces twice the anomalous fields than in the upper hemisphere, the smallest fields were measured in the plane of the quadcopter structure.

Figure 8 shows the number of frequencies in the range of up to 25 kHz, which is generated by one electric power part (ESC + 80W BLDC motor) of the small multicopter [22]. Engine speed was increased in the range from 40% to 100% power with the 10% increments. By sensing the magnetic field with a suitable inductive magnetometer, it is therefore possible to detect various flying objects in the immediate vicinity of buildings even in a fully autonomous mode.

4 Conclusion

Based on the experimental results, it can be stated that the measurement of magnetic fields in buildings and/or around them is a suitable complementary system in the solution of tasks of interior and exterior security. All the basic manifestations of human activity tested could be monitored in magnetic field and they caused sufficient change in the sensed signals to be identified as a disturbance of the area of interest, thus triggering the alarm.

Acknowledgments. This work was supported by the following grants: KEGA 052TUKE-4-2018, APVV-17-0184, APVV-18-0248 and KEGA 013TUKE-4/2020 Transfer of knowledge from the research of resources using the augmented reality into the educational process in the field of safety of technical systems.

References

1. Tumanski, S.: Handbook of Magnetic Measurements. CRC Press (2011). ISBN 978-1-4398-2951-6
2. Pačaiová, H., Sinay, J., Turisová, R., Hajduová, Z., Markulik, Š: Measuring the qualitative factors on copper wire surface. Measurement **109**, 359–365 (2017). https://doi.org/10.1016/j.measurement.2017.06.002
3. Sinay, J., Pačaiová, H., Oravec, M.: Present state of machinery safety assessment. In: Annals of DAAAM for Proceedings of the International DAAAM Symposium (2005)
4. Turisová, R., Svetlík, J.: The structural design of the basic module for construction of modular machines. Appl. Mech. Mater. **613**, 434–440 (2014). https://doi.org/10.4028/WWW.SCIENTIFIC.NET/AMM.613.434
5. Balážiková, M., Sinay. J., Dulebová, M., Markulik, Š., Kotianová, Z.: Measurement of low-frequency noise during CNC machining and its assessment. Measurement **119**, 190–195 (2018). ISSN 0263–2241
6. Balážiková, M., Andrejiová, M., Wysoczańská, B.: Methodology for risk assessment of noise effect at workplace. MM Sci. J. 3426–3430 (2019). ISSN 1803-1269
7. Ovseník, Ľ, Turán, J., Huszaník, T., Oravec, J., Kováč, O., Oravec, M.: Image encryption algorithm with plaintext related chaining. Comput. Inform. **38**, 647–678 (2019). https://doi.org/10.31577/cai_2019_3_647
8. Pačaiová, H., Oravec, M., Šmelko, M., Lipovský, P., Forraj, F.: Extra low frequency magnetic fields of welding machines and personal safety. J. Electr. Eng. **69**, 493–496 (2018). https://doi.org/10.2478/jee-2018-0084

9. Oravec M., Kulikov, A., Jurc, D., Kulikova, O., Vahovsky, J., Abraham, M.: Analysis of the effect of non-ionizing radiation in the area of magnetism under the limit of 1 Hz generated by means of transport from the perspective of neurobehavioral effects. In: 2019 IEEE 17th World Symposium on Applied Machine Intelligence and Informatics (SAMI), pp. 89–94 (2019). https://doi.org/10.1109/SAMI.2019.8782759
10. Šmelko, M., et al.: Low frequency magnetic fields and safety. Acta Phys. Pol. A **137**, 693–696 (2020). https://doi.org/10.12693/APhysPolA.137.693
11. Turisová, R., Sinay, J., Pacaiova, H., Kotianová, Z., Glatz, J.: Application of the EFQM model to assess the readiness and sustainability of the implementation of I4.0 in Slovakian companies. Sustainability **12**, 5591 (2020). https://doi.org/10.3390/su12145591
12. Veľas, A., Kutaj, M., Ďurovec, M.: Influence of changing the parameters of the camera system on video-based motion detection. In: 2017 International Carnahan Conference on Security Technology (ICCST), pp. 1–5 (2017). https://doi.org/10.1109/CCST.2017.8167829
13. Velas, A., Svetlík, J.: Design of equipment to improve safety on roads. In: Proceedings of 20th International Scientific Conference. Transport Means, pp. 614–617 (2016)
14. Kutaj, M., Jasenčák, Ľ., Veľas, A., Boroš, M.: The design of a testing device for passive infrared motion detectors. Saf. Secur. Eng. **174**, 457–463 (2018). https://doi.org/10.2495/SAFE170411, ISBN978-1-78466-211-0, ISBN978-1-78466-212-7
15. Hudák, J., Blažek, J., Praslička, D., Mikita, I., Lipovský, P., Gonda, P.: Sensitivity of VEMA-04.1 magnetometer. J. Electr. Eng. **61**, 28–31 (2010). ISSN 1335-3632
16. Praslicka, D., Lipovský, P., Hudak, J., Šmelko, M.: Estimation of multichannel magnetometer noise floor in ordinary laboratory conditions. Acta Phys. Pol. A **131**, 1123–1125 (2017). https://doi.org/10.12693/APhysPolA.131.1123
17. Kliment, T.; Lipovský, P.; Draganová, K., Moucha, V.: Comparison of speed and accuracy of chosen magnetometer calibration algorithms in the presence of external interference. In: New Trends in Signal Processing (NTSP), pp. 85–89 (2018). ISBN 978-80-8040-546-5. ISSN 1339-1445
18. Kliment, T.; Praslicka, D.; Lipovský, P.; Draganová, K., Zavodsky, O.: Calibration of magnetometer for small satellites using neural network. Acta Physica Polonica A **131**, 1129–1131 (2017). ISSN 0587-4246
19. Lipovský, P., et al.: Relax-type magnetometer with direct optocoupler relaxation. Acta Phys. Pol. A **137**, 681–683 (2020). https://doi.org/10.12693/APhysPolA.137.681
20. Hudak, J., et al.: Fluxgate sensors based on magnetic microwires for weak magnetic fields measurement. J. Electr. Eng. **66**, 153–156 (2015). ISSN 1335-3632
21. Lipovský, P., Čverha, A., Hudak, J., Blazek, J., Praslicka, D.: Noise characteristics of microwire magnetometer. Acta Phys. Pol. A **126**, 384–385 (2014)
22. Blažek, J., Lipovský, P., Heško, F., Repčík, D.: Electromagnetic image of small UAV in very low frequency range. J. Electr. Eng. **69**, 438–441 (2018). https://doi.org/10.2478/jee-2018-0069.ISSN1335-3632
23. Stodola, P., Drozd, J., Šilinger, K., Hodický, J., Procházka, D.: Collective perception using UAVS: autonomous aerial reconnaissance in a complex urban environment. Sensors **20**, 2926 (2020)
24. Mazal, J., Stodola, P., Procházka, D., Kutěj, L., Ščurek, R., Procházka, J.: Modelling of the UAV safety manoeuvre for the air insertion operations. In: Hodicky, J. (ed.) MESAS 2016. LNCS, vol. 9991, pp. 337–346. Springer, Cham (2016). https://doi.org/10.1007/978-3-319-47605-6_27
25. Stodola, P.: Optimization methods for position of unattended ground monitoring systems. In: Proceedings of the 20th International Scientific Conference. Transport Means (2006)

Safety and Productivity Enhancement Through Ergonomics Development (SPEED) at the Embassy in the Philippines

Alma Maria Jennifer Gutierrez(✉) and Rosemary Seva

Industrial & Systems Engineering Department, De La Salle University, 2401 Taft Avenue, Malate, Manila, Philippines
alma.gutierrez@dlsu.edu.ph

Abstract. The ergonomics audit conducted at the Embassy in the Philippines was done in order to identify ergonomic hazards in the office and propose practical solutions to these problems. The result of the audit is used as input to educate employees about ergonomics principles and how these can be used to prevent musculoskeletal disorders. A total of 27 people mostly female (89%) responded to the symptom survey. Fifty-nine percent (59%) of the respondents complained of discomfort. The most complained body parts are the low back, hand/wrist, shoulder and neck. The values obtained from the audit showed that illuminances in some areas exceed recommended values. In order to address the problems identified, recommendations were given for implementation by employees and management. Ergonomics training was conducted for the purpose of educating employees how to prevent common musculoskeletal disorders experienced in the workplace.

Keywords: Safety · Productivity · Ergonomics

1 Introduction

The office is a worker's second home. A great percentage of a person's waking time is spent in this environment so it is crucial to keep it safe, convenient, and stimulating. The modern workplace changes every day as a result of the introduction of new technologies, equipment, and furnishings. However, employees may not be knowledgeable of how the office environment affects their personal well-being.

Many workers use furniture available in the office without a conscious effort to fit them to their needs. As such, they may experience discomforts that they sustain for a significant period of time.

In order to address these occupational ergonomics issues in the workplace, this study wants to achieve the following specific objectives:

1. Document the occurrence of musculoskeletal problems in the workplace and the perceived causes of these problems.
2. Identify ergonomic hazards in the workplace by conducting an audit.
3. Provide practical recommendations for preventing ergonomic hazards.

R. S. Goonetilleke et al. (Eds.): AHFE 2021, LNNS 273, pp. 244–251, 2021.
https://doi.org/10.1007/978-3-030-80713-9_32

2 Audit Methodology

2.1 Ergonomic Symptom Survey

Many tasks in the office have the potential for ergonomics hazard. The purpose of the symptom survey is to:

- Identify areas of discomfort in the body of employees
- Determine the type and extent of discomforts experienced by employees.
- Recognize employees' perception of the causes of these discomforts
- Determine actions taken by employees to remedy these discomforts

The results of the survey were summarized and analyzed to identify prevalent problems in the workplace. Such information was used to unearth possible causes related to workstation design.

2.2 Illumination Measurement

A sampling of offices and work areas was done to measure illumination. Illumination measurements were taken in order to know if sufficient lighting is provided to all workstations. Working in dim light can cause eye strain because the person struggles to maintain a clear vision despite of the condition. On the other hand, the presence of glare whether direct or indirect can cause headaches, eye strain, and fatigue that can severely affect the productivity and morale of workers.

Illumination was measured using a lux meter. The lux meter measures the amount of light falling on a surface and the value obtained is compared to established standards found in literature.

During the audit, the lux meter is placed on top of the area of interest (i.e., writing and reading area) so as to measure the illumination at that point.

2.3 Postural Analysis

Workers in offices usually work more than 5 h a day in a sitting position. The long working hours can result in many WMSD's such as back pain, wrist pain, and neck pain. In order to diagnose possible causes of these disorders a postural analysis was conducted to determine deviations from neutral postures.

Several pictures were taken while the employees are working. Areas of interest in the pictures are the following: wrists, upper and lower arm, upper and lower back, legs and neck.

2.4 Workstation Analysis

Workstations currently used by workers were also analyzed to determine if they are fitted to the workers. The Embassy employs mixed races who have significant anthropometric differences especially with respect to height. It is not known if the workers have been trained to adjust their workstation and if the range of adjustment conforms to their zone of comfort.

Workstation analysis was done by taking measurements of the chairs and tables that were found in the sampled areas and compare them with Filipino anthropometric data available in literature. Measurements were taken using a measuring tape.

3 Audit Results

3.1 Symptom Survey

A total of 27 people mostly female (89%) responded to the symptom survey. Fifty-nine percent (59%) of the respondents complained of discomfort.

The most complained body parts are the low back, hand/wrist, shoulder and neck as can be seen from Fig. 1. These are the body parts that are closely associated to poor and static posture while typing for long periods of time.

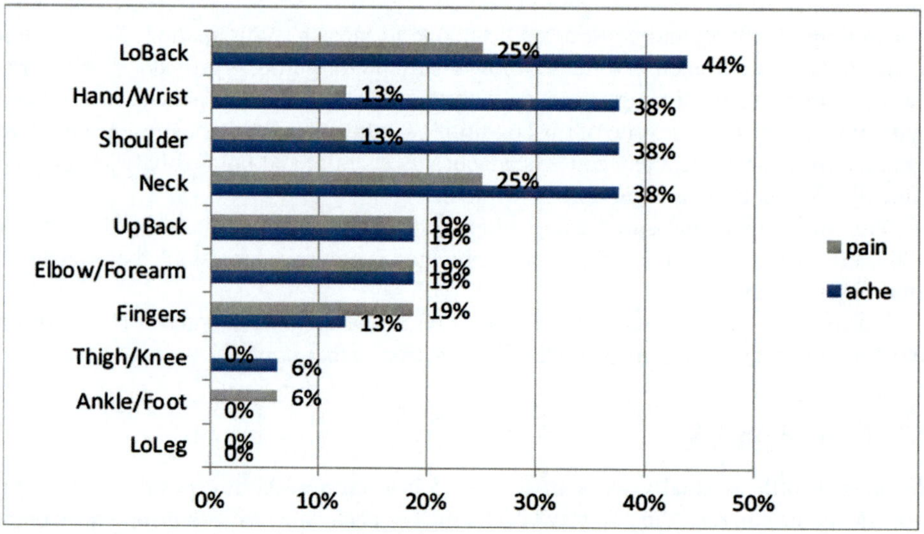

Fig. 1. Summary of complained body parts

The body parts complained usually ache (44%) and experience pain (25%) which last for a day and sometimes for a week. One person had experienced hand and wrist pain for more than 5 years. She had been with the embassy for seven years and might have been acquired from computer work. The most cited problem is low back pain (44%), hand and wrist pain, shoulder pain, and neck pain (38%) as can be seen from Fig. 1.

Those who complained of discomfort attributed their problems to poor continuous typing (31%), prolonged sitting (25%), poor posture, and prolonged use of mouse (19%). There are two employees, however, that have existing health condition that are not related to their work.

Thirty-one percent (31%) of those who complained went to company or personal doctor to seek treatment. Treatment provided by the doctors generally helped the employees relieve their discomfort. Those who did not seek treatment thought that there is no need to go to the doctor to relieve their pain so they self-medicated or rested to alleviate their discomfort.

3.2 Illumination Analysis

Several offices and open areas were sampled, and the following illumination values were obtained as shown in Table 1. These values were compared to recommended illuminance values proposed by [1].

Table 1. Recorded illuminances

Workstation condition	Range of illuminance (lux)
Without windows	180–374
Windows with shades	175–423
Windows w/shades (rolled up)	265–596

The values obtained from the audit showed that illuminances in some areas exceed recommended values. Some areas are well illuminated especially if the window shades are rolled up. There are areas where window shades are not used because employees want to enjoy the view outside the office. However, not all people in the same area welcome excessive illumination.

Glare

Close observation of the workstations revealed several sources of glare. Glare is an interference of visual perception due to a very bright light source. Glare can cause discomfort or be disabling. Discomfort glare still enable people to see information needed to accomplish their daily activities but interferes with perception. Disabling glare decreases contrast and reduces visibility.

Windows are sources of high intensity light on a bright day. There are a number of areas that have windows and some employees observed do not use their shades to prevent glare. Those interviewed are aware that they are supposed to use the shades in the morning, but they do not want to do so because they enjoy the view. Using the shades is a personal decision but employees should be aware that excessive light can cause headache and can affect their sense of perception.

3.3 Postural Analysis

It was observed that more than half of the workers assume awkward postures while working. Some examples of awkward postures observed are shown for illustrative purposes:

1. Extended arm

Most employees interviewed use computers for more than 5 h per day although they take constant breaks. Based on actual observation some employees use their keyboards with an awkward posture such as the one shown in Fig. 2 below. The location of the mouse is too far from their body causing them to extend the dominant arm while working. If this position is sustained for a long time, shoulder pain will be experienced. The shoulder exerts force to maintain this position and if this is repeated several times within the day, it can cause stress on the shoulder.

Fig. 2. Extended arm

2. Deviated posture of lower arm

The picture below shows that the middle of the lower arm rests on the table. This is an indication that the keyboard is placed a little bit high making the arms unable to rest on the arm rest provided by the chair and maintain a 90-degree angle between the upper and lower arms. This position may cause the wrist to also have an awkward posture.

Fig. 3. Deviated posture of lower arm

Moreover, it can be seen that the lower arm comes in contact with the edge of the table that can compress blood flow (Fig. 3).

3. Leg obstructions

One worker was observed to be in an awkward posture due to the small cabinet that is blocking his thigh. The cabinet prevents him from sitting comfortably while typing because his legs bump the cabinet (Fig. 4).

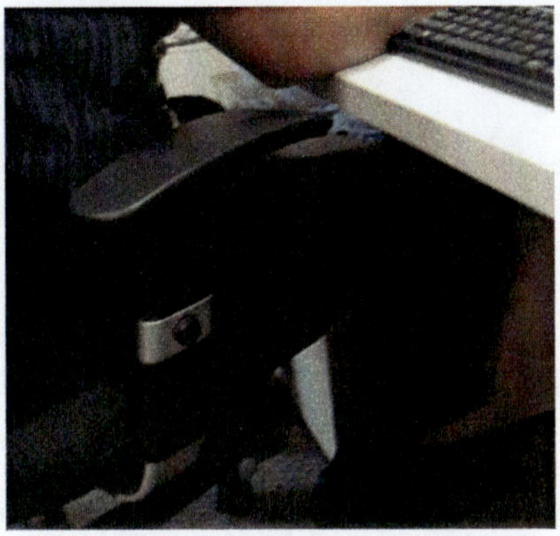

Fig. 4. Leg obstruction

3.4 Workstation Analysis

Ergonomic Chair

The backrest width of the chair is adequate to accommodate big employees in the embassy. It is also made of mesh fabric which ventilates the back.

Although this chair has a headrest, not many people interviewed use them because the height is not adjustable. As such, only employees whose head can rest comfortably on the headrest can use them. Those that are too tall or too short turn it around because it will only hurt the back of their heads.

The height is too high to be reached by the smallest employees of the embassy even at its lowest setting. This would force small users to dangle their feet or use the chair casters to rest their feet.

Some short employees do not lower the height of their tables. In order to reach the table, they adjust their chair to maximum height so they can reach the keyboard and type comfortably. The problem, however, is dangling of the feet and accumulation of pressure under the thigh. After keeping this position for several hours, employees may experience cramps in the legs.

The arm rest height is also too high for short employees. As a result, the arms maintain an awkward position when they use the armrest. The tendency is to raise elbow together with the shoulders. Such position can cause shoulder pain.

Although the chairs are adjustable in many dimensions, only a few employees know how to make proper adjustments or bother to do so. The employees have also not been trained on proper posture while working so they are not able to make appropriate adjustments to their workstation.

Table
Tables located in offices have heights that are adjustable. This is a very good feature because it can allow tall users to make adjustments to suit their height. Adjustment of the height is also very easy because each table has a crank that is easy to manipulate. However, although table heights maybe adjusted, they have not made any adjustments due to the following reasons:

- They do not feel the need to adjust because they adjust their chairs
- They do not know that the tables can be adjusted
- They are not aware of the effect of table height on good working posture

4 Recommendations

The results of the ergonomics audit illustrated the major problems experienced by employees in their workplace and the causes of these problems. A pro-active approach of problem solving through proper education was proposed. If employees are adequately trained about the proper ways of solving usual ergonomics problems in the office, management is unloaded the burden of constantly checking an employee's workplace for ergonomic hazards. Employees can also benchmark on good practices of other people that generate immediate improvement.

Recommendations enumerated identify points of action, reasons, and responsibilities. They address the causes of problems identified in the audit.

4.1 Ergonomic Training

It is believed that the best way to prevent ergonomic hazards in the office is to train employees how to recognize these hazards and their effects to overall well-being. Ergonomics training was conducted for the purpose of educating employees how to prevent common musculoskeletal disorders experienced in the workplace.

4.2 Illumination

Employees should prevent glare by doing the following:

- Use window shades in the morning to control excessive illumination in their workstation.
- Ensure that their work area is beside a light source. If an existing light source is too bright and within their field of view, they should rearrange their workstation or ask the building property section to decrease the number of lamps.

4.3 Workstation

The chairs that are available in the Embassy should be changed in the future. The back-rests are not adjustable vertically so the employees cannot set it to the appropriate location that is comfortable. The embassy employs anthropometrically diverse populations, so it is important to purchase furniture that has acceptable range of adjustment [2].

4.4 Job Characteristics

The Embassy should impose a mandatory rest break and exercise after two hours of continuous office work. Area supervisors should make sure that they follow the schedules. If this presents a problem in terms of monitoring, the Embassy should schedule rest and exercise through an automatic computer program that mandates them to stop working for a few minutes.

References

1. Kroemer, K., Kroemer, H., KroemerElbert, K.: Ergonomics: How to Design for Ease and Efficiency. Prentice-Hall, Upper Saddle River (2002)
2. del Prado-Lu, J.L.: Anthropometric measurement of Filipino manufacturing workers. Int. J. Ind. Ergonomics **37**, 497–501 (2007)

Wearable Sensing in Physical Ergonomics and Safety

Functional Data Representation of Inertial Sensor-Based Torso-Thigh, Knee, and Ankle Movements During Lifting

Sol Lim[1(✉)] and Clive D'Souza[2]

[1] Department of Systems and Industrial Engineering,
The University of Arizona, 1127 E. James E. Rogers Way, Tucson, AZ 85721, USA
lims@arizona.edu

[2] Department of Industrial and Operations Engineering, University of Michigan,
Ann Arbor, MI, USA
crdsouza@umich.edu

Abstract. This study examined the goodness-of-fit of using a sigmoid function to characterize time-series angular displacement trajectories during two-handed anterior lifting. Twenty-six participants performed two-handed anterior lifting with a low (4.5 kg) vs. high (22.7 kg) load at floor vs. knee lifting height. A sigmoid function with three parameters was fit to the torso-thigh included angle, knee flexion-extension (F-E), and ankle F-E angles in the sagittal plane obtained from body-worn inertial sensors. Mean $\pm$ SD RMSE between measured vs. fitted trajectories were $3.6 \pm 2.9°$, $3.9 \pm 4.2°$, and $2.7 \pm 2.8°$ for the torso-thigh included angle, knee F-E, and ankle F-E angles, respectively. Findings suggest that the sigmoid function adequately describes the trajectory shape of two-handed lifting kinematics. Functional representations facilitate data aggregation and feature extraction in large time-series datasets encountered in inertial-based motion analysis and machine learning applications.

Keywords: Lifting kinematics · Curve-fitting · Functional data · Wearable sensing

1 Introduction

Occupational lifting poses a significant risk to low back disorders [1]. Wearable inertial sensing provides new opportunities for continuous, real-time assessment of physical risk factors associated with work-related lifting. This is a critical step in the design of targeted ergonomic interventions for injury prevention.

Recent studies also demonstrate the potential for combining inertial sensor-based data with statistical prediction (i.e., machine learning; ML) to infer relevant lifting parameters, e.g., load level, lifting height [2]. The predictive performance of such ML algorithms relies on extracting representative kinematic features from continuous time-series sensor data as predictor variables for characterizing lifting conditions. Commonly

R. S. Goonetilleke et al. (Eds.): AHFE 2021, LNNS 273, pp. 255–260, 2021.
https://doi.org/10.1007/978-3-030-80713-9_33

used features or predictors include point estimates (e.g., mean, range of motion, minimum, maximum, trajectory kurtosis) of angular displacement, velocity, and acceleration at the torso, hip, knee, ankle and measures of inter-joint/segment coordination to distinguish different lifting task parameters [3, 4]. However, point estimates fail to adequately capture the shape features of the overall time-series trajectory. Functional data representations are useful for this purpose. The objective is to represent the time-series data with a function containing a fixed and usually small number of parameters in vector space. Another benefit of functional representation is data reduction or aggregation whereby a simple function can capture the essence of a large set of time-series measurement data. Curve-fitting methods have also been used in characterizing hip and knee postural angles during lifting tasks, including hyperbolic tangent function [5].

This study examined the use of sigmoid functions to represent the overall shape of kinematic time-series data during two-handed anterior lifting. Specifically, we analyzed the goodness of fit of a sigmoid function to characterize the torso-thigh included angle, knee flexion-extension (F-E) angle, and ankle F-E angles in the sagittal plane during two-handed lifting. The effects of load level and lifting height on the goodness of fit were statistically assessed.

2 Methods

An experiment was conducted wherein 15 men and 11 women performed 2 repetitions of two-handed sagittal lifting at two load levels, i.e., low (4.5 kg) vs. high (22.7 kg) and from two workstation heights (floor vs. knee). Three inertial sensors (80 Hz; Biostamp RC, mc10 Inc.) attached on the upper back (T6), right thigh, and right shank (Fig. 1) were used for computing continuous angular displacement between the torso-thigh, at the knee joint (between thigh and shank), and the ankle joint (shank relative to vertical), respectively. The sensors were attached by aligning one of the sensor-axis (i.e., x-axis) with the proximal-distal axis of the body segment. The angles were referenced to an upright vertical posture using a set of reference posture angles obtained by averaging data from 30s of quiet standing.

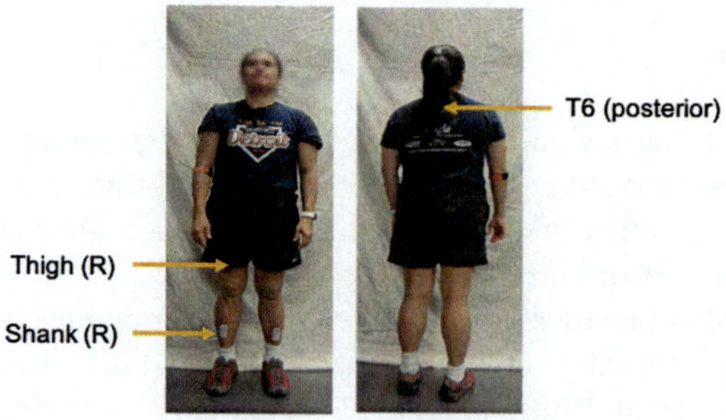

Fig. 1. Images depicting the three anatomical locations for inertial sensors attachment, namely, at T6, thigh (right), and shank (right).

3 Functional Data Representation

First, the continuous angular displacement data were segmented to the start-end of each lifting trial using a threshold-based criteria for T6 angular displacement and velocity. Next, the data were time-normalized (to 100%) to account for between-subject variability in lifting duration. The time-normalized trajectories were aligned by translating the data in magnitude such that the end of each trajectory data was set to zero, which corresponds to an upright standing posture. For each measured trial, the aligned angle data for the torso-thigh, knee, and ankle were fit to a sigmoid function in MATLAB R2019a. The sigmoid function takes the general form:

$$y(x) = \frac{a}{1 + e^{-b\times(x-c)}} \tag{1}$$

with three parameters that can be interpreted as follows,

a: the range of angular displacement from lift initiation to upright standing,
b: stiffness, which reflects the rate of rising/extension, and
c: the percent time corresponding to the midpoint of angular displacement.

Goodness-of-fit was evaluated by computing Root-Mean-Square Error (RMSE) values between the measured vs. fitted curves for each lift trial. Three 2 x 2 repeated measures ANOVA were performed on the RMSE values obtained for each of the three postural angles to examine differences by load level and lifting height. Significant main and interaction effects ($p < 0.05$) were examined using Bonferroni post-hoc tests. Statistical analyses were performed in R v.3.3.1 [6].

4 Results

A sigmoid function was fit to the torso-thigh, knee, and ankle angles for 208 lifting trials (= 26 participants × 2 repetitions × 2 height condition × 2 load condition). Table 1 summarizes the mean ± standard deviation (SD) values for the fitted function parameters, *a*, *b*, and *c* stratified by workstation height and load level condition.

Figure 2 shows the mean (±SD) measured vs. fitted angular displacement trajectories across all participant trials while lifting loads from the floor vs. knee height levels. The combined RMSE across all trials had a 90th percentile value of 10.4°. Table 2 summarizes the mean ± SD RMSE between measured and fitted curves across trials stratified by lifting height and load level. The mean RMSEs between measured vs. fitted curves generally were low between 2–3° across the different posture angles. RMSEs for the knee and ankle F-E angles were influenced by workstation height. The RMSEs for knee F-E angle were significantly greater at the floor vs. knee height condition. Conversely, RMSEs for the ankle F-E angle were significantly greater at the knee vs. floor height condition.

Table 1. Mean ± SD values for the fitted function parameters *a*, *b*, and *c* from Eq. (1) stratified by different workstation heights and hand load levels (n = 208).

Parameters	Floor height		Knee height	
	Low load (n = 52)	High load (n = 52)	Low load (n = 52)	High load (n = 52)
1. Torso-Thigh				
a: range	168.4 ± 18.5	169.5 ± 23.3	88.5 ± 17.4	99.0 ± 18.4
b: stiffness	−0.079 ± 0.013	−0.085 ± 0.024	−0.077 ± 0.011	−0.083 ± 0.02
c: midpoint	51.4 ± 10.0	52.8 ± 13.2	49.2 ± 8.0	51.6 ± 12.4
2. Knee				
a: range	112.0 ± 30.0	123.5 ± 34.2	33.4 ± 33.0	47.4 ± 31.5
b: stiffness	−0.081 ± 0.017	−0.085 ± 0.047	0.144 ± 0.637	−0.077 ± 0.21
c: midpoint	47.4 ± 13.5	49.1 ± 13.9	63.3 ± 22.0	50.3 ± 19.8
3. Ankle				
a: range	31.3 ± 9.0	39.1 ± 13.5	8.3 ± 9.8	13.1 ± 11.2
b: stiffness	−0.102 ± 0.024	−0.102 ± 0.038	0.347 ± 0.923	−0.05 ± 0.241
c: midpoint	46.0 ± 16.7	46.4 ± 16.9	69.5 ± 29.0	51.4 ± 28.5

Table 2. Mean (±SD) RMSE (in degrees) between the measured vs. fitted angular displacements across all lifting trials stratified by different workstation heights and hand load levels (n = 208). Letter notations in the superscript indicate significant pairwise comparison by row at a Bonferroni-adjusted $p < 0.05$.

Angle	Floor height		Knee height	
	Low load (n = 52)	High load (n = 52)	Low load (n = 52)	High load (n = 52)
Torso-Thigh	4.9 ± 4.6	4.8 ± 4.4	1.8 ± 0.8	2.6 ± 1.6
Knee	5.0 ± 5.1^{a}	4.6 ± 5.1^{a}	2.5 ± 2.7^{b}	3.3 ± 3.9^{b}
Ankle	1.7 ± 1.5^{a}	2.3 ± 3.0^{a}	4.1 ± 3.9^{b}	2.5 ± 3.0^{b}

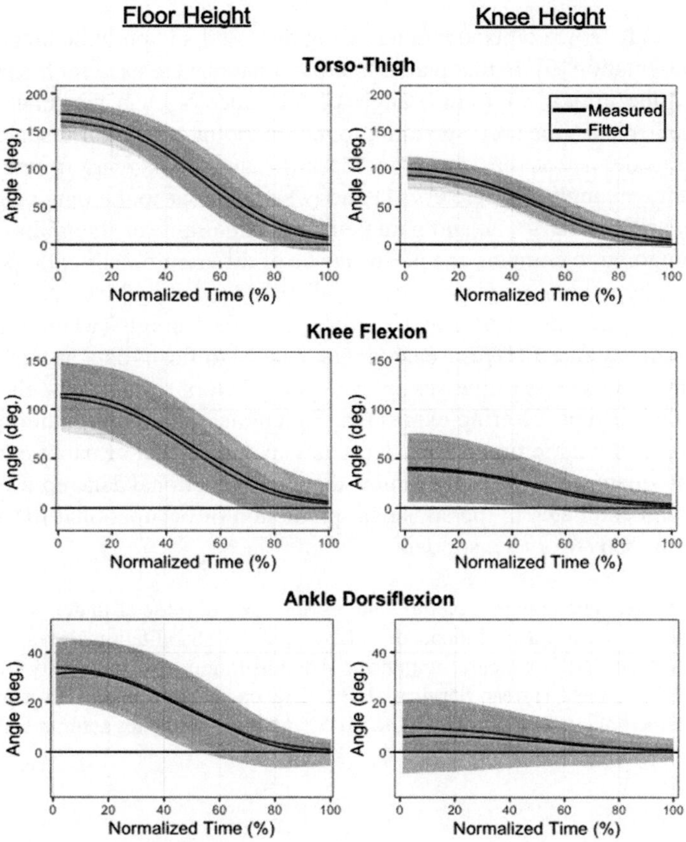

Fig. 2. Mean (±SD) measured (black lines) vs. fitted (blue lines) angular displacement trajectories across all participant trials while lifting loads from the floor (left-panel, n = 104) vs. knee (right-panel, n = 104) height levels. Trajectories are normalized to each detected lifting duration.

5 Discussion and Conclusions

Findings from this study suggest that the sigmoid function yields an adequate description of the characteristic s-shaped angular displacement data for the torso-thigh included angle, knee F-E, and ankle F-E angles during two-handed anterior lifting. The proposed method is applicable when the time-series data have similar shape and are aligned to the terminal posture. The three parameters of the sigmoid representation provides sufficient shape control to capture differences in lifting kinematics at two load levels, i.e., low (4.5 kg) vs. high (22.7 kg), and at two workstation heights (floor vs. knee). This was reflected in relatively low RMSE values across task conditions.

A prior study by Boston et al. [5] demonstrated use of the hyperbolic tangent function for representing knee F-E angles during isodynamic lifting trials. The sigmoid function used in our study is a more generic form of the hyperbolic tangent. Although not presented in detail, the average R^2 of 0.992 for the knee F-E angle (measured vs. fitted curves) across all lifting trials in our study was similar to the $R^2 = 0.991$ for the knee F-E in the previous study by [5] on isodynamic lifting that used a hyperbolic tangent function for data representation [5]. In that paper, four parameters were used for the curve fitting, including (1) the initial (minimum) angle of the trajectory, (2) the final (maximum) angle, (3) the percent time corresponding to the midpoint of angular displacement, and (4) the rise time defined as the time required for the angle to increase from 12 to 88% of the total change in angle. Parameters (1) and (2) are similar to the range parameter 'a' in our study, i.e., $a = (2) - (1)$, and parameter (3) is the same as the midpoint time (c). Analysis is ongoing to compare the performance of different functional representations (e.g., sigmoid, hyperbolic tangent, polynomial) suited to lifting kinematics.

On average, each lifting trial had 113.5 ± 37.2 data points (when collected with sampling frequency of 80 Hz) for each angle data. Significant data and feature reduction is required to use such time-series data as predictors for ML models to estimate the occupational lifting-lowering exposures. Functional data representations provide a method for reducing large time-series datasets to a reduced set of parameters in vector space. In subsequent work, we will examine the use of functional data representations vs. traditional point estimates in the statistical prediction of occupational lifting-lowering exposures from body-worn sensor data.

Acknowledgments. The contents of this manuscript were developed under a grant from the National Institute on Disability, Independent Living, and Rehabilitation Research (NIDILRR 90IF0094-01). NIDILRR is a Center within the Administration for Community Living (ACL), Department of Health and Human Services (HHS). The contents of this manuscript do not necessarily reflect the policy of NIDILRR, ACL, HHS, and you should not assume endorsement by the Federal Government.

References

1. Marras, W.S., et al.: Biomechanical risk factors for occupationally related low back disorders. Ergonomics **38**(2), 377–410 (1995). https://doi.org/10.1080/00140139508925111
2. Lu, M.-L., Barim, M.S., Feng, S., Hughes, G., Hayden, M., Werren, D.: Development of a wearable IMU system for automatically assessing lifting risk factors. In: Duffy, V.G. (ed.) HCII 2020. LNCS, vol. 12198, pp. 194–213. Springer, Cham (2020). https://doi.org/10.1007/978-3-030-49904-4_15
3. Yehoyakim, M., Bellefeuille, S., Côté, J.N., Plamondon, A.: Relationship between leg and back strength with inter-joint coordination of females during lifting. Int. J. Ind. Ergon. **56**, 32–40 (2016). https://doi.org/10.1016/j.ergon.2016.08.013
4. Burgess-Limerick, R., Abernethy, B.: Toward a quantitative definition of manual lifting postures. Hum. Factors **39**(1), 141–148 (1997). https://doi.org/10.1518/001872097778940632
5. Boston, J.R., Rudy, T.E., Mercer, S.R., Kubinski, J.A.: A measure of body movement coordination during repetitive dynamic lifting. IEEE Trans. Rehabil. Eng. **1**(3), 137–144 (1993). https://doi.org/10.1109/86.279263
6. Ihaka, R., Gentleman, R.: R: A Language for Data Analysis and Graphics. J. Comput. Graph. Stat. **5**(3), 299–314 (1996). https://doi.org/10.1080/10618600.1996.10474713

BIONIC: Custom Sensors for Risk Assessment and Training of Older Workers

Alberto Ferreras Remesal(✉), Juan Fernando Giménez Pla, Purificación Castelló Mercé, Salvador Pitarch Corresa, Raquel Marzo Roselló, and Mercedes Sanchís Almenara

Instituto de Biomecánica de Valencia, Universidad Politécnica de Valencia, Edificio 9C. Camino de Vera s/n, 46022 Valencia, Spain
{alberto.ferreras,juanfernando.gimenez,puri.castello, salvador.pitarch,raquel.marzo,mercedes.sanchis}@ibv.org

Abstract. The ageing of working population is becoming a significant challenge that affects both the health and safety of workers and the productivity and competitiveness of companies.

The BIONIC project aims to develop a body sensor system to monitor the worker, in order to accurately obtain the level of risk associated with each of the tasks performed. The system consists of a platform for real-time risk alerting and continuous coaching, enabling the design of workplace interventions adapted to the needs and fitness levels of specific ageing workforce.

For the preparation of this system it has been necessary to develop a specific model for ergonomic evaluation and orientation, as the actual methods are not adapted to the specifics characteristics of older workers. This model has to consider both the changes related with age and the actual functional fitness of each individual worker, in order to develop an effective assessment system. To do that, we propose an individualized analysis that allows to compare the characteristics of the person with those of the job and, in this way, detect the degree of adjustment between both.

Keywords: Ergonomics assessment · Real-time risk alerting · Older workers · Ageing · Workplace design · Workplace adaptation · Work ability

1 Introduction

The ageing of population in Western societies has a marked impact on social reality and its effects affect many areas, including work. Ageing is a relevant aspect for companies, since the percentage of workers over 50 is increasing and in many countries lengthening of working life above 65 is being considered. This reality affects both the health and safety of workers and the productivity and competitiveness of companies. Ageing at work is conditioned by the work ability, which is determined by the interaction between the working conditions and the characteristics of the workers.

R. S. Goonetilleke et al. (Eds.): AHFE 2021, LNNS 273, pp. 261–269, 2021.
https://doi.org/10.1007/978-3-030-80713-9_34

1.1 The BIONIC System

BIONIC is a European project[1] that aims to develop a body sensor system to monitor the worker, in order to accurately obtain the level of risk associated with each of the tasks performed. The system consists of a holistic, unobtrusive, autonomous and privacy preserving platform for real-time risk alerting and continuous coaching, enabling the design of workplace interventions adapted to the needs and fitness levels of specific ageing workforce.

The overall system is based on the concept of a cloud platform, two different Body Sensor Networks (BSN), one for using it at work and one for being used at home. The interaction between the worker and the BSN is done via an Edge device that acts as the interface between the cloud platform and the BSN. Through the applications, the end users (older workers, doctors and managers) can retrieve their processed data from the platform. A gamification layer is available for the workers that recommends exercises to help prevent injuries or to counteract unilateral stress of some tasks that were performed throughout the workday (Fig. 1).

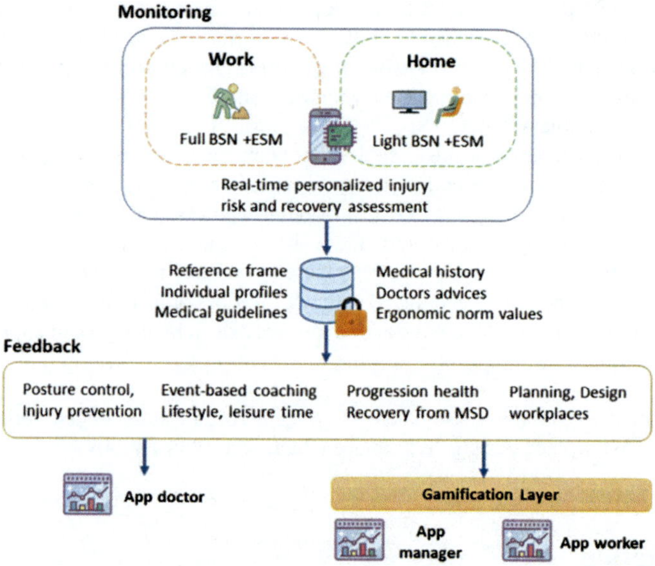

Fig. 1. Overview of the BIONIC system.

The BSN is based on loosely integrated sensors fitted into everyday or work clothing with dynamic monitoring of overall body posture. The system will use kinematic data (e.g. body postures) in conjunction with kinetic, physiological and environmental sensor data related to the situation of the workers and their surroundings. These data will be used to monitor worker's fatigue and to calculate the ergonomic risks that are present

[1] The project has received funding from the European Union's Horizon 2020 research and innovation program under grant agreement No826304.

during the realization of tasks. To do that, validated ergonomic risk assessment methods will be implemented in the system.

1.2 Ergonomic Risk Assessment Methods

The ergonomics assessment methods make possible the identification and evaluation of the risk factors present at the workplaces in order to put forward redesign options to reduce the risk to an acceptable level for workers [1].

Assessment is based on identifying situations at the workplace which generate extensive physical or environmental loads on the workers. The key aspects identified are the different postures of body segments (neck, trunk, arms, hands, legs) while doing the tasks. Other factors are also considered, such as: time of exposure, repetitiveness of movements, loads carried or forces exerted. The aspects are normally identified by an expert observation at the workplace or visualizing video recordings of the tasks. The methods are applied to analyze these factors and generate a risk level. This risk indicator can be used to identify harmful conditions and to prioritize the risk factors that have to be modified. OWAS [2] and the standard ISO 11226 [3] are examples of these kind of methods and they have been selected to be implemented at the BIONIC system. In addition to the automatic analysis and interpretation of the information, the main advantage of the implementation of these methods in BIONIC is that the coding of postures and movements can be done objectively and in real time through the BSN.

Nevertheless, these ergonomic methods are addressed to general population, so they don't consider that some workers can have limited or different capabilities, for example the reduction of physical fitness related to age.

1.3 Functional Changes of the Ageing Workforce

Physical, physiological and psychological abilities are expected to change as people age. There is a biological plausibility along with support from studies showing a modification in the physical and cognitive capabilities among older persons [4]. Understanding the implications ad scope of these changes it is necessary to perform an accurate detection of ergonomic problems.

A summary of the main changes includes physical, physiological and psychosocial aspects [5]:

- Physical Changes: As people age, they begin to lose strength, flexibility, balance, sight, reaction time and speed, hearing, manual dexterity and feedback, and body fat [6, 7].
- Physiological Changes: Ageing leads to decrease in maximum oxygen intake, rising systemic blood pressure, earlier onset of fatigue and greater susceptibility to extreme temperatures [8].
- Psychosocial Changes: As workers age, they have different shift preferences, training and learning styles and sometimes tend to experience disenfranchisement and disengagement with their work [5, 9].

Many of these changes can have a major impact on the day-to-day work of employees, since a lot of physical, mental and sensorial skills are required in the workplace. Studies indicate that aged workers suffer more serious but less frequent workplace injuries than younger workers and that musculoskeletal disorders are often the result of a failure to match the work-based requirements of a task to the functional status of workers [10].

Nevertheless, these changes seem to be uneven and with a non-linear evolution. Although clear trends in the ageing process can be demonstrated, the rate of ageing remains highly individual [11]. Many factors e.g. genetics, lifestyle or past diseases can have an influence on functional aspects.

In any case, it seems that no stereotype about older workers is likely to be true for all, or even for most older workers, particularly, the belief that chronological age is the most important determinant of health [12].

1.4 Alternatives for the Age-Sensitive Ergonomic Assessment

As seen above, a linear relationship of biomechanical factors and age seems to be insufficient. Several indicators e.g. age, personal fitness, gender and working conditions may result in in a better estimation of an ergonomic risks.

Therefore, it is necessary to know which is the actual functional physical fitness of each individual worker, in order to develop an effective assessment system.

There are only a few methods that consider the real capacity of workers. One of the most known proposals to do that is the Work Ability Index (WAI) [13]. The work ability index gives an idea of individuals' perceived work ability and involves a questionnaire complemented by an interview; it can also provide an indication of the potential for disability in the future or early retirement. This approach is not useful for the ergonomic analysis as it is founded on subjective data collected by a self-assessed questionnaire and there is no direct comparison between specific demands and specific capacities [1].

The Job Matching Methods are another alternative to solve these issues. They are based on the analysis of both work and worker features using similar criteria and common assessment levels to facilitate their comparison and interpretation of results. Job matching methods are intended mainly for workers with functional limitations [14], but they can be also be used with older workers to identify if there are functional limitations associated with age that may be causing mismatches with the work requirements (e.g. Ergo + 50 method [1]).

This Job Matching procedure is the approach used to implement at the BIONIC system. We propose an individualized analysis that allows to compare the characteristics of the person with those of the job and, in this way, detect the degree of adjustment between both. This way, in addition to the ergonomic assessment, if the worker has issues related to their physical fitness, the resultant mismatches will allow to prioritize the existing risks in accordance to each individual's situation.

2 Results: An Age-Adapted Ergonomic Risk Model

2.1 Model Overview

The age-adapted ergonomic risk model is based on the ergonomic analysis of the tasks, the functional assessment of the older worker and the comparison between both aspects (Fig. 2).

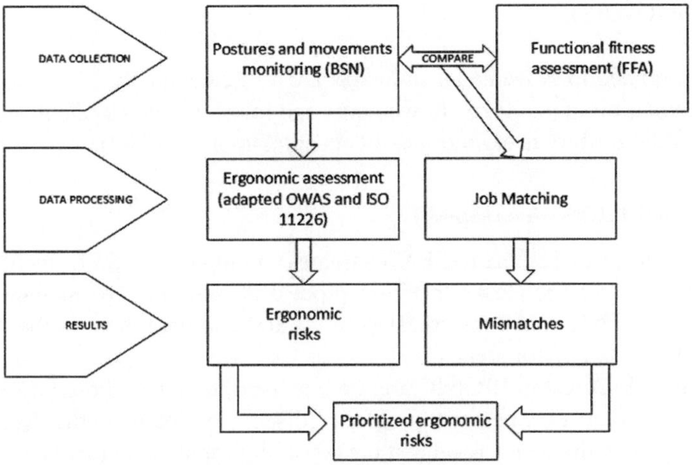

Fig. 2. BIONIC age-adapted ergonomic risk model.

The model starts collecting data from the user through a functional fitness assessment (FFA) Data from the work activities (postures and movements) are also gathered through the BSN. The second step is the data processing: the postures and movements are used to detect the ergonomic risks and, when compared with the FFA, to detect the mismatches between work performance and worker fitness. Mismatches are used to detect age-specific issues and also to prioritize the ergonomic risks.

2.2 Ergonomic Assessment

Postures and movements performed by the worker are recorded automatically through the Body Sensor Network. These postures and movements are used primarily to detect ergonomic risks but also to be compared with the individual capacity.

The ergonomic risks are calculated using OWAS [2] and the standard ISO 11226 [3]. Both methods are based on the codification of body positions while the worker is performing the tasks:

- Neck/head posture (ISO)
- Trunk posture (OWAS and ISO)
- Arms posture (OWAS and ISO)
- Legs posture (OWAS and ISO)
- Load/force (OWAS)

Variables related to movements and loads are also considered.

The combination of positions, movements and loads is used to calculate a risk level (OWAS) or if the posture is ergonomically acceptable or not (ISO).

2.3 Functional Fitness Assessment

A functional capacity assessment (FFA) evaluates an individual's adjustment to perform activities related to their job tasks. The FFA process is used to figure out the individual's health status, and body functions and structures, and compare them to the demands of the job and the work environment.

The goals of FFA test in BIONIC are, on one hand, to detect those physical aspects in which older workers may perceive problems (discomfort, pain or other issues) and, on the other hand, to modulate the results of the ergonomic analysis. Therefore, the FFA for BIONIC includes the same items as those considered at the ergonomic analysis: neck, trunk, arms, leg, load.

The items to collect at BIONIC FFA have been structured in a questionnaire formed by a list of 27 items that are assessed using the following scale:

A. Normal/No problem referred
B. Minor problems or discomfort
C. Moderate/high problems or discomfort

The FFA is applied at the beginning of the assessment, and can be done in the context of a medical interview or through a test using the BSN.

2.4 Individual Adaptation – Adjustment Between Ergonomic and Functional Fitness Assessment

The age-adapted model proposed for BIONIC is a matching procedure. The specific comparison uses, on one hand, the worker capacity values (the results of the FFA questionnaire) and, on the other hand, the variables collected for the ergonomic analysis (measured by the BSN during work performance) (Fig. 2).

The system will compare each variable in the fitness questionnaire (*Fitness value*) with the data that is measured in real time from the ergonomic analysis (*Work measurement*). The result of the comparison indicates whether the situation is adequate or if there is a mismatch that may increase the existing ergonomic risk. Depending on the level of fitness, mismatches are classified into different grades:

- Slight mismatch
- Moderate mismatch (Table 1).

Table 1. BIONIC Matching rules.

FFA value	Working condition	Situation
A – Normal	YES	Potential ergonomic risk
A – Normal	NO	Adequate situation
B – Minor	YES	Slight mismatch - increase of ergonomic risk
B – Minor	NO	Adequate situation
C – Moderate	YES	Moderate mismatch - increase of ergonomic risk
C – Moderate	NO	Adequate situation

2.5 System Implementation

The implementation of the age-adapted ergonomic risk model at the BIONIC system will include the following steps:

- The worker will perform the FFA test before using the system for the first time. The data collected at the FFA test will be added to the user profile.
- The BIONIC system will collect in real-time the postures and movements done by the worker while performing the tasks. This information will be processed to obtain the following information:
  - Ergonomic risks (through the application of OWAS and ISO11226).
  - Mismatches (through the comparison between the data of FFA test and the posture/movement measurements.
- The results will be shown to BIONIC users (workers, doctors, managers) through the corresponding apps (Fig. 3).

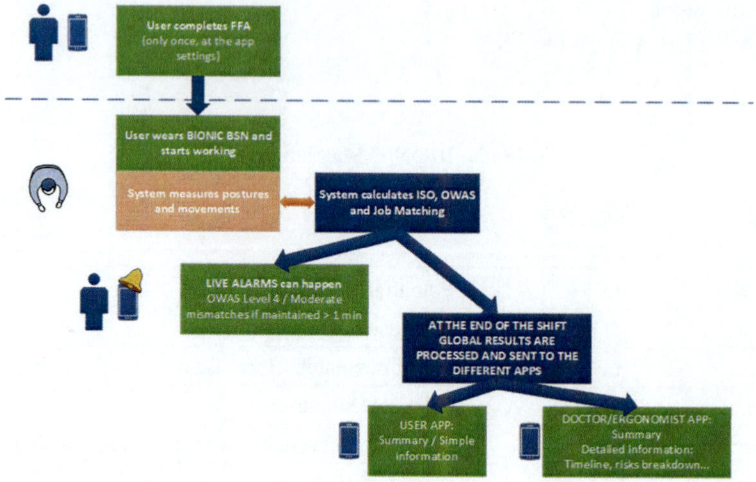

Fig. 3. Age-adapted model implementation at the BIONIC system.

3 Conclusions and Further Work

An age-adapted model has been developed to prioritize the ergonomic risks that are related to the working tasks. The model is based on the following components:

- A standard ergonomic risks assessment, based on well-known procedures (OWAS and ISO 11226) that have been selected and adapted to the measurement possibilities of the BIONIC system.
- A functional fitness assessment, integrated in the user profile, to detect the perception of the user about their health status.
- A matching procedure to compare ergonomic risks and functional fitness status. This comparison will allow to detect the mismatches between task execution and worker physical status. Mismatches can be used to modulate and prioritize the ergonomic risks to the real condition of the specific user.

This model will be tested during a 6-month pilot study in 2021 that will involve the participation of more than 40 older workers in the construction and manufacturing sectors.

References

1. Remesal, A.F., et al.: Ergo + 50: Ergonomic Assessment Methodology Aimed at Older Workers. In: Goossens, R.H.M., Murata, A. (eds.) AHFE 2019. AISC, vol. 970, pp. 27–36. Springer, Cham (2019). https://doi.org/10.1007/978-3-030-20145-6_3
2. Mattila, M., Karwowski, W., Vilkki, M.: Analysis of working postures in hammering tasks on building construction sites using the computerized OWAS method. Appl. Ergon. **24**(6), 405–412 (1993). https://doi.org/10.1016/0003-6870(93)90172-6

3. Delleman, N.J., Dul, J.: International standards on working postures and movements ISO 11226 and EN 1005-4. Ergonomics **50**(11), 1809–1819 (2007). https://doi.org/10.1080/00140130701674430
4. World Health Organization: World report on ageing and health. World Health Organization (2015)
5. Ergonomics and the Aging Workforce - How to Improve Workplace Design. ErgoPlus (2019)
6. Brunner, F., Schmid, A., Sheikhzadeh, A., Nordin, M., Yoon, J., Frankel, V.: Effects of aging on Type II muscle fibers: a systematic review of the literature. J. Aging Phys. Act. **15**(3), 336–348 (2007). https://doi.org/10.1123/japa.15.3.336
7. Kowalski-, K.M., Steiner, L.J., Schwerha, D.J.: Safety considerations for the aging workforce. Saf. Sci. **43**(10), 779–793 (2005). https://doi.org/10.1016/j.ssci.2005.08.014
8. Bittles, A.H., Collins, K.J.: The biology of human ageing. Cambridge University Press (1986). https://doi.org/10.1017/CBO9780511753350
9. Fozard, J.L.: Sensory and cognitive changes with age. Mobil. Transp. Elder. 1–61 (2000)
10. Silverstein, M.: Meeting the challenges of an aging workforce. Am. J. Ind. Med. **51**(4), 269–280 (2008). https://doi.org/10.1002/ajim.20569
11. Menard, D., Stanish, W.D.: The aging athlete. Am. J. Sports Med. **17**(2), 187–196 (1989). https://doi.org/10.1177/036354658901700208
12. Benjamin, K., Sally, W.: Facts and misconceptions about age, health status and employability. Health and Safety Laboratory Buxton, Derbyshire (2005)
13. Tuomi, K., Juhani, I., Antti, J., Lea, K., Arto, T.: Work ability index. Institute of Occupational Health Helsinki (1994)
14. Ferreras, A., et al.: Ergonomic adaptation of workplaces for people with disabilities: methodology and examples. In: Challenges for Assistive Technology. AAATE 2007 Conference Proceedings, pp. 136–140 (2007)

Using Deep Learning Methods to Predict Walking Intensity from Plantar Pressure Images

Hsing-Chung Chen[1,2], Sunardi[1,3], Yih-Kuen Jan[4,5,6], Ben-Yi Liau[7], Chih-Yang Lin[8], Jen-Yung Tsai[9], Cheng-Tsung Li[9], and Chi-Wen Lung[4,10](✉)

[1] Department of Computer Science and Information Engineering, Asia University, Taichung, Taiwan
[2] Department of Medical Research, China Medical University Hospital, China Medical University, Taichung, Taiwan
[3] Department of Mechanical Engineering, Universitas Muhammadiyah Yogyakarta, Yogyakarta, Indonesia
[4] Rehabilitation Engineering Lab, University of Illinois at Urbana-Champaign, Champaign, IL, USA
[5] Kinesiology and Community Health, University of Illinois at Urbana-Champaign, Champaign, IL, USA
[6] Computational Science and Engineering, University of Illinois at Urbana-Champaign, Champaign, IL, USA
[7] Department of Biomedical Engineering, Hungkuang University, Taichung, Taiwan
[8] Electrical Engineering, Yuan Ze University, Chung-Li, Taiwan
[9] Department of Digital Media Design, Asia University, Taichung, Taiwan
[10] Department of Creative Product Design, Asia University, Taichung, Taiwan
cwlung@asia.edu.tw

Abstract. People with diabetes are recommended to perform exercise such as brisk walking to maintain their health. However, a fast walking speed can increase plantar pressure, especially at the forefoot and rearfoot areas, thereby increasing the risk of diabetic foot ulcers (DFU). The deep learning model can identify plantar pressure patterns for an early detection of DFU when performing various intensities of exercise. Therefore, this study aimed to identify differences in walking speeds to the plantar pressure response using deep learning methods, including Resnet50, InceptionV3, and MobileNets. The deep learning models were used to classify the plantar pressure images of healthy people walking on a treadmill. The design consisted of three walking speeds (1.8 mph, 3.6 mph, and 5.4 mph). Through 5-fold cross-validation, accuracy, and robustness, the Resnet50 model had a better performance compared to the other two models in the image classification with a mean F1 score of 0.8646 and a standard deviation of 0.0466. The results indicated that the Resnet50 model can be used to analyze plantar pressure images for assessing risks of DFU.

Keywords: Diabetic foot ulcers · Resnet50 · InceptionV3 · MobileNets

R. S. Goonetilleke et al. (Eds.): AHFE 2021, LNNS 273, pp. 270–277, 2021.
https://doi.org/10.1007/978-3-030-80713-9_35

1 Introduction

DFU sufferers are estimated to be 15–25% among people with diabetes [1]. People with diabetes are required to have proper activities to maintain their health. However, people walk with a significant variation in speed causes a more rapid response of blood flow to the plantar skin, which increases the risk of DFU [2]. Variations in walking speed (1.8, 3.6, and 5.4 mph) are at risk of developing DFU, where the higher the walking speed, the lower the plantar tissue stiffness [3]. Increased plantar pressure during weight-bearing exercise activities is a significant factor in the development of DFU [4]. People walked faster, the resulting pressure values were higher, and the average peak plantar pressure values were in the heel and forefoot areas [5, 6].

Foot plantar pressure is a 2-dimensional pattern widely used to identify the research object's health condition. The plantar pressure sensor is a set of sensor arrays that can detect the pressure level and is usually integrated into the footwear [7]. Deep learning based gait detection used plantar pressure distribution data during walking to classify the three disorders of hemiplegic, Parkinson's, and sensory ataxic gait [8]. The machine learning method based on plantar pressure data has been used to classify five outpatient activities: flat walking, decreasing inclines, sloping inclines, descending stairs, and walking upstairs to obtain an effective walking pattern [9]. The depth learning methodology is based on image data and numerical value data of plantar pressure to classify the type of foot deformation (concave, normal, and flat) using the arch index method [10]. Deep learning is used as the research method because it can perform accurate plantar pressure results [11].

This is a pilot study to investigate the effect of different walking intensities on the response to the plantar pressure distribution using deep learning methods. Results can be the basis for early detection system for walking speed in people who are at risk of developing diabetic foot ulcers.

2 Methodology

2.1 Participants and Instruments

In this study, subjects were 8 healthy males who had not suffered leg injuries in the last six months. The research objectives and the plantar pressure data collection procedure were explained to each subject. The demographic data of the participants were as follows: age 31.0 ± 7.2 years, body weight 68.4 ± 9.2 kg, body height 168.0 ± 6.1 cm, and body mass index (BMI) 24.2 ± 3.0 kg/m^2. All data collection was performed using plantar pressure measurements from the F-scan system (Tekscan, South Boston, MA, USA). The learning model uses a deep learning model that is executed on a personal computer system window 10 with a 64-bit operating system and $\times$64-based processor, Intel (R) Core (TM) i7-10700K CPU @ 3.80 GHz 3.79 GHz, 32 GB RAM, GPU (NVIDIA Ge-Force RTX 3080), 10 GB display memory (VRAM). Python 3.6 has been used in learning environments for training and model testing.

Participants who wear sensor shoes before the experiment is first calibrated to stand on one leg (right foot and left foot) to ensure that the right and left foot pressure is balanced. Procedure for collecting the first plantar pressure data, the subjects were asked to stand on both feet, followed by standing on one leg alternately for 5 s. Second, subjects who walked on a treadmill with variations in the speed of slow (1.8 mph), moderate (3.6 mph), and fast (5.4 mph) each speed with a duration of 2 min with 5 min of rest between each experiment. Each subject was asked to walk on a treadmill with a different walking speed.

2.2 Data Preparation

The measurement device's experimental output results were numerical values of plantar pressure in a comma-separated values (CSV) file type. MATLAB R2019b (Mathworks, Inc., Natick, MA, USA) is used to divide each subject's walking speed into 30 steps and convert the plantar pressure data from a CSV file into an image data file (jpg) with an image intensity.

Preprocessing data is an essential step in preparing and transforming data before being used in model training. Normalization is a common technique used during data processing in which numeric column values are changed to have a uniform scale when the data samples vary in range [12]. Because the resulting numeric plantar pressures value has a wide range of values in one column, namely between 0 and 255 psi, it is crucial to normalize the data into a value scale with a range of 0 – 1 before being used for model training. This way is to make sure the data is distributed correctly while constructing a predictive model. In this study, raw plantar pressure data were carried out at the data preprocessing stage before being trained in the deep learning model. In the next data preprocessing stage, the raw data image size is 120 × 42 pixels from the pressure measuring device (Fig. 1A). The training dataset input requirement is the image size of 224 × 224 pixels for training models of deep learning. These resized images also use a background padding process for decreasing image distortion (Fig. 1B). The final process is labeling the plantar pressure data for slow, moderate, and fast walking speeds (Fig. 1C).

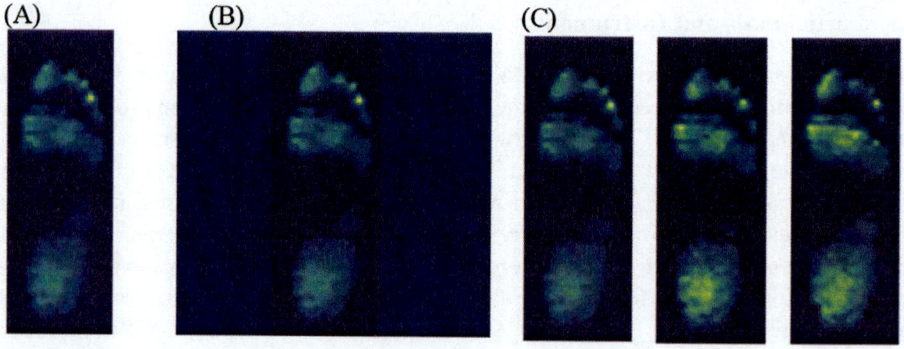

Fig. 1. (A) Raw data image (120 × 42), (B) Data image resize (224 × 224), (C) Labeling of image data from left to right was slow, moderate, and fast walking speed.

2.3 Proposed Model

The flowchart in Fig. 2 shows the training model in this study. Firstly, collect plantar pressure image data from 8 subjects. Secondly, save it in a CSV file with a metric size of 120×42. Each subject walk at three walking speeds (1.8, 3.6, and 5.4 mph), each subject consisted of 30 plantar images/walking speed. Thus, the total amount of plantar data collected from 8 subjects $\times$ 3 walking speed $\times$ 30 steps is 720 plantar images.

Thirdly, at the stage of data preprocessing progress for training models. The dataset is normalized first, and then the image data is changed from the plantar image size from 120×42 to 224×224 so that the data fits the model's needs. In the fourth step, the plantar dataset used in model learning divided by a ratio of 70% (6 subjects = 540 plantar data) was used for model training data and 30% (2 subjects = 180 plantar data) for model validation purposes. This method of dividing datasets in ratios is standard in adopting machine learning/deep learning models for biomedical prediction purposes and solving problems in other fields [13].

The fifth stage is the model training process. Transfer learning (TL) is a method that uses a model that has been trained on a dataset to solve other similar problems by using it as a starting point, modifying and updating its parameters to fit the new dataset. The TL model will help reduce the amount of data used, the calculation process, and the calculation time when the model training process is carried out [14]. TL models for walking speed classification have been used in this study, such as Resnet50, InceptionV3, and MobileNets [10, 14], as this model is one of the best-performing transfer learning models (trained models) widely used by researchers for image classification.

In the sixth stage, the learning dataset used for model training is based on the K-fold cross-validation method [10], which 5-fold cross-validation is commonly used for model evaluation and improving model performance in general. Cross-validation is used mostly on small, composed datasets [9]. Evaluation of the training model is carried out to determine each model's performance in classifying walking speed using plantar pressure validation data. Finally, each model's performance results are measured by the F1 score with the formula:

$$\text{F1 score} = (2 * \text{TP})/(2 * \text{TP} + \text{FP} + \text{FN}) \tag{1}$$

(worst value: 0; best value: 1) F1 ranges in [0,1], where the minimum is reached for TP = 0, that is, when all the positive samples are misclassified, and the maximum for FN = FP = 0, that is for perfect classification.

3 Results

The results of the graph of each model training show the accuracy and validation (Resnet50 in Fig. 3A, InceptionV3 in Fig. 3B, and MobileNets in Fig. 3C). A blue line presents each model's training accuracy, and a red line represents the accuracy of the validation. In contrast, the line X-axis represents the epoch's number, and Y-axis represents the model accuracy magnitude. In this study, the validation accuracy value for the Resnet50 model (0.9053), the InceptionV3 model (0.8113), and the MobileNets

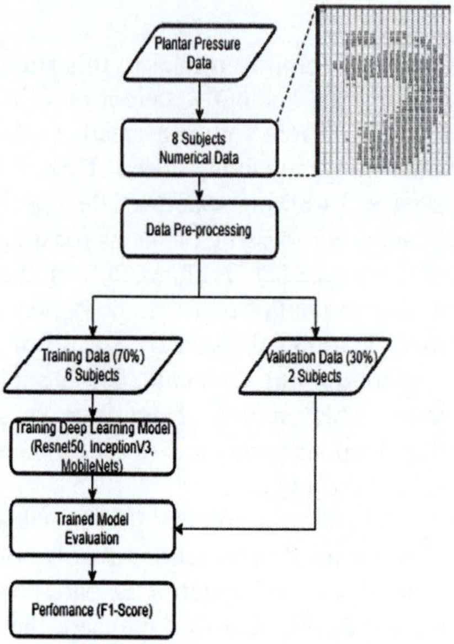

Fig. 2. Flowchart training model.

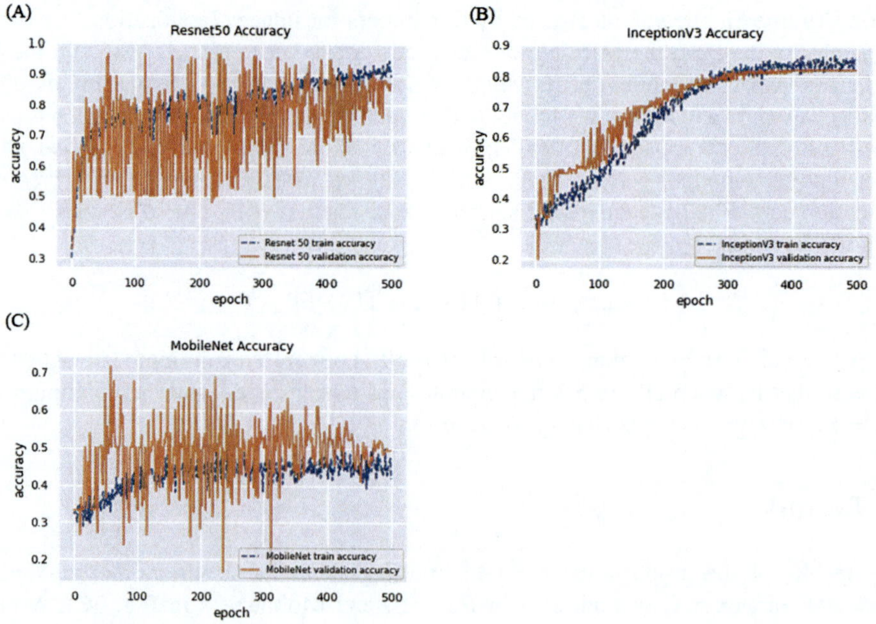

Fig. 3. Results of training accuracy and validation accuracy. (A) Resnet50 model, (B) InceptionV3 model, (C) MobileNets model.

model (0.4484) with the model training process using the respective learning rate parameters ($1e^{-2}$, $1e^{-4}$, $1e^{-3}$), batch size respectively (64, 32, 64), all models using epoch 500.

The trained model used each subset and was evaluated for test data by calculating the mean F1 score. The results of the performance of each model are presented. The three boxplots show the F1 score with 5-fold cross-validation, the first and third quartiles are shown a horizontal line above and below the box (standard deviation), the centerline in the box is the median value (Resnet50: 0.8748, InceptionV3: 0.7843, and MobileNets: 0.4298) or called the second quartile, whereas the model maximum values (0.9053, 0.8113, and 0.4484 respectively) and the minimum values (0.7849, 0.7384, and 0.4261 respectively) are represented by the outer horizontal lines along the vertical lines in the box (Fig. 4 and Table 1).

Table 1 provides information about all models' numerical results seen from statistical parameters such as the mean, standard deviation, median, minimum, and maximum values based on the five F1 scores. The highest model performance is achieved in the Resnet50 model with a mean statistical parameter value of 0.8646, a standard deviation of 0.0466, a minimum value of 0.7849, a median value of 0.8748, and a maximum value of 0.9053.

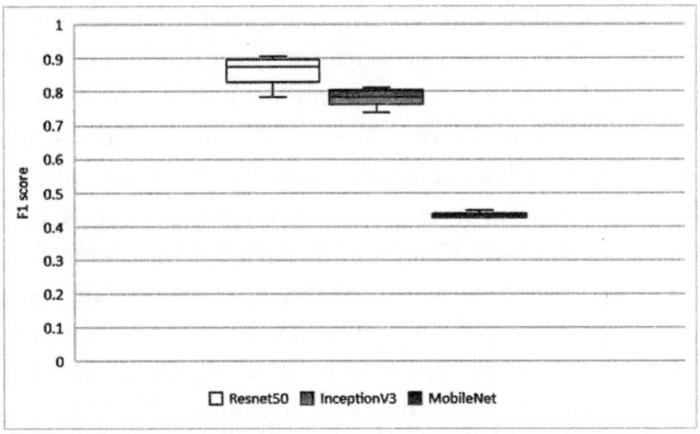

Fig. 4. Performance results of all models.

Table 1. Numerical results of the statistical parameters of the F1 score in the 5-fold cross validation.

	Resnet50	InceptionV3	MobileNets
Mean	0.8646	0.7823	0.4320
Std	0.0466	0.0269	0.0093
Min	0.7849	0.7384	0.4261
Median	0.8748	0.7843	0.4298
Max	0.9053	0.8113	0.4484

4 Discussion

The trend of the training line and the validation line coincide with each other. The model is shown that there is no over-fitting or under-fitting of the model when it is executed. Resnet50 model and InceptionV3 model show a reasonably good performance with a mean F1 score obtained by 0.8646 and 0.7823. The accuracy value provides information on the accuracy of a model, or it can be interpreted that the higher the accuracy means the higher the performance of a model [15]. Parameters related to the training process, including learning rate, batch size, and the number of epochs, were adjusted. By setting the right parameters, the training process can be achieved optimally.

The higher the accuracy of a model, the closer the box position is to the F1 score value 1. It shows the higher the model robustness, the narrower the distance between the outer values. The numerical results of the statistical parameters of the F1 score using 5-fold cross-validation are described. The highest average value is achieved using the Resnet50 model. However, in this model, the standard deviation value between InceptionV3 and MobileNets models. In this analysis, the Resnet50 model is the best model for classifying plantar pressure images in walking speed variations.

The Convolution Neural Network (CNN) principle is the basis for the neural network of the Resnet50, InceptionV3, and MobileNets models. What distinguishes each model is only on each model's architecture a function extractor layer and a hidden layer of fully connected neurons. The basic principle of feature extraction is by shifting the kernel in the image. The features extracted and studied in the same group have a high misclassification rate in different plantar images [10]. Plantar pressure image data trained with the MobileNets model showed relatively lower performance than the Resnet50 and InceptionV3 models. The problem of the MobileNets model is more challenging to extract plantar image data than the Resnet50 and InceptionV3 models.

The MobileNets model has the lowest standard deviation value compared to Resnet50 and inceptionaV3. It is because the MobileNets model is not sensitive to deviations or differences that occur in plantar images. while the Resnet50 model has the highest standard deviation, value compared to MobileNets and InceptionV3 because the Resnet50 model uses an optimized maximum depth value to prevent overfitting so that it has a high level of sensitivity to the differences that occur in each plantar pressure image with different walking speeds.

The limitation of the study is to obtain plantar pressure dataset from a subject with a consistent quality plantar pressure distribution pattern for each walking speed [10]. The plantar data is often within the criteria for the classification of walking speed. The resulting data conditions have a somewhat ambiguous effect on features, resulting in less-than-optimal performance. Although the deep learning model is effective in classifying the walking speed via plantar pressure images, future work will require a suitable plantar pressure data preprocessing method and a method of combining multiple superior models to achieve better performance than the current model.

5 Conclusion

The results of this study demonstrated that the deep learning model has potential to classify plantar pressure images from three walking speeds with a mean F1 score of 0.8646 using Resnet50, 0.7823 using InceptionV3, and 0.4320 using MobileNets. This study indicate that the Resnet50 model has the highest accuracy performance than InceptionV3 and MobileNets on analyzing plantar pressure distribution images.

Acknowledgments. The Authors wish to express gratitude to Mr. Fityanul Akhyar, M.Sc. and Mr. Prayitno, M.T. for their assistance. This study was supported by the Ministry of Science and Technology of the Republic of China (MOST 108-2221-E-241-008, MOST-108-2221-E-468-018 and MOST 110-2218-E-468-001-MBK), Kuang Tien General Hospital and Hungkuang University (HK-KTOH-109-04), and Asia University Hospital and China Medical University Hospital (ASIA-107-AUH-09).

References

1. Armstrong, D.G., Boulton, A.J., Bus, S.A.: Diabetic foot ulcers and their recurrence. N. Engl. J. Med. **376**, 2367–2375 (2017)
2. Wu, F.L., Wang, W.T., Liao, F., Elliott, J., Jain, S., Jan, Y.K.: Effects of walking speeds and durations on plantar skin blood flow responses. Microvasc. Res. **128**, 103936 (2020)
3. Lung, C.-W., Wu, F.-L., Zhang, K., Liau, B.-Y., Townsend, R., Jan, Y.-K.: Using elastographic ultrasound to assess plantar tissue stiffness after walking at different speeds and durations. Appl. Sci. **10**(21), 7498 (2020)
4. Bus, S., et al.: Footwear and offloading interventions to prevent and heal foot ulcers and reduce plantar pressure in patients with diabetes: a systematic review. Diabetes/Metab. Res. Rev. **32**, 99–118 (2016)
5. Burnfield, J.M., Few, C.D., Mohamed, O.S., Perry, J.: The influence of walking speed and footwear on plantar pressures in older adults. Clin. Biomech. **19**, 78–84 (2004)
6. Pataky, T.C., et al.: New insights into the plantar pressure correlates of walking speed using pedobarographic statistical parametric mapping (pSPM). J. Biomech. **41**, 1987–1994 (2008)
7. Ramirez-Bautista, J.A., Huerta-Ruelas, J.A., Chaparro-Cárdenas, S.L., Hernández-Zavala, A.: A review in detection and monitoring gait disorders using in-shoe plantar measurement systems. IEEE Rev. Biomed. Eng. **10**, 299–309 (2017)
8. Potluri, S., Ravuri, S., Diedrich, C., Schega, L.: Deep learning based gait abnormality detection using wearable sensor system. In: 2019 41st Annual International Conference of the IEEE Engineering in Medicine and Biology Society (EMBC), pp. 3613–3619. IEEE (2019)
9. Nguyen, N.D., Bui, D.T., Truong, P.H., Jeong, G.-M.: Classification of five ambulatory activities regarding stair and incline walking using smart shoes. IEEE Sens. J. **18**, 5422–5428 (2018)
10. Chae, J., Kang, Y.J., Noh, Y.: A deep-learning approach for foot-type classification using heterogeneous pressure data. Sensors (Basel) **20**(16), 4481 (2020)
11. Lee, C.-H., Li, L.-T.: Cost-effective person identity recognition based on plantar pressure images. In: 2020 IEEE International Conference on Consumer Electronics-Taiwan (ICCE-Taiwan), pp. 1–2. IEEE (2020)
12. Grus, J.: Data science from scratch: first principles with python. O'Reilly Media (2019)
13. Hashi, E.K., Zaman, M.S.U., Hasan, M.R.: An expert clinical decision support system to predict disease using classification techniques. In: 2017 International Conference on Electrical, Computer and Communication Engineering (ECCE), pp. 396–400. IEEE (2017)
14. Wu, Y., Wu, Q., Dey, N., Sherratt, S.: Learning models for semantic classification of insufficient plantar pressure images. Int. J. Interact. Multimedia Artific. Intell. **6**, 51–61 (2020)
15. Yang, F., Li, F., Zhang, K., Zhang, W., Li, S.: Influencing factors analysis in pear disease recognition using deep learning. Peer-to-Peer Network. Appl. 1–13 (2020)

Machine Learning-Based Pre-impact Fall Detection and Injury Prevention for the Elderly with Wearable Inertial Sensors

Xiaoqun Yu, Jaehyuk Jang, and Shuping Xiong(✉)

Human Factors and Ergonomics Lab, Department of Industrial and Systems Engineering, Korea Advanced Institute of Science and Technology (KAIST), 291 Daehak-ro, Yuseong-gu, Daejeon 34141, Republic of Korea
{xiaoqunyu,jaejang0113,shupingx}@kaist.ac.kr

Abstract. Falls are the leading cause of death and non-fatal injuries among older people, thus pre-impact fall detection that detects a fall before body-ground impact is of crucial significance. 32 young subjects performed different types of falls and daily activities, and their motion data was recorded by a wearable inertial sensor to establish a large-scale motion dataset. Five commonly used machine learning algorithms were applied and compared thoroughly in terms of accuracy and practicality for pre-impact fall detection. Results showed that in terms of sensitivity, specificity and lead time, both support vector machine (SVM: 99.77%, 93.10%, 362 ± 150 ms) and random forest (RF: 100%, 92.90%, 357 ± 145 ms) achieved better results than other 3 models. SVM showed a much shorter latency (66 ms) than RF (1047 ms) running in a microcontroller. Those findings suggest that SVM has the highest potential to be embedded into a wearable sensor based system to provide real-time fall protection for the elderly.

Keywords: Pre-impact fall detection · Machine learning · Wearable sensor

1 Introduction

Falls among the elderly are a major health and safety issue in our society due to rapid demographic aging. Yearly fall rates are between 30% for those over 65 years old to 50% for those aged over 85 [1]. Twenty to thirty percent of falls result in moderate to serious injuries such as hip fracture, being the leading cause of injury-related hospitalization in older people [2]. Due to the high prevalence and severe consequences of the fall, it was regarded as "Global Burden of Disease" by WHO [3]. Therefore, it is critical to prevent falls and fall-related injuries for the elderly.

Many research works have been devoted to post-fall detection which is expected to identify the occurrence of fall events thus to avoid unnecessary losses caused by "long lie" [4]. As the name implies, this approach is reactive since the fall accident has happened so that fall-induced injuries cannot be prevented. Hence, some researchers have shifted their focus of algorithm & system design from post-fall detection to pre-impact fall detection, which is aimed to detect the fall before the body hits against ground

R. S. Goonetilleke et al. (Eds.): AHFE 2021, LNNS 273, pp. 278–285, 2021.
https://doi.org/10.1007/978-3-030-80713-9_36

rather than after body-ground impact. Compared with post-fall detection, this method is proactive because it has good potential to enable on-demand fall protective devices such as airbags to prevent or minimize the injuries from impact falls [5].

Over the past decade, various algorithms were proposed based on wearable inertial sensors due to their low-cost, miniature and no space constraints. While many studies utilized simple threshold-based algorithms for the pre-impact fall detection, they lack the generalizability for different types of falls and populations. Thus, threshold-based algorithms are difficult for practical applications. For example, Bourke, O'Donovan [6] utilized vertical velocity as a single threshold and achieved 100% of sensitivity and specificity based on the data with 6 types of activities of daily life (ADLs) and 4 types of falls from 5 young subjects. However, when another group of researchers applied the same threshold with modified values based on the data from 25 young subjects with 7 types of ADLs and 5 types of falls, they only acquired 80% of sensitivity [7]. Fewer studies have attempted to apply machine learning algorithms to detect pre-impact fall and most of them only implemented a single type of machine learning algorithm based on a small dataset with limited types of falls and activities of daily living (ADLs) [8, 9]. In addition, practical requirements of those algorithms (such as efficiency aside from accuracy) for the fall protection system were seldom considered.

Therefore, the objective of this study is to explore the feasibility of different machine learning algorithms for pre-impact fall detection based on a large-scale motion dataset and propose guidelines for the algorithm selection when designing practical wearable fall protection systems for the elderly.

2 Methods

2.1 Participants and Apparatus

Thirty-two young male subjects (age: 24.9 ± 3.7 years, height: 174.0 ± 6.3 cm, weight: 69.3 ± 9.5 kg) participated in this study. All subjects were in heathy conditions and none of them had any musculoskeletal diseases or injuries. All participants gave consent for the experimental protocol that was approved by University Institutional Review Board (IRB No: KH2020–068).

A 9-axis inertial sensor, LPMS-B2 (LP-RESEACH Inc., Tokyo, Japan) was used for recording motion data at a sampling rate of 100 Hz. The senor communicated with a Raspberry Pi 4 as a host PC via a Bluetooth connection and it was placed on each participant's low back (pelvis, Fig. 1a). Figure 1b illustrates a representative time series data of 3-axis acceleration from the sensor for a forward fall when trying to sit down. In addition, Raspberry Pi High Quality Camera was connected to Raspberry Pi 4 to record the motion video synchronously at a frequency of 90 Hz based on a self-developed GUI program.

2.2 Experimental Tasks and A Motion Dataset

All the subjects performed 21 types of ADLs and 15 types of falls (Table 1). Most of ADLs and fall tasks were adopted from SisFall dataset [10] while some ADLs frequently

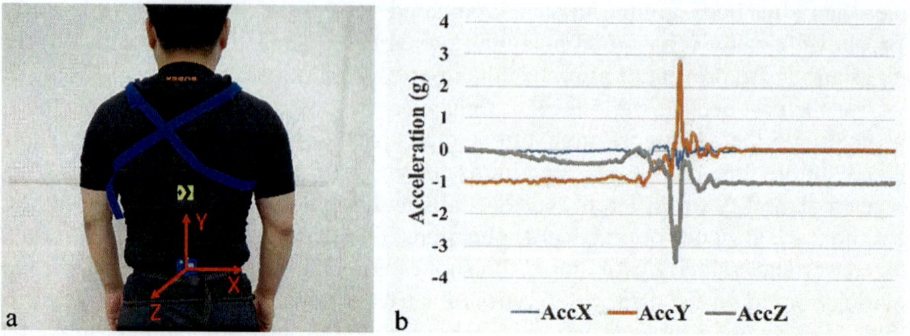

Fig. 1. (a) Inertial sensor position and 3D coordinate system; (b) 3-axis acceleration data of a forward fall when trying to sit down

observed in old Korean's daily life such as sitting to the ground and getting up, sitting to the sofa with inclining to the back support and getting up were also considered. Except for the static tasks (D01, D11, D12, and D17) which only required 1 trial, all other tasks performed 5 times. For the safety reason, subjects were instructed to perform fall activities on a mattress of 15 cm in thickness with wearing protective knee pads. This newly created motion database offers most various types of ADLs covering from low-intense motions, high-dynamic behaviors even to near-fall activities and falls compared with existing public datasets [11].

2.3 Data Preparation and Labeling

In order to detect pre-impact fall in real time, sliding window technique was applied on the motion file with window size of 0.5 s and step size of 0.1 s [9]. Each window was then labeled as non-fall or fall class in order to train supervised machine learning models.

For the ADL tasks, all the windows were labelled as non-fall class. While for the fall tasks which usually include transition period from preceding ADLs to fall, they are more complicated to label. Similar to an earlier study [12], we divided the fall task into 3 phases: pre-fall (performing ADLs), falling (descending to the ground) and post-fall. Post-fall phase was removed because it is the state after the fall impact. For the windows containing at least 40% of readings in falling phase, they were labeled as fall class, whereas the rest windows were labeled as non-fall class.

2.4 Feature Extraction

In this study, a comprehensive set of motion features were extracted from each window based on previous studies [13, 14]. Those features can be divided into 3 categories: acceleration features, angular velocity features and orientation features. The magnitude of acceleration was utilized to extract acceleration features. Except commonly used features which include mean, variance and Root Mean Square (RMS), advanced ones are listed as follows:

Table 1. Experimental tasks of 21 types of ADLs and 15 types of falls

No	Activity	Trials
D01	Stand	1(30 s)
D02	Stand, slowly bend at knees, tie shoe lace, and get up	5
D03	Pick up an object from the floor	5
D04	Gently jump (try to reach an object)	5
D05	Stand, sit to the ground, wait a moment, and get up with normal speed	5
D06	Walk normally with turn (4 m)	5
D07	Walk quickly with turn (4 m)	5
D08	Jog normally with turn (4 m)	5
D09	Jog quickly with turn (4 m)	5
D10	Stumble while walking	5
D11	Sit on a chair	1(30 s)
D12	Sit on the sofa (back is inclined to the support)	1(30 s)
D13	Sit down to a chair normally, and get up from a chair normally	5
D14	Sit down to a chair quickly, and get up from a chair quickly	5
D15	Sit a moment, trying to get up, and collapse into a chair	5
D16	Stand, sit to the sofa, inclined to the support, and get up normally	5
D17	Lie on the bed	1(30 s)
D18	Sit a moment, lie down to the bed normally, and get up normally	5
D19	Sit a moment, lie down to the bed quickly, and get up quickly	5
D20	Walk upstairs and downstairs normally (5 steps)	5
D21	Walk upstairs and downstairs quickly (5 steps)	5
F22	Forward fall when trying to sit down	5
F23	Backward fall when trying to sit down	5
F24	lateral fall when trying to sit down	5
F25	Forward fall when trying to get up	5
F26	lateral fall when trying to get up	5
F27	Forward fall while sitting, caused by fainting	5
F28	lateral fall while sitting, caused by fainting	5
F29	Backward fall while sitting, caused by fainting	5
F30	Vertical(forward) fall while walking caused by fainting	5
F31	Fall while walking, use of hands to dampen fall, caused by fainting	5
F32	Forward fall while walking caused by a trip	5

(*continued*)

Table 1. *(continued)*

No	Activity	Trials
F33	Forward fall while jogging caused by a trip	5
F34	Forward fall while walking caused by a slip	5
F35	Lateral fall while walking caused by a slip	5
F36	Backward fall while walking caused by a slip	5

- Zero-Crossing Rate (ZCR): the number of sample which is over the mean of the window;
- Absolute Difference (ABSDIFF): sum of the absolute difference between each sample and the mean of the window divided by the number of samples;
- First 5-FFT Coefficients: the first 5 of the fast Fourier transform (FFT) coefficients;
- Spectral Energy (SE): the sum of the squared FFT coefficients divided by the number of samples.

For the angular velocity features, same set of features were applied to the magnitude of angular velocity. The orientation features were computed from pitch and roll angles of the sensor. Similarly, mean, standard deviation, RMS, ZCR, ABSDIFF and SE were considered for both of pitch and roll angles. Finally, a total of 34 features were formed for the supervised machine learning models.

2.5 Pre-impact Fall Detection Using Machine Learning Algorithms

Five commonly used machine learning algorithms were investigated to detect the pre-impact fall with our dataset: support vector machine (SVM), random forest (RF), decision tree (DT), k-nearest neighbors (KNN) and logistic regression (LR). Followed by the general guideline, 80% of data (26 subjects) was set for training and the rest 20% of data (6 subjects) was used for testing purpose. Before the training, hyper-parameter tuning for each of the algorithm was implemented by random and grid search methods.

The algorithm performance is evaluated by different metrics including sensitivity, specificity, lead time and latency. Sensitivity and specificity are calculated by Eqs. 1 and 2, respectively.

$$\text{Sensitivity} = \frac{TP}{TP + FN}. \tag{1}$$

$$\text{Specificity} = \frac{TN}{TN + FP}. \tag{2}$$

where TP (true positive) denotes the fall file in which the window being classified as fall for the first time allocates in falling phase; FN (false negative) is the fall file in which all windows are inferred as non-fall; TN (true negative) is the ADL file in which all the windows are classified as non-fall; and FP (false positive) is the ADL file in which there is at least 1 window is misclassified as fall. Lead time is defined as the time interval between fall detection moment and body-ground impact moment. In addition, latency is specified as the computation time for each inference based on the trained model. In this study, machine learning models implemented by the Scikit-learn library [15] were trained on Windows PC while latency test was conducted on a microcontroller (Raspberry Pi 4) considering the practical applications.

3 Results and Discussion

Table 2 summarizes the performances of 5 machine learning algorithms comprehensively in terms of accuracy related measures (sensitivity, specificity and lead time) which only depend on the algorithms per se and practical measures (latency) which also rely on the hardware. Overall, all the machine learning models showed reasonable performance after conducting hyper-parameter tuning. In terms of accuracy related measures, SVM (99.77%, 93.10%, 362 ± 150 ms) and RF (100%, 92.90%, 357 ± 145 ms) performed better than DT (97.97%, 94.48%, 279 ± 150 ms), KNN (97.97%, 94.67%, 373 ± 151 ms) and LR (94.82%, 91.91%, 231 ± 133 ms). With respect to practical measures, SVM had a short latency of 66.0 ms running in a microcontroller, which was much lower than RF (1047.1 ms) and KNN (391.5 ms) but comparable as DT (62.4 ms) and LR (63.3 ms).

Table 2. Summary of pre-impact fall detection performances of 5 machine learning algorithms

Algorithm	Sensitivity (%)	Specificity (%)	Lead time (ms)	Latency (ms)
SVM	99.77	93.10	362 ± 150	66.0
RF	100	92.90	357 ± 145	1047.1
DT	97.97	94.48	279 ± 150	62.4
KNN	97.97	94.67	373 ± 151	391.5
LR	94.82	91.91	231 ± 133	63.3

For accuracy related measures, it is understandable that non-linear classifiers (SVM, RF, DT and KNN) outperformed the linear classifier (LR) since LR assumes the data is linearly (or curvy linearly) separable in space. However, some features of ADL could be very similar to the fall, therefore, those features are less likely to be linearly separable. Both LR and DT have a shorter lead time which indicates those two models could only detect the fall in the later stage of falling (when fall features are very obvious to be distinguished from ADL features). In other words, shorter lead time may result in lack of time to enable fall protection devices (such as wearable inflatable airbags) to prevent fall-induced injuries. This is because they do not have good prediction power due to their simple working mechanisms [16].

Both SVM and RF performed well in accuracy related measures, which was generally consistent with the previous studies [17, 18]. Since the consequences of missing detection of the fall events were much more serious than false alarms, DT and KNN underperformed compared with SVM and RF due to their lower sensitivities. For SVM, it tries to find a hyperplane which maximizes the margin between two classes but also allows some flexibility to achieve a good balance between increasing the distance of decision boundary to classes and maximizing the number of samples correctly classified (soft margin). In this study, Gaussian radial basis function (RBF) kernel was selected to map the input features to higher dimensional space where a linear hyperplane can be easily found based on the hyper-parameter tuning. Due to the soft margin strategy and kernel trick, the robustness and great performance of SVM for classification have also been validated in different kinds of real-life applications[16]. As for RF, it is an ensemble algorithm by building multiple decision trees (i.e. a forest) and final classification is determined by the result of major votes. Since each tree is only trained on random subset of features from a bootstrapped sample of the data, RF is less likely to be overfitting and more robust (not depending on some specific features) than DT.

The latencies of deployed machine learning models for pre-impact fall detection were rarely reported in the literature since most of the algorithms only run on the Desktop PC [8, 9, 17]. However, for practical applications, those algorithms are usually embedded into a microcontroller with low-performance CPU. Therefore, latencies are not negligible in such device. Due to the very short period of falling (around 800 ms) [19], low latency is mandatory for the algorithm running in the microcontroller to detect the fall before the body hits to the ground. In this study, RF and KNN models showed much higher latencies on the microcontroller (Raspberry Pi 4) compared with other models, which were infeasible for the real-time applications because their latencies were longer than their lead time. This was expected as latencies were determined by the computational complexity of algorithms. The computational complexity of RF is highly related to the number of trees in the algorithm. Because there were 1200 trees in this study from the result of hyper-parameter tuning, the computational cost was very high. Whereas for the KNN model, the computational cost is proportional to the number of samples in the training set since it will loop through every training sample and compute the distance between the training sample and new observation to make the classification. The size of training set was relatively big in this study, which resulted in higher latency of the model.

4 Conclusion

In this paper, we have investigated the feasibility of five commonly used machine learning algorithms for pre-impact fall detection based on a newly established large-scale motion dataset. The algorithm accuracy and practicality for pre-impact fall detection and injury prevention have been compared comprehensively. In terms of accuracy related measures (sensitivity, specificity and lead time), SVM (99.77%, 93.10%, 362 $\pm$ 150 ms) and RF (100%, 92.90%, 357 $\pm$ 145 ms) were comparable but better than DT (97.97%, 94.48%, 279 $\pm$ 150 ms), KNN (97.97%, 94.67%, 373 $\pm$ 151 ms) and LR (94.82%, 91.91%, 231 $\pm$ 133 ms). However, when deploying the model into a microcontroller (Raspberry Pi 4), the latency of SVM (66.0 ms) was much shorter than RF's (1047.1 ms). Those findings suggest that among those five machine learning algorithms, SVM has highest potential to be embedded into a wearable inertial sensor based system to provide real-time pre-impact fall detection so that protective devices can be triggered timely to prevent fall-induced injuries for the elderly.

Acknowledgments. This work was funded by the KAIST End-Run Project and partially supported by the Basic Science Research Program of the National Research Foundation of Korea [NRF2017R1C1B2006811] and the China Scholarship Council.

References

1. Lord, S.R., et al.: Falls in Older People: Risk Factors and Strategies for Prevention, 2nd edn. Cambridge University Press, Cambridge (2007)
2. Bergen, G., Stevens, M.R., Burns, E.R.: Falls and fall injuries among adults aged≥ 65 years—United States, 2014. Morb. Mortal. Wkly Rep. **65**(37), 993–998 (2016)
3. Rehm, J., et al.: Global burden of disease and injury and economic cost attributable to alcohol use and alcohol-use disorders. The Lancet. **373**(9682), 2223–2233 (2009)
4. Vallabh, P., Malekian, R.: Fall detection monitoring systems: a comprehensive review. J. Ambient. Intell. Humaniz. Comput. **9**(6), 1809–1833 (2017). https://doi.org/10.1007/s12652-017-0592-3
5. Hu, X., Qu, X.: Pre-impact fall detection. Biomed. Eng. Online **15**(1), 61 (2016)
6. Bourke, A.K., O'Donovan, K.J., Ólaighin, G.: The identification of vertical velocity profiles using an inertial sensor to investigate pre-impact detection of falls. Med. Eng. Phys. **30**(7), 937–946 (2008)
7. Sabatini, A.M., et al.: Prior-to-and post-impact fall detection using inertial and barometric altimeter measurements. IEEE Trans. Neural Syst. Rehabil. Eng. **24**(7), 774–783 (2015)
8. Shan, S., Yuan, T.: A wearable pre-impact fall detector using feature selection and support vector machine. In: IEEE 10th International Conference On Signal Processing Proceedings. IEEE (2010)
9. Aziz, O., et al: The effect of window size and lead time on pre-impact fall detection accuracy using support vector machine analysis of waist mounted inertial sensor data. In: 2014 36th Annual International Conference of the IEEE Engineering in Medicine and Biology Society, EMBC 2014 (2014)
10. Sucerquia, A., López, J.D., Vargas-Bonilla, J.F.: SisFall: A fall and movement dataset. Sensors **17**(1), 198 (2017)
11. Casilari, E., Santoyo-Ramón, J.-A., Cano-García, J.-M.: Analysis of public datasets for wearable fall detection systems. Sensors **17**(7), 1513 (2017)
12. Musci, M., et al.: Online fall detection using recurrent neural networks on smart wearable devices. IEEE Trans. Emerg. Top. Comput. 1 (2020). https://doi.org/10.1109/TETC.2020.3027454
13. Incel, O.D.: Analysis of movement, orientation and rotation-based sensing for phone placement recognition. Sensors **15**(10), 25474–25506 (2015)
14. Lee, J, Kim, J.: Energy-efficient real-time human activity recognition on smart mobile devices. Mobile Inf. Syst. **2016**, 1–12 (2016)
15. Pedregosa, F., et al.: Scikit-learn: machine learning in python. J. Mach. Learn. Res. **12**, 2825–2830 (2011)
16. Hastie, T., Tibshirani, R., Friedman, J.: The Elements of Statistical Learning: Data Mining, Inference, and Prediction. Springer Science & Business Media (2009)
17. Kim, T.H., et al.: Machine learning-based pre-impact fall detection model to discriminate various types of fall. J. Biomech. Eng. **141**(8) (2019)
18. Qiu, H., et al.: Application of wearable inertial sensors and a new test battery for distinguishing retrospective fallers from non-fallers among community-dwelling older people. Sci. Rep. **8**, 16349 (2018)
19. Tao, X., Yun, Z.: Fall prediction based on biomechanics equilibrium using Kinect. Int. J. Distrib. Sens. Netw. **13**(4), 1550147717703257 (2017)

Workload Assessment Methods and Techniques

A Pilot Study on the Use of Changes in Facial Features to Assess Physical Workload in Real-Time

Qian Zhang(✉) and Lora Cavuoto

Department of Industrial and Systems Engineering,
University at Buffalo, Buffalo, NY 14226, USA
{qzhang42,loracavu}@buffalo.edu

Abstract. An understanding of the level of physical workload experienced by an operator is important for designing workplace interventions that reduce injury risk. Existing methods for physical workload assessment can be disruptive to workers, are highly sensitive to the environment, and may not be feasible to implement in real-time on the manufacturing shop floor. The goal of this study was to investigate the potential of using changes in facial features as a method for real-time physical workload assessment. Four participants performed an assembly task during two sessions completed on different days, while measures of ratings of perceived exertion (RPE), performance and changes of facial features were collected. Results showed that facial features, such as head movement, left and right inner eyebrow point and bottom point of the lip had significant changes over time. Future work should investigate these changes further for the potential to assess physical workload in real-time onsite manufacturing.

Keywords: Facial features · Physical workload assessment · Repetitive exertions

1 Introduction

The nature of manufacturing work often involves low load and highly repetitive tasks. Low load and highly repetitive tasks, such as placing and assembly, may lead to an accumulation of physical workload. Highly repetitive movement, short cycle times, and a lack of task variety or breaks during assembly tasks has been shown to lead to work-related musculoskeletal disorders in neck, shoulder and wrist in the long term [17]. In the short term, the accumulation of physical workload may lead to physical fatigue, which can negatively affect attention [8]. Real-time physical workload assessment can support the monitoring of operator status for designing interventions to release physical fatigue and reduce injury risks.

Various strategies have been proposed to measure physical workload. Subjective rating scales, such as ratings of perceived exertion (RPE) [2] and visual analogue scales (VAS) [12], are popular to assess physical workload in both in-lab and field studies. These scales are easy to implement onsite via self-reporting from individuals with only

R. S. Goonetilleke et al. (Eds.): AHFE 2021, LNNS 273, pp. 289–296, 2021.
https://doi.org/10.1007/978-3-030-80713-9_37

paper and pen [16]. However, these rating scales may not be capable of providing continuous assessment of physical workload, since self-reporting only allows implementation at time intervals and any questioning may disrupt an operator's task performance [1]. Physiological measurements can provide real-time and continuous assessment of physical workload. Heart rate monitoring devices can capture cardiovascular load, an indicator of both physical and mental workload [9]. At the muscle level, electromyography (EMG) has been widely used for measurement of muscle activities [3]. Both heart rate and EMG may not be feasible to measure physical workload onsite in the workplace due to sensitivity to the working environment and the intrusive nature of the measurement [23]. The use of motion capture and inertial sensing can quantify kinematic features related to physical workload, via tracking the coordinates and acceleration of body segments and joints overtime [14, 18]. This approach is limited by fixed cameras for motion capture systems that only allow for tracking within specific areas and accurately can be negatively influenced by other materials present in the environment. Inertial sensing does not require fixed cameras, however, studies reported that kinematic parameters processed from inertial sensors did not change significantly with increasing physical workload [4]. Additionally, inertial sensors are limited by the same concern of intrusiveness as physiological monitoring. Overall, existing measurement approaches may not be feasible in order to effectively monitor physical workload in real-time at the workplace, considering 1) disruption and extra mental workload towards individuals via self-reporting, 2) discontinuous data acquisition, 3) limitations on workspace, 4) high sensitivity towards environment, and 5) intrusiveness toward the human operators.

Changes of facial features have been proposed for real-time assessment of physically demanding tasks. Timme and Brand found the mouth openness was significantly related to exercise intensity in incremental cycling task [24]. Asadi et al. pointed out the average pixel movement of lips and eyes increased with incremental intensity of hand grip task [1]. Khanal et al. reported increasing eye blink frequency from 8 times/min to 26 times/min at the beginning and end of incremental cycling tasks [13]. Miles et al. indicated increasing head movement at the occurrence of exhaustion from individuals in cycling task [20]. However, there is a lack of literature on the potential use of this approach for real-time onsite physical workload assessment in manufacturing. Thus, the purpose of this pilot study is to investigate the potential of using changes in facial features as a method for real-time physical workload assessment for a constrained physical task. The research questions include:

1. Are measured changes in facial features consistent for participants performing the same task during separate sessions?
2. How do facial features change over time during a repetitive, low-load task?
3. Are changes of facial features related to ratings of perceived exertion (RPE)?
4. How do changes of facial features correlate with performance?

2 Method

2.1 Participants

Four participants (3 males, 1 female; age, 29 ± 2 years) were recruited from the university population for pilot data collection. The study was approved by the University

Institutional Review Board and all participants provided written informed consent. Each participant completed two sessions of the assembly task on different days with at least a one-week interval. In total, eight datasets were captured for this pilot study.

2.2 Task Design

Each participant was asked to complete an assembly task, where s/he screwed a nut over a bold as fast as possible and as many times as possible, with his/her dominant hand, until exhaustion. The bolt was held on an assembly box (length 0.4 m × width 0.4 m × height 0.3 m) and located at the center of its upper surface. The assembly box was placed at the center of a workstation (length 0.8 m × width 0.5 m × height 0.7 m). The participant stood close to the workstation with their toes next to the edge of the workstation. S/he was asked to initiate the task posture with elbow flexion at 90°, which was confirmed by the researcher with a manual goniometer. The shoulder, forearm, and wrist position were adjusted to reach the bolt placed at fixed height (1.1 m from the ground). After adopting the posture, the participant was required to maintain the posture, without any interruption or breaks until the end of task, in order to induce a sustained physical workload. No arm support was provided to participants, to make the task similar to conditions experienced in the field.

2.3 Procedures

A repeated measures study design was used to investigate test-retest reliability of changes of facial features. There was at least one-week interval between the two sessions completed by each participant to provide sufficient time to recover from the first session. For each session, a camera (GoPro, Inc., San Mateo, California) was set at 30 cm away from the participant aligned to eye height. Another camera (Canon, Japan) was set to capture the participant performance, counting the number of finger movement during assembly task. Participants were instructed to maintain the same strategy task performance strategy over time, screwing up and down the nut with thumb, index and middle fingers, to minimize the influence of changes in strategy on workload and performance. After starting the assembly task, ratings of perceived exertion (RPE) of the arm, shoulder, back and whole body were collected via self-report every two minutes using the Borg CR-10 scale [2], with low (0, very light effort) and high (10, maximal effort) anchor points. Participants continued the task to exhaustion.

3 Data Processing

3.1 Facial Landmarks

Fourteen facial landmarks, corresponding to the features of the eyes, eyebrows, mouth, and head (Fig. 1), were calculated on the basis of Euclidean distance between 3D coordinates of the landmark for each frame relative to the first frame of task recording. The window of 10 s was selected right before the time point of self-reported RPE for every 2 min of the task. For example, there were 18 windows for the participant who continued the assembly task for 36 min.

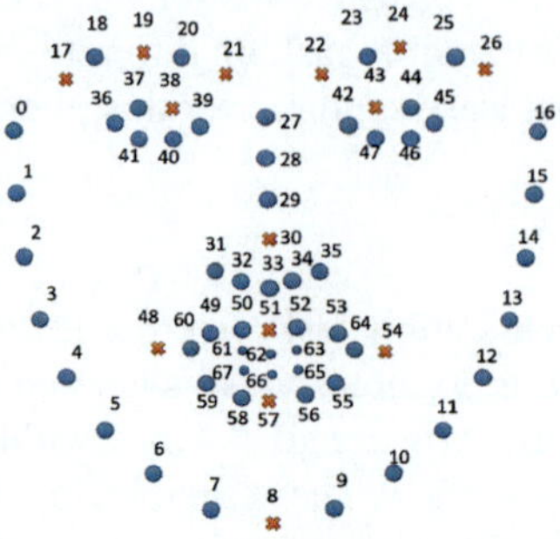

Landmarks	Features
8	Jaw
17	Right Outer Point of Eyebrow
19	Right Middle Point of Eyebrow
21	Right Inner Point of Eyebrow
22	Left Inner Point of Eyebrow
24	Left Middle Point of Eyebrow
26	Left Outer Point of Eyebrow
30	Head Movement
38	Right Eye Up Lid
43	Left Eye Up Lid
48	Right Corner of Lip
51	Lip Top Point
54	Left Corner of Lip
57	Lip Bottom Point

Fig. 1. Selected features with facial landmark numbers (selected facial landmarks are marked with an "X")

3.2 Pixel Movement

The implementation of 3D coordinates from 2D videos was realized by Convolutional Experts Constrained Local Model (CE-CLM) [28]. The Euclidean distance between the 2D coordinates of two landmarks may change with head movement in depth (e.g., nodding), even though there is no relative movement between the two landmarks. To account for the influence of head movement, the coordinate of each landmark was first subtracted by its coordinate at the first frame at the beginning of the task and then the head movement was subtracted. The head movement was captured by the coordinate of landmark 30 (Fig. 1) subtracted from the position at the first frame from the beginning of the task. The calculation is shown in Eq. 1,

$$P_{fn\_i} = \left(P_{n_i} - P_{i0}\right) - \left(P_{n_{head}} - P_{head_0}\right) \tag{1}$$

where $P_{fn_i} - \left(X_{fn_i}, Y_{fn_i}, Z_{fn_i}\right)$, 3D coordinates of landmark i in window n after subtracting head movement.

$P_{n_i} - \left(X_{n_i}, Y_{n_i}, Z_{n_i}\right)$, 3D coordinates of landmark *i* in window *n*.

$P_{i0} - \left(X_{i0}, Y_{i0}, Z_{i0}\right)$, 3D coordinates of landmark *i* at first frame at beginning of assembly task.

$P_{n_{head}} - \left(X_{n_{head}}, Y_{n_{head}}, Z_{n_{head}}\right)$, 3D coordinates of head in window *n*.

$P_{head_0} - \left(X_{head_0}, Y_{head_0}, Z_{head_0}\right)$ 3D coordinates of head at first frame at beginning of assembly task.

3.3 Data Analysis

Changes in RPE, performance, and average movement of selected facial features from the first to last window were analyzed using paired t-tests. Pearson product-moment correlation was used to assess the relationships between the facial feature changes and the RPE and task performance. Significance was set at $p < 0.05$.

4 Results

RPE scores for the arm, shoulder, and whole body increased significantly from the first to the last window (RPE$_{arm}$ $t = -2.99$, $p = 0.02$, RPE$_{shoulder}$ $t = -12.27$, $p < 0.001$,

RPE$_{wholebody}$ $t = -6.42$, $p < 0.001$). Mean maximum RPE of arm, shoulder, back and whole body were 5.63, 8.63, and 7.13, with standard deviations (SD) 2.50, 2.00, and 3.00, respectively. Performance, as the number of finger movements when screwing the nut, decreased 12.5% from the first to the last window.

4.1 Changes of Facial Features at the Start and End of the Task

Average pixel movement of facial features observed from the windows at the start and end of assembly task are shown in Table 1. The 2D locations of the facial landmarks at the start and end frame of the assembly task for each session are shown in Fig. 2. No major trends of the directions of changes of the facial features were observed.

Table 1. Paired t-test on RPE, %HRR, performance, and average pixel movement facial features of first and last window (across sessions). * indicates $p < 0.05$

	1st		Last		t-value	p-value
	Mean	S.D	Mean	S.D		
RPE$_{arm}$	2.88	1.64	5.63	2.50	−2.99	0.02*
RPE$_{shoulder}$	1.50	0.93	8.63	2.00	−12.27	<0.001*
RPE$_{back}$	0.88	0.64	2.63	2.07	−2.33	0.05*
RPE$_{wholebody}$	1.25	0.89	7.13	3.00	−6.42	<0.001*
Performance (# finger movements)	36.13	3.48	31.63	4.17	3.13	0.02*
Facial features (pixels)						
Jaw	7.58	2.33	11.85	5.71	−2.05	0.08
Right outer point of eyebrow	7.58	2.92	12.68	5.86	−2.65	0.03*
Right middle point of eyebrow	6.52	3.38	11.33	4.02	−3.08	0.02*
Right inner point of eyebrow	5.28	2.92	8.74	2.73	−2.76	0.03*
Left inner point of eyebrow	4.39	2.04	7.87	2.89	−3.04	0.02*
Left middle point of eyebrow	4.09	1.02	8.19	4.13	−2.82	0.03*
Left outer point of eyebrow	3.44	1.25	9.40	3.70	−3.69	0.01*
Head movement	14.80	4.06	27.78	8.94	−3.19	0.02*
Right eye upper lid	5.84	2.57	9.84	3.93	−2.71	0.03*
Left eye upper lid	3.26	1.20	7.10	3.26	−2.96	0.02*
Right corner of lip	4.87	1.21	6.97	2.67	−1.86	0.11
Left corner of lip	5.32	1.70	8.02	3.72	−1.84	0.11
Lip top point	3.02	1.01	4.39	0.94	−2.99	0.02*
Lip bottom point	4.34	1.43	6.86	2.09	−3.10	0.02*

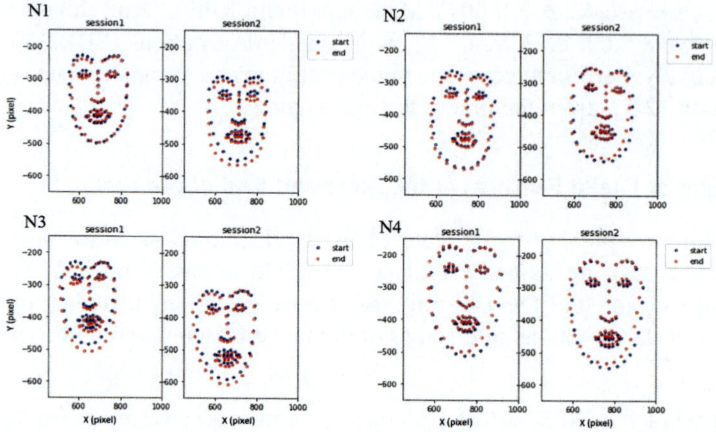

Fig. 2. Facial landmarks with 2D coordinates at the start and end of the assembly task

4.2 Correlation Between Changes of Facial Features and RPE, Performance

There were significant positive correlations between the average pixel movement of inner, middle and outer point of right and left eyebrow and the arm, shoulder, back, and whole body RPE scores ($r = 0.19$–$0.29, p < 0.01$). similarly, the average head movement were significantly correlated with RPE scores ($r = 0.38$–0.56, $p < 0.001$), as well as and right corner, top and bottom point of lip ($r = 0.32$–0.59, $p < 0.001$). Significant correlations were also found between average pixel movement of right and left inner eyebrow point (right: $r = -0.19, p = 0.03$; left: $r = -0.20, p = 0.01$), head movement ($r = -0.22, p = 0.01$) and lip bottom ($r = -0.17, p = 0.05$).

5 Discussion

In this study, head movement, left and right inner eyebrow, and lip bottom point increased significantly from first to last window close to the start and end of the assembly task, and showed significant correlation with RPE score and performance. The results of head and eyebrow movement changes are consistent with previous research that described increasing facial landmark movement may be correlated with increasing physical workload [1, 10, 20] Miles et al. found the facial region of eyebrow and head movement increased from low to high exercise intensity cycling task [20]. Asadi et al. reported that average pixel movement of the eyebrow and head increased highest at 100% maximum volume contraction (MVC) with increasing physical workload, through increasing grip exertions [1]. Haque et al. also indicated severe head vibration from individuals when exhaustion occurred in grip exertion [10].

However, there were no significant changes of jaw and lip corner, even though average pixel movement of the lip bottom point was found to increase significantly. Different findings were reported in previous studies. Significantly greater activity of the jaw and mouth was report with increasing physical workload induced by cycling intensity [1, 20]. Lip corner from both sides pulled up significantly when physical intensity was increased by percent of maximal weight lifted during an arm curl exercise [25]. The non-significant changes of jaw and mouth in this study may be due to the lower metabolic demand induced by this low-load task. In heavy intensity exercise, such as incremental cycling, breath strategy may change from nose to mouth in order to increase the air flow and absorb more oxygen to supply the energy [24].

Additionally, no major trend of general changes of facial features were observed from the start to the end of task based on 2D coordinates (Fig. 2). One possible explanation may be the individual difference in facial expression of emotions [6]. Different trends of changes of facial features may result from participants' different emotions at the end of task, such as relaxed (almost finishing the task) or nervous (still maintaining the speed of finger movement though exhaustion).

The present study discussed the potential use of changes of facial features in real-time physical workload assessment, while additional work remains before further investigating facial feature movement as a reliable and valid measurement to physical workload and translating this method into workspace. Current results of changes of features were demonstrated based on a small sample size and may not indicate the variability across the population and workspaces. Also, with more data collection from participants, intraclass correlation should be analyzed, in order to investigate the test-retest reliability of facial feature changes within subjects and between different sessions. Additionally, the task of constantly screwing the nut may only represent a limited subset of actual assembly tasks in the workspace. Extra features of tasks should be considered in future with types of postures (e.g. sitting), movements (e.g. push and place), so as to investigate the effectiveness of changes of facial features as physical workload measurement across various conditions in manufacturing.

References

1. Asadi, H., Zhou, G., Lee, J.J., Aggarwal, V., Yu, D.: A computer vision approach for classifying isometric grip force exertion levels. Ergonomics **63**(8), 1010–1026 (2020)
2. Borg, Gunnar: Borg's Perceived Exertion and Pain Scales. Human kinetics (1998)
3. Bosch, T., et al.: The effect of work pace on workload, motor variability and fatigue during simulated light assembly work. Ergonomics **54**(2), 154–168 (2011)
4. Brandt, M., et al.: Effects of a participatory ergonomics intervention with wearable technical measurements of physical workload in the construction industry: cluster randomized controlled trial. J. Med. Internet Res. **20**(12), e10272 (2018)
5. de Morree, H.M., Marcora, S.M.: Frowning muscle activity and perception of effort during constant-workload cycling. Eur. J. Appl. Physiol. **112**(5), 1967–1972 (2012)
6. Dhall, A., et al.: Static facial expression analysis in tough conditions: data, evaluation protocol and benchmark. In: 2011 IEEE International Conference on Computer Vision Workshops (ICCV Workshops). IEEE (2011)
7. Elliott, D., Charlton, L.: A preliminary investigation into the use of the emotional contagion effect in the exercise environment. Compr. Psychol. **5**, 2165222816634038 (2016)
8. Faber, M., et al.: Cognition-enhanced assembly sequence planning for ergonomic and productive human–robot collaboration in self-optimizing assembly cells. Prod. Eng. **11**(2), 145–154 (2017)
9. Goodie, J.L., et al.: Validation of polar heart rate monitor for assessing heart rate during physical and mental stress. J. Psychophysiol. **14**(3), 159 (2000)

10. Haque, M.A., et al.: Facial video-based detection of physical fatigue for maximal muscle activity. IET Comput. Vis. **10**(4), 323–330 (2016)
11. Huang, D.-H., et al.: Frowning and jaw clenching muscle activity reflects the perception of effort during incremental workload cycling. J. Sports Sci. Med. **13**(4), 921 (2014)
12. Jensen, M.P., et al.: The Measurement of clinical pain intensity: a comparison of six methods. Pain **27**(1), 117–126 (1986)
13. Khanal, S.R., et al.: Classification of exercise intensity by using facial expression analysis. In: 2018 Second International Conference on Computing Methodologies and Communication (ICCMC). IEEE (2018)
14. Lee, G., et al.: Higher physical workload risks with notes versus laparoscopy: a quantitative ergonomic assessment. Surg. Endosc. **25**(5), 1585–1593 (2011)
15. Lee, W., Migliaccio, G.C.: Physiological cost of concrete construction activities. construction innovation (2016)
16. Li, G., Buckle, P.: Current techniques for assessing exposure to work-related musculoskeletal risks, with emphasis on posture-based methods. Ergonomics **42**(5), 674–695 (1999)
17. Lutz, T.J., et al.: The use of mirrors during an assembly task: a study of ergonomics and productivity. Ergonomics **44**(2), 215–228 (2001)
18. Madgwick, S.O., et al.: Estimation of IMU and MARG orientation using a gradient descent algorithm. In: 2011 IEEE International Conference on Rehabilitation Robotics. IEEE (2011)
19. Maman, Z.S., et al.: A data-driven approach to modeling physical fatigue in the workplace using wearable sensors. Appl. Ergon. **65**, 515–529 (2017)
20. Miles, K.H., et al.: Facial feature tracking: a psychophysiological measure to assess exercise intensity? J. Sports Sci. **36**(8), 934–941 (2018)
21. Mo, S., Yang, W., Wang, G., Liao, Q.: Emotion Recognition With Facial Landmark Heatmaps. In: Ro, Y.M., et al. (eds.) MMM 2020. LNCS, vol. 11961, pp. 278–289. Springer, Cham (2020). https://doi.org/10.1007/978-3-030-37731-1_23
22. Peruzzini, M., Pellicciari, M.: A framework to design a human-centred adaptive manufacturing system for aging workers. Adv. Eng. Inform. **33**, 330–349 (2017)
23. Tankisi, H., et al.: Standards of instrumentation of EMG. Clin. Neurophysiol. **131**(1), 243–258 (2020)
24. Timme, S., Brand, R.: Affect and exertion during incremental physical exercise: examining changes using automated facial action analysis and experiential self-report. PloS ONE **15**(2), e0228739 (2020)
25. Uchida, M.C., et al.: Identification of muscle fatigue by tracking facial expressions. PloS ONE **13**(12), e0208834 (2018)
26. Yang, H., Zhao, G., Zhang, L., Zhu, N., He, Y., Zhao, C.: Real-time emotion recognition framework based on convolution neural network. In: Pan, J.-S., Li, J., Tsai, P.-W., Jain, L.C. (eds.) Advances in Intelligent Information Hiding and Multimedia Signal Processing. SIST, vol. 157, pp. 313–321. Springer, Singapore (2020). https://doi.org/10.1007/978-981-13-9710-3_33
27. Yu, Y., et al.: Estimating construction workers physical workload by fusing computer vision and smart insole technologies. In: ISARC. Proceedings of the International Symposium on Automation and Robotics in Construction. IAARC Publications (2018)
28. Zadeh, A., et al.: Convolutional experts constrained local model for 3D facial landmark detection. In: Proceedings of the IEEE International Conference on Computer Vision Workshops (2017)

sEMG and Postural Analysis for Biomechanical Risk Assessment in a Banknotes Printing Process

Lorenzo Fiori, Alessio Silvetti(✉), Antonella Tatarelli, Alberto Ranavolo, and Francesco Draicchio

DIMEILA, INAIL Research Area, Monte Porzio Catone (Rome), Italy
{lo.fiori,al.silvetti,a.tatarelli,a.ranavolo, f.draicchio}@inail.it

Abstract. The purpose of this paper is to assess the biomechanical overload risk of some tasks that are typical of the printing industry, by means of surface electromyography and postural analysis software.

In the first task of manual loading, muscle activation percentage duration of the Bicepses and Tricepses were similar for similar duty cycles. The Erector Spinae muscles had higher %MVC sustained for a greater percentage of the duty cycle. In the second task of aeration and transfer, Bicepses were activated for most of the duty cycle with a low %MVC between 0 and 5%. Differently, Triceps and Erector Spinae muscles had muscle activations with higher %MVC and higher percentage duration. In the third task, the muscular activity achieved the highest values in the Bicepses, while the activity of the Erector Spinae and Triceps muscles was less significant. Finally, in the fourth task, the compressive force values at the L5/S1 level found by 3DSSPP software ranged from 1072 N to 1863 N. Still at the L5/S1 level, shear forces ranged from 263 N to 310 N.

In the observed conditions, the used methods found no significant biomechanical overload risk in the studied tasks. The %MVC values within the cycle were all below the threshold proposed by ACGIH. The force values at level L5/S1 estimated with 3DSSPP software were also less than the 3400 N threshold limit value proposed by NIOSH for compressive forces and less than the 700 N limit proposed for shear forces by Gallagher in his review.

One of the observed activities at a faster working pace could increase the biomechanical risk. This is the case of the manual loading of the offset printing machine. In fact, the sheet board with the sheets to be loaded was positioned frontally to the printing machine loading area, thus forcing the operator to a 180° rotation.

Keywords: Ergonomic · Musculoskeletal disorders · Muscle fatigue · 3DSSPP

1 Introduction

Despite the high degree of automation in the typographic industry some tasks are still performed manually. In terms of biomechanical risk, these activities can be assessed

R. S. Goonetilleke et al. (Eds.): AHFE 2021, LNNS 273, pp. 297–304, 2021.
https://doi.org/10.1007/978-3-030-80713-9_38

using standardized and validated protocols [1, 2]. These methods can, however, be supplemented with electrophysiological and biomechanical investigations [3, 4] in order to achieve a more detailed view of the workers' workload.

Several studies have investigated musculoskeletal disorders (MSDs) in the printing industry. A first epidemiological study [5] showed that printing is one of the five industries with the highest prevalence rate of MSDs of the upper limbs in women.

Another epidemiological study [6], showed that among the various industries, the printing industry has a high prevalence rate of MSDs concerning wrists and fingers.

Very interesting is the analysis carried out by the Occupational Safety and Health Administration (OSHA). Starting from the task analysis, OSHA identified, for each working stage, the riskiest tasks and suggested possible corrective actions for reducing the biomechanical overload risk (https://www.osha.gov/SLTC/etools/printing/index.html). Moreover, Burke [7] reported a case study for possible actions in industrial printing plants, aimed at reduce the biomechanical risk in the task of rewinding paper following a jam.

This study investigated tasks selected following a preliminary investigation, namely the following:

1) Manual loading of sheets in the offset printing machine, from a pallet positioned at 180° with respect to the loading area (Task A).
2) Aeration of freshly printed sheets and their transfer to a new pallet. This task was carried out to prevent the freshly printed sheets from sticking to each other due to the presence of ink (Task B).
3) Transfer of the printed sheets from the pallet to the jogger for subsequent cutting (Task C).
4) Handling boxes weighing 10 kg from the roller conveyor to the pallet for shipping, at the end of the machining process (Task D).

This paper aims at assessing the biomechanical overload risk of the aforesaid tasks, by using myoelectric parameters combined with biomechanical models. To our knowledge there are no paper that used these tools in print industry.

Moreover, a deepen biomechanical characterization would help to identify appropriate ergonomic interventions.

2 Material and Methods

Analyses were carried out on three workers selected for their experience in the investigated tasks. No participants had any self-reported musculoskeletal injuries or disorders in the past 12 months.

Table 1 reports the anthropometric data of the workers who took part in the study.

In three of the four tasks (Task A, Task B, Task C), surface electromyography (sEMG) signals were recorded on both sides of the Brachial Biceps, Brachial Triceps and Erector Spinae muscles.

The sEMG signals were recorded by means of an electromyography system (FreeEMG, BTS SpA) equipped with 8 wireless surface probes at a sampling frequency

Table 1. Anthropometric data of the subjects involved in the study

Subject	Tasks	Height (cm)	Weight (kg)	Age (years)
Subject 1	Task A	179	71	47
Subject 2	Task B	179	80	47
Subject 3	Tasks C and D	165	67	41

of 1 kHz. These electromyography probes were equipped with an instrumentation amplifier with a common-mode rejection ratio of 100 dB and a Hamming filter with cut-off frequencies of 10 and 400 Hz. The probes were positioned by means of disposable pregelled electrodes Ag/AgCl (H124SG, Kendall Arabic, Donau, Germany) in accordance with the recommendations of the Atlas of Muscle Innervation Zones [8]. The electromyography signals were processed using MATLAB R2017B software (vers. 9.3.0, MathWorks, Natick, MA, USA). First, the signal was filtered in the frequency band of interest [20–450 Hz] using a digital filter and Butterworth 9th-order passband to reduce motion artifacts (electrode-to-skin) and other components of the high frequency noise. Subsequently, to extract the muscular activity profile, the signal was rectified and then filtered using a Butterworth 3rd-order low-pass filter with a cut-off frequency of 10 Hz, thus obtaining the linear envelope. The sEMG signals were normalized with respect to maximum voluntary contraction (MVC). MVC acquisitions were performed according to SENIAM instructions [9].

The percentage duration of muscle activations within the duty cycles (%DC) were calculated from the electromyography signals recorded for Tasks A, B, and C. These values were compared with the limits recently proposed by ACGIH [10]. The limits proposed by ACGIH (Fig. 1) are based on a mathematical model of muscle fatigue proposed by Potvin [11].

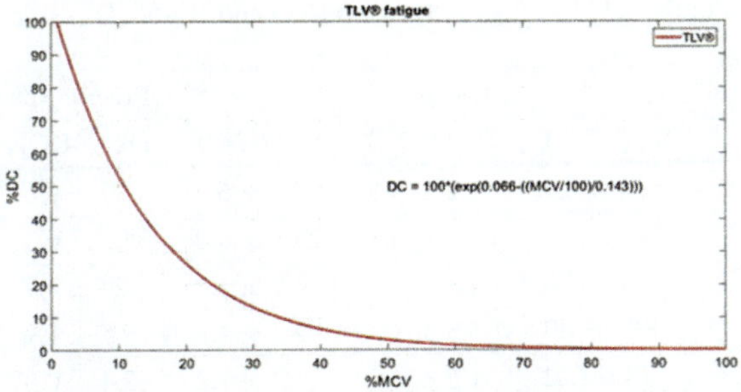

Fig. 1. The image shows the percentage duration limits of muscular activations within the duty cycles according to the mathematical model proposed by Potvin and used by ACGIH as a threshold limit for muscle fatigue in healthy subjects (red curve).

Task D was analyzed by means of 3DSSPP software [12–14], which allowed a load prediction of shear and orthogonal forces at the L5-S1 joint level. In particular, were analyzed those transfers that, the workers judged as the most demanding, i.e., those on the penultimate (ninth) and in the last (tenth) row of the pallet.

To assess the daily workload required by Task D, the handled weight was compared with the table of the cumulative mass per day, which refers to the handling distance indicated in ISO 11228-1 [1].

3 Results

3.1 sEMG Results

The workers' muscular activity during Task A, B and C are reported in Tables 2, 3 and 4. The tables show the average values (±SD) of %DC in which MVC percentages were comprised between 0% and 30%, discretized into 5% bands. The results relating to percentages higher than 30% with respect to the MVC have been not reported because occasional.

Task A
In Task A the right Biceps was active for most of the duty cycle (60.7%), with a %MVC usually between 0 and 5 % of the MVC, and for 23.8 % of the DC with an activation between 5 and 10% of the MVC. For the Erector Spinae muscles a muscular activation was recorded between 0 and 10% of the MVC sustained for two thirds of the DC (63.3% right Erector Spinae and 66.8% left Erector Spinae), as well as a significant activation between 10 and 15 % of the MVC sustained for 24% of the DC on the right side and 17.9% of the DC on the left side. Finally, the Tricepses showed activation levels ranging between 0 and 10% of the MVC for three quarters of duty cycle duration (75.7% of the DC for the right side and 81.9% of the DC for the left side). Levels of 25-30% of the MVC were achieved for a significant period only by the left Biceps (2.4% of the DC).

Table 2. MVC percentage values (discretised into 5% bands) and their relative percentage duration of activation in the DC for Task A for Left (Sx) and Right (Dx) side

	Biceps		Triceps		Erector Spinae	
%MVC	Dx (%DC)	Sx (%DC)	Dx (%DC)	Sx (%DC)	Dx (%DC)	Sx (%DC)
0–5%	60.7 ± 5.2	39.9 ± 6.7	52.2 ± 5.6	58.1 ± 6.2	27 ± 9.6	34.9 ± 12.8
5–10%	23.8 ± 4.7	25.5 ± 3.6	23.5 ± 6.3	23.8 ± 4.0	36.3 ± 6	31.9 ± 6.5
10–15%	9.7 ± 2.3	14 ± 2.8	14.3 ± 3.7	9.2 ± 1.4	24 ± 4.7	17.9 ± 4.9
15–20%	3.3 ± 1	8.5 ± 1.9	5.3 ± 3	3.9 ± 1.3	9.5 ± 1.9	8.3 ± 2.3
20–25%	1 ± 0.3	5.1 ± 1.2	1.6 ± 0.5	2 ± 1.2	2.4 ± 0.8	3.9 ± 1.9
25–30%	0.6 ± 0.5	2.4 ± 0.6	0.8 ± 0.4	1.1 ± 0.9	0.6 ± 0.4	1.9 ± 0.8

Task B
In task B the Bicepses were active for most of the duty cycle (82.2% the right Biceps and 85.9% the left Biceps) with a %MVC between 0 and 10%. For the Erector Spinae

muscles, a muscular activation was recorded between 0 and 10% of the MVC sustained for over two thirds of the DC (70.1% right Erector Spinae and 70.5% left Erector Spinae); relevant was also the activation between 10 and 15% of the MVC sustained for 20.1% of the DC on the right side and of 18.8% of the DC on the left side. Finally, the Tricepses, like the Erector Spinae muscles, showed activation levels ranging between 0 and 10% of the MVC for over two thirds of the duty cycle duration (73.7% of the DC for the right side and 68.9% of the DC for the left side). Levels of 25–30% of the MVC were achieved for a relevant period only from left Triceps (2% of the DC).

Table 3. MVC percentage values (discretised into 5% bands) and their relative percentage duration of activation in the DC for Task B for Left (Sx) and Right (Dx) side

	Biceps		Triceps		Erector Spinae	
%MVC	Dx (%DC)	Sx (%DC)	Dx (%DC)	Sx (%DC)	Dx (%DC)	Sx (%DC)
0–5%	63.2 ± 5.2	66.6 ± 3.9	45.2 ± 6.9	37.8 ± 5.2	37 ± 5.1	42.9 ± 7.4
5–10%	19 ± 1.9	19.3 ± 2.6	28.5 ± 2.3	31.1 ± 3.9	33.1 ± 3	27.6 ± 4
10–15%	7.7 ± 1.4	7.1 ± 1.1	13.5 ± 2.5	14.8 ± 2.1	20.1 ± 3.6	18.8 ± 2.6
15–20%	3.7 ± 1	3.2 ± 0.8	6.1 ± 1.4	6.6 ± 1.2	7.3 ± 2.1	7.1 ± 1.9
20–25%	2 ± 0.7	1.7 ± 0.6	2.9 ± 0.8	3.3 ± 0.8	1.7 ± 0.7	2 ± 0.6
25–30%	1.1 ± 0.4	0.8 ± 0.4	1.7 ± 0.6	2 ± 0.6	0.4 ± 0.2	0.9 ± 0.5

Task C

In Task C the Bicepses and Tricepses were active for most of the duty cycle, between 68.6% and 77.6% of the DC, for MVC percentages between 0 and 5%. The Erector Spinae muscles were the most involved (35.4% right Erector Spinae and 43.4% left Erector Spinae) for activation levels between 5 and 10% of the MVC. Activation was relevant between 10 and 15% of the MVC (16.8% of the DC for right Erector Spinae and 17.9% of the DC for left Erector Spinae). All activation levels over 20% of the MVC were not relevant, except for the right Erector Spinae, which was active for 3.7% of the DC.

3.2 3DSSPP Results

Table 5 shows the values of the compression and shear forces at the level of the L5-S1 lumbosacral joint. The values obtained for compressive forces ranged from a minimum of 1042 N to a maximum of 1863 N, and those for shear forces ranged from 263 N and 310 N.

Figure 2 is an example of the reconstruction made with 3DSSPP software. Specifically, it shows the results of the handling a 10 kg box, with the box positioned rear in the tenth row.

Table 4. MVC percentage values (discretised into 5% bands) and their relative percentage duration of activation in the DC for Task C for Left (Sx) and Right (Dx) side

	Biceps		Triceps		Erector Spinae	
%MVC	Dx (%DC)	Sx (%DC)	Dx (%DC)	Sx (%DC)	Dx (%DC)	Sx (%DC)
0–5%	77.6 ± 4.3	74.3 ± 2.4	74.7 ± 3.1	68.6 ± 5.4	34.2 ± 6	34.4 ± 3.9
5–10%	12.3 ± 2.6	16.2 ± 2.2	13.6 ± 2.0	18.6 ± 4.2	35.4 ± 4.4	43.3 ± 4.1
10–15%	5.2 ± 1.2	4.4 ± 0.6	5 ± 1	5.6 ± 1.2	16.8 ± 2.1	17.9 ± 1.7
15–20%	2 ± 0.9	1.7 ± 0.2	2.2 ± 0.3	2.7 ± 0.6	8.2 ± 0.8	4 ± 0.8
20–25%	1 ± 0.6	1.2 ± 0.3	1.5 ± 0.3	1.3 ± 0.3	3.7 ± 0.4	0.3 ± 0.2
25–30%	0.4 ± 0.2	0.7 ± 0.3	1 ± 0.2	0.8 ± 0.3	1.3 ± 0.5	0.03 ± 0.05

Table 5. The table shows the values (in Newton) of compression and shear forces at the lumbosacral joint level, for each of the examined frames

L5-S1 sacral zone		
Position	Compression (N)	Shear (N)
9th row left rear	1772	306
9th row front left	1338	304
9th row right rear	1863	310
9th row right front	1042	291
10th row left rear	1609	305
10th row front left	1072	270
10th row right front	1189	263
10th row right rear	1482	308

3.3 ISO 11228-1 Results

For assessing the biomechanical load throughout the entire working day for Task D, the handling distance and frequency limits proposed by the ISO 11228-1 for the cumulative mass per day were used. The reference limit values that are closer to the working conditions observed are those of 10000 kg/8 h, 7200 kg/h and 120 kg/min for a handling distance of 1 m.

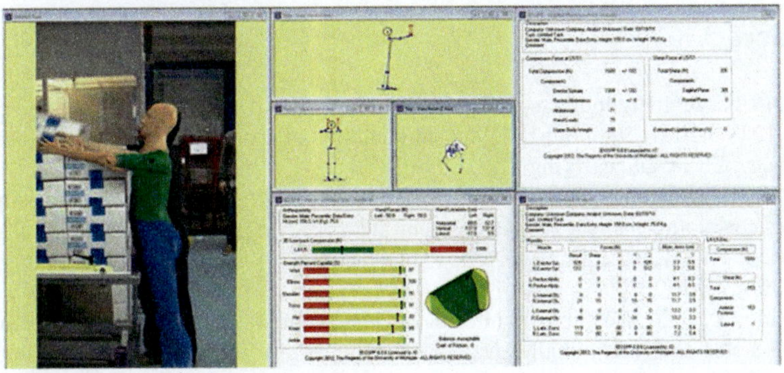

Fig. 2. The image shows the 3DSSPP reconstruction of the task of handling on the pallet

The observed work pace was four pallets per work shift. This frequency corresponds to the lifting of 160 boxes of 10 kg each per shift, i.e., reaching a cumulative mass handled per day of 1600 kg/8 h, 200 kg/h and 30 kg/min.

4 Discussion

In the four investigated tasks, our results revealed non-significant biomechanical risk levels.

However, these results must refer to the peculiar type of observed production. In fact, different results could be found for the same tasks if performed for the handling of sheets of other sizes or at a different pace and in a different working environment.

The muscular activity analysis in Tasks A, B, and C, recorded by sEMG, has shown that for all the investigated muscles, the MVC percentages were maintained for duty cycle percentage duration below the TLV proposed by ACGIH.

Even the study of pallet handling (Task D) with 3DSSPP revealed no significant levels of biomechanical overload risk. The force values at the L5-S1 level were estimated between 1042 and 1863 N, and therefore below the 3400 N limit proposed by NIOSH [15]. Furthermore, the shear forces, between 263 and 310 N, were far from the 700 N threshold limit value proposed by Gallagher in his review [16].

At the observed work pace, Task D was below the limit values even with respect to the cumulative mass per day proposed by ISO 11228-1. At the observed frequencies, the handled kilos were lower than those proposed by the standard for a handling distance up to 4 m.

However, during the inspections we carried out, some conditions were observed that at a faster working pace could increase the biomechanical risk. This is the case of the manual loading of the offset printing machine. In fact, the pallet was positioned frontally to the printing machine loading area, thus forcing the operator to a 180° trunk rotation. Positioning the pallet at 90° with respect to the loading zone, as in most recent equipment, would allow a risk reduction.

References

1. ISO/DIS 11228-1 Ergonomics—Manual handling—Part 1: Lifting, lowering and carrying
2. ISO 11228-3:2007 Ergonomics—Manual handling—Part 3: Handling of low loads at high frequency: lifting and carrying
3. Alberto, R., Draicchio, F., Varrecchia, T., Silvetti, A., Iavicoli, S.: Wearable monitoring devices for biomechanical risk assessment at work: current status and future challenges-a systematic review. Int. J. Environ. Res. Publ. Health **15**(9), pii: E2001 (2018). https://doi.org/10.3390/ijerph15092001
4. Ranavolo, A., et al.: Wearable monitoring devices for biomechanical risk assessment at work: current status and future challenges-a systematic review. Int. J. Environ. Res. Publ. Health **15**(11) (2018)
5. Roquelaure, Y., et al.: Epidemiologic surveillance of upper-extremity musculoskeletal disorders in the working population. Arthritis Rheum. **55**(5), 765–778 (2006)
6. Yu, W., et al.: Work-related injuries and musculoskeletal disorders among factory workers in a major city of China. Accid. Anal. Prev. **48**, 457–463 (2012)
7. Burke, G.: Using ergonomics in the printing industry. AAOHN J. **56**(1), 41–42 (2008)
8. Barbero, M., Merletti, R., Rainoldi, A.: Atlas of Muscle Innervation Zones: Understanding Surface Electromyography and Its Applications. Springer, Milano (2012). https://doi.org/10.1007/978-88-470-2463-2
9. Hermens, H.J., Freriks, B., Merletti, R., et al.: European recommendations for surface ElectroMyoGraphy (SENIAM). 2000b Ed CLUT
10. ACGIH: TLVs and BEIs, pp. 202–204 (2015). ISBN 978-1-607260-77-6
11. Potvin, J.: Predicting maximum acceptable efforts for repetitive tasks: an equation based on duty cycle. Hum. Fact. **54**(2), 175–188 (2012)
12. Chaffin, D.B., Erig, M.: Three-dimensional biomechanical static strength prediction model sensitivity to postural and anthropometric inaccuracies. IIE Trans. **23**(3), 215–227 (1991)
13. Chaffin, D.B.: Biomechanical modeling for simulation of 3D static human exertions. In: Computer Applications in Ergonomics, Occupational Safety and Health. Elsevier Publishers, B.V., Amsterdam (1992).
14. Chaffin, D.B., Andersson, G.B.J., Martin, B.J.: Chapter 6 of Occupational Biomechanics, 4th edn. Wiley, Hoboken (2006)
15. Waters, T., Putz-Anderson, V., Garg, A. (eds.): Applications Manual for the Revised Lifting Equation, (DHHS/NIOSH) Pub. No. 94-110. CDC, NIOSH, Cincinnati (1994)
16. Gallagher, S., Marras, W.S.: Tolerance of the lumbar spine to shear: a review and recommended exposure limits. Clin. Biomech. **27**(10), 973–978 (2012)

Diagnostics of the Stress State by the Method of Pupillography

Isaeva Oksana(✉) and Boronenko Marina

Yugra State University, 16 Chekhov street, 628012 Khanty-Mansiysk, Russia

Abstract. Existing systems for recognizing the emotional state of a person draw conclusions from facial expressions, galvanic reactions of the skin, electrocardiogram and pulse, analysis of sound waves, vibraimage. In this case, a person can deceive a car. Emotions greatly affect pupil size. This is due to the unambiguous uncontrolled reaction of the human body to the information received. In this way, the reliability of emotion detection can be increased. Our goal: to develop a method for remote diagnostics of a person's stress state using the pupillography method. In accordance with the set goal, conditions were found under which test objects create the same illumination of the eye surface; a technique has been developed that increases the accuracy of tracking the center of attention in the absence of limiting head movements; a calibration method has been selected, relative to which the stress state of a person can be determined.

Keywords: Pupillometry · Emotion · Test objects

1 Introduction

To determine the stressful state, you can use facial expressions, pulse, vibraimage. At the same time, a person's ability to restrain mimic expression of emotions can be attributed to the disadvantages of the method for determining emotions by facial expressions. The non-contact heart rate measurement method is too sensitive to the amount of movement. Emotional control by sound waves does not work if the person is silent. Vibraimage identification does not register small changes in the psychophysical state. Therefore, all these methods must be used together. An uncontrolled reaction of the human body to heard (seen) information, which manifests itself in a change in the size of the pupils, can expand the capabilities of such contactless systems for determining a stress state. Pupillograms are actively used for diagnostics in ophthalmology, neuropathology, and narcology [1]. Real-time diagnostics of mental disorders has been developed [2]. As a result of a comparative analysis of the shape of the pupils of the subjects with samples of the pupils of people with known mental disorders, thereby diagnosing the disease. It is also known that the diameter of the pupils is associated with theta, the brain rhythm [3], which is a reliable marker of stress [4], and, if necessary, is used as an indicator of object recognition by individuals [5]. It has also been established that there is a relationship between microsaccades and attention with cognitively modulated brain potentials, pupil diameter and psychophysical state of a person [6]. The peculiarity of remote diagnostics

R. S. Goonetilleke et al. (Eds.): AHFE 2021, LNNS 273, pp. 305–312, 2021.
https://doi.org/10.1007/978-3-030-80713-9_39

of stress (stress) state is that the respondents do not have any special expensive equipment. Therefore, our goal was to develop a technique that allows remotely, without specialized equipment, to monitor changes in the size of the pupils and assess the emotional state (tension, stress).

2 Experimental Methods and Techniques

As you know, a person involuntarily directs the most attention to what is most significant. Therefore, a person who turns to a psychologist for help must have a topic that especially excites him. Any information, especially new information related to this topic, is significantly important for a person. In this regard, our research has the following logic. Mental, depressive or stressed state of a person is accompanied by a change in his emotional response to a stimulus. A person's emotional reaction leads to a change in the size of the pupils to one degree or another. With a collection of images and videos on the most common mental health topics, pupillographic studies can be done. To achieve this goal, it is necessary to solve the following tasks: development of a methodology to determine the presence of a stress state; checking the technical ability to track the track of attention without special equipment.

During the receipt of meaningful information, depending on the depth of interest, there is a concentration of attention or even an emotional reaction arises. In order for the images to arouse increased human attention, a survey was conducted (according to the results, χ^2 was calculated). Anyone took part in the survey. In the future, it is planned to involve doctors and patients with psychological disorders for research. To determine the emotional response of a person to the presented stimulus, a presentation was made from images that received an overwhelming number of votes. The greatest emotional response (χ^2) was received by images that to one degree or another evoke the emotion "fear". Each stimulus was presented once. A total of 9 images were presented, each slide was shown for 5 s. At the beginning of the presentation, there was always a calibration slide. To check the equivalence of the eye-tracking results in the absence of a rigid coordinate connection between the recorder and the head, we used two video cameras. One was attached to a pupillographic module, the second was located separately (Fig. 1a). Since the distance between the camera and the observed scene is much greater than the focal length of the optical system, we can assume that the image is built in its focal plane (Fig. 1b). The man sat on a chair. The stimulus material was demonstrated from a monitor located at a distance, at which the fluctuations of illumination (Fig. 1c) from it are insignificant [7].

After the end of the experiment, the subject was asked to assess the emerging emotional state when viewing images. All the subjects were volunteers, they are satisfied with the results, they have no complaints. The video files with images of the pupils were analyzed in the free software Fi-Ji, an improved and extended distribution of ImageJ [8], which integrates many plug-ins that allow full scientific analysis of the image. After the preparation of the video files, the pupil was outlined in each frame. Results visualization and statistical analysis were performed using the OriginPro 2019 software. When analyzing pupillograms, the data were cleaned from artifacts caused by blinking.

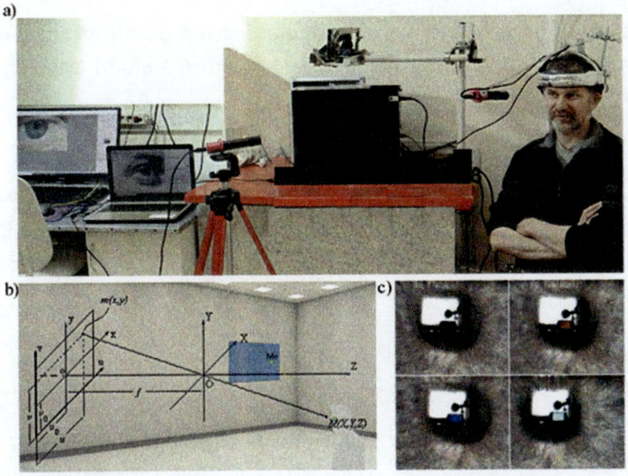

Fig. 1. a) Observation of pupillary reaction with two video cameras; b) Camera standard coordinate system; c) Change in the size of the pupil under the influence of light of different intensities (all other things being equal).

2.1 Comparative Analysis of Eye-Tracking of the Pupillographic Module and a Separate Video Camera

To check the identity of the tracks received by the video camera 1 - fixed on the head and the video camera 2 – "free", that is not having a rigid coordinate connection with the head, the following experiment was carried out. The first astro-ZWO-120 video camera is located at a distance of 40 cm from the plane of the eyes. The choice of the brand of the second camcorder (Canon SX620HS, 25X optical zoom) is due to its prevalence. The second video camera is independent, (free), mounted on a tripod 250 cm from the plane of the eyes. The video was filmed at 30 fps. For the obviousness of the results obtained, the subjects had to look at a black dot (6 mm diameter). To study the effectiveness of the technique for eliminating track distortions (while viewing a slide with a dot), you need to shift and turn your head without taking your eyes off the point. The internal time of the cameras is synchronized to thousandths of a second. Camcorders recorded the pupillary response during the slide show. The area of the pupil, the coordinates of its center of mass, and the coordinates of the center of mass of the monitor's reflection on the cornea were measured in each frame of the video files. Since all attention was focused on the point, the coordinates of the center of attention in each frame should be grouped near one center. The smaller the spread, the more accurate the result. However, the tracks of the spotlight before the adjustment are a complex trajectory (Fig. 2a, b). Therefore, without correction, the use of such a technique is impossible. An increase in the accuracy of determining the coordinates of the center of attention is carried out by moving from a coordinate system associated with a stationary head to a coordinate system associated with a moving center of the pupil. For this, an image of the monitor reflected in the pupil was isolated on the obtained video frames. Then the center of mass of the selection was tracked in each frame and its coordinates were determined. The coordinates of the center of mass of the monitor reflection in the pupil and the pupil diameter measured in each frame were used to construct pupillograms and oculograms. After the performed actions, both tracks are localized in a small area, corresponding to the observed point (Fig. 2c).

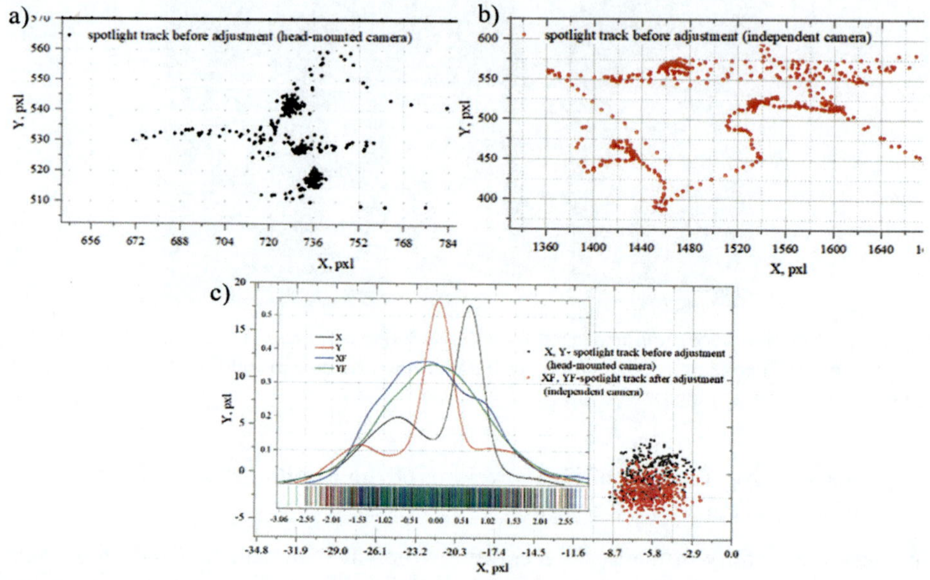

Fig. 2. Tracks of attention before correction, obtained using: a) a video camera with a rigid connection with the head; b) free video camera; c) Statistical: Histogram + Rug/Distribution + Rug and localization of the track of attention around the point.

Thus, visually the results are the same. If the numerical results coincide, then the technique can be used for remote control of the pupillary response to a stimulus. Checking the data for normality gave a negative result, despite the fact that the sample sizes are large enough. The standardization of variables also did not give the desired result. Therefore, nonparametric criteria were used. Let us formulate testable hypotheses. Null Hypothesis: The samples come from the same population. Alternative Hypothesis: The samples come from different populations. At the 0.01 level, the populations are not significantly different. Thus, we can assume that remote eye tracking according to the developed methodology gives a reliable result.

2.2 Controlling Unplanned Light Changes

For successful recognition of the stressed state of a person, among other parameters, it is necessary to measure the size of the pupils. The area of the pupil is more resistant to change than the diameter. As you know, the size of the pupils is strongly influenced by the fluctuation of illumination. Therefore, if the distance to the demonstration monitor is less than the permissible [9], then the stimulus material (images) should be converted to an equal brightness form. Luma flattening is done in ImageJ by creating a stack of all used images and normalizing local contrast. It also calculates the total number of pixels of images with a certain brightness. Figure 3 shows the statistics of the total brightness of the images before and after the alignment procedure.

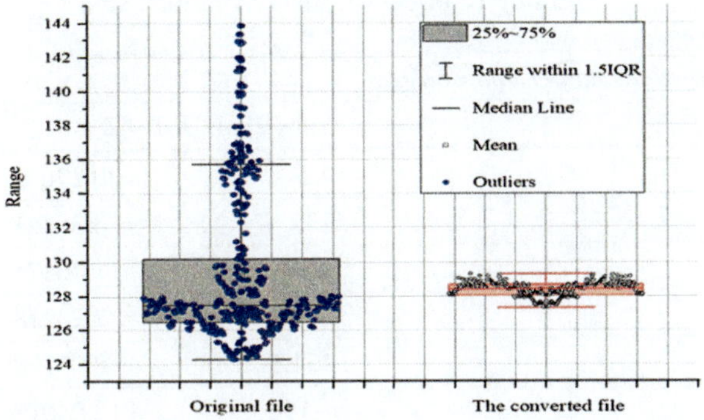

Fig. 3. Descriptive Statistics of images brightness (test objects).

Since the differential threshold of the light sensitivity of the eye is 1–1.5%, the images obtained fully satisfy the given condition. In this case, it is considered that the glow of the entire surface of the monitor is uniform. Additional control of illumination can be carried out on the area of the skin around the eyes. With the pupil illumination E_1, the pupil area S_1, the average brightness of the pixels included in the selected area of the skin near the pupil in grayscale I_1. Then, with an unplanned change in illumination E_2, the area of the pupil will become S_2. The brightness of the skin I_2. It is known that the transformation of light by the photo matrix is linear. Therefore, the brightness of the image is proportional to the illumination of the photomatrix. Then the change in illumination in $E_2 \backslash E_1 = I_2 \backslash I_1$ times is taken into account by introducing a correction factor $k = I_1 \backslash I_2$. In this case, the illumination in the room where the experiments are carried out must be uniform and meet the requirements of regulatory documents.

3 Emotional Response Experiments

More than 100 people took part in the experiments. The best 30 were selected from the total number of video files received (typical number of blinks). Video filming was carried out on a ZWO-120 video camera at a speed of 30 fps; images to which an emotional reaction is most likely were used as a stimulus material. The images were selected according to the previously described method. Among other images, a calibration slide was required. After the experiment, the survey showed that some respondents were tense. Their files were tagged. After processing and analysis, the pupillograms were compared with the described feelings of the subjects. The results obtained were subjected to statistical analysis.

All received pupillograms contain reactions to different test objects. At the same time, a tense (emotional) state manifested itself on a specific slide. If the pupillograms belong to one person, then the analyzed groups of pupillograms are dependent. In this case, for coherent samples, statistical hypotheses are tested by Friedman ANOVA (the results are presented in Table 1.

Table 1. Friedman ANOVA results.

Wilcoxon-Nemenyi-McDonald-Thompson Test	Sum Rank Diff	Z	Prob	Sig
"3 Gy" "2 Gy"	4.5	0.27643	0.99968	0
"3 Gy" "h"	136	8.35441	0	1
"3 Gy" "7"	−191.5	11.76375	0	1
"3 Gy" "1 Gy"	−36.5	2.24218	0.51031	0
"2 Gy" "h"	131.5	8.07798	0	1
"2 Gy" "7"	−196	12.04018	0	1
"2 Gy" "1 Gy"	−41	2.51861	0.38644	0
"h" "7"	−327.5	20.11817	0	1
"h" "1 Gy"	−172.5	10.59659	0	1
"7" "1 Gy"	155	9.52157	0	1

Sig equals 1 indicates that the difference of the means is significant at the 0.01 level. Sig equals 0 indicates that the difference of the means is NOT significant at the 0.01 level.

Results of comparing pupillary response using the Friedman ANOVA test in groups of pupillograms of a person without emotional response (h), emotional outburst (7) and response on the calibration slide showed that the levels are significantly different. Therefore, on any pupillogram it is possible to distinguish from 1 to 4 zones (Fig. 4).

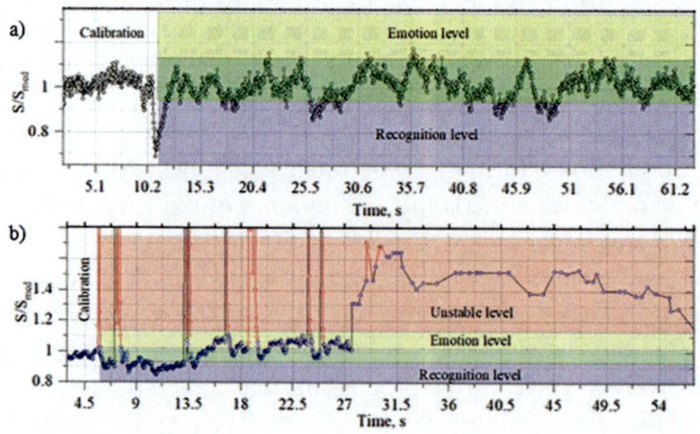

Fig. 4. a) Pupillogram of a mentally stable reaction; b) Pupillogram with the manifestation of an emotional reaction, indicating mental stress.

In Fig. 4a) the pupillogram was taken without the use of special equipment. All experienced emotions are in the "green" zone, everything is normal. Figure 4b) a pupillogram from experiments with a pupillographic module, laboratory studies. The method for obtaining pupillograms is the same, the stimulus material was static images. Several stimuli contained photographs of insects (pupillogram in the yellow emotional zone). Viewing a cute spider corresponds to a pupillogram in the instability zone (red level). This is a pupillogram of a girl who is afraid of spiders (phobia). After the experiment, the girl reported exactly this. Red peaks are not counted, this is blinking.

Thus, by measuring the width of the zone of the calibration pupillogram, and marking it on the entire pupillogram (to make an individual adjustment of the pupillographic module), it becomes possible to quickly determine the stress state of a person.

4 Discussion

Obtaining pupillograms in the context of remote experiment implementation has its own nuances. The technique needs to be adjusted taking into account the fact that the user does not have any special equipment. The measurements will be carried out under the following conditions: there is no infrared illumination, the head is not fixed, the stimulus material is shown from a laptop screen from a close distance, the registration of pupils is carried out without a magnifying lens. However, if the spatial resolution of video cameras is weak, then the procedure becomes useless. An alternative option could be a smartphone video camera. The distance between the screen surface and the person's face is determined by the convenience of the respondents. Usually, a person places his head parallel to the screen, the projection is almost in the center. In this case, the distance was 1 m. At this distance, the change in luminous flux when changing frames has a significant effect on the size of the pupils. In order to eliminate this problem, the video files used as stimulus material were converted to equal brightness.

The method of taking into account the influence of head rotation on the coordinates of the center of attention allows eye-tracking without additional devices. From the given tracks it becomes clear that if you do not correct the coordinates of the center of attention, then the conclusion will be false. In general, the intensity of the evoked emotions on a scale from 1 to 10 did not exceed 4, where 10 is the strongest emotion that can be. However, it can be assumed that for a girl with a phobia, the level of emotion was quite high, most likely a threshold. Therefore, when developing a stimulus material, one should remember about possible extreme unexpected reactions of respondents.

5 Main Results

The problem of determining the concentration of a subject's attention to a stimulus was recently solved by measuring the microsaccadian dynamics of the subject's eye movement [10]. It has also been shown that pupil dilation reliably reflects the decision-making process and there is a noticeable pupil response when a target is detected, even in the presence of a distraction [11]. In our research, we attempt to expand the scope of application of synchronized puppilography and oculography to identify a person's stressed state by their excessive emotionality.

In accordance with the purpose of the research, the stimulus images were converted to equal brightness. A technique has been developed that increases the accuracy of tracking the center of attention (determining the angle of displacement) in the absence of limitation of head movements. The developed techniques for constructing pupillograms were tested in the absence of a rigid limitation of head movements (solid angle of 30 degrees), specialized equipment and infrared illumination. Pupillograms were analyzed offline. A method of calibration has been chosen, in relation to which it is possible to identify a stressed mental state of a person. The developed technique can be used for remote diagnostics. The results also contribute to the development of machine learning and artificial intelligence systems; open up new opportunities for early diagnosis of mental disorders of a person and the study of his psychophysical state remotely.

Acknowledgments. The study was carried out with the financial support of the Russian Foundation for Basic Research in the framework of the research project 18-47-860018 p_a.

References

1. Pashkov, A.A., Dakhtin, I.S., Kharisova, N.S.: Electroencephalographic biomarkers of experimentally induced stress. Bull. South Ural State Univ. Ser. Psychol. **10**(4), 68–82 (2017)
2. Ohayon, J.: Pupil distortion measurement and psychiatric diagnosis method. U.S. Patent Application No. 15/853, 899
3. Mesin, L.: Estimation of complexity of sampled biomedical continuous time signals using approximate entropy. Front. Physiol. **9**(710) (2018). (Edited by, Mesin, L.)
4. Sapountzis, P.: Neural signatures of attention: insights from decoding population activity patterns. Front. Biosci. **23**, 221–246 (2018). (Edited by, Sapountzis, P., Gregoriou, G.G.)
5. Otero-Millan, J.: System and method for using microsaccade dynamics to measure attentional response to a stimulus: Patent US 9,854,966; assignee: Dignity Health 14/359,235; filed 23.11.12 (2013). pub. WO2013/078462.–10. (Edited by, Otero-Millan, J., Macknik, S.L., Martinez-Conde, S.)
6. Boronenko, M.P., Zelensky, V.I., Kiseleva, E.S.: Application of waves of attention as a marker of hidden intentions. Natl. Psychol. J. **2**(34), 88–98 (2019)
7. Bobrova, D., Bikberdina, N., Boronenko, M.: Investigation of the color effect of the test object on the pupil response. J. Phys. Conf. Ser. **1124**(3), 031016 (2018). IOP Publishing
8. Oksana, I., Marina, B.: Application of ImageJ program for the analysis of pupil reaction in security systems. J. Phys. Conf. Ser. **1519**, 012022 (2020)
9. Bobrova, D., Boronenko, M., Bikberdina, N.: The empowerment of biometric systems of psychophysical research. J. Phys. Conf. Ser. **1410**(1) (2019). IOP Publishing
10. Otero-Millan, J., et al.: Modeling the triggering of saccades, microsaccades, and saccadic intrusions. Front. Neurol. **9**(346), 1–12 (2018)
11. Strauch, C., Greiter, L., Huckauf, A.: Pupil dilation but not microsaccade rate robustly reveals decision formation. Sci. Rep. **8**(1), 1–9 (2018)

Effectiveness of Reduced Work Pace to Decrease the Risk of Work-Related Musculoskeletal Disorders in a Chicken Slaughterhouse

Diogo Cunha dos Reis(✉), Adriana Seara Tirloni, and Antônio Renato Pereira Moro

Federal University of Santa Catarina, Florianópolis, SC, Brazil
adri@tirloni.com.br, renato.moro@ufsc.br

Abstract. The aim of this study was to evaluate the risks in relation to workers' repetitive movements of the upper limbs, as well as analyze the effects of a reduced work pace on the risk levels in a chicken slaughterhouse. The study was conducted in a slaughterhouse with 3,200 workers, who were divided into two shifts, where 300,000 chickens were slaughtered daily. The OCRA checklist was used to assess 10% of the total workforce during work tasks. The average OCRA checklist score was 18.2 ± 4.8 (moderate risk). Considering the five risk categories proposed by the OCRA method, 2 tasks were deemed high risk (9.5%), 15 presented moderate risk (71.5%) and 4 exhibited low risk (19%). By conducting simulated interventions, it was possible to reduce the risk of work-related musculoskeletal disorders to very low levels in 19 of the 21 tasks by only decreasing the work pace (−39.8 ± 16.8%).

Keywords: Risk assessment · Ergonomics · Work pace · Slaughterhouse · OCRA

1 Introduction

In recent years, Brazil has been consolidating a leading role in the world's meat production and exports [1]. This position was achieved due to the abundance of land, favorable climate, and fertile soil to produce food, along with the companies' ability to overcome challenges [2]. Most Brazilian chicken meat is processed manually (knives) or by means of specific machines, considering that the cuts are the main product of this industry [1]. Therefore, slaughtering and meat processing work involve high loading intensities and cyclic repetitive muscle actions of the upper limbs, and thus implies an elevated risk of work-related musculoskeletal disorders (WMSDs) [3, 4].

The OCRA method resulted from a consensus document developed by the Technical Committee on Musculoskeletal Disorders of International Ergonomics Association (IEA) endorsed by the International Commission on Occupational Health (ICOH) [5]. This method was established to evaluate workers' exposure to tasks featuring many upper-limb injury risk factors (lack of recovery periods, repetitiveness, force requirement, awkward postures and movements, and "additional factors") [6].

R. S. Goonetilleke et al. (Eds.): AHFE 2021, LNNS 273, pp. 313–319, 2021.
https://doi.org/10.1007/978-3-030-80713-9_40

Although a series of recent studies have evaluated the risk of WMSDs in chicken slaughterhouses [7–12], intrinsic characteristics of each production plant have been verified due to the particularities of the production mix. According to the production mix, different work activities are carried out, and consequently, different occupational risks are present.

Considering these assumptions, the aim of this study was to evaluate the risks in relation to repetitive movements of the workers' upper limbs of, as well as analyze the effects of a reduced work pace on the risk levels in a chicken slaughterhouse.

2 Method

The local Human Research Ethics Committee approved the procedures for this study, in accordance with the Declaration of Helsinki.

Data collections were performed in a Brazilian slaughterhouse with 3,200 workers, who were divided into two shifts, where 300,000 chickens were slaughtered daily. In order to measure the risks related with repetitive movements of the upper limbs, 10% of the workforce was assessed while performing their work tasks, using the OCRA checklist [6]. The evaluated tasks were considered individually, regardless of whether they were part of job rotation schemes. A sampling of 10 cycles for each task was recorded with a Sony® HDR-XR160 digital camcorder camera.

From the results of the OCRA checklist, simulated interventions were performed to achieve borderline risk levels by reducing the work pace for each of the evaluated activities. The scores for each risk factor were weighted in proportion to the proposed work pace reductions.

Descriptive statistics (mean, standard deviation and percentage) were used, as well as the Student t-test (SPSS 17.0; $p \leq 0.05$), in order to compare the risk between the sides of the worker's body.

3 Results and Discussion

The 21 investigated work tasks (homogeneous groups) were from the following sectors: packing (10), cutting (9), chiller (1), reception (1).

In the analyzed industry, the work shift was 08 h 48 min with six rest breaks of 10 min each, thus, the net duration of the repetitive work was classified in the range of 421 to 480 min ("duration" multiplier 1). For the risk factor "recovery", the multiplier 1.05 was used. The scores of the other risk factors (force, frequency of technical actions, posture with stereotyped movements, and "additional factors") were assigned according to the features of each task and the operating mode of each worker.

The occupational repetitive actions performed by workers were 63.3 ± 20.2 per minute, representing 9 points on the OCRA scale (0- to 10-point scale). (Table 1). Many studies have also shown a high frequency of technical actions in poultry slaughterhouses, with averages between 63.7 and 79.8 [8–13]. Kilbom [14] recommends that workers must not exceed 25–33 actions/min. to prevent tendon injuries, since higher rates offer insufficient breaks for fatigue recovery between contractions (micropauses).

The work in chicken processing industries is repetitive, strenuous and exposes workers to overuse injuries [15]. Upon arriving at the slaughterhouse, live chickens are received and then introduced to a production line that requires workers to hang, slaughter, pluck, clean, eviscerate, cut, pack and box chicken parts at a fast pace. Additionally, the workers clean and repair equipment, assemble boxes and move packaged chicken pallets.

The average OCRA checklist score was 18.2 ± 4.8 (moderate risk), and there was no significant difference ($p = 0.165$) between the scores on the right upper limb (18.1 ± 4.9) and left upper limb (17.9 ± 2.9). A number of studies in Brazilian slaughterhouses have also found a moderate risk level [7, 8, 10, 11]. In an Italian poultry slaughterhouse, Colombini and Occhipinti [6] found that 22.4% of workers exposed to moderate risk tasks (an average of 20 points on the OCRA Checklist) were diagnosed with WMSDs (based on clinical evaluations and complementary medical examinations).

Considering the five risk categories of the OCRA method, 2 tasks were deemed high risk (9.5%), 15 presented moderate risk (71.5%) and 4 exhibited low risk (19%). Despite some studies have found that high-risk activities predominate in slaughterhouses in Brazil (56.5%) [7], Italy (90%) [13] and Iran (67%) [16], many recent studies indicate that moderate-risk activities predominate in Brazilian slaughterhouses [8–11]. It is speculated that the reduction of high-risk activities, found in these studies, may be related to the promulgation of the Brazilian Regulatory Standard n° 36 (NR-36) [17]. This standard defines parameters for the evaluation, monitoring and control of occupational risks in meat processing industries [17]. Among the parameters established by the NR-36, the mandatory inclusion of rest breaks directly influences the outcomes of the OCRA checklist [17]. When meeting these parameters, the exposure time for repetitive activities was reduced throughout the Brazilian meat processing industry. Hence, in the OCRA checklist analyzes, the "recovery" score is reduced, which then diminishes the risk of WMSDs as well.

Founded on epidemiological data regarding WMSDs, the authors of the OCRA method used statistical procedures (regression analysis) to define hypotheses for the prevalence of WMSDs according to the occupational conditions [6]. Specific percentages were determined for each level of the WMSDs incidence. As an example of the studies that originated the method, Colombini and Occhipinti [6] found an incidence of 47.7% of WMSDs in meat deboning workers, who were classified with high risk (28 points) on the OCRA checklist. In this way, most of the workers evaluated in the present study had a probability of developing WMSDs between 10.8 and 21.5% (moderate-risk tasks).

Based on the assumption that there is a predominance of activities with highly repetitive movements of the upper limbs [8–12, 18], some studies have recommended that a reduced work pace decreases the risk of WMSDs in chicken slaughterhouses [8–12]. Therefore, simulated interventions were performed, reducing the work pace to reach very low risk levels on the OCRA checklist. By conducting these simulated interventions, it was possible to diminish the risk of WMSDs to very low levels in 19 of the 21 tasks by only decreasing the work pace (−39.8 ± 16.8%). Two of the tasks ("turn the boxes" and "close the boxes") were unachievable because of the high demand for strength required to perform these tasks. Similarly, previous studies also executed simulated interventions to lower the risk by reducing the work pace (−42.1 ± 14.5%; −

Table 1. The OCRA Checklist risk assessment and simulations to reduce the risk by decreasing the work pace.

Tasks	Current situation				Simulations for risk reduction			
	Units/min	TA/min	OCRA score	Risk level	Units/min	TA/min	OCRA score	Risk level
Closing boxes	8.0	56.0	32.0	5	#	#	#	#
Meat transfer – bowl to box – secondary packing	6.0	54.0	28.5	5	#	#	#	#
Boning leg	2.5	75.0	21.0	4	1.0	30.0	11.0	2
Boneless leg screening	17.1	85.7	20.0	4	7.1	35.0	11.0	2
Packing boneless breast – 20 kg	1.9	70.6	20.0	4	1.0	35.0	11.0	2
Trimming boneless leg	8.5	102.3	20.0	4	2.9	35.0	11.0	2
Trimming breast	11.8	58.8	19.0	4	6.0	30.0	11.0	2
Packing boneless leg – automatic scale	3.7	98.1	19.0	4	1.5	40.0	11.0	2
Placing breast on the tray	3.3	88.3	19.0	4	1.5	40.0	11.0	2
Packing boneless leg – manual scale	4.5	90.2	19.0	4	2.0	40.0	11.0	2
Weighing and packing boneless breast – 7.5 kg	3.3	66.7	17.0	4	2.0	40.0	11.0	2
Standardizing breast – IQF (with knife)	17.6	52.9	17.0	4	12.0	36.0	11.0	2
Re-hanging chicken – Stork/cone	15.0	45.0	16.0	4	10.0	30.0	11.0	2
Secondary packaging – leg	*	59.0	16.0	4	*	40.0	11.0	2
Repositioning sassami on the mat – IQF	*	57.0	16.0	4	*	40.0	11.0	2

(*continued*)

Table 1. (*continued*)

Tasks	Current situation				Simulations for risk reduction			
	Units/min	TA/min	OCRA score	Risk level	Units/min	TA/min	OCRA score	Risk level
Hanging live chickens	13.0	39.1	15.0	4	10.0	30.0	11.0	2
Interfold legs	*	62.0	15.0	4	*	40.0	11.0	2
Re-hanging chicken – post chiller	13.6	40.9	14.0	3	10.0	30.0	11.0	2
Sealing packages – boneless leg	2.3	56.0	13.5	3	1.7	40.0	11.0	2
Repositioning breast on the mat – IQF	*	42.5	13.0	3	*	39.0	11.0	2
Re-hanging chicken – cone	10.0	30.0	12.0	3	8.6	26.0	11.0	2
Average	7.0	63.3	18.2	4	4.2	35.6	11.0	2
Standard-deviation	5.7	20.2	4.8	–	4.0	4.8	0,0	–

Risks: 5-high; 4-medium; 3-low; 2-very low; 1-acceptable; TA-technical actions; *Task with variable work rate; # The task needs to be restructured due to the high force requirement

44.9 ± 13.7%; −48.5 ± 11.8%; −38.8 ± 4.8% and −50.4 ± 7.7%, respectively) and were effective in most cases (24/26, 28/30, 15/15, 33/35 and 46/47 tasks, respectively), except for those activities with an excessive demand for strength [8–12].

4 Conclusion

Considering the results of this study, it is possible to conclude that:

- Most of the analyzed work activities were classified as moderate risk, predisposing the workforce to a WMSDs incidence between 10.76 and 21.51%;
- The risk of developing WMSDs is similar for both sides of the body;
- Simulated interventions, reducing the working pace, showed the effectiveness of this organizational measure to lower the risk of WMSDs in most of the analyzed tasks;
- Further studies are necessary to verify whether the results of the present study can be generalized to other slaughterhouses.

Lastly, to lessen the risk of WMSDs in chicken slaughterhouses, it is suggested to implement several organizational measures: reduce the work pace, increase the workforce in each task, take rest breaks every hour, implement an efficient job rotation (between activities with different biomechanical demands), keep knives sharp (avoid unnecessary effort), and monitor the risk level of work tasks using quantitative tools, such as the OCRA checklist.

References

1. Brazilian Association of Animal Protein. Annual Report 2020. http://www.abpa-br.com.br/relatorios. Accessed 17 Jan 2021
2. Nääs, I.A., Mollo Neto, M., Canuto, S.A., Walker, R., Oliveira, D.R.M.S., Vendramento, O.: Brazilian chicken meat production chain: a 10-year overview. Rev. Bras. Cienc. Avic. **17**, 87–93 (2015)
3. Punnett, L., Wegman, D.H.: Work-related musculoskeletal disorders: the epidemiologic evidence and the debate. J. Electromyogr. Kinesiol. **14**, 13–23 (2004)
4. Magnusson, M., Ortengren, R., Andersson, G.B., Petersén, I., Sabel, B.: An ergonomic study of work methods and physical disorders among professional butchers. Appl. Ergon. **18**, 43–50 (1987)
5. Colombini, D., Occhipinti, E., Delleman, D., Fallentin, N., Kilbom, A., Grieco, A.: Exposure assessment of upper limb repetitive movements: a consensus document. In: Karwowski, W. (ed.) International Encyclopedia of Ergonomics and Human Factors. Taylor & Francis, New York (2001)
6. Colombini, D., Occhipinti, E.: Risk analysis and Management of Repetitive Actions: A Guide for Applying the OCRA System. CRC Press, New York (2016)
7. Reis, D.C., Reis, P.F., Moro, A.R.P.: Assessment of risk factors of musculoskeletal disorders in poultry slaughterhouse. In: Azeres, P., et al. (eds.) SHO 2015, 1st edn, vol. 1, pp. 294–296. SPOSHO, Guimarães (2015)
8. Reis, D.C., Ramos, E., Reis, P.F., Hembecker, P.K., Gontijo, L.A., Moro, A.R.P.: Assessment of risk factors of upper-limb musculoskeletal disorders in poultry slaughterhouse. Procedia Manuf. **3**, 4309–4314 (2015)
9. Reis, D.C., Tirloni, A.S., Ramos, E., Moro, A.R.P.: Assessment of risk factors of upper-limb musculoskeletal disorders in a chicken slaughterhouse. Jpn. J. Ergon. **53**, 458–461 (2017)
10. dos Reis, D.C., Tirloni, A.S., Ramos, E., Dias, N.F., Moro, A.R.P.: Risk assessment of repetitive movements of the upper limbs in a chicken slaughterhouse. In: Bagnara, S., Tartaglia, R., Albolino, S., Alexander, T., Fujita, Y. (eds.) IEA 2018. AISC, vol. 825, pp. 323–329. Springer, Cham (2019). https://doi.org/10.1007/978-3-319-96068-5_35
11. dos Reis, D.C., Tirloni, A.S., Moro, A.R.P.: Risk assessment of upper limb musculoskeletal disorders in a poultry slaughterhouse. In: Arezes, P.M. (ed.) AHFE 2019. AISC, vol. 969, pp. 265–272. Springer, Cham (2020). https://doi.org/10.1007/978-3-030-20497-6_24
12. dos Reis, D.C., Tirloni, A.S., Moro, A.R.P.: Effects of reduced work pace on the risk of developing upper-limb musculoskeletal disorders in a poultry slaughterhouse. In: Karwowski, W., Goonetilleke, R.S., Xiong, S., Goossens, R.H.M., Murata, A. (eds.) AHFE 2020. AISC, vol. 1215, pp. 87–94. Springer, Cham (2020). https://doi.org/10.1007/978-3-030-51549-2_12
13. Colombini, D., Occhipinti, E.: Results of risk and impairment assessment in groups of workers exposed to repetitive strain and movement of the upper limbs in various sectors of industry. Med. Lavoro **95**, 233–246 (2004)
14. Kilbom, A.: Repetitive work of the upper extremity: part II - the scientific basis (knowledge-base) for the guide. Int. J. Ind. Ergon. **14**, 59–86 (1994)
15. Cartwright, M.S., et al.: The prevalence of carpal tunnel syndrome in Latino poultry processing workers and other Latino manual workers. J. Occup. Environ. Med. **54**(2), 198–201 (2012)
16. Mohammadi, G.: Risk factors for the prevalence of the upper limb and neck work-related musculoskeletal disorders among poultry slaughter workers. J. Musculoskelet. Res. **15**, 1250005 (2012)

17. Brasil: Norma Regulamentadora 36. Segurança e Saúde no Trabalho em Empresas de Abate e Processamento de Carnes e Derivados, Ministério do Trabalho e Emprego, Portaria MTE n. 555, de 18 de abril de 2013, Diário Oficial da União (2013)
18. Sundstrup, E., Jakobsen, M.D., Jay, K., Brandt, M., Andersen, L.L.: High intensity physical exercise and pain in the neck and upper limb among slaughterhouse workers: cross-sectional study. BioMed. Res. Int. 218546 (2014)

Job Analysis and Ergonomic Design

Collaborative Workspace – Concept Design of an Interactive System for Total Airport Management

Mandra Bensmann[1], Alicia Lampe[1], Thomas Hofmann[1](✉), and Steffen Loth[2]

[1] Hochschule Osnabrück, Industrial Design, Sedanstraße 60, 49076 Osnabrück, Germany
{mandra.bensmann,alicia.lampe,t.hofmann}@hs-osnabrück.de
[2] German Aerospace Center, Institute of Flight Guidance, Lilienthalplatz 7, 38108 Braunschweig, Germany
steffen.loth@dlr.de

Abstract. The future of aviation should be safe, efficient, measurable, collaborative, fair and transparent. However, only a few attributes can be fulfilled under the given circumstances at airports at present. The German Aerospace Center (DLR) is developing concepts regarding the implementation of Collaborative Decision Making and Performance Based Management at airports. In cooperation with the University of Applied Sciences Osnabrück, a design concept is developed which deals with the implementation of an interactive visualization module for a control center to support decision making in a multi-stakeholder environment at airports. The approach and first results will be discussed and plans for further studies are deduced.

Keywords: Performance Based Management · Total Airport Management · Collaborative Decision Making · User Experience Design · Design Thinking · Human Computer Interface

1 Introduction

Airports nowadays are organized by a multi-stakeholder environment, in which every stakeholder works with an individual Human Computer Interface (HCI) system and with technology that is not state-of-the-art to decide about daily business operations. The status quo results in a lack of communication among the different stakeholders and causes a time-consuming decreased system performance. This effects a disproportionate amount of work, which leads to significant mental distraction for the users. Critical situations might be detected too late or not at all [1]. Due to the complex structure, fluctuation and incomplete flows of data information, the integration of land and air side processes at the airport is difficult and limits the view of overall processes for all parties. In addition, all individual parties are caught in a so-called *coopetition* [2, 3], since both common and individual interests must be enforced. To this day, there is no

M. Bensmann and A. Lampe—These authors contributed equal to this work.

R. S. Goonetilleke et al. (Eds.): AHFE 2021, LNNS 273, pp. 323–330, 2021.
https://doi.org/10.1007/978-3-030-80713-9_41

common basis for discussion in order to be able to collaboratively make profound and sustainable decisions in the interests of all parties. Consequently, the potential of the airport infrastructure and capacities cannot be fully exploited.

The German Aerospace Center (DLR) developed the concepts of Total Airport Management (TAM) and Performance Based Airport Management (PBAM), that address these problems. Regarding these concepts and in collaboration with the University of Applied Sciences Osnabrück, one responsive concept for an onsite- and remote-control solution is being developed, which focus on the design and implementation of an interactive visualization module for an airport operation center (APOC) to support collaborative decision making depending on the size and infrastructure of an airport.

The overarching objective is to initiate, provide and support communication as well as to collect, analyze and link data information to enhance the general efficiency, safety, sustainability and success of the institution airport.

On the basis of the already existing TAM/PBAM concepts and a far-reaching interdisciplinary market and trend research, different solution variants were developed, from which an exemplary design concept is worked out. The result is a prototypical ideation for a responsive interface design as well as an ergonomic workplace suggestion of physical control center including emerging technologies for the period 2025 to 2030.

2 Background

Airports have increasingly developed from simple infrastructure providers to high performance companies. To further improve operations and provide high quality of services for passengers and airlines, different concepts have been developed.

While A-CDM [1] as a first approach is based on providing and sharing more accurate data between different stakeholders at defined milestones DLR together with Eurocontrol introduced the idea of TAM [5] as a method for reliably improving airport operations. TAM covers the cooperation of stakeholders based on operating strategies in a pre-tactical time horizon and the collaborative decision-making process at airports. Work has been done regarding interface design and cooperation of stakeholders as well as user requirements for future APOC operators in [6] and [7] by DLR. Currently research and industry partners develop and test solutions for TAM in the context of SESAR, the Single European Sky ATM Research Program [8].

As airport performance is likely to be rated against Key Performance Indicators (KPI) in the future, the concept of PBAM [3] has been introduced, combining the existing concepts of TAM and the general approach of Performance Based Management (PBM).

Within the TAM/PBAM concept the idea of an APOC is described. Either defined as a physical facility with a common location and infrastructure or a virtual remote system, it acts the central coordination environment. Stakeholders are connected with each other and provided with a common situation presentation. Actual flight data from the airport operational database as well as the measurements and results from the PBAM systems can deliver a full picture of the current status and the compliance with the performance targets. In addition, weather information or information about disturbances in the ATM system can be used to forecast situations and implications to the airport.

There is no defined standard or best practice available yet how these APOCs should be implemented. Challenges exist how stakeholders can be supported at best with the display of the huge amount of different data as well as to quickly identifying critical situations and their reasons.

3 Design Goal

The aim was to develop an interactive visual system at airports for a new control center in the sense of TAM and PBAM and to develop visualization options for displaying larger amounts of data. The basis for this is a topological map and current traffic data of Hamburg Airport.

The system was designed to recognize and display critical situations and bottlenecks. For this purpose, suitable parameters must be selected for the visualization, which are to be defined and abstracted as visualization elements. Individual elements are then selected as examples for implementation and suitable visualization techniques. The functionality of the implementation and validation of the selected elements will be demonstrated for use. Due to the complex scope of TAM and PBAM, this paper mainly addresses air side processes. Land based processes are not considered and need to be worked out in the future.

4 Design Approach

4.1 Design Process

The challenge from the design perspective lies in the design of an HCI that is optimally adapted to the needs of humans as users and offers creative problem-solving approaches to complex tasks (user-centered design). In iterative procedures, the phases of the design process are evaluated under the aspect of design thinking methods [9] and DIN EN ISO 13407, 9241 [10]. The iterative phases in this work are shown in Fig. 1. Thereby, both strictly methodical and freely explorative investigations were conducted to work scientifically substantiated as well as to understand the users' experiences and thinking [11].

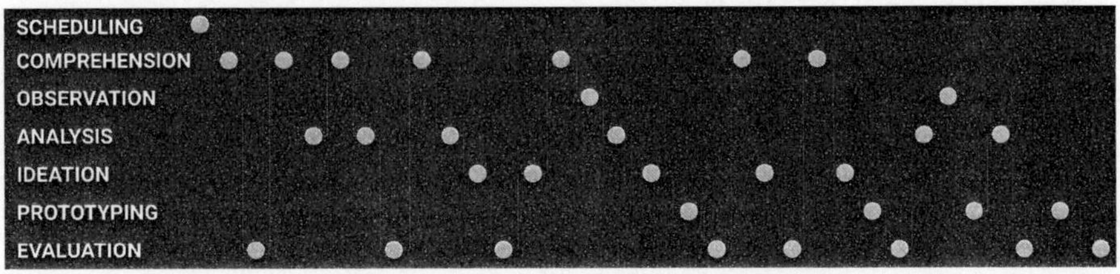

Fig. 1. Iterative model of the design phases in this work

The DLR has researched initial ideas and implementation options for designing a holistic system in internal trials [12]. In addition to a far-reaching market and trend analysis, future considerations are formed. Further inspirations followed from science fiction movies, literature and video games of various genres, as well as futuristic design concepts. Likewise, interdisciplinary industries and their handling of emergency situations, here using the example of an emergency room, were examined. Based on the previous research and experience of former projects [13, 14], Scientific Personas were identified, which form the main user group of APOC. The four most important stakeholders are airport management, air traffic control, ground handling, airlines whose areas of responsibility are clearly defined [12]. As a result, they represent different interests which, in terms of user-centred design, place different demands on the workstation and the HCI of the control center arise. Despite the clearly separated areas of responsibility, the parties involved are interdependent and interrelated through contracts and the indispensable exchange of information for smooth operations at the airport.

Following the research, problem areas were defined, resulting in lists of requirements for the design, users and system of the control center. Based on this, different variants for concept solutions were developed and sorted using morphological boxes. Subsequently, evaluation matrices were used to determine a solution variant as a design concept.

The analysis shows that a control center is essential for every airport, but they have different requirements for the new type of workstation, as not all of them have the same starting point and there are significant differences in terms of financial resources, size and infrastructure of airports. For this reason, the HCI must be able to be used responsively on different end devices and the control center must be flexible and expandable [5].

The HCI should be intuitive, even if it is assumed to be used by professionals, so that the user experience is as pleasant as possible, because the users are confronted with immensely fluctuating and complex data sets [14]. These large amounts of data have to be abstracted and processed in a target-oriented way to keep the user's attention at all times. The introduction of playful elements under the gamification approach can also promote work motivation and motivation for teamwork [1].

4.2 Design Concept

Based on the previous research and analysis, different modes could be identified that have specific requirements regarding HCI and interaction with the system. On the one hand, users *independently* perform everyday actions with the system, and on the other hand, there is a need for *collaborative* decision making in certain situations. In a collaborative workspace, direct communication (face-to-face) is the most effective because it makes it easier to circumvent misunderstandings [12]. The design solution is a stand-alone system that can be connected to the systems already in place at the airport. By implementing a situational adaptive behavior to the HCI, communication between stakeholders will be initiated and supported, and events will be categorized depending on their priority and scope.

The visualization of only relevant data and an integrated airport overview might most certainly strengthen situational awareness with regard to air side processes and all possible activities, as users are not confronted with a flood of data and all information shown are comparable and comprehensible. For this purpose, the visualization of the general performance targets creates an optimized basis for planning and discussion.

Because of the different initial conditions at airports, two solutions are developed to meet the presumed requirements for this design concept. The solution of an on-site workstation implies a shared use of a physical space as a control center. The users communicate directly with each other. The solution for a remote workplace assumes that the users are spatially separated from each other at different locations and can only communicate via virtual and digital channels (Fig. 2).

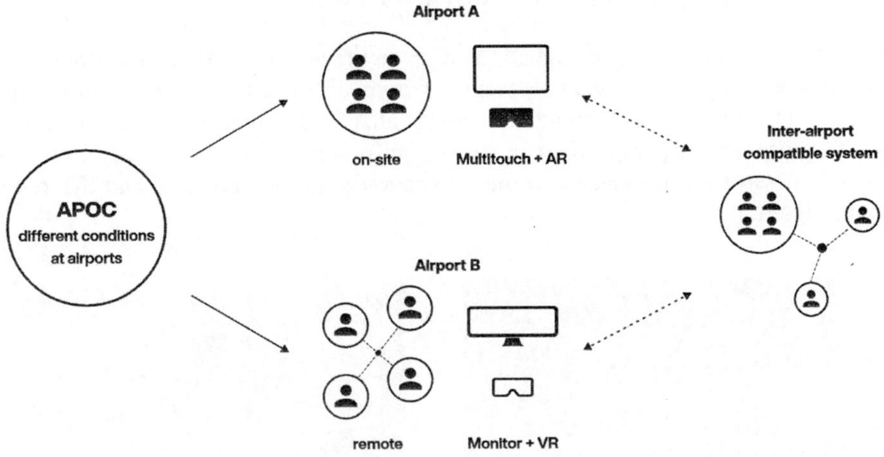

Fig. 2. On-Site and Remote solution for APOC

Data has to be prepared individually for the respective stakeholders and represented on a common overview. In order to meet the individual needs of the users, personalized widgets are used to supplement the common overview with specific elements, such as target definitions at different Performance Levels or the location of parameters such as average speeds on certain runway sections or the allocation of parking positions.

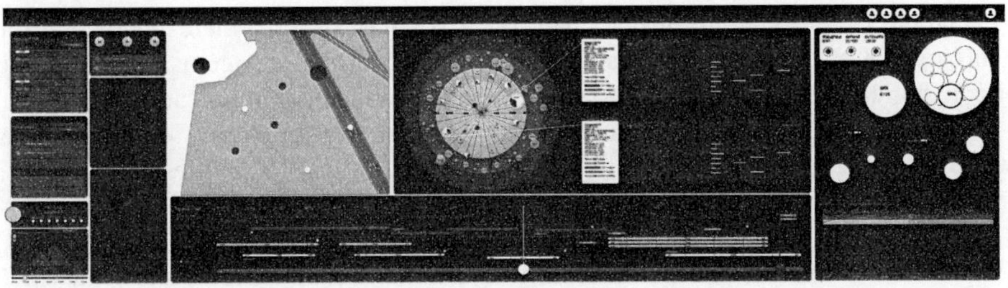

Fig. 3. Exemplary Arrangement of Widgets created in Adobe XD

Figure 3 shows a first intermediate result of the design process. Through collaboration with experts from DLR and University of Applied Sciences Osnabrück, these were iteratively adapted to the needs of the scientific personas to represent the entire spectrum of possible uses of the design elements. It turned out that any change within the airport system always affects the appearance of the entire interface. These fluctuating changes must be perceived directly by the users, which is why repetitive data sets are helpful. As a result, redundant information is loaded context-specifically and visualized depending on it. In addition, critical situations are marked by warnings so that users can react immediately.

5 Simulation Based on User Journeys

To see the application of the designed interface in the user context, user journeys were created based on two personas in the form of storyboards and limited interactional click dummies. The interplay of methodical (storyboards) and exploratory approaches (role plays from the user's perspective) in the design process made it possible to test the responsiveness of the interface and the interaction with the hardware and the system (Fig. 4, Fig. 5).

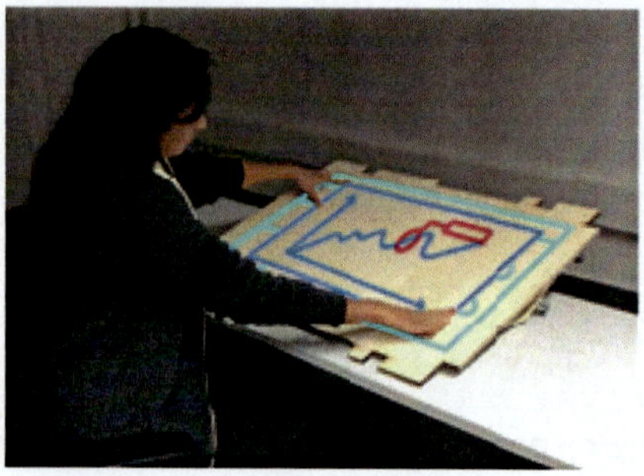

Fig. 4. Role play from user's perspective

Fig. 5. Storyboards

Persona A works at a larger airport and represents the ground handler agent in APOC. In this scenario, all representatives are physically present and therefore share common space in the control center.

Persona B works as an airport agent at a smaller airport where all responsible agents remain at their usual workstations and need a digital device as a suitable tool for collaborative decision-making to simplify and establish communication.

Both personas work with the same interface, which has structural similarities in the form of same design elements. However, due to their different activities and locations, it differs in terms of content, depth of information, and preparation.

With the help of augmented reality (persona A) and virtual reality (persona B), it is possible to interact and work flexibly in 3-dimensional space to show the direct interdependency of parameters and interactions via geometrical shapes and thus make it presumably more accessible for users. Therefore, conferences and interactions with the system can take place in digital and virtual space so that miscommunications are reduced [12] and collaborative solutions can be developed.

6 Discussion

The conceptualization of an interactive and responsive visualization system to support TAM provides an idea for the possible implementation of a superimposed workspace, that reduces the existing problems of non-communication and miscommunication at the airport [12].

For the first time, designers were involved in the development process as support for visualizing the concept around the use cases TAM and PBAM which had the advantage that HCI were examined from a development perspective and adapted to users' needs. The scenario-based design approach by using storyboards ensured that the developed concept is placed in an application-related context through simulation, which meant that the usability of the concept can be evaluated and adapted repetitively.

The next steps in the project require a more detailed elaboration of the prototypical click dummies into real interfaces. This would require usability testing under realistic circumstances to identify vulnerabilities and get feedback from the real user group [15]. Future technologies beyond AR/VR will bring unknown opportunities in communication, handling of complex data sets and the overall structures at airports that are not considered in this paper. In addition, new regulations and guidelines in national and international air traffic might cause changes in the concept. In further development, it should be investigated how the decisions made in an APOC can be communicated to other parties involved at the airports, e. g. via wearables. A combination of the solution approaches is possible, both within and across airports, as the concept is variably compatible due to its responsiveness and flexible transferability and can be adapted to the prevailing circumstances.

For the targeted period of 2025 to 2030, the elaborated design concept with its remote and on-site workplace solutions for an interactive visualization module is considered feasible and reasonable, both from the system and user perspective. The merger of interdisciplinary institutions for the research of HCI can be seen as novel and enriching in this context and provide new impulses for future research.

References

1. Kumar, J.: Gamification at work: designing engaging business software. In: Marcus, A. (ed.) DUXU 2013. LNCS, vol. 8013, pp. 528–537. Springer, Heidelberg (2013). https://doi.org/10.1007/978-3-642-39241-2_58
2. Bengtsson, M., Kock, S.: Coopetition in Business Networks—to Cooperate and Compete Simultaneously. Ind. Mark. Manage. **29**, 16 (2000)
3. Helm, S.: Advancing total airport management - an introduction of performance based management in the airport context. In: ATRS World Conference (2015)
4. Eurocontrol, A-CDM. https://www.eurocontrol.int/concept/airport-collaborative-decision-making
5. Günther, Y., et al.: Total Airport Management (Operational Concept & Logical Architecture). In: Meier, C., Eriksen, P. (2006)
6. Jipp, M, Schaper, M., Günther, Y., Papenfuss, A.: Ecological interface design and its application to total airport management. In: IEEE Conference on System, Man, and Cybernetics, 09–12 October 2011, Anchorage, USA (2011)
7. Papenfuss, A.: Konzeption einer anzeige zur unterstützung von verhandlungsprozessen: methodik und erste validierungsergebnisse. In: Kooperative Arbeitsprozesse DGLR-Bericht 2009-02, 2009-0, pp. 107–123, Braunschweig (2009)
8. SESAR PJ04-TAM. https://www.sesarju.eu/projects/tam

9. HPI School of Design Thinking - Hasso-Plattner-Institut. https://hpi.de/school-of-design-thinking/design-thinking/was-ist-design-thinking.html
10. Deutsche Akkreditierungsstelle. https://www.dakks.de/content/leitfaden-usability
11. Hofmann, T., Bergner, J.: Air Traffic Control HCI - how to handle very special user requirements. In: Proceedings of the 8th International Conference on Applied Human Factors and Ergonomics (AHFE 2017), Los Angeles (2017)
12. Günther, Y.: Airport2030, HAP 3 (AP 3.1) Leitstandkonzept (2014)
13. Hofmann, T., et al.: DFS ConceptDesk - an approach to an intuitive UX for air traffic control. In: Proceedings of the 8th International Conference on Applied Human Factors and Ergonomics (AHFE 2017), Los Angeles (2017)
14. Hofmann, T., Syndicus, M., Bergner, J.: AirTrafficControl HMI - Herausforderungen der Integration mentaler Modelle spezieller Nutzer in den Designprozess. In: Proceedings of the Frühjahrskongress der Gesellschaft für Arbeitswissenschaft e.V., Dresden (2019)
15. Barnum, C.M.: Usability Testing Essentials: Ready, Set…Test! Morgan Kaufmann, Burlington (2011)

Worker Satisfaction of Job Rotations in Brazilian Poultry Slaughterhouses: A Cross-Sectional Study

Adriana Seara Tirloni(✉), Diogo Cunha dos Reis, and Antônio Renato Pereira Moro

Federal University of Santa Catarina, Florianópolis, Santa Catarina, Brazil
adri@tirloni.com.br

Abstract. This cross-sectional study aimed to analyze the poultry slaughterhouse workers' satisfaction of job rotations, as well as verify the association of job rotation schemes and bodily discomfort perception. Workers from two Brazilian poultry slaughterhouses were selected randomly ($n = 235$) and interviewed about job rotations and bodily discomfort. The binary logistic regression between the job rotation schemes ($\leq$1h and $>$1h) and the bodily discomfort perception were applied ($p \leq 0.05$). Most workers performed job rotations ($n = 226$ workers; 96.2%) with 2–9 tasks, and of these, 96.5% were satisfied with this work organization. The main reasons for this satisfaction were activities' diversification (60.9%), decreased tiredness (56.2%), reduced monotony (17.4%) and alternation between standing and sitting (11.1%). Additionally, of those who performed rotations, 94 (41.6%) workers felt bodily discomfort. There was not a significant association ($p = 0.759$) between the task durations that made up the rotations ($\leq$1h and $>$1h) and the workers' bodily discomfort perception.

Keywords: Job satisfaction · Job rotation · Slaughterhouse · Ergonomics · Worker

1 Introduction

In Brazil, there is a Regulatory Standard (NR-36) that recommends job rotation implementation in slaughtering and meat processing industries, as it is a preventive measure to reduce the workers' exposure to various risks involved in slaughterhouses [1].

According to OSHA [2], one example of administrative solutions used effectively by poultry processors is the rotation schedule to address high-risk tasks or to minimize exposure to cold. However, there are limitations regarding its implementation in Brazilian slaughterhouses, since the job rotation schemes should obey the hygienic-sanitary aspects, the workers' pay scale and the proximity between the tasks [3].

It is unanimous that the job rotation design should promote rotating jobs that use different muscle groups and reduce the postural requirements, in order to work in more neutral postures [1, 2]. Despite this, a systematic review demonstrated weak evidence for job rotation effectiveness when implemented in manufacturing industries, regardless

R. S. Goonetilleke et al. (Eds.): AHFE 2021, LNNS 273, pp. 331–337, 2021.
https://doi.org/10.1007/978-3-030-80713-9_42

of the rotation target, to prevent and control upper-limb work-related musculoskeletal disorders (UL-WMSDs) and to reduce exposure to physical factors [4].

On the other hand, studies have shown workers' satisfaction with the task rotation schemes of different industries [5–7], though, only one study was found that verified the satisfaction of slaughterhouse workers [8]. Nevertheless, this research did not address the relationship between the duration of the task rotation intervals and the perception of workers' bodily discomfort. Tirloni et al. [9] demonstrated that, on average, 71.5% of workers of three poultry slaughterhouses felt bodily discomfort, but it was not associated with job rotations.

Workers who produced car seats in Québec perceived that performing job rotations decreased monotony [10]. This result was also established in the study by Dias and Moro with poultry slaughterhouse workers [8]. Nonetheless, Dias and Moro [8] observed that the main reason for the workers' satisfaction was activity diversification in relation to the change of pace, the use of hand tools and/or the work environment (62.4%).

As stated by the Occupational Repetitive Action (OCRA) method [11], when the tasks of a job rotation are >1 h, the prediction of obtaining UL-WMSDs is greater than <1 h. However, the question arises: "Is the bodily discomfort perception of slaughterhouse workers associated with the task durations that made up the rotations? Thus, this study aimed to analyze the poultry slaughterhouse workers' satisfaction of job rotations, as well as verify the association of job rotation schemes and bodily discomfort perception.

2 Method

This research was approved by the local Committee of Ethics in Research with Human Beings, protocol nº 2098/2011. Workers from two Brazilian poultry slaughterhouses participated in this cross-sectional study. In one slaughterhouse, there were 326 employees in which 111,000 chicken were slaughtered per day and in another, 2,000 employees slaughtered 180,000 chicken per day.

Workers were selected randomly, totaling 235 workers from seven sectors. Data were collected with women (31.6 ± 10.2 years) and men (29.6 ± 10.6 years) who worked in artificially cold and natural environments.

Employees were asked about the number and duration of the tasks that made up the job rotation, their satisfaction with the job rotation scheme and reasons, decreased physical effort due to job rotations, along with the presence of bodily discomfort. Workers were able to mention more than one reason to justify their satisfaction or dissatisfaction with the task rotation. Moreover, those who did not perform job rotations were asked why they did not do them.

Observing the hygienic-sanitary aspects, the employer must implement job rotations within a daily schedule that attends at least one of the following situations [1]: (1) alternating work positions, such as standing and sitting postures, (2) alternating muscle groups required, (3) switching with non-repetitive tasks, (4) reducing postural demands, decreasing more frequent static and dynamic efforts, (5) alternating tasks where environmental exposure is more comfortable, (6) reducing carrying, handling and lifting of loads (mass) and (7) decreasing monotony.

Based on these guidelines, the health and safety team of the slaughterhouses predetermined the job rotation schemes, in which the work intervals lasted 30, 50, 100, 120 and/or 210 min. Due to problems during data collection, information on bodily discomfort were not collected from all workers. Therefore, 176 workers responded about discomfort and were classified into two groups of job rotation schemes: $\leq$ 1h (30 and 50 min) and $>$1h (100 min).

The descriptive statistics were used in the analysis of worker satisfaction and task rotation characteristics. The binary logistic regression between the job rotation schemes and the bodily discomfort perception were applied ($p \leq 0.05$).

3 Results

The results showed that most workers performed job rotations (n = 226 workers; 96.2%) with 2–9 tasks, and of these, 96.5% were satisfied with this work organization (Table 1).

The main reasons that caused the satisfaction of these workers were activity diversification (65.6%), decreased tiredness (60.6%), reduced monotony (18.8%) and alternation between standing and sitting (11.9%) (Table 2). It is noteworthy that 84.1% of workers perceived that job rotations reduced physical demand.

Dissatisfied workers (n = 8) that performed job rotations reported having relationship difficulties with other employees (n = 1) and the fact that they did not want to do a rotation activity that they considered "bad" (n = 5), two workers did not answer.

The workers who stated that they do not perform job rotations, cited that it is because there is a lack of workers in the sector to rotate, that they would like to do it to rest and that they do not know why it is not performed. Additionally, there are few workers who do one specific task, the other workers do not change, and finally, due to a shoulder problem, making it impossible to carry out some activities.

Most workers felt bodily discomfort (52.4%), but of those who performed rotations, 41.6% of workers felt discomfort (n = 94). However, there was not a significant association ($p = 0.759$) between the task durations that made up the rotation ($\leq$ 1 h and $>$1 h) and the workers' bodily discomfort perception (OR = 0.90; 95% CI 0.46 to 1.76).

4 Discussion

A job rotation system with different tasks is created to reduce exposure to any single risk factor and to allow body parts to either rest completely, work at slower rates, use less force, or work in more neutral postures [2]. As in the current study, Dias and Moro [8] found that most workers of a poultry slaughterhouse were satisfied to rotate (95%), although not all workers have identified effort reduction in at least one task in the rotation scheme (77.3%).

Corroborating with the present study (65.6%), Dias and Moro [8] also confirmed that the most prevalent reason justifying the workers' satisfaction was activity diversification in relation to change of work pace, the use of hand tools and/or the alteration of the work environment (62.4%). Dias et al. [3] verified that the alternation of sitting and standing posture, postural requirement reduction and monotony were met by most sectors in job rotation schemes of a poultry slaughterhouse.

Table 1. Work organizational factors of slaughterhouses and social-demographic characteristics of workers.

Variables	n	%
Slaughterhouse *		
Slaughterhouse A	91	38.7
Slaughterhouse B	144	61.3
Gender*		
Female	130	55.3
Male	105	44.7
Length of time working at the company*		
<1 year	75	31.9
1.1 to 5 years	88	37.4
>5 years	72	30.7
Work environments*		
Natural	98	41.7
Artificially cold	137	58.3
Job rotation *		
Yes	226	96.2
No	9	3.8
Number of tasks in job rotations		
2	126	55.8
3	53	23.4
4	15	7.1
5	13	5.7
≥6	18	8.0
Total	226	100.0
Intervals of the job rotation schemes		
≤1 h	129	73.3
>1 h	47	26.7
Total	176	100.0
Bodily discomfort		
Yes	97	52.4
No	88	47.6

* Variables with 235 workers; Bodily discomfort (n = 185)

Table 2. Reasons reported by workers to justify satisfaction with job rotation schemes.

Satisfaction reasons	n	%
Activity diversification (work pace, knife use and environment)	143	65.6
Decreased tiredness	132	60.6
Reduced monotony	41	18.8
Alternations of work positions, such as standing and sitting postures	26	11.9
Prevention of pain onset	18	8.3
Reduced static and dynamic efforts, task with displacement	7	3.2
Did not know	4	1.7

(n = 218 workers satisfied with job rotation schemes)

One study found that nurses' job rotations had a positive influence on job satisfaction [6], an optimal sequence in task rotation aimed at maximizing production levels and worker satisfaction [5]. According to Zawiah and Taha [12], job rotation, work method, training and goal setting indicated strong correlation with job satisfaction.

The task alternation where environmental exposure is more comfortable is a recommendation given by NR-36 for the implementation of job rotations [1]. Assessments have shown that slaughterhouse workers had finger temperatures below acceptable ($\leq$24 C°) [13–15], despite wearing an average of three overlapping gloves [15, 16] and felt cold in their hands (61%) [16]. Besides, a research of 925 poultry slaughterhouse workers established a significant association between perception of bodily discomfort and cold perceptions (OR = 2.05; 95% CI 1.44 to 2.91) [9]. However, in this paper, no worker mentioned that he/she liked the rotation by changing the environment (natural to artificially cold and vice versa). Perhaps this was because they must be observing the hygienic-sanitary aspects for implementation of job rotations in this economic sector, which can make changing the environment unfeasible.

In the present study, few workers (n = 5) declared that they were dissatisfied with the task rotation, as there was a "bad" task in the rotation scheme that they performed. Guimarães et al. [5] applied Learning Curve (LC) modeling and the analysis revealed that workers did not present significant differences in learning rate and final performance when rotating between tasks of different complexities. The authors suggest the use of LC modeling will define an optimal sequence in task rotations aimed at maximizing production levels and worker satisfaction. Conversely, the researchers did not comment on the interval duration of the task rotations. In contrast, Dias et al. [17] applied the OCRA method, verifying that the median risk score for developing WMSD was significantly lower in the "with job rotation – tasks <1 h" condition (18.6) when compared to the "job rotation – tasks >1 h" (19.4) ($p < 0.001$) and the "without job rotation" (19.0) ($p = 0.038$) conditions.

Jorgensen et al. [18] indicated that 42.7% of the contacted companies in the Midwest US manufacturing sector used job rotations. Furthermore, the authors cited that major limitations to successful implementation of job rotations were the rotation of individuals with medical restrictions, decreased product quality and lack of jobs to rotate. In the results of this study, only one worker did not perform a job rotation because she had a shoulder disorder. Conforming to OSHA [2], all rotations should be monitored for employee reports of symptoms to assure that the rotation is not aggravating a problem. Therefore, in the set up a job rotation system, employers should consider the nature and extent of exertions and the body parts used for each task in order to prevent the development of musculoskeletal disorders.

Several studies evaluated the work pace in poultry slaughterhouses in various tasks and checked the high number of technical actions per minute, 77.0 $\pm$ 22.5 [19], 75.5 $\pm$ 23.9 [17] and 69.1 $\pm$ 13.3 (high frequency) [20]. In this way, job rotation schemes can be a preventive measure to reduce the work pace in slaughterhouses and the risk of developing WMSD, if the tasks of these schemes decrease the physical and psychological demands of workers.

As reported by Dias et al. [3], it is difficult to implement efficient rotations due to particularities of work in slaughterhouses such as: tasks with similar musculoskeletal requirements, paces imposed by machines, inability to perform rotations between different sectors (health and occupational constraints), the predominance of tasks with moderate and high risks, hindering the distribution of risks between the rotation tasks. This was proven when evaluating 36 tasks in a productive area of a poultry slaughterhouse, as Dias et al. [17] found that 34 tasks were repetitive and 2 non-repetitive.

Therefore, it is suggested that the health and safety team of the slaughterhouses carry out studies in order to analyze the risk of all tasks, identifying which ones have the lowest risk of illness and/or meet the requirements of NR-36, which can be inserted in the job rotation scheme and, consequently, reduce occupational risks and increase worker satisfaction.

5 Conclusion

It was concluded that most workers liked the job rotations implemented by the slaughterhouses. In addition, several job rotation conditions required in NR-36 were present in the work organization of slaughterhouses: alternating work positions (standing and sitting) and the solicited muscle groups; reduced postural demands, efforts, and monotony. Future research may be performed to investigate the relationship between bodily discomfort and organizational, environmental conditions, as well as the risks present in each job rotation task.

References

1. Brasil. Ministério do Trabalho. Norma Regulamentadora NR 36 - Segurança e saúde no trabalho em empresas de abate e processamento de carnes e derivados. Portaria MTE no. 555, de 18 de abril de 2013. Brasília – DF (2013)
2. OSHA: Prevention of musculoskeletal injuries in poultry processing [Internet]. United States of America (2013). https://www.osha.gov/Publications/OSHA3213.pdf
3. Dias, N.F., Tirloni, A.S., Reis, D.C., Moro, A.R.P.: Effect of job rotation on the risk of developing UL-WMSDS in poultry slaughterhouse workers. Braz. J. Poult. Sci. **21**(2), eRBCA-2018-0843 (2019)

4. Padula, R.S., Comper, M.L.C., Sparer, E.H., Dennerlein, J.T.: Job rotation designed to prevent musculoskeletal disorders and control risk in manufacturing industries: a systematic review. Appl. Ergon. **58**, 386–397 (2017)
5. Guimarães, L.B., Anzanello, M.J., Renner, J.S.: A learning curve-based method to implement multifunctional work teams in the Brazilian footwear sector. Appl. Ergon. **43**(3), 541–547 (2012)
6. Ho, W.-H., Chang, C.S., Shih, Y.-L., Liang, R.-D.: Effects of job rotation and role stress among nurses on job satisfaction and organizational commitment. BMC Health Serv. Res. **9**(1), 8 (2009)
7. Kuijer, P.P.F.M., van der Beek, A.J., van Dieën, J.V., Visser, B., Frings-Dresen, M.H.W.: Effect of job rotation on need for recovery, musculoskeletal complaints, and sick leave due to musculoskeletal complaints: a prospective study among refuse collectors. Am. J. Ind. Med. **47**(5), 394–402 (2005)
8. Dias, N.F., Moro, A.R.P.: Análise da satisfação de trabalhadores sobre o rodízio de atividades em um abatedouro de aves. In: XXXVIII Encontro Nacional de Engenharia de Produção, pp. 1–12. ABEPRO, Maceió (2018)
9. Tirloni, A.S., Reis, D.C., Borgatto, A.F., Moro, A.R.P.: Association between perception of bodily discomfort and individual and work organisational factors in Brazilian slaughterhouse workers: a cross-sectional study. BMJ Open. **9**, e022824 (2019)
10. Ouellet, S., Vézina, N.: L'implantation de la rotation de postes: un exemple de démarche préalable. PISTES **5**(2), 3322 (2003)
11. EPM. MAPPING WITH CHECKLIST. Free software in English (2013). http://www.epmresearch.org/a57_free-software-in-english.html. [cited 27 December 2020]
12. Zawiah, S., Taha, Z.: The effect of job organizational factors on job satisfaction in two automotive industries in Malaysia. J. Hum. Ergol. **36**(2), 63–68 (2007)
13. Ramos, E., Reis, D.C., Tirloni, A.S., Moro, A.R.P.: Thermographic analysis of the hands of poultry slaughterhouse workers exposed to artificially cold environment. Proced. Manuf. **3**, 4252–4259 (2015)
14. Tirloni, A.S., Reis, D.C.D., Ramos, E., Moro, A.R.P.: Thermographic evaluation of the hands of pig slaughterhouse workers exposed to cold temperatures. Int. J. Environ. Res. Public Health. **14**, 838 (2017)
15. Tirloni, A.S., Reis, D.C.D., Dias, N.F., Moro, A.R.P.: The use of personal protective equipment: finger temperatures and thermal sensation of workers' exposure to cold environment. Int. J. Environ. Res. Public Health. **15**, 2583 (2018)
16. Tirloni, A.S., dos Reis, D.C., Dias, N.F., Moro, A.R.P.: Evaluation of worker satisfaction with the use of hand tools in a poultry slaughterhouse. In: Goonetilleke, R.S., Karwowski, W. (eds.) AHFE 2018. AISC, vol. 789, pp. 476–488. Springer, Cham (2019). https://doi.org/10.1007/978-3-319-94484-5_49
17. Dias, N.F., Tirloni, A.S., Reis, D.C., Moro, A.R.P.: Risk of slaughterhouse workers developing work-related musculoskeletal disorders in different organizational working conditions. Int. J. Ind. Ergon. **76**, 102929 (2020)
18. Jorgensen, M., Davis, K., Kotovski, S., Aedla, P., Dunning, K.: Characteristics of task rotation in the midwest US manufacturing sector. Ergonomics **48**(15), 1721–1733 (2005)
19. dos Reis, D.C., Tirloni, A.S., Moro, A.R.P.: Effects of reduced work pace on the risk of developing upper-limb musculoskeletal disorders in a poultry slaughterhouse. In: Karwowski, W., Goonetilleke, R.S., Xiong, S., Goossens, R.H.M., Murata, A. (eds.) AHFE 2020. AISC, vol. 1215, pp. 87–94. Springer, Cham (2020). https://doi.org/10.1007/978-3-030-51549-2_12
20. Tirloni, A.S., Reis, D.C.D., Tirloni, S.F., Moro, A.R.P.: Exertion perception when performing cutting tasks in poultry slaughterhouses: risk assessment of developing musculoskeletal disorders. Int. J. Environ. Res. Public Health. **17**, 9534 (2020)

Physical Ergonomics Design and Evaluation of Civil Aircraft Cockpit Control Devices

Xinyang Zhu(✉), Hongyu Zhu, Zhefeng Jin, and Yinbo Zhang

COMAC ShangHai Aircraft Design and Research Institute, No. 5188 Jinke Road, Pudong New Area, Shanghai, China
{zhuxinyang,zhuhongyu,jinzhefeng,zhangyinbo}@comac.cc

Abstract. Physical ergonomics evaluation of civil aircraft cockpit control devices has become increasingly important for aircraft engineers when designing pilot-centered cockpit. This paper puts forward the principles for physical ergonomics design of control devices, as well as a physical ergonomics evaluation method for cockpit control devices based on flight scenarios in virtual environment to demonstrate how digital human modeling software can be used effectively for such evaluation. Three digital pilot models are generated from anthropometric database, representative of the shortest, average and tallest Chinese pilot population, and interfaced with digital prototype of cockpit control devices in RAMSIS software for physical ergonomics design analysis of the control devices. The analysis mainly includes the control device's appearance, range of movement, operating force, and the analysis results can be used to achieve design iteration of cockpit control devices.

Keywords: Physical ergonomics · Control devices · Civil aircraft · Flight scenarios

1 Introduction

The current state-of-the-art in human-centered design evaluation is the application of digital modeling and simulation [1]. With this advanced technology, control devices can be evaluated in virtual digital prototype with digital human models. This evaluation can help us identify potential issues or defects in the physical ergonomics design of control devices. That allows engineers to ergonomically optimize the control devices early in the design process, when changes can be made much more efficiently than in the physical design stage.

Physical ergonomics of control device refers to the interaction relationship between the physical characteristics of the control device and the ergonomic characteristics of the human. This relationship mainly includes the comfort, force, range of human operating control devices. The physical ergonomics design of control devices must consider the pilot's physical parameters, perception characteristics, behavior characteristics and performance limits. The physical ergonomics evaluation of the control device can make the design of the control device more suitable for the user's ergonomic characteristics.

R. S. Goonetilleke et al. (Eds.): AHFE 2021, LNNS 273, pp. 338–345, 2021.
https://doi.org/10.1007/978-3-030-80713-9_43

With the help of digital human model and simulation technology, this paper proposes a method for evaluating the physical ergonomics design of cockpit control device based on flight scenarios, and applies this method to evaluate the design of the cockpit rudder and brake pedal. The evaluation mainly includes the pedal appearance, neutral pedal position, rudder pedal's range of movement, brake pedal's range of movement and operating force. The evaluation results can be used to effectively optimize control devices' structure and layout, and thus improve the efficiency of the integrated design of the civil aircraft cockpit.

2 Physical Ergonomics Design Requirements for Control Devices

Well-designed control devices can ensure both functional and comfortable accommodation of the pilots. However, the variations in pilot body sizes means that the structural design and layout of control devices may not meet the optimal needs of the entire target population. According to the regulation requirements of 14 CFR, CS, or CCAR 25.777(c), the design and layout of the control device should meet the operational needs for the flight crew with a height from 5′2″ (158 cm) to 6′3″ (190 cm). The appearance, range of movement, and operating force design of the control device should comply with the following design requirements.

(a) The appearance and size design of the control device should consider the operating posture of any crew member with a height of 158cm to 190cm;
(b) The range of movement of the control device should consider the operational accessibility requirements of any crew member with a height of 158cm to 190 cm;
(c) The operating force design of the control device should consider the operating capability requirements of any crew member with a height of 158cm to 190 cm.

3 Methodology

In this chapter, we proposes the analysis tool and method to evaluate the physical ergonomics design of the control device based on the flight scenarios, and shows an example using the rudder and brake pedal for an analysis.

3.1 Ergonomics Evaluation Tool

RAMSIS is a computer-aided human body digital system for occupant simulation [2]. The powerful posture prediction models of RAMSIS can predict seated postures of the pilot in cockpit. It enables the user to calculate possible pilot postures automatically based on a task description and eliminates arbitrary manipulation of digital manikins by the user [3]. RAMSIS's ergonomics analysis tools include: human body modeling, posture calculation, reachability analysis, comfort analysis, distance analysis, joint angle, maximum operating force calculation, etc.

3.2 Ergonomics Evaluation Steps

Taking the CAD data of cockpit control device as the object of evaluation, the virtual simulation evaluation process of physical ergonomics of control device based on RAMSIS is mainly comprised of the following five steps:

(a) Test manikins creation: Statistical analysis shows that the height of the human body is approximately of Gaussian distribution. The height of the 50th percentile males can represent the "average" of most human bodies. Therefore, the design and layout of the control device should be evaluated with RAMSIS manikins representing 158cm females (the shortest Chinese pilots), 173cm males (50th percentile males of Chinese pilots), 190cm males (the tallest Chinese pilots);
(b) Flight scenarios selection: Flight scenario refers to the expected behavior (or function) of the aircraft in a combination of crew (pilot and flight attendant), external environment (atmospheric, radio, terrain, electromagnetic, etc.) and internal conditions (normal or malfunction). By using the flight scenarios development method proposed by Hongyu Zhu [4], a flight scenarios database can be constructed, and relevant flight scenarios that include using of the control device can be found by searching the database. These scenarios can be used as target scenarios to evaluate the physical ergonomics design of the control device. Considering the occurrence frequency of the flight scenarios, the workload of the flight crew, and the complexity of the crew task under a flight scenario, a typical flight scenario can be chosen from these flight scenarios;
(c) Behavior extraction: Crew task is the purpose, action or corresponding behavior that the flight crew needs to achieve in a specific flight scenario. Pilot's behavior can be extracted by analyzing the crew task;
(d) Task description: The task description in RAMSIS's task editor is generally defined by users based on his subjective judgment of pilot behavior. However, task description based on flight scenarios can more accurately define interactions between a digital human model and a CAD environment, and reflect interactions between real pilots and cockpits as realistically as possible. The analysis of the crew task and pilot's behavior under the flight scenarios selected from the database can be used as the reference for task description. With many manikins available, by activating the required skin points on the manikins, including heel points, pedal points, and grasping points, one can thereby set the constraints between the manikin's skin points and the CAD environment in task editor;
(e) Postures calculation and analysis: RAMSIS automatically calculates the most probable or "average" flying posture of the manikin under task description. The resulting postures analyses could provide information for possible improvements of the control device design.

3.3 Ergonomics Evaluation of Rudder and Brake Pedal Unit

The rudder and brake pedal unit is used for rudder control and brake control of the aircraft. The evaluation process is as follows:

(a) Test manikins creation: Creating three manikins which include 190cm males (the tallest Chinese pilots), 173cm males (50th percentile males of Chinese pilots), 158cm females (the shortest Chinese pilots). All manikins will have medium sitting height and medium waist circumference. The manikins are shown in Fig. 1;

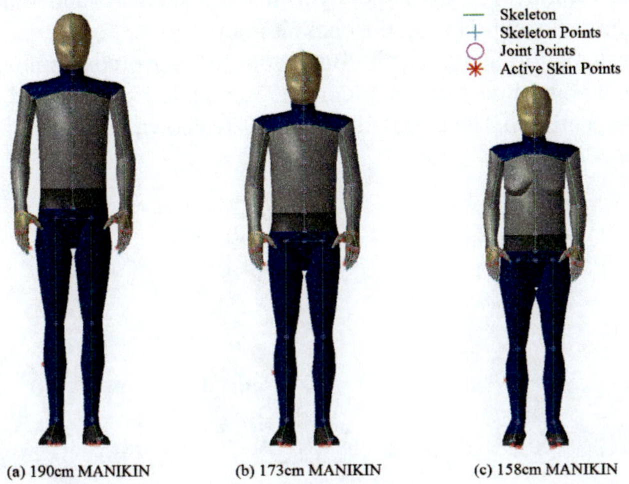

Fig. 1. Graphical representation of the test sample of RAMSIS manikins

(b) Flight Scenarios selection: In the flight scenarios database (considering only basic flight scenarios), search for the crew tasks which include the usage of rudder pedal control. The pilot uses the steering and braking control function of the pedal on the ground, and uses the yaw control function in flight. The pilot's control of the pedal on the ground can be divided into taxiing and ground take-off stage. In the ground take-off stage, the pilot's workload is higher compared to other flight stages. The pilot not only has to pay attention to the change in the speed of the aircraft and the display information, but also has to control the side stick, throttle lever and rudder pedal. Thus, a normal take-off scenario which time, environmental, and status dimensions are normal is chosen as the typical scenario to analyze the pilot's behavior;

(c) Behavior extraction: In normal take-off flight scenario, the crew task is to keep the aircraft in the center of the runway and decide whether to abort the take-off before the take-off decision speed. The corresponding pilot behavior (taking the left seat pilot as an example) is to maintain a line of sight straight ahead and control the rudder pedals with the feet to adjust the aircraft's rolling direction and use the brakes if necessary. The left hand is held on the side stick to control the elevator for rotation of the aircraft, and the right hand is held on the throttle lever on the central pedestal to control engines' power;

(d) Task description: After obtaining the analysis result of the pilot's behavior in the typical flight scenario of the control device's usage, the relationship of the manikin and the 3D environment can be set in the task editor. The task constraints between the manikin and the 3D environment include: 1) Constrain the eyes of the pilot at the level of Design Eye Point (DEP); 2) Constrain the line of sight straight ahead; 3) Constrain the shoe sole surface to be parallel to the plane of the pedals; 4) Constrain the left hand's grasping point at the side stick; 5) Constrain the right hand's grasping point at the throttle lever; 6) Fix pelvis with no tilt sideways and long axis rotation; 7) Constrain the heel points on the cockpit floor;

(e) Postures calculation and analysis: Based on the above constraints, calculate the rudder pedal control postures of 190cm, 173cm male manikins and 158cm female manikin respectively. The calculation results are shown in Fig. 2.

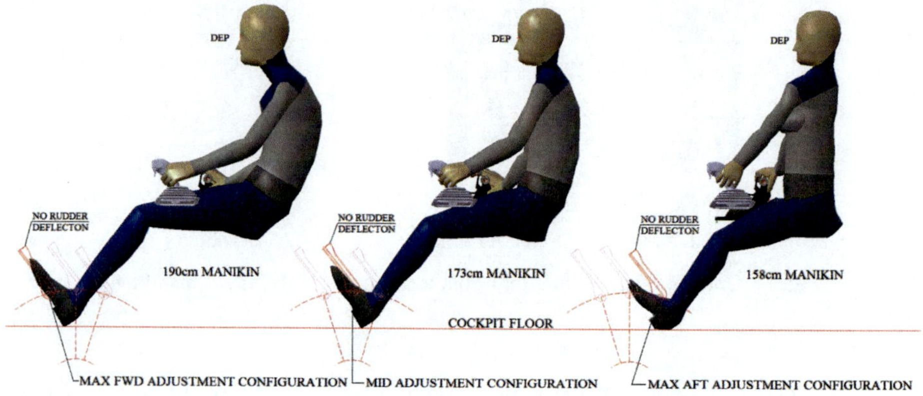

Fig. 2. Determining the preferred neutral position for the rudder pedals

4 Analysis and Discussion

Based on the method puts forward in this paper, the dimension of the pedal, the adjustment range of neutral pedal position, the rudder pedal's range of movement, the brake pedal's range of movement, and the control force can be evaluated.

4.1 External Dimension Evaluation

The length of the pedal should be able to adapt to pilots with different foot lengths. If the pedal is too long, the braking torque will be too large. If the pedal is too short, the pilot's feet will go beyond the top of the pedal. According to the feet length of the human body of different heights in GB-10000–88, considering the thickness of the shoes and the operating gap, a reasonable length and width of the pedal can be defined and verified by digital human models.

4.2 Neutral Pedal Position Adjustment Range Evaluation

The cockpit offers adjustable rudder pedals. Pilots can choose neutral rudder pedal position according to their needs. According to the calculation result in 3.3(e), the tallest, the medium, and the shortest manikins' preferred neutral pedal position can be determined. The predicted foot positions are basically consistent with the designed position of the max forward adjustment, mid adjustment, max afterward adjustment of the rudder pedal. Meanwhile, the angle value of each joint of the manikin can be obtained, and used to confirm that it is within the comfortable value range of the human joint angle. The result of comparison shows that the human body is a comfortable posture, so the adjustment range of the neutral pedal position is reasonable.

4.3 Rudder Pedal's Range of Movement Evaluation

When the rudder pedal is fully deflected, the full forward rudder deflection and the full afterward rudder deflection define the maximum range of movement of the rudder pedal. It is assumed that each manikin would "pick" the neutral rudder pedal position that was the closest to the predicted foot position. Constraining the sole of the manikin's foot at the right and left rudder pedal respectively. The calculation result of each manikin's posture is shown in Fig. 3. Each manikin is able to fully depress the rudder pedal, and the rudder pedal's range of movement is reasonable.

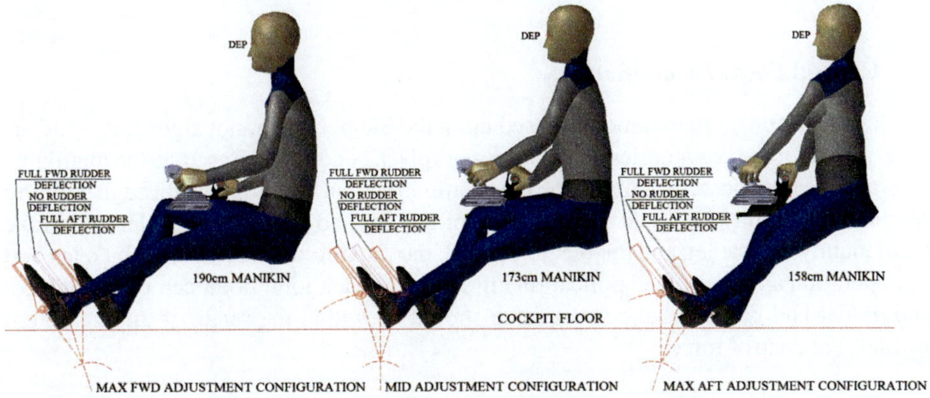

Fig. 3. Manikins fully depressing the rudder pedal

4.4 Brake Pedal's Range of Movement Evaluation

When the brake pedal is fully depressed, the pedals will deflect a certain angle based on the pedal rotation axis. This angle defines the maximum range of movement of the brake pedal. The brake control requires that every pilot must be able to fully apply the brakes. There is a concern that shorter pilots may not be able to fully apply the brakes. It is assumed that pilots do not slide back and forth or move up and down in the seat while applying the brakes. Each manikin's posture when fully depressing the upper part of both brake pedals is calculated under the constraint that the H point is fixed to the coordinates of the most probably seating posture in flight, see Fig. 4. The resulting posture analyses indicated that the shortest female could be able to fully apply the brakes. Thereby, the brake pedal's range of movement is reasonable.

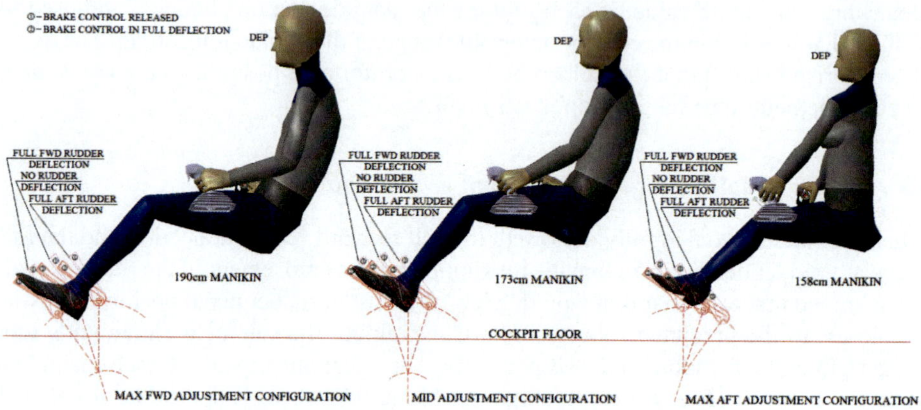

Fig. 4. Manikins fully depressing the brake pedal

4.5 Control Force Evaluation

The RAMSIS force computation is based upon the Siemens-Burandt algorithm which is widely used for the ergonomics analysis of workplace conditions. Based on the manikin's "average" seating posture in flight, the maximum force can be computed through defining the manikin's joint chain and the direction of the applied force. The rudder and brake pedal mainly use the leg and the feet to control, the maximum force of the pilot's leg and feet operation and the force application efficiency of each joint point can be calculated. The rudder and brake pedal control force should be within the range of the maximum leg and feet control force.

5 Conclusion

With digital human models (such as RAMSIS), a physical ergonomics evaluation method of cockpit control devices based on flight scenarios is proposed in this paper. The interaction between a digital human model and a CAD environment can be defined accurately based on selected flight scenarios, thereby reflecting real pilots' behavior as realistically as possible under a specific crew task. Taking the rudder and brake pedal as an example, the evaluation process of the physical ergonomics design is shown. The analysis of the evaluation result allows engineers to ergonomically optimize the control device at the digital stage in the design process, when changes can be made much more efficiently than in the physical design stage.

References

1. Naumann, A., Röetting, M.: Digital human modeling for design and evaluation of human-machine systems. MMI-Interktiv **12**, 27–35 (2007)
2. Vogt, C., Mergl, C., Bubb, H.: Interior layout design of passenger vehicles with RAMSIS. J. Hum. Factors Ergon. Manuf. Serv. Ind. **15**(2), 197–212 (2010)
3. Van der Meulen, P.A., Diclemente, P.: Ergonomic evaluation of an aircraft cockpit with ramsis 3d human modeling software (2001)
4. Zhu, H., Meng, H., Lu, S., Bao, G.: The flight scenarios development method for cockpit design and evaluation of civil aircraft. In: Stanton, N. (ed.) AHFE 2020. AISC, vol. 1212, pp. 603–609. Springer, Cham (2020). https://doi.org/10.1007/978-3-030-50943-9_77

Investigation of Anatomical Shape of Thumb of de Quervain's Tenosynovitis Patients

Eunice Wai-si Tam[1], Joanne Yip[1](✉), Kit Lun Yick[1], Sun Pui Ng[2], and Christian Fang[3]

[1] Institute of Textiles and Clothing, The Hong Kong Polytechnic University, Hung Hom, Hong Kong
eunice.ws.tam@connect.polyu.hk, {joanne.yip,tcyick}@polyu.edu.hk
[2] Hong Kong Community, The Hong Kong Polytechnic University, Hung Hom, Hong Kong
zerance.ng@cpce-polyu.edu.hk
[3] Department of Orthopaedics and Traumatology, The University of Hong Kong, Queen Mary Hospital, Pok Fu Lam, Hong Kong
cfang@hku.hk

Abstract. Patients with de Quervain's tenosynovistis (dQt) experience pain on the radial side of their injured wrist. One of the conservative treatments prescribed for this hand condition is splinting. The fit of the splint is a critical element that may affect the treatment compliance rate and wear comfort of patients. To fabricate a splint with well-fitting properties, exploring the anatomical shape of the hand of dQt patients is essential. In this study, 16 female dQt patients are recruited for three-dimensional hand scanning and investigation of their thumb curvatures. The measured angles which show the different curvatures along the thumb are discussed.

Keywords: De Quervain's tenosynovitis · Anatomical · Thumb · Three-dimensional · Splinting treatment

1 Introduction

De Quervain's tenosynovitis (dQt) is a common and painful hand condition that is mainly caused by repetitive hand movement, particularly motions that involve twisting the wrist and extending the thumb. dQt is in general more prevalent among females than males with 1.3% of women who suffer from dQt versus 0.5% of men [1]. Others at higher risk of dQt include women in their postpartum period or those in their forties to sixties [2]. Patients with dQt may experience pain, tenderness and soreness along the radial side of their hand due to inflammation in the first dorsal compartment [3].

Splinting is one of the non-invasive methods used to treat dQt. Wearing a splint facilitates the healing process of the inflamed tendons and sheaths by supporting the thumb and minimizing hand movement. There are mainly two types of splints for dQt: rigid and soft splints. The majority of the former are fabricated with thermoplastic materials and designed for wear during nighttime [4]. Since hand activities could be

R. S. Goonetilleke et al. (Eds.): AHFE 2021, LNNS 273, pp. 346–352, 2021.
https://doi.org/10.1007/978-3-030-80713-9_44

minimized at home after work hours, it is recommended that patients wear the rigid splint at night and even when they are sleeping so that the injured wrist could also be protected during sleep. Soft splints are for daytime wear as patients may need to carry out a small range of wrist and thumb movements during different daily life activities that require a certain degree of flexibility. Soft splints are mainly fabricated with neoprene and fabrics. Since there is less constraint of hand movement with the use of softer materials, these splints are usually recommended for dQt patients who have more mild symptoms [5].

Patients with dQt are to wear their splint for long hours in order to sufficiently heal their hand. In a study that compared the effectiveness of bandaging versus splinting as a treatment, the patients had to wear either the splint or bandage for at least six hours per day to determine their efficacy [6]. The long hours of wear mean that regardless whether the splint is rigid or soft, the fit of the splint on the hand is the most important factor. Wearing a splint with a poor fit may lead to discomfort, skin irritation and formation of pressure ulcers. Therefore, to develop a functional splint that is well fitting and accommodates the shape of the hand, a good understanding of the anatomical shape of the hand is critical. This paper investigates the anatomical shape of the hand of dQt patients, especially focusing on the curvature along the radial side of the wrist to the proximal phalanx of the thumb.

2 Methodology

2.1 Participants

In this study, 16 female dQt patients are recruited with 5 who are between 18 and 45 years old, and 11 who are 46 and older. All are right-handed and their right hand is officially diagnosed with dQt by a professional clinician. The majority have experienced pain in the injured area for 3 months to approximately 2 years. However, 1 of the patients has suffered from wrist pain for more than 5 years.

2.2 Hand Scanning and Measuring Process of Thumb Curvature

The fabricated shape and efficacy of a medical splint are greatly affected by the shape of the human hand. To explore the anatomical shape of the hand of those afflicted with dQt, three-dimensional (3D) images of the right hand of the subjects are first taken by placing the hand in a neutral resting position, as shown in Fig. 1 [7]. The distal phalanx of the thumb is gently placed onto the middle phalanx of the index finger. The 3D contours of the positioned hand were then captured by using an Artec Eva handheld scanner. Afterwards, the 3D hand images were imported into Artec Studio software for image processing.

The images of the hand of each patient were divided into 6 specific areas to produce the cross-section curves. The distance between two adjacent cross-section curves is 15 mm. The specific measured areas include the proximal phalanx (PP), metacarpophalangeal (MCP) joint, metacarpal (MC) bone, carpometacarpal (CMC) joint, carpals (C) and radius (R) of the hand; see Fig. 2. The cross-section curves along the midline of

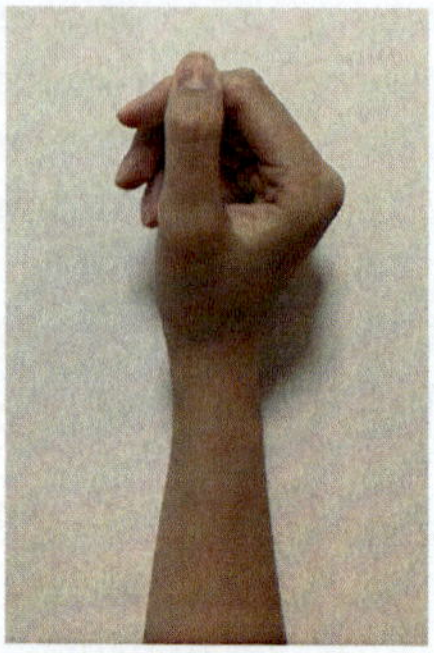

Fig. 1. Neutral resting position of hand

thumb are marked vertically to create the points of intersection. Three angles (α, θ and β) are formed by extending 2 straight lines that are 5 mm in length from the point of the intersection at both sides respectively; see Figs. 3a and 3b. The three angles were measured in order to determine the degree of curvature at the different areas. A smaller angle indicated a larger curvature at the location.

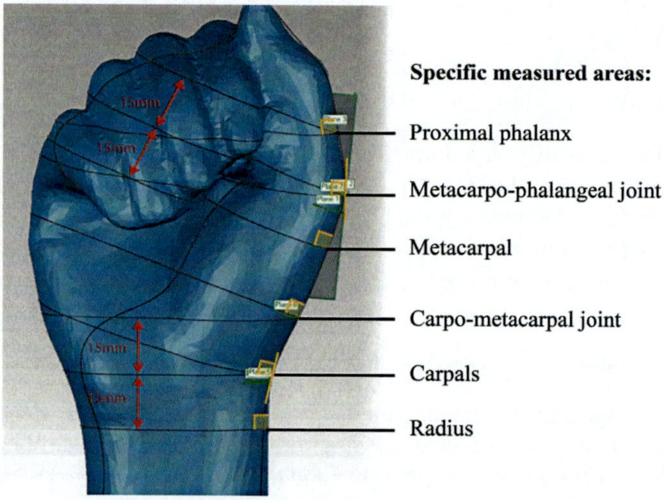

Fig. 2. Location of cross-section curves

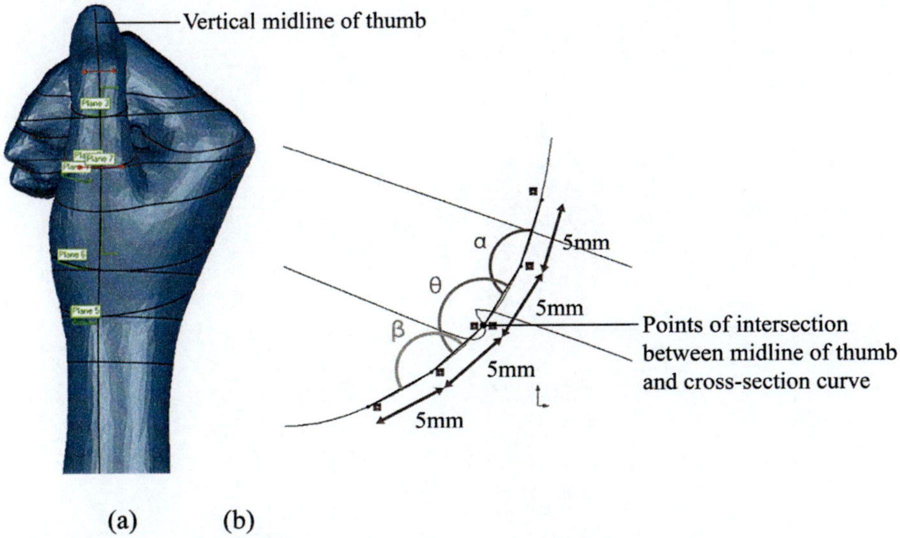

Fig. 3. (a) Midline of thumb marked vertically, and (b) angles α, θ and β along cross-section curve

3 Result and Discussion

Three angles, α, θ and β, along the 6 cross-section curves were measured correspondingly and are presented in Fig. 4. The PP has the largest curvature in comparison to the other hand areas. The mean values of the α, θ and β angles at the PP, MCP joints, MC bones, CMC joints, C, and R of the hand are 151.3° (SD = 9.4), 155.3° (SD = 4.3) and 153.0° (SD = 5.8); 159.6° (SD = 6.0), 162.5° (SD = 4.6) and 157.7° (SD = 3.9); 171.3° (SD = 2.7), 169.7° (SD = 6.7) and 162.5° (SD = 3.7); 172.1° (SD = 6.1), 169.1° (SD = 5.1) and 163.5° (SD = 5.4); 169.4° (SD = 3.9), 168.6° (SD = 5.2) and 162.1° (SD = 6.1); and 164.4° (SD = 3.8), 162.8° (SD = 5.3) and 159.8° (SD = 5.5), respectively. The area with the greatest curvature is found at the α angle of the PP, whereas the lowest curvature is found at the α angle of the CMC joint.

The changes in curvature from the PP to R of the hand along each angle are presented in Fig. 5. As the arc of the measured area passes through the distal end points of the two straight lines via the vertex of the angle, a smaller angle indicates a larger curvature. The curvatures along the α angle from the PP to the CMC joint of the hand are obviously reduced, and gradually increase from the CMC joint to the R of the hand. On the other hand, the curvatures along the θ and β angles from the PP to the R of the hand show less variation.

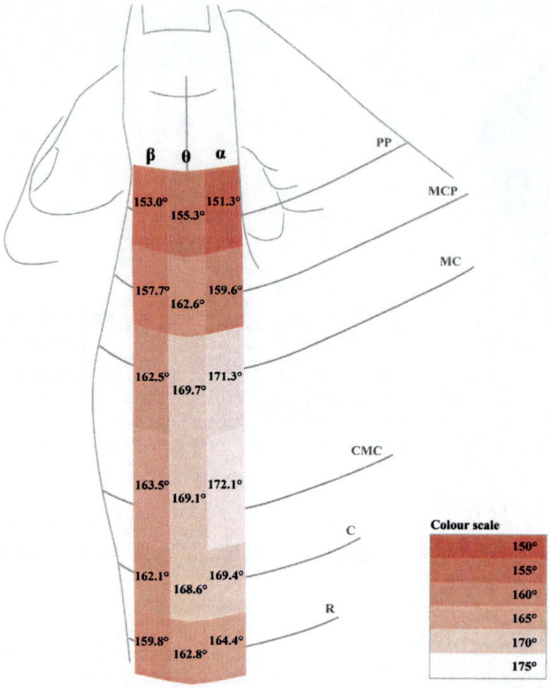

Fig. 4. Mean values of α, θ and β at six specific positions

With larger curvatures at the PP and MCP joints of the hand, clinicians may need to pay more attention to these areas when molding the thermoplastic sheet for the splint or noting the properties of the softer materials used for the splint pattern. The fluctuation of the curvatures along the α angle may cause issues when constructing a well-fitting splint. The measured angles in this study clearly reflect the curvature of each specific area along the radial side of the patient's hand. Clinicians can have a better sense of the anatomical shape of the patient's hand more quickly and easily by referring to measured angles at specific hand areas instead of merely visually evaluating the shape of the hand as an informal means of measurement. Furthermore, constructing a splint simply after observation and subjective evaluation may lead to an improper fit. By scanning the 3D images of patient hands, the curvatures of the targeted parts of the hand can be measured accurately with the use of computer software. For dQt patients, the radial side of their hand can be extracted and investigated for better construction of thumb supports and splints.

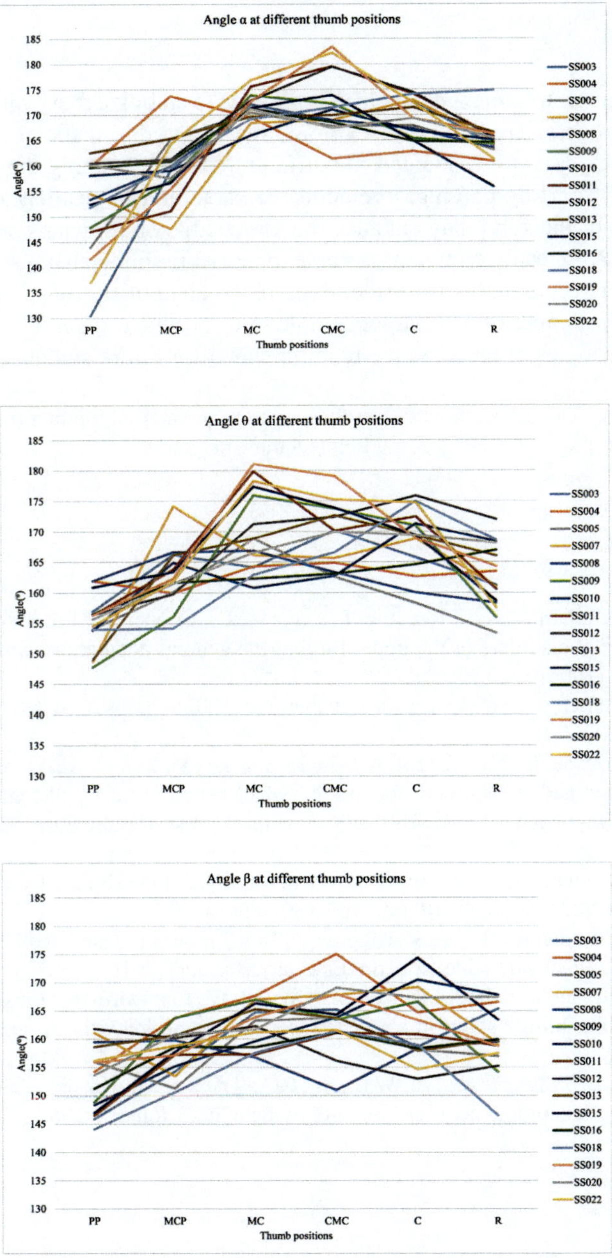

(a)
(b)
(c)

Fig. 5. Changes in curvature from PP to R of hand along (a) α, (b) θ and (c) β angles

4 Conclusion

Splinting as a form of treatment can help dQt patients to stabilize their injured wrist and aid the healing process. This study has explored the ergonomic shape of the thumb and wrist of dQt patients, and investigated the different curvatures along the radial side of their hand by using part-by-part measurements based on three angles. The quantification of the curvatures of the hand may enhance the knowledge of clinicians on the shape of the human hand. As a result, clinicians are able to fabricate splints that better support the thumb with a more accurate fit. To further explore the anatomical shape of the hands of dQt affected patients, recruiting a larger sample size for image scanning and collecting more data on the angles of the hand are recommended for future studies.

Acknowledgments. This research work is fully supported by internal funding from the Institute of Textiles and Clothing of the Hong Kong Polytechnic University.

References

1. Walker-Bone, K., Palmer, K.T., Reading, I., Coggon, D., Cooper, C.: Prevalence and impact of musculoskeletal disorders of the upper limb in the general population. Arthrits Care Res. **51**(4), 642–651 (2004)
2. Satteson, E., Tannan, S.C.: De Quervain Tenosynovitis (2020). https://www.ncbi.nlm.nih.gov/books/NBK442005/
3. Goubau, J.F., Goubau, L., Van Tongel, A., Van Hoonacker, P., Kerckhove, D., Berghs, B.: The wrist hyperflexion and abduction of the thumb (what) test: a more specific and sensitive test to diagnose de Quervain tenosynovitis than the eichhoff's test. J. Hand Surg. (Eur. Vol.) **39**(3), 286–292 (2014)
4. Chow, K.K.C.: Splint for de Quervain disease: a new design. Hong Kong J. Occup. Ther. **19**(1), 31–32 (2009). https://doi.org/10.1016/S1569-1861(09)70042-7
5. Saunders, R., Astifidis, R., Burke, S., Higgins, J., McClinton, M.: Hand and Upper Extremity Rehabilitation: a Practical Guide, 4th edn. Elsevier, Missouri (2016)
6. Jongprasitkul, H., Suputtitada, A., Kitisomprayoonkul, W., Pintawiruj, K.: Elastic Bandage vs. neoprene thumb stabilizer splint in acute de Quervain's tenosynovitis. Asian Biomed. **5**(2), 263–267 (2011)
7. Lin, W.T., Kuo, L.C., Liu, H.Y., Wu, W.L., Su, F.C.: The three-dimensional analysis of three thumb joints coordination in activities of daily living. Clin. Biomech. **26**(4), 371–376 (2011)

Human Characteristics and Influencing Factors

Using Ultrasound to Assess Microchambers and Macrochambers Tissue Properties After Walking at Different Speeds and Durations

Wei-Cheng Shen[1], Yih-Kuen Jan[2,3,4], Chi-Wen Lung[2,5], Hsin-Chieh Chen[6], Cheng-Tsung Li[5], Jian-Guo Bau[6], and Ben-Yi Liau[6(✉)]

[1] Department of Digital Media Design, Asia University, Taichung, Taiwan
[2] Rehabilitation Engineering Lab, University of Illinois at Urbana-Champaign, Illinois, USA
[3] Kinesiology and Community Health, University of Illinois at Urbana-Champaign, Illinois, USA
[4] Computational Science and Engineering, University of Illinois at Urbana-Champaign, Illinois, USA
[5] Department of Creative Product Design, Asia University, Taichung, Taiwan
[6] Department of Biomedical Engineering, Hungkuang University, Taichung, Taiwan
byliau@hk.edu.tw

Abstract. Exercise has been shown to improve health in people with diabetes. However, various walking speeds and durations in daily life may increase plantar pressure, thereby increasing risk for diabetic foot ulcers. Plantar thickness and stiffness have been demonstrated to be associated with plantar pressure. The pads under the heel and metatarsal heads include microchambers and macrochambers for absorbing impact forces during weight-bearing activities. The objective of this study was to investigate the effect of different walking intensities on the changes of biomechanical properties of plantar soft tissues. The healthy participants were tested in a 3 × 2 factorial design, including three walking speeds (1.8, 2.7, and 3.6 mph) and two durations (10 and 20 min). B-mode ultrasound images were obtained at the first metatarsal head to quantify plantar soft tissue thickness and stiffness before and after walking. Our results showed that the walking duration factor causes a significant effect on the macrochambers thickness. In addition, the speed factor caused a downward trend of the macrochambers stiffness with a lower walking speed.

Keywords: Diabetic foot ulcer · Ultrasound · Weight-bearing · Metatarsal heads

1 Introduction

The plantar soft tissue in diabetes had a risk of developing ulcers, it was the most recognized complication in people with diabetes mellitus (DM) [1]. In diabetes tissue glycation, the fat pad increased collagen fibril density causes it to stiffen risk of developing foot ulcers [2]. The American Diabetes Association (ADA) recommends that people with DM daily exercise or at least not allowing more than 2 days to elapse between

R. S. Goonetilleke et al. (Eds.): AHFE 2021, LNNS 273, pp. 355–363, 2021.
https://doi.org/10.1007/978-3-030-80713-9_45

exercise sessions have 150 min/week of physical activity [1]. An appropriate intensity of exercise can provide necessary physical stress to maintain tissue health. However, if the intensity is too low, it would decrease the tolerance of tissues to subsequent stresses, or if the intensity too high, it would lead to tissue injury [3]. But most weight-bearing physical may decrease tissue thickness and increased plantar pressure [4]. Other studies showed the increasing pressure demonstrated to be a major risk factor for diabetic foot ulcers, measure the stiffness in plantar soft tissue has been used to detect the endured pressure at plantar tissue [5, 6].

Plantar soft tissue stiffness and thickness are critical biomechanical variables to understand stress concentrations that may contribute to tissue injury [7]. The plantar soft tissue of diabetic was generally increased the stiffness and thickness decrease [8]. it can be caused by damaged motor neurons of the foot musculature or blood flow reduction [9]. The research revealed that walking speed is one of affected skin blood flow increase [10], and make it decrease the stiffness[11]. Also, soft tissue thickness decrease makes functional on difficult to cushioning capacities, cause the rising pressure [12].

Soft tissue is a multilayer embodying, anisotropic, and viscoelastic properties to withstand large structural deformations [13]. This multilayer tissue include microchambers and macrochambers [6]. Microchambers contain predominantly elastic fibers and equal amount of collagen and elastic fibers are identified in the macrochambers [14]. The microchambers function to maintain most of the macrochamber tissue beneath the plantar and prevent excessive deformation of the deep pad structure [15]. The macrochambers play a significant role in the resiliency, i.e., the tissue's property to recover its shape after deformation and are supposed to play a cushioning role in the walking [3].

This study can provide a foundation to understand the effect of different walking exercises including different time duration and different speeds on the changes of mechanical property of plantar tissue in people at risk for foot ulcers. This study is to observe the effect of different walking intensities on the microchambers and macrochambers thickness and stiffness properties of plantar soft tissue to the best of our knowledge. Thus, initial work is essential to analyze healthy people's reactions, which can provide a foundation for understanding the effect of diabetes on the response. Therefore, the current study aimed to investigate the effect mechanism of different walking speeds and durations on microchambers and macrochambers thickness and stiffness of plantar soft tissue properties, from multiple skin surface indentation tests in non-diabetics.

2 Method

2.1 Plantar Soft Tissue

Ultrasonography can provide information on the soft tissue structures in their plantar soft tissue thickness [11, 16]. Also it can be used to assess soft tissue stiffness [17].In this study, skin epidermal plus dermal thickness was obtained using a 12 MHz B-mode ultrasound (ArtUs EXT-1H, Milano, Italy). Images (7 mm depth, 9 mm width) were acquired using a linear ultrasound. A gel pad was used to control the contact interference area between the probe and skin. Using a custom-built motor indenter device, loadcell and B-mode ultrasound images were measured at the first metatarsal heads to hold the similar pressure value quantify plantar soft tissue thickness after walking. Thickness and stiffness are measures with the same equipment. It includes a gel pad, ultrasound probe, loadcell, and stepper motor. Gel pad to limit the area of ultrasound contact with the skin, ultrasound probe for soft tissue image RF data, the acquired load cell signal was recorded using LabVIEW software (DAQ, National Instruments Austin, TX, USA), and stepper motor can monitor pressure value, and a stepping motor to regulate the advancing speed and distance of ultrasound probe (Fig. 1).

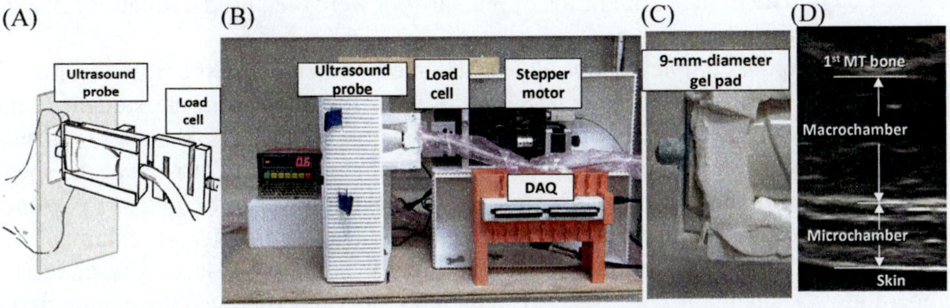

Fig. 1. (A) Experimental at the first metatarsal head, (B) Indentation system, (C) The ultrasound probe with gel pad, (D) Macrochamber and microchamber beneath 1st Metatarsal head.

The majority of skin structural deformation of the foot sole occurred within the first metatarsal [18], choose first metatarsal for the experimental location. A 9-mm-diameter indenter gel pad probe was used in this study. A load cell linear actuator was assembled on a load cell holder. We controlled the linear displacement of the indenter using a step motor driver. The rate was calculated by using the maximum deformation and the plantar pressure force to obtain Young's modulus obtained by motor applying cyclic compressions on the first metatarsal head [3, 12].

To quantify the elastic properties of soft tissues, we used the effective Young's modulus (E). It is a traditional material constant for response to the stiffness of soft tissue property [19].To extract effective Young's modulus E, the equation was defined as below:

To quantify the elastic properties of soft tissues, we used the effective Young's modulus (E). It is a traditional material constant for response to the stiffness of soft tissue property [19].To extract effective Young's modulus *(E)*, the equation was defined as below:

$$\boldsymbol{E} = \frac{(1-\boldsymbol{v}^2)}{2\boldsymbol{a}\cdot \boldsymbol{k}(\boldsymbol{v},\boldsymbol{a}/\boldsymbol{h})}\frac{\boldsymbol{P}}{\boldsymbol{w}} \tag{1}$$

Where $\boldsymbol{v}$ is Poisson's ratio; $\boldsymbol{a}$ is the indenter radius; $\boldsymbol{k}$ is a scaling factor dependent on the Poisson's ratio, indenter radius (4.5 mm), and soft tissue thickness; $\boldsymbol{h}$ is the soft tissue thickness; $\boldsymbol{P}$ is the force of pressure loading (indentation); $\boldsymbol{w}$ is the depth of indentation. Generally, 0.45 is set as the Poisson's ratio for biological soft tissues [20]. The $\boldsymbol{k}$ value was obtained from the information extracted from the publication of Hayes and colleagues [21].

2.2 Data Acquisition

A 3 × 2 factorial design, including 3 speeds (1.8 mph, 2.7 mph, and 3.6 mph) and 2 durations (10 and 20 min), was used in this study. A total of 6 walking protocols was tested in this study. The participant received the 1.8 mph protocol in the first week, the 2.7 mph protocol in the second week, and the 3.6 mph protocol in the third week. The order of duration (10 and 20 min) was randomly assigned. Each protocol was separated by 7 ± 2 days. The tissue strain ratios of plantar foot were then used to quantify plantar thickness and stiffness.

Each participant completed a provided informed written consent before the study's onset. Healthy participants between 18 and 28 were recruited from the student at the Asia University. The exclusion criteria included active foot ulcers, diabetes, vascular diseases, hypertension, and inability to walk 20 min independently or walk at the speed of 3.6 mph independently. Subjects were free from recent lower limb injury or presence of lower limb pain, as well as any sensitivities to adhesives or gels. All examinations were performed in the Rehabilitation Engineering Lab at the Asia University. and room temperature was maintained at 24 ± 2 °C. Our laboratory has validated the plantar tissue feasibility of the test plan [11, 22, 23].

2.3 Experimental Procedures

For each participant, all experiment process was done during a single laboratory visit lasting approximately one hour. Set the plantar postures for control of the foot dorsum and set relax 15-min by lifting replacing the left foot onto the sit while relaxing, then neutral for the foot was stand on the ultrasound tube box. The operator adjusts each participant's left plantar position as required to access the necessary plantar sole place test. All participants were reminded to relax all lower limb muscles to avoid any confounding effects of active muscle-mediated skin structural deformation. The same operator performed specific components of the experimental procedure across all subjects to reduce variability.

2.4 Statistical Analysis Method

The significant differences in each parameter among the three speeds (1.8, 2.7, and 3.6 mph) were tested using one-way analysis of variance (ANOVA). The paired sample *t*-test was used to evaluate the variation of two durations (10 and 20 min) and the between the speeds. It examines the interaction between the speed and duration factors on macrochambers and microchambers stiffness thickness. All statistical tests were performed using SPSS 22 (IBM, Somers, NY, USA) at the significance level of 0.05.

3 Results

Among the one-way factors of speed, there were no significant difference exist between different speeds. However, thickness and stiffness of the macrochamber trend to downward more than that of the microchamber. Including the walking speed rate at the same time factor, the thickness of the macrochamber is lower at 2.7 mph than that at 1.8 and 3.6 mph (Table 1). Moreover, the stiffness of the soft tissue continues to decrease with walking time and speed. Regarding the walking duration, there is a significant difference in the plantar macrochambers thickness between 10 min and 20 min in 3.6 mph (109 ± 8.4% v.s. 121.7 ± 6.1%, $P = 0.047$) (Table 2 and Fig. 2).

Table 1. Measure the walking speed of thickness and stiffness.

Duration		Speed			One-way ANOVA	Fisher LSD Post hoc		
		1.8 mph (Mean ± SE)	2.7 mph (Mean ± SE)	3.6 mph (Mean ± SE)	*P* value	1.8 mph vs. 2.7 mph	1.8 mph vs. 3.6 mph	2.7 mph vs. 3.6 mph
Thickness (%)								
Macro	10 min	107.5 ± 6.1	95.5 ± 13.2	109.0 ± 8.4	0.576	0.404	0.912	0.348
	20 min	116.2 ± 4.6	107.3 ± 9.4	121.7 ± 6.1	0.384	0.395	0.594	0.182
Micro	10 min	92.0 ± 14.5	82.4 ± 10.7	98.4 ± 14.5	0.546	0.370	0.333	0.939
	20 min	97.1 ± 9.7	90.8 ± 7.0	92.6 ± 16.7	0.792	0.598	0.928	0.538
Stiffness (%)								
Macro	10 min	134.0 ± 35.3	92.9 ± 28.8	89.5 ± 27.7	0.949	0.755	0.843	0.909
	20 min	128.0 ± 36.7	84.6 ± 59.8	118.9 ± 56.4	0.827	0.569	0.904	0.652
Micro	10 min	99.1 ± 27.4	76.8 ± 22.2	73.0 ± 35.4	0.694	0.652	0.406	0.695
	20 min	110.6 ± 25.6	135.3 ± 48.2	88.0 ± 37.4	0.692	0.659	0.685	0.404

Note: %, normalized with before excise value. Data was presented as mean ± standard errors.

4 Discussion

This study demonstrated that the 3.6 mph walking durations (10 and 20 min) significantly affected plantar tissue thickness and the walking speed of 1.8 and 2.7 mph did not and walking at 2.7 mph for 10 min resulted in a decreased thickness macrochambers tissue. It showed a trend that plantar tissue stiffness after walking at 2.7 and 3.6 mph for the same duration was lower compared to walking at 1.8 mph. This finding is significant because walking durations of 20 min can increase plantar tissue thickness compared to slow walking durations of 10 min. This study suggests people at risk for foot ulcers should walk at the speed (3.6 mph) rather than walk slowly (1.8 mph) and keep more duration (20 min).

The result of this study found 2.7 mph walking 10 min may decrease the plantar thickness. The thickness decrease represents the macrochambers attenuate to shape deformation ability [3], this phenomenon might indicate that the soft tissue universal received more pressure. This phenomenon disappears after 20 min of exercise.

Table 2. Measure the duration of thickness and stiffness.

Parameter	Layer	Speed	Duration		Paired t-test
			10 min	20 min	P value
			(Mean ± SE)	(Mean ± SE)	
Thickness (%)	Macro	1.8 mph	107.5 ± 6.1	116.2 ± 4.6	0.093
		2.7 mph	95.5 ± 13.2	107.3 ± 9.4	0.079
		3.6 mph	109.0 ± 8.4	121.7 ± 6.1	**0.047***
	Micro	1.8 mph	92.0 ± 14.5	97.1 ± 9.7	0.655
		2.7 mph	92.9 ± 28.8	76.8 ± 22.2	0.686
		3.6 mph	89.5 ± 27.7	73.0 ± 35.4	0.690
Stiffness (%)	Macro	1.8 mph	134.0 ± 35.3	99.1 ± 27.4	0.167
		2.7 mph	92.9 ± 28.8	76.8 ± 22.2	0.686
		3.6 mph	89.5 ± 27.7	73.0 ± 35.4	0.690
	Micro	1.8 mph	128.0 ± 36.7	110.6 ± 25.6	0.801
		2.7 mph	84.6 ± 59.8	135.3 ± 48.2	0.339
		3.6 mph	118.9 ± 56.4	88.0 ± 37.4	0.464

Note: %, normalized with before excise value. Data was presented as mean ± standard errors

Our previous study demonstrated that the walking speeds significantly affected plantar tissue stiffness and the walking durations did not [11], combined with the findings from this study, the trend of stiffness also supports the research. The other study showed the decreased stiffness macrochambers tissue represents means increased cushioning capacities [12]. And this study showed that walking duration was significantly affected plantar tissue thickness at 3.6 mph, this may indicate that in the walking duration (10 and 20 min), the thickness of the soft tissue changed can affect the plantar more significantly than the stiffness.

Several issues still have to be clarified in the future work. The stress increase caused by wearing shoes or different shoe-wearing habits is different. In this study, the subjects themselves are accustomed to walking shoes, which should be limited in the future. Besides, for the subjects exhibiting different walking posture at 3.6 mph speed, the walking posture may change the soft tissues' compression conditions, but this depends on the subjects' own exercise habits. It should consider that how to reduce the difference between the subjects' moving postures. In this study, we developed a method for quantifying the plantar foot's mechanical properties and analyzed the differences in skin and fat mechanical properties with speed and duration. This study suggests that the possibility of the soft tissue buffer capacity attenuation is related to mechanical property changes in the walking duration time. This study can be used to help preserve healthy plantar tissues in diabetic patients.

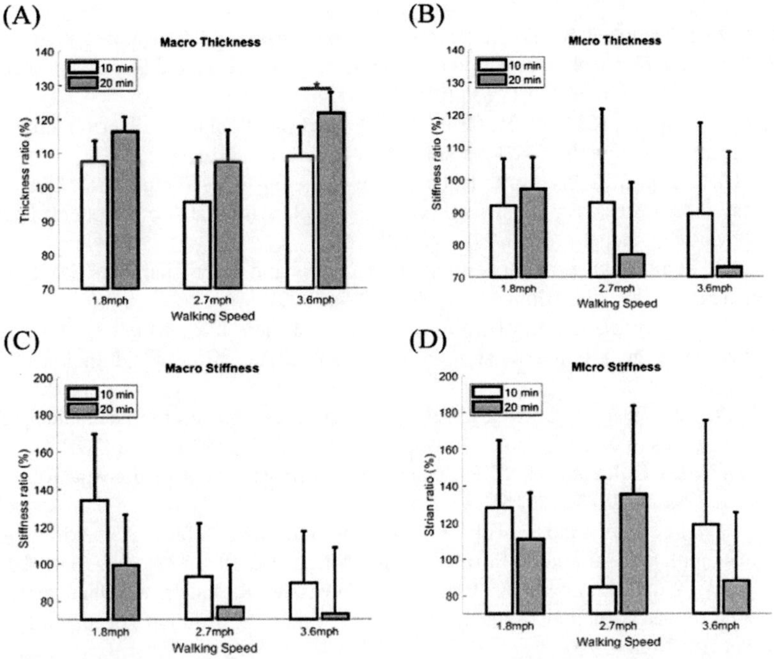

Fig. 2. Percentage change of soft tissue (A) thickness of macrochambers after walking, (B) thickness of microchambers after walking, (C) stiffness of macrochambers after walking, (D) stiffness of microchambers after walking.

5 Conclusions

The results of this study demonstrated that the walking duration (10 and 20 min) significantly affected plantar macrochambers tissue thickness and the walking speed (1.8, 2.7, and 3.6 mph) did not. Our results showed that the walking duration factor causes a significant main effect on the macrochambers thickness. In addition, the speed factor caused a downward trend of the macrochambers stiffness. This study suggests that people at risk for foot ulcers should walk at speeds at 3.6 mph with duration 20 min.

Acknowledgments. This study was supported by the Ministry of Science and Technology of the Republic of China (MOST 108-2221-E-241-008, MOST-108-2221-E-468-018), Hungkuang University and Kuang Tien General Hospital (HK-KTOH-109-04), and Asia University Hospital and China Medical University Hospital (ASIA-107-AUH-09).

References

1. Colberg, S.R., et al.: Physical activity/exercise and diabetes: a position statement of the American diabetes association. Diabetes Care **39**, 2065–2079 (2016)
2. Wearing, S.C., Smeathers, J.E.: The heel fat pad: mechanical properties and clinical applications. J. Foot Ankle Res. **4**, 1 (2011)

3. Mueller, M.J., Maluf, K.S.: Tissue adaptation to physical stress: a proposed "physical stress theory" to guide physical therapist practice, education, and research. Phys. Ther. **82**(4), 383–403 (2002). https://doi.org/10.1093/ptj/82.4.383
4. Gooding, G., Stress, R., Graf, P., Grunfeld, C.: Heel pad thickness: determination by high-resolution ultrasonography. J. Ultrasound Med. **4**, 173–174 (1985)
5. Sun, J.-H., Cheng, B.K., Zheng, Y.-P., Huang, Y.-P., Leung, J.Y., Cheing, G.L.: Changes in the thickness and stiffness of plantar soft tissues in people with diabetic peripheral neuropathy. Arch. Phys. Med. Rehabil. **92**, 1484–1489 (2011)
6. Hsu, C.-C., et al.: Diabetic effects on microchambers and macrochambers tissue properties in human heel pads. Clin. Biomech. **24**, 682–686 (2009)
7. Garcia, C.A., Hoffman, S.L., Hastings, M.K., Klaesner, J.W., Mueller, M.J.: Effect of metatarsal phalangeal joint extension on plantar soft tissue stiffness and thickness. Foot **18**, 61–67 (2008)
8. Chao, C.Y., Zheng, Y.-P., Cheing, G.L.: Epidermal thickness and biomechanical properties of plantar tissues in diabetic foot. Ultrasound Med. Biol. **37**, 1029–1038 (2011)
9. Aumiller, W.D., Dollahite, H.A.: Pathogenesis and management of diabetic foot ulcers. J. Am. Acad. PAs **28**, 28–34 (2015)
10. Wu, F.L., Wang, W.T., Liao, F., Elliott, J., Jain, S., Jan, Y.K.: Effects of walking speeds and durations on plantar skin blood flow responses. Microvasc. Res. **128**, 103936 (2020)
11. Lung, C.-W., Wu, F.-L., Zhang, K., Liau, B.-Y., Townsend, R., Jan, Y.-K.: Using elastographic ultrasound to assess plantar tissue stiffness after waking at different speeds and durations. Appl. Sci. **10**, 7498 (2020)
12. Gwak, Y., Kim, J., Lee, K.M., Koo, S.: Increase of stiffness in plantar fat tissue in diabetic patients. J. Biomech. **107**, 109857 (2020)
13. Browm, L.A.: A scanning electron microscope study of the effects of uniaxial tension on human skin. Br. J. Dermatol. **89**, 383–393 (1973)
14. Buschmann, W.R., Jahss, M.H., Kummer, F., Desai, P., Gee, R.O., Ricci, J.L.: Histology and histomorphometric analysis of the normal and atrophic heel fat pad. Foot Ankle Int. **16**, 254–258 (1995)
15. Hsu, C.-C., Tsai, W.-C., Wang, C.-L., Pao, S.-H., Shau, Y.-W., Chuan, Y.-S.: Microchambers and macrochambers in heel pads: are they functionally different? J. Appl. Physiol. **102**, 2227–2231 (2007)
16. Hsu, C.-C., et al.: Effects of aging on the plantar soft tissue properties under the metatarsal heads at different impact velocities. Ultrasound Med. Biol. **31**, 1423–1429 (2005)
17. Lee, S.Y., Joo, S.Y., Cho, Y.S., Hur, G.Y., Seo, C.H.: Effect of extracorporeal shock wave therapy for burn scar regeneration: A prospective, randomized, double-blinded study. Burns (2020)
18. Smith, S.G., Yokich, M.K., Beaudette, S.M., Brown, S.H., Bent, L.R.: Effects of foot position on skin structural deformation. J. Mech. Behav. Biomed. Mater. **95**, 240–248 (2019)
19. Liau, B.Y., Chen, C.L., Jan, Y.K., Chiu, H.Y., He, P.S., Lung, C.W.: A comparative study of the effects of electrical stimulation and intermittent compressive forces on soft tissue mechanical properties. In: Goonetilleke, R., Karwowski, W. (eds.) Advances in Physical Ergonomics and Human Factors. AHFE 2017. Advances in Intelligent Systems and Computing, vol. 602, pp. 89–97. Springer, Cham (2018). https://doi.org/10.1007/978-3-319-60825-9_11
20. Zheng, Y.P., Mak, A.F.T.: Effective elastic properties for lower limb soft tissues from manual indentation experiment. IEEE Trans. Rehabil. Eng. **7**, 257–267 (1999)
21. Hayes, W.C., Keer, L.M., Herrmann, G., Mockros, L.F.: A mathematical analysis for indentation tests of articular cartilage. J. Biomech. **5**, 541–551 (1972)

22. Lung, C.W., Hsiao-Wecksler, E.T., Burns, S., Lin, F., Jan, Y.K.: Quantifying dynamic changes in plantar pressure gradient in diabetics with peripheral neuropathy. Front. Bioeng. Biotechnol. **4**, 54 (2016)
23. Liau, B.-Y., Wu, F.-L., Lung, C.-W., Zhang, X., Wang, X., Jan, Y.-K.: Complexity-based measures of postural sway during walking at different speeds and durations using multiscale entropy. Entropy **21**, 1128 (2019)

Analysis and Application of Influencing Factors of Mirror Drawing Ability

Minxia Liu[1(✉)], Yu Gu[2], Jiping Lu[1], Lin Gong[1], and Qing Xue[1]

[1] Beijing Institute of Technology, 5 South Zhongguancun Street, Haidian District, Beijing 100081, China
liuminxia@bit.edu.cn

[2] Beijing Railway Signal Company Limited, No. 456 Sicun Langfa Huangcun, Daxing District, Beijing 102613, China

Abstract. This article analyzed the factors affecting the mirror drawing ability through the experimental data collected by the mirror drawing experiment. The specific research selected the following five aspects for analysis: gender, personality, the intersection of gender and personality, the difficulty degree of skills, and the influence between left and right hands. Using statistical analysis methods to analyze the experimental data. Gender and personality have no significant effect on the formation of skills. The time of mirror drawing and the number of errors are not significant in the intersection of gender and personality. However, skill difficulty (the difficulty degree of graphics) has an impact on the formation of skills. The process of skills formation is a process of skill transfer. Based on this, mirror-drawing graphics with different difficulty degrees, which could be as a basis for evaluating the ability of personnel to transfer skills. Through the analysis of experimental data, the dominant hand has a certain positive transfer effect on the non-dominant hand. The process of mirror drawing is a process of representation training. Participants processed the experience and skills gained in the process of practicing mirror drawing with the dominant hand in the brain, and then applied to the process of non-dominant hand mirror drawing, which improved their non-dominant hand mirror drawing ability and realized the migration of mirror drawing ability. In fact, the practice process had played the role of representation training. Therefore, use the method of representation training to promote skill transfer. For example, we can establish correct technical action representations to improve the quality of actions and speed up the automation of actions. In addition, in the post-epidemic era, use network virtual experiments for representation training to promote the transfer of students' experimental skills. Moreover, there are some other applications.

Keywords: Movement skills · Mirror drawing ability · Influences

1 Introduction

Mirror drawing experiment is a commonly used psychological experiment. D.Starch pioneered the mirror painting experiment in 1910. The mirror instrument has become

M. Liu—The project number supporting this study is 2017YFB1102802.

R. S. Goonetilleke et al. (Eds.): AHFE 2021, LNNS 273, pp. 364–371, 2021.
https://doi.org/10.1007/978-3-030-80713-9_46

the most commonly used psychological experimental instrument for studying motion learning. The device works by asking the subject to look at a figure through a mirror and draw on the figure with a pen. Reflected by the mirror, compared with the original figure, the figure in the mirror is upside down but unchanged from left to right, so the subjects often draw in the opposite direction of the usual upper and lower strokes, which brings some difficulties to the beginners, but the subjects will quickly adapt to the original customary drawing method through training. The results demonstrate the positive transfer effect from one hand to the other.

2 Experimental Data Collection

We collected data include 27 women and 23 men. Before analyzing the influencing factors of skill formation, it is necessary to consider the relationship between gender and personality. If there is a strong relationship, use gender to approximate the personality. If there is no relationship, we cannot approximate a substitute.

2.1 The Correlation Between Personality and Gender

Using the SPSS to analysis data and the analysis results showed in the Table 1.

Table 1. The correlation between gender and personality

The correlation			
		Gender	Character
Gender	Pearson correlation	1	.010
	Significance (bilateral)		.924
	N	100	100
Character	Pearson correlation	.010	1
	Significance (bilateral)	.924	
	N	100	100

It can be seen from the table that the correlation between gender and personality is 0.01, indicating that the correlation between personality and gender is very low. Taking the significance level $\alpha = 0.05$, 0.924 is far greater than α, which proves that there is no significant difference between personality and gender factors. Therefore, we can discuss the influence of gender and personality on skill formation respectively in the following paragraphs.

2.2 The Impact of Gender on Skill Formation

According to the data, we can know the mean value about male and female on the complete time and the error time. The results are in the Table 2 and Table 3.

Table 2. The complete time mean value between gender and skill formation

Time	Male	Female
I -type	48.57417	53.62514
Flower-type	61.51274	56.53256

Table 3. The error time mean value between gender and skill formation

Error times	Male	Female
I -type	6.869565	6.148148
Flower-type	15.34783	7.148148

Only in terms of descriptive statistics, personality has some influence on skill formation, but not much. Extroverts drew more slowly when drawing simple shapes, and more quickly when drawing complex shapes. Extroverts made more mistakes than introverts did.

When personality and graph are controlled unchanged (Note: in SPSS analysis, 1 represents male, extroverted and I-type, and 0 represents female, introverted and flower type), using the SPSS to analyze the correlation of drawing time and error number, which the results are as follows in the Table 4.

Table 4. Correlation and significance of gender on drawing time and error number

The correlation					
Control variables			Gender	Time	Error times
Character & Graphics	Gender	Pearson correlation	1.000	−.001	.247
		Significance (bilateral)		.994	.014
		df	0	96	96
	Time	Pearson correlation	−.001	1.000	.066
		Significance (bilateral)	.994		.517
		df	96	0	96
	Error times	Pearson correlation	.247	.066	1.000
		Significance (bilateral)	.014	.517	
		df	96	96	0

The correlation between gender and drawing time is -0.001, indicating that the male painting time is short, but the value is close to 0, indicating that the correlation is small or even irrelevant. Taking the significance level $\alpha = 0.05$, 0.994 is far greater than α, proving that there is no significant difference in drawing time in terms of gender factors.

Similarly, the correlation between gender and error is 0.247, indicating that the number of male errors is more, but the value does not reach 0.3, proving that the correlation is very small. Taking the significance level α = 0.05, 0.014 is greater than α, proving that the number of errors has no significant difference in gender factors. Conclusion: Although gender has no significant effect on skill formation, the effect of gender on error number is greater than that of drawing time.

2.3 The Character Influence on Skill Formation

Take the same methods with above, we can analyze the character influence on skill formation. The results showed in the Table 5 and Table 6.

Table 5. The complete time mean value between character and skill formation

Time	Outgoing	Introverted
I -type	52.36204	49.95218
Flower-type	57.45011	60.57132

Table 6. The error time mean value between characterand skill formation

Error times	Outgoing	Introverted
I -type	6.607143	6.318182
Flower-type	11.25	10.5

Only in terms of descriptive statistics, personality has some influence on skill formation, but not much. Extroverts drew more slowly when drawing simple shapes, and more quickly when drawing complex shapes. Extroverts made more mistakes than introverts did.

When gender and graphics remain unchanged through SPSS control, the correlation table of personality to drawing time and number of errors is as follows in the Table 7.

From the table that the correlation between personality and drawing time is -0.008, indicating that extroverts spend less time drawing, but the correlation is very low. Taking the significance level α = 0.05, 0.935 is far greater than α, which proves that there is no significant difference in the drawing time on personality factors.

Similarly, the correlation between personality and the number of errors is 0.026, indicating that the number of errors in males is more, but the correlation is also very low. Taking the significance level α = 0.05, 0.801 is greater than α, it proves that there is no significant difference in the number of errors in gender factors.

Conclusion: Personality has no significant effect on skill formation.

Table 7. Correlation and significance of personality on drawing time and error number

The correlation					
Control variables			Time	Error times	Character
Gender & Graphics	Time	Pearson correlation	1.000	.068	− .008
		Significance (bilateral)		.504	.935
		df	0	96	96
	Error times	Pearson correlation	.068	1.000	.026
		Significance (bilateral)	.504		.801
		df	96	0	96
	Character	Pearson correlation	− .008	.026	1.000
		Significance (bilateral)	.935	.801	
		df	96	96	0

2.4 Skills Migration

Previous research has shown that skill formation has little to do with gender or personality. We analyzed the influence of gender and personality on skills transfer. Use the following formulas to computer the migration effect.

$$C = \left((A - B)/A\right) \times 100 \tag{1}$$

$$\text{The migration effect} = \left(C_x - C_y\right) \tag{2}$$

Where A and B represent the painting time of the first and second time with the non-dominant hand respectively.

Controlling the gender and personality respectively to get their correlation and the skills transfer effect. The results are in the Table 8 and Table 9.

Table 8. The correlation between personality and migration effect

The correlation				
Control variables			Migration effect	Character
Gender	Migration effect	Pearson correlation	1.000	.096
		Significance (bilateral)		.512
		df	0	47
	Character	Pearson correlation	.096	1.000
		Significance (bilateral)	.512	
		df	47	0

Table 9. The correlation between gender and migration effect

The correlation				
Control variables			Migration effect	Gender
Character	Migration effect	Pearson correlation	1.000	− .227
		Significance (bilateral)		.116
		df	0	47
	Character	Pearson correlation	− .227	1.000
		Significance (bilateral)	.116	
		df	47	0

From the above data, we can see that personality has almost nothing to do with the effect of skill transfer, while gender has a certain relationship with the effect of skill transfer, but the relationship is not large enough. Among them, male's skill transfer effect is worse than female's.

Analyzing the skills transfer effect of men and women. Using SPSS data analysis function, obtained the statistical description of skill transfer effect of male and female. The dominant hand of male and female has a certain positive transfer effect on the non-dominant hand, and gender and other factors effected the movement transfer of left and right hand. For example, the transfer effect of male is worse than that of female.

2.5 Experimental Conclusion

The data in the experiment showed that the difficulty of skills had a great influence on the formation of skills. The more difficult the skills the subjects had to train, the more times they made mistakes during the training. Moreover, skill formation has little to do with gender or personality.

The subjects had the right hand as the dominant hand. In the experiment, due to the habit of the right hand, the experiment took a long time and made many mistakes. However, the left hand is not the dominant hand, which has little influence on the drawing. Moreover, the subjects first experimented with the right hand, then the left hand. The right hand exerted influence on the left hand through many practices, so that the subjects could get the learning experience of the right hand in the process of practice. The spatial position of the image in the mirror is opposite to that of the actual image, which results in the weakening of the visual control and the enhancement of the motion control. Meanwhile, the left hand's habitual low does not affect the left hand's motion sense, so the exercise of the right hand has a positive transfer effect on the left hand.

3 Interpretation of the Results of the Mirror Drawing Experiment

3.1 The Conditions Under Which Skills Transfer Occurs

Because there is a common rule between different skills, so in the mirror painting experiment, the dominant hand can transfer the skill to the non-dominant hand. In the mirror painting experiment, we need to operate in the same direction as the image inside the mirror, which is called image training. The purpose of imagery training is to make the subjects clearly realize and recall that the skills used in the operation with the dominant hand can be transferred to the non-dominant hand. Continuous reinforcement and repetition of this transfer will improve the operation level and ability of non-dominant hand, and reduce the difference of operation level between dominant hand and non-dominant hand.

3.2 Fundamentals of Neurology

Previous studies have shown that the basis of motor task and motor imagery is through the role of nerves. Action imagery means that there is no obvious body movement, only the motor imagination in the brain. Action task and action imagery are regarded as "functional equivalence". Brain imaging studies show that finger; hand and foot, movements can activate the auxiliary motor area, premotor area and the primary motor cortex of cerebellum. Due to the functional equivalence of action task and action representation, it can be considered that the corresponding brain regions are also activated in the experiment of non-dominant hand mirror painting, which can produce positive transfer effect.

4 Application of Positive Migration

The transfer of action skills to non-handedness is a very useful transfer in real life. Taking basketball players as an example, if a right-hander can effectively transfer his layup and shooting skills from his right hand to his left hand, his personal strength will be greatly improved. Through the above experimental analysis, we know that the repeated practice of the dominant hand will have a positive effect on the movement ability of the non-dominant hand. When athletes improve the athletic ability of the non-dominant hand, they can train through the above principles, to obtain a substantial improvement in athletic ability.

4.1 Original Shooting Training

Since the invention of basketball in 1891, the scientific shooting training method has been the shortcut for all basketball fans to pursue progress. At the initial stage, basketball-shooting training was carried out by repeatedly training sharp hand shooting. At that time, basketball players did not realize the importance of two-hand attack. Therefore, in the first 30 years of the invention of basketball, basketball fans still use one-handed attack as the main means of attack.

4.2 The Development of Two-Handed Attack and the Initial Training Method

With the development of basketball, the importance of left and right hand flexible attack is gradually reflected in the game. The famous player Larry Bird is relying on its strong court dominance, the left and right hand flexible attack techniques and tactics thoroughly written into the training subjects of basketball players. In the initial training of left-hand and right-hand attack, athletes often use the way of "long time practice of good hand shooting rest long time practice of non-good hand shooting", which enables athletes to develop the offensive ability of non-good hand. With the progress of training, the athletes and coaches at that time found a problem: the offensive ability of the non-handedness of the athletes was greatly improved in the first three months of training, and they often could not make satisfactory progress in the later training.

4.3 Using the Principle of Mirror Painting for Shooting Training

After 1990, basketball players have developed a new training method based on the positive transfer law of the dominant hand to the non-dominant hand: the dominant hand and the non-dominant hand cross shooting. Based on the analysis of the experimental data of mirror painting, we can draw the following conclusions: the repeated training of the dominant hand will have a positive transfer effect on the non-dominant hand. However, through the analysis of the experimental data, we can know that this positive transfer effect cannot make the non-dominant hand's movement ability completely improve to the same level as the dominant hand. Through the cross practice of the dominant hand and the non-dominant hand, we can repeatedly use the positive transfer to improve the sports ability of the non-dominant hand, and gradually improve the sports ability of the non-dominant hand.

4.4 Conclusion

From the mirror drawing experiment, we can know that the positive transfer of dominant hand to non-dominant hand exists objectively, and this positive transfer can be fully applied in various sports. In addition to basketball, in the process of table tennis, tennis, badminton and other sports, we can also use the transfer function to strengthen training and improve the level of athletes' hands. However, it is difficult for the non-dominant hand to achieve the same level as the dominant hand. Therefore, in the process of training, through the "repeated" and "short-term" cross training, we should constantly play the positive transfer role of the dominant hand to the non-dominant hand, so that the sports ability of the non-dominant hand tends to the sports ability of the dominant hand.

The training of mirror painting ability can be applied in the related physical education, especially in basketball. Appearance can promote the formation of non-dominant hand skills of basketball players, and has a positive effect on the development of inferior brain. In addition, imagery training can be used to eliminate the interference of negative transfer.

Implementing Participatory Ergonomics Among Indigenous Women of Ecuador to Preserve Ancestral Customs and Knowledge

Alejandra Lascano, Thalia San Antonio(✉), Fernando Urrutia, and Maria Augusta Latta

Universidad Técnica de Ambato, Ambato, Ecuador
{am.lascano,t.sanantonio,fernandourrutia,mariaalatta}@uta.edu.ec

Abstract. Severe spine deformations are common among elderly indigenous women of Ecuadorian highlands mainly dedicated to agricultural labors. Participatory Ergonomics (PE) is used to reduce or eliminate the incidence of musculoskeletal disorders. Two demanding daily activities of these women were evaluated with MREBA and 3DSSPP, then the ergonomic recommendations were discussed with the ERGO group, and their opinions regarding the proposals were used to modify them according to the women traditions and customs. The PE is an effective instrument to adjust the recommendations, in the case of this study to be respectful of the customs and traditions of the intervened community. However, the ingrained customs suggest that the modifications should be implemented gradually, especially for older women who have the most compromised musculoskeletal health. A major detected problem is the great magnitude of the loads these women lift and transport, tied to their backs (sacks or children) since they exceeded by far the recommended levels.

Keywords: Participatory ergonomics · Indigenous women · Spine deformations

1 Introduction

Agriculture is a crucial part of economic growth and poverty reduction. However, it is one of the activities with the highest rate of health risks, especially in countries where this activity is not technologically enhanced and is performed with manual tools that require great physical effort, which leads to important musculoskeletal disorders (MSD) [1]. Another relevant aspect is that agricultural work requires the frequent adoption of uncomfortable postures for extended periods of time resulting in pain in different overexerted parts of the body; reports indicate that pain occurs more frequently in areas, such as the neck, lumbar area, and knees [2]. In addition to the above, in developing countries, it is common for agricultural work to be performed by vulnerable groups, such as children, women, and the elderly, in unfavorable weather conditions [3].

The aim of ergonomic studies is to assess the musculoskeletal risks of different work activities and to propose improvement measures that can be oriented at changing the

R. S. Goonetilleke et al. (Eds.): AHFE 2021, LNNS 273, pp. 372–379, 2021.
https://doi.org/10.1007/978-3-030-80713-9_47

design of manual handling operations, activities, objects, tools, and the work environment, so that tasks are adapted to the characteristics of the workers performing them [4]. Interventions, as well as the above-mentioned modifications, can include the redesign of equipment and tools, the improvement of work practices, and the purchasing of new equipment that allows workers to perform tasks with reduced physical loads [5].

In Ecuador, specifically in the highland's region, agricultural work is performed very rudimentarily and mostly by children and women of all ages, including seniors. These activities entail forced postures, repetitive movements, carrying high loads on the back, and long journeys over uneven terrain. All these conditions, along with precarious healthcare, have resulted in high-risk ergonomic assessments in female workers [6, 7].

Participatory Ergonomics (PE) allows the development of ergonomics intervention programs with the involvement of all the studied actors in order to improve the work environment conditions and to reduce safety and health risks [8]. It is based on the assumption that workers are the experts and that, if they are provided with knowledge, skills, tools, resources and incentives, they are the most suitable ones to identify and analyze problems, as well as to propose and apply effective solutions to reduce injury risk and MSD in work activities [9]. Because of these characteristics, PE is considered as the most suitable methodology to implement changes with indigenous women, so that they accept said changes and that their traditions are respected.

There is a growing interest in the development of PE in different countries in the world. Several companies worldwide are interested implementing PE. For instance, the National Institute of Occupational Safety and Health (NIOSH) considers PE to be a useful strategy for preventing musculoskeletal disorders caused by work tasks [10]. Studies show a lack of results on the application of PE in Latin America, a warning that must be considered, since the region faces big problem-solving challenges when it comes to ergonomics [11].

Since the main objective of PE is to achieve the direct participation of the involved individuals with a focus on obtaining favorable results and, therefore, an improvement in the quality of life of the participants, the additional effect of PE is to favor the empowerment of the community by having its members participate in decision-making and in actions in favor of their health [12]. Since PE involves, among others, improvements in working stations, it is worth mentioning the difference between its two approaches. Taylor's approach is based on the study of the body movements of humans, which are necessary to perform a task, as well as on measuring the time devoted to each one of these movements. The macroergonomics approach is based on reducing biomechanical and cognitive risks, and it considers social, organizational, and psychological factors of work [13].

In Ecuador, the Ministry of Public Health (MSP, in Spanish *Ministerio de Salud Pública*) and the Ministry of Economic and Social Inclusion (MIES, in Spanish *Ministerio de Inclusión Económica y Social*) have quality-of-life improvement policies and programs in place, addressed to the elderly in rural areas [14]. The aim of this work is to apply PE via the facilitators of MIES in elderly indigenous women of Quisapincha, Tungurahua, Ecuador, since this group shows important MSD [7] that have been assessed using the modified REBA (MREBA) method for the hard daily work of these women [15].

2 Materials and Methods

This study is performed in 6 groups of elderly people from indigenous communities. The MIES is in charge of creating the groups through its qualified staff, referred to as facilitators, who are in charge of organizing the groups and performing workshops with cultural, sport, leisure, and health activities. Each group is assisted by a facilitator that visits the community twice a week. The main purpose of these groups is to carry out activities in order to entertain the elderly and keep them active, thus improving their quality of life.

The ergonomic assessment was carried out using the MREBA method, which is modified for the specific work carried out by these human groups. From the assessment of different tasks, it was determined that the most demanding ones are: loading and transporting sacks containing agricultural products and tasks performed while carrying a child on the back [6], as well as tasks performed without the collaboration of another person making them more demanding. One of the characteristics of the two assessed

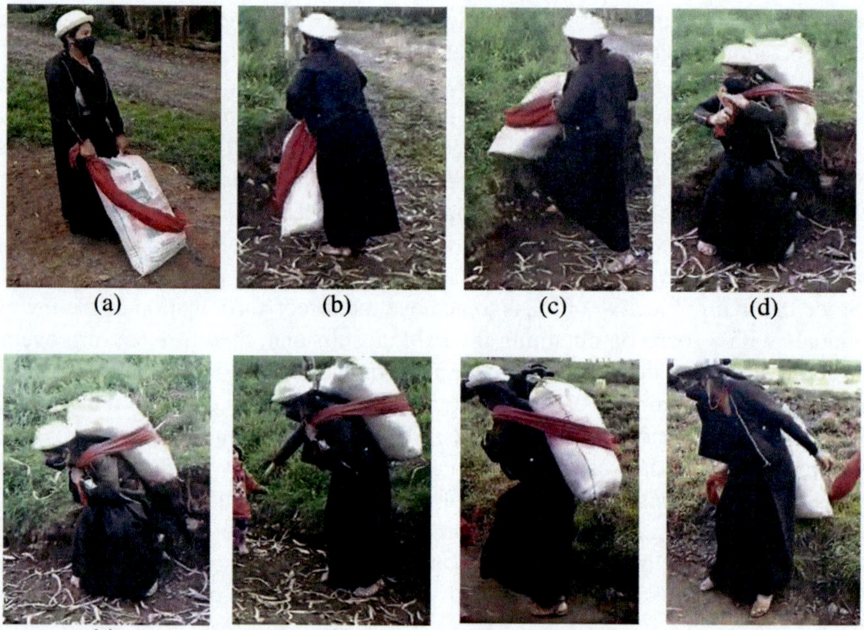

(a) (b) (c) (d)

(e) (f) (g) (h)

Fig. 1. Task 1: loading, transfer and unloading the sack of agricultural products a) a standing woman drags the sack towards her body, b) she surrounds the sack with her shawl and lifts it with her arms c) she places the sack on a higher area of the terrain by pushing it up with her leg d) she locates the sack at the level of her back and fastens it with her shawl e) she bends and accommodates the sack on her back; she then ties the ends of her shawl around her chest f) she stands up with the sack on her back and keeps her chest tilted forward g) she walks forward with her chest tilted forward h) she unloads the sack from her back by untying and putting away the shawl.

activities is the fact that women use knitted shawls called *chalinas* to tie things to their backs. Figure 1 describes the assessed tasks in detail.

The group ERGO was formed to apply EP. It is defined as a working group that encourages the participation of all actors; at a company level, it is composed of workers [1]. In this study, this group is made up by MIES facilitators. From the results of task assessments using the MREBA, recommendations were made, and they were analyzed and improved based on: the experience of facilitators with their elderly groups, the feasibility of the implementation, the possibility of adaptation to the cultural, social, economic, and demographic environment of the intervention group. Once the recommendations were adjusted, the facilitators conducted dissemination workshops for the elderly, focused on minimizing the risks of developing MSD. From a Participatory Ergonomics perspective, the best way of reaching this objective is by transferring knowledge and ergonomics methods from the expert to the worker, so the latter learns to identify problems before they lead to an accident or injury [4] (Fig. 2).

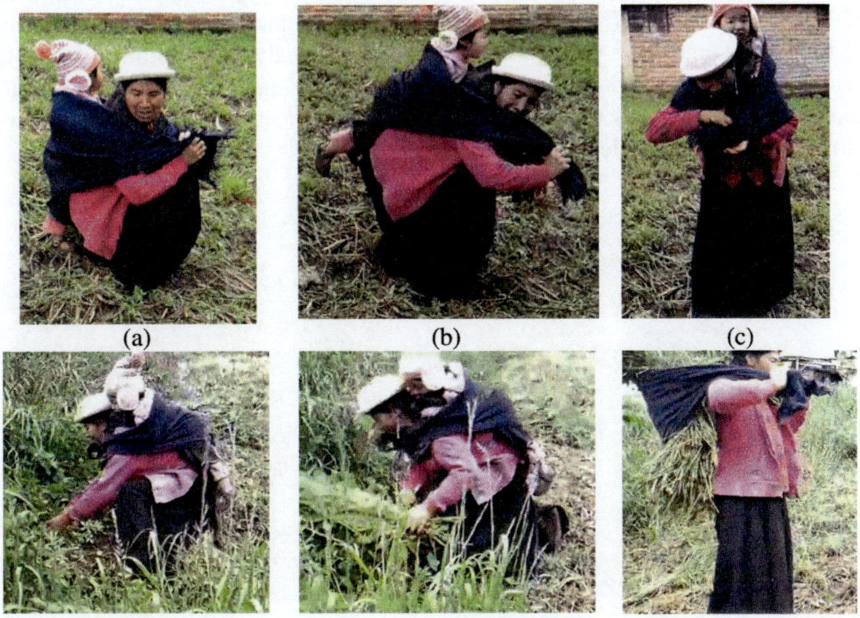

(a) (b) (c) (d) (e) (f)

Fig. 2. Task 2: cutting grass with an infant secured on the back a) a woman in squatting position places the child against her back and places the shawl around them b) she tilts her torso forward and accommodates the child on her back c) she stands up and ties the ends of her shawl at chest level d) she crouches down to cut the grass e) she repeats the cutting movement with her right hand; she ties the cut grass with her left hand and rotates left to pile the cut grass up. She does all these movements in a squatting position f) she unloads the child and loads the tied-up grass on her back, using the same shawl.

3 Results

The two assessed tasks were selected by considering their precariousness, frequency, and postural load. Likewise, in order to choose the most critical posture within each task, the magnitude of the deviation from neutral postures and the weight of the load were considered. All this is focused on analyzing the woman-task interface based on the necessary force for executing the task and the way in which the load adjusts to the body.

3.1 Task Analysis Using 3DSSPP

Figure 3 shows a summary of the results from analyses conducted with the 3DSSPP in order to evaluate the critical postures of each task. This analysis allows the 3D simulation of postures, as well as of anthropometric parameters and external loads associated with the critical postures of the assessed tasks. Results show that the L4/L5 disc compression force is 908 lb for task 1 and 813 lb for task 2; therefore, both exceed the reference value of 770 lb established by NIOSH. These values indicate that lumbar pain is three times more likely in women that perform these tasks in the analyzed conditions. In the case of load transportation (sack/child), frequency, time, and load magnitude are significant factors; however, analyses indicate that weight is the most determining factor for the occurrence of musculoskeletal disorders.

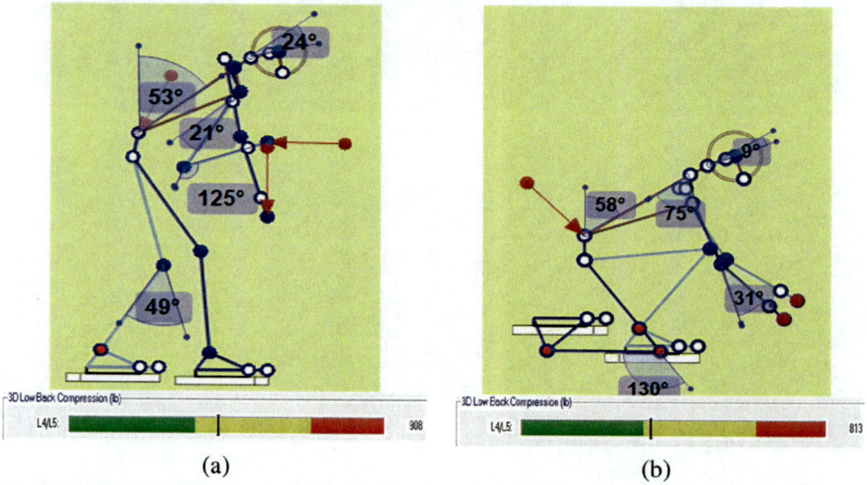

Fig. 3. Summary of the results of the 3DSSPP analysis of critical postures in analyzed tasks (a) Sack loading, transportation, and unloading, (b) Grass cutting with a child tied to the back.

3.2 Task Analysis Using MREBA

The results of applying the MREBA method to the critical postures of the analyzed tasks show that, for both tasks, the risk level is 4: Very High. It is the result of the exposure to inadequate postures, as postural load has extremely harmful effects of the musculoskeletal system. The main causes of this high risk in the case of both tasks are: high trunk tilting values, neck flexing and bending, lateral trunk bending, and a very heavy load on the back; additionally, only for task 1: knee flexing, lateral trunk and neck bending and other body parts that remain static for over 1 min.

3.3 List of Recommendations Following PE Application

After the meetings with the ERGO group composed of MIES facilitators to discuss the results of the task analyses and once the corresponding recommendations were made, the following ones were chosen to be disseminated to the elderly people in the groups:

- Use aids for the transportation of heavy loads, such as wheelbarrows, or perform this task in pairs.
- Carry children on your backs as few as possible, especially if they can walk.
- If a task demands forced postures, such as crouching down, these postures must be adopted without loads on the back.
- If you need to carry a child on your back, try to fasten them in a symmetrical manner.
- Loads over 10 kilos must be transported for the shortest time possible and slopes must be avoided.
- When performing repetitive movements, breaks must be taken to prevent postural overload.
- Whenever possible, elevate the working area to the waist level in order to minimize trunk flexing and twisting.

4 Conclusions

PE is a very useful tool to make adjustments to the recommendations that ergonomics experts propose for improving working conditions and protecting workers from MSD, especially in cases such as the ones presented in this study, where the affected population performs tasks according to traditions and knowledge passed down from generation to generation.

While the proposed adjustments are not aimed at reducing loads and modifying postures until they no longer pose a health risk, they do allow the reduction of current negative effects. In addition, once they have been analyzed with and explained to the workers, there is a greater likelihood that they will be implemented.

The physical health of indigenous women is particularly important, as they use their physical strength to perform the tasks with which they financially support themselves and their families; therefore, all the initiatives undertaken in order to protect them are of great value to the affected communities.

Acknowledgments. This work is being funded by (1) Dirección de Investigación y Desarrollo (DIDE), of Universidad Técnica de Ambato under the project: UTA-CONIN-2020-0298-R "*Desarrollo de un sistema de capacitación para la protección de salud física de las usuarias* *de chumbi mediante la creación de contenidos utilizando instrumentos pedagógicos adaptados a su cultura y entorno*" and (2) Corporación Ecuatoriana para el Desarrollo de la Investigación y la Academia (CEDIA) under the project: CEPRA XIII-2019-14, *Llankak Warmikunata Kamachikkuna (Protec the working woman).*

References

1. Kee, D., Haslam, R.: Prevalence of work-related musculoskeletal disorders in agriculture workers in Korea and preventative interventions. Work **64**(4), 763–775 (2019)
2. Gangopadhyay, S., Das, B.B., Ghosh, T., Das, T., Ghoshal, G.: Un estudio de ergonomía sobre el malestar relacionado con la postura entre los trabajadores agrícolas preadolescentes de Bengala Occidental, India. Revista Internacional de Seguridad y Ergonomía en el Trabajo **11**(3), 315–322 (2005)
3. Naeini, H.S., Karuppiah, K., Tamrin, S.B., Dalal, K.: Ergonomía en la agricultura: un enfoque en la prevención de trastornos musculoesqueléticos relacionados con el trabajo (WMSD). Revista de Ciencias Agrícolas y Ambientales **3**(2), 33–51 (2014)
4. Rost, K.A., Alvero, A.M.: Participatory approaches to workplace safety management: bridging the gap between behavioral safety and participatory ergonomics. Int. J. Occup. Saf. Ergon. **26**, 194–203 (2018)
5. Capodaglio, E.M.: Ergonomía participativa para la reducción de la exposición musculoesquelética de los trabajadores de mantenimiento. Revista internacional de seguridad y ergonomía en el trabajo, pp. 1–11 (2020)
6. San Antonio, T., Larrea, A., Lascano, A., Urrutia, J.: Ergonomic evaluation of daily activities of women from a community in the Ecuadorian highlands. In International Conference on Applied Human Factors and Ergonomics. Springer, Cham, pp. 102–110 (2018)
7. Antonio, T.S., Latta, M.A., Naranjo, M.T., Guevara, D.: Postural analysis in older females from an indigenous community of ecuador that uses a Chumbi as part of their traditional outfit. In: Kalra, J., Lightner, N. (eds.) Advances in Human Factors and Ergonomics in Healthcare and Medical Devices. AHFE 2020. Advances in Intelligent Systems and Computing, vol. 1205, pp. 350–356. Springer, Cham (2020). https://doi.org/10.1007/978-3-030-50838-8_47
8. Wahyuni, D., Panjaitan, N., Budiman, I., Manurung, E.D.: La intervención de la ergonomía participativa en las actividades de mejora de la carga repetitiva de camiones para productos de agua mineral. In: IOP Conference Series: Ciencia e ingeniería de materiales. Publicación de IOP, vol.1003, no. 1, p. 012076 (2020)
9. Sundstrup, E., Seeberg, K.G.V., Bengtsen, E., Andersen, L.L.: A systematic review of workplace interventions to rehabilitate musculoskeletal disorders among employees with physical demanding work. J. Occup. Rehabil. **30**(4), 588–612 (2020). https://doi.org/10.1007/s10926-020-09879-x
10. García, A.M., Gadea, R., Sevilla, M.J., Genís, S., Ronda, E.: Ergonomía participativa: empoderamiento de los trabajadores para la prevención de trastornos musculoesqueléticos. Rev. Esp. Salud Publica **83**(4), 509–518 (2009)
11. Lorca, O.P.: Ergonomía participativa: Aproximación a su estudio e implementación a través de la evidencia científica. Rev. Ergon. Invest Desar. **1**(1), 147–157 (2019)
12. World Health Organization. Promoción de la Salud. Glosario. Ginebra (1998)
13. Broday, E.E.: Ergonomía participativa en el contexto de la Industria 4.0: una revisión de la literatura. Cuestiones teóricas en la ciencia de la ergonomía, pp. 1–14 (2020)

14. Políticas integrales de salud para el adulto mayor, Ministerio de Salud Pública. https://aplicaciones.msp.gob.ec/salud/archivosdigitales/documentosDirecciones/dnn/archivos/Pol%C3%ADticas%20Integrales%20Salud%20Adulto%20mayor1.pdf
15. Lascano, A., Antonio, T.S., Larrea, A., Ciaccia, M.: Ergonomic methods adaptation for risk evaluation associated to musculoskeletal disorders in elderly indigenous women of the ecuadorian highlands. In: Karwowski, W., Goonetilleke, R.S., Xiong, S., Goossens, R.H.M., Murata, A. (eds.) AHFE 2020. AISC, vol. 1215, pp. 266–273. Springer, Cham (2020). https://doi.org/10.1007/978-3-030-51549-2_35

Prediction Model of One-Handed Pull Strength in the Sagittal Plane

Hailiang Wang[1(✉)], Mian Yan[2], and Da Tao[3]

[1] School of Design, The Hong Kong Polytechnic University, Hung Hum, Hong Kong, China
hailiang.wang@polyu.edu.hk

[2] School of Intelligent Systems Science and Engineering, Jinan University (Zhuhai Campus), Zhuhai, China
yanmian@jnu.edu.cn

[3] Institute of Human Factors and Ergonomics, College of Mechatronics and Control Engineering, Shenzhen University, Shenzhen, China
taoda@szu.edu.cn

Abstract. Understanding individuals' capability of one-handed pull strength is crucial in manual handling task design as one-handed pulling is frequently conducted in daily life. The present study was designed to develop prediction models of one-handed pull strength from anthropometrics and body-joint angles. One hundred Chinese adults were recruited and instructed to provide their maximum one-handed pull strength in the sagittal plane. Sagittal-plane photographs were taken for measuring three joint angles (i.e., trunk angle, knee angle, and ankle angle). T-tests, ANOVAs, and stepwise multiple regression analysis were conducted for data analysis. Five prediction models were developed with the adjusted R^2 values ranging from 0.621 to 0.818 (all $ps < 0.001$). Significant predictors were reported and discussed. The findings contribute to physical ergonomics and human factors by providing prediction models for reference values of one-handed pull strength of a population, further facilitating safety designs of tasks involved one-handed pulling (144 words).

Keywords: Physical ergonomics · One-handed pull strength · Prediction · Sagittal plane

1 Introduction

Mismatching between individuals' strength capability and task strength requirement can lead to musculoskeletal disorders for workers in manual material handlings. It is of importance to have and use reference strength data when designing the tasks. One-handed pulling is a frequently performed motion in many jobs and daily activities (e.g., pulling luggage out of a car's trunk or objects from a running conveyor in manufacturing facilities). To facilitate the design of tasks involved one-handed pulling, some researchers have conducted attempts to build reference data on one-handed pull strength [1–4]. For example, Lin and colleagues [2] collected static one-handed pulling strength from 84

R. S. Goonetilleke et al. (Eds.): AHFE 2021, LNNS 273, pp. 380–387, 2021.
https://doi.org/10.1007/978-3-030-80713-9_48

Americans and built a normative dataset on the populational strength capability of one-handed pulling from four heights (24 inches, 30 inches, waist height, and above-shoulder height). Or and colleagues [1] utilized the same apparatus and experimental protocol that Lin et al. used [2] to test a Chinese population and established a normative dataset of one-handed pulling strength. However, creating a normative strength dataset is time consuming and costly, especially for population-level data [5]. A prediction model can be beneficial to the industry community by providing reference strength data for task design. Previous studies have reported some influencing factors on pull strength, for example, gender [4], age [6], body mass [7], hand-handle interface [8], friction between floor [9], and pulling heights and directions [1, 2]. These findings make it possible to develop prediction models for one-handed pull strength. However, only few prediction models were published in the literature. For example, Garg and Chaffin [10] established a biomechanical computerized model of hand forces, but the model was only for seating posture. Voorbij and Steenbekkers [6] created a prediction model of one-handed pull strength, but they only had one predictor – 'age'. To the best of our knowledge, no prediction model of one-handed pull strength was developed for Chinese population.

Therefore, the present study was designed to develop and validate prediction models of one-handed pull strength using a Chinese population. In order to well control the experimental environment, the hand-handle interface, floor friction, and pulling direction (i.e., the sagittal plane) were all remained unchanged during the experiment. Based on the literature [4, 6, 7], age, gender, body weight, and stature were selected as the potential predictors. Moreover, considering potential body twists in the pulling process, three main body-joint angles (trunk angle, knee angle, and ankle angle) were examined in the development of prediction models of one-handed pull strength.

2 Methodology

2.1 Participants

One hundred Chinese adults participated in the study, including 10 males and 10 females from each of five age groups (18–24, 25–34, 35–44, 45–54, and 55–64 years old). All participants self-reported heathy and free of musculoskeletal injuries at the time of conducting this study. Written consent forms were obtained from each participant.

2.2 Apparatus

Rulers and an electric weight scale were used to measure the participants' stature and body weight, respectively. One-handed pull strength was evaluated using the test frame in Fig. 1. Four load cells (Model AG100, Scaime S.A.S., France) were mounted to the aluminum cross bar whose height was adjustable. When the participants pulled the cylindrical aluminum handle (diameter = 4 cm) mounted to the center of the cross bar, a computer-based application developed using MATLAB recorded the values of the four load cells and calculate the pull strength as a vector sum of the orthogonal components [1, 2]. The test frame was placed on a plywood slip-resistant platform (friction coefficient $\mu = 0.83$). A camera was located on the right side of the participants to capture their sagittal-plane photographs. An angle measuring software (Image Tool v3.0) was used to measure the participants' trunk angle, knee angle, and ankle angle according to the published measurement protocol [10].

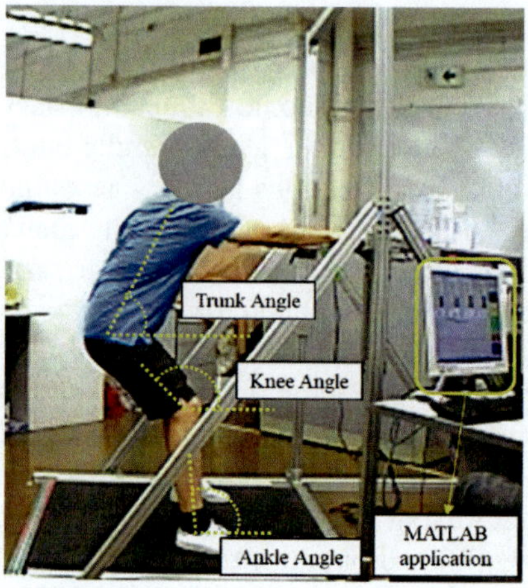

Fig. 1. Illustration of the test platform and three joint-angles.

2.3 Procedure

Two trained research assistants first explained the study protocol to the participants and collected their written consent form, following by a questionnaire for demographic information (e.g., gender and age) and the measurements of body weight and stature using weight scale and ruler, respectively. In the measurement of one-handed pull strength, the participants were asked to: 1) stand on the platform facing to the handle in front of the body (see Fig. 1); 2) pull the handle towards the chest using their domain hand; 3) build up their strengths, without any jerk, in the first two seconds and hold the maximum for the following three seconds. Pulling process for each trial was accompanied with a slogan "ramp up, ramp up, hold, hold, hold, relax". The greater strength value of the two replications was used for data analysis. A sagittal-plane photograph was taken at the time of the second "hold". A 2-min rest (longer if needed) was given between trials.

2.4 Data Analysis

Descriptive analysis was conducted to analyze the means and standard deviations (SDs) of anthropometrics. T-tests were used to evaluate the difference between two genders in anthropometrics and one-handed pull strength. One-way ANOVAs were performed to examine the effects of age-group and pulling height. The 100 participants were divided into two subsets: subset-1 comprising 80 participants (8 males and 8 females randomly selected from each of the 5 age groups) and subset-2 comprising the remaining 20 participants (10 males and 10 females). Stepwise multiple regression analysis was conducted for each pulling height, with age, gender, body weight, stature, trunk angle, knee angle, and ankle angle as potential predictor variables. The standard error of the estimate (SEE) and the mean absolute percentage error (MAPE) were calculated for validating the prediction models through the two equations below [11]. The smaller the SEE and/or MAPE values, the higher prediction accuracy and better validity [12].

$$\text{SEE} = \sqrt{\frac{\sum(Y - \bar{Y})^2}{(n-2)}} \tag{1}$$

$$\text{MAPE} = \frac{100\%}{n}\sum_{1}^{n}\left|\frac{Y - \bar{Y}}{Y}\right| \tag{2}$$

where Y is the observed value, $\bar{Y}$ is the predicted value, and n is the measurements.

3 Results and Discussion

Table 1 lists the means and SDs of age, body weight, and stature of the 100 participants. The results of t-tests showed significant differences between the two genders in body mass, $t(98) = 6.235$, $p < 0.001$, and stature, $t(98) = 9.696$, $p < 0.001$, but not age, $t(98) = 0.035$, $p = 0.972$.

Table 1. Anthropometric information (means and SDs) of the 100 participants.

Characteristics	18–24		25–34		35–44		45–54		55–64	
	M	F	M	F	M	F	M	F	M	F
Age (years)	21.5 (2.1)	20.9 (2.1)	27.6 (2.3)	28.0 (2.5)	39.2 (3.0)	39 (2.4)	50.3 (2.9)	50.1 (2.0)	58.5 (3.4)	58.6 (2.9)
Body weight (kg)	63.7 (7.9)	53.6 (7.4)	69.5 (8.8)	52.6 (6.3)	74.7 (15.9)	58.9 (8.1)	71.8 (7.9)	55.6 (7.1)	62.8 (7.5)	60.3 (13.1)
Stature (cm)	175.1 (3.7)	163.0 (7.0)	174.6 (5.4)	161.3 (4.7)	172.5 (5.8)	162.0 (6.3)	169.8 (6.6)	158.7 (6.2)	167.7 (5.5)	156.1 (4.4)

Table 2 shows the means and SDs of maximum one-handed pull strength when the participants pull in the sagittal plane. The difference in one-handed pull strength between females (172.6 N) and males (241.4 N) was significant, $t = -6.704$, $p < 0.001$, consistent with previous studies [2, 5]. The ANOVA results showed no significant difference in one-handed pull strength among the five age groups for females ($F = 0.421$, $p = 0.793$) but significant difference was found for males ($F = 3.870$, $p = 0.009$). The results also presented that pulling height significantly affected the one-handed pull strength for females ($F = 46.833$, $p < 0.001$) and males ($F = 103.185$, $p < 0.001$), in line with previous studies [2, 5].

The results of stepwise multiple regression analysis showed that the predictors varied for different pulling heights (see Table 3). Age is a negative predictor for one-handed

Table 2. Means (SDs) of maximum one-handed pull strength (in N) of the 100 participants.

Pulling height	Age group	Female	Male
20 inches	18–24	210.8 (36.6)	337.0 (122.1)
	25–34	221.3 (50.7)	378.1 (99.6)
	35–44	223.5 (67.0)	367.2 (63.9)
	45–54	197.3 (59.6)	285.1 (61.3)
	55–64	235.82 (95.4)	269.6 (96.5)
	Total	217.7 (63.6)	327.4 (98.1)
32 inches	18–24	183.0 (39.7)	302.5 (91.9)
	25–34	209.4 (45.1)	312.6 (90.7)
	35–44	196.5 (59.9)	327.0 (57.2)
	45–54	196.7 (68.2)	258.6 (67.0)
	55–64	182.87 (58.4)	229.4 (64.0)
	Total	193.7 (53.9)	286.0 (81.2)
44 inches	18–24	151.3 (25.6)	239.0 (37.8)
	25–34	168.5 (28.9)	249.4 (51.5)
	35–44	175.4 (46.3)	267.2 (51.8)
	45–54	157.7 (43.8)	219.4 (57.8)
	55–64	162.48 (63.9)	190.0 (54.8)
	Total	163.1 (42.9)	233.0 (55.9)
56 inches	18–24	134.8 (29.4)	196.5 (33.6)
	25–34	149.0 (44.3)	233.5 (92.3)
	35–44	159.2 (35.2)	215.1 (47.5)
	45–54	132.9 (24.1)	175.0 (42.0)
	55–64	162.06 (63.7)	155.3 (37.6)
	Total	147.6 (41.8)	195.1 (59.7)
68 inches	18–24	126.8 (41.4)	162.0 (37.0)
	25–34	132.2 (65.3)	200.8 (88.6)
	35–44	143.4 (31.5)	173.0 (43.1)

(*continued*)

Table 2. (*continued*)

Pulling height	Age group	Female	Male
	45–54	142.9 (39.4)	152.3 (64.1)
	55–64	158.9 (63.9)	140.5 (31.2)
	Total	140.8 (49.4)	165.7 (58.3)
Total	18–24	161.3 (46.2)	247.4 (96.4)
	25–34	176.1 (57.8)	274.9 (104.4)
	35–44	179.6 (55.6)	269.9 (87.9)
	45–54	165.5 (54.7)	218.1 (75.6)
	55–64	180.4 (73.5)	197.0 (75.9)
	Total	172.6 (58.3)	241.4 (93.0)

pull strength when pulling at the two lower heights, i.e., 20 inches and 32 inches. Stature is only a negative predictor when pulling at the highest height (68 inches). One possible reason could be that the higher-stature participants may attempt to pull the handle in a horizontal direction but cannot make use of their body weight, as compare to those with lower stature. Trunk angle plays as a positive predictor for the lowest height but a negative predictor for the two highest heights. For the pulling from a high height, a smaller trunk angle may denote that the participants bend their body and tried to make full use of their body weight for the pulling. Weight, ankle angle, and knee angle are all positive predictors for the one-handed pull strength from the five pulling height. The adjusted R^2 of the five models varied from 0.621 (for 20 inches) to 0.88 (for 44 inches) with all p values < 0.001.

Table 3. Results of the stepwise multiple regression analysis for maximum one-handed pull strength (n = 80) for each pulling height (in inches).

Height	(Constant)	Age	Gender	Weight	Stature	KA	AA	TA	Adjusted R^2	F, *p-value*
20	−913.3	−1.8	0	4.6	0	2.4	5.8	2.4	0.621	26.9, *< 0.001*
32	−752.7	−0.8	0	3.2	0	3.0	4.2	0	0.772	67.9, *< 0.001*
44	−643.2	0	0	2.6	0	2.6	3.6	0	0.818	119.2, *< 0.001*
56	−549.1	0	0	2.0	0	3.1	3.4	−1.2	0.809	84.9, *< 0.001*
68	77.2	0	0	2.0	−3.4	4.0	3.9	−3.2	0.716	40.8, *< 0.001*

The subset-2 data (n = 20) were used for the model validation by calculating the SEE and MAPE values (see Table 4). The results showed that when predicting one-handed pull strength for 20 inches, both SEE value and MAPE value were the largest, i.e., 65.2 and 23.5%, respectively; when predicting for 44 inches, both SEE and MAPE were the smallest, i.e., 29.0 and 10.8%, respectively. One possible reason could be that pulling strength from 20 inches is more dependent to body weight and pulling posture (e.g., body twisting), which may make the prediction more sensitive.

Table 4. Means (SDs) of the actual strength and predicted strength for the subset-2 (n = 20) and the validation results (i.e., SEE and MAPE values) for each pulling height (in inches).

Height	Actual strength	Predicted strength	SEE	MAPE
20	280.8 (118.4)	291.5 (76.9)	65.2	23.5%
32	237.8 (90.0)	224.3 (69.2)	50.3	17.3%
44	195.9 (59.3)	191.4 (46.4)	29.0	10.8%
56	161.7 (41.3)	150.9 (44.2)	35.7	14.5%
68	132.6 (32.0)	141.8 (42.8)	39.4	22.0%

4 Conclusion

A normative dataset of one-handed pull strength was established for designing manual handling tasks that involve one-handed pulling. Based on the dataset, five prediction models were developed with the adjusted R^2 values ranging from 0.621 to 0.818 (all *ps* < 0.001), among which body weight, knee angle, and ankle angle are three positive predictors regardless of pulling heights. Low SEE values and low MAPE values denote a good validity of the models. The findings contribute to physical ergonomics and human factors by providing prediction models for reference values of one-handed pull strength of a population, further facilitating safety designs of tasks involved one-handed pulling.

Acknowledgments. The authors would like to thank Dr. Calvin K.L. Or and Dr. J.H. Lin for the experimental design and data collection.

References

1. Or, C., et al.: Normative data on the one-handed static pull strength of a Chinese population and a comparison with American data. Ergonomics **59**(4), 526–533 (2015)
2. Lin, J.-H., McGorry, R.W., Maynard, W.: One-handed standing pull strength in different postures: normative data. Appl. Ergon. **44**(4), 603–608 (2013)
3. Chaffin, D.B., et al.: Volitional postures during maximal push/pull exertions in the sagittal plane. Hum. Factors **25**(5), 541–550 (1983)

4. Cheng, T.-S., Lee, T.-H.: Human pulling strengths in different conditions of exertion. Percept. Mot. Skills **98**(2), 542–550 (2004)
5. Wang, H., Lin, J.-H., Or, C.K.L.: Prediction of maximum static grip strength in a standing posture and with preferred grip span in a Chinese sample. IISE Trans. Occup. Ergon. Hum. Factors **7**(2), 71–80 (2019)
6. Voorbij, A.I.M., Steenbekkers, L.P.A.: The composition of a graph on the decline of total body strength with age based on pushing, pulling, twisting and gripping force. Appl. Ergon. **32**(3), 287–292 (2001)
7. Ayoub, M.M., McDaniel, J.W.: Effects of operator stance on pushing and pulling tasks. AIIE Trans. **6**(3), 185–195 (1974)
8. Fothergill, D.M., Grieve, D.W., Pheasant, S.T.: The influence of some handle designs and handle height on the strength of the horizontal pulling action. Ergonomics **35**(2), 203–212 (1992)
9. Haslam, R.A., et al.: Maximum acceptable loads for pushing and pulling on floor surfaces with good and reduced resistance to slipping. Saf. Sci. **40**(7–8), 625–637 (2002)
10. Garg, A., Chaffin, D.B.: A biomechanical computerized simulation of human strength. AIIE Trans. **7**, 1–15 (1975)
11. de Myttenaere, A., et al.: Mean absolute percentage error for regression models. Neurocomputing **192**, 38–48 (2016)
12. Portney, L.G., Watkins, M.P.: Foundations of Clinical Research: Applications to Practice. Pearson Prentice Hall, New Jersey (2009)

Cross-Cultural Decision Making

Examining the Cultural Differences of Users' Characteristics Between the United States and Japan Related to User Interface Design

Toshihisa Doi(✉) and Atsuo Murata

Graduate School of Natural Science and Technology, Okayama University, 3-1-1 Tsushimanaka, Kita-ward, Okayama, Japan
tdoi@okayama-u.ac.jp, murata@iims.sys.okayama-u.ac.jp

Abstract. This study aims to grasp the differences in user characteristics related to user interface (UI) design from the perspective of cultural differences and examine how to approach UI design. This study mainly focused on the cultural differences between the United States and Japan to examine UI design that takes cultural differences into account. Based on the following six cultural differences between the United States and Japan, differences in user characteristics related to UI design were examined, and how to approach UI design considering cultural differences was examined: (1) short-term or long-term orientation, (2) individualism or totalism, (3) having control of the environment or acceptance of a given environment, (4) preferring change or being risk-averse and seeking for stability, (5) emphasizing the value of time or the value of human relation, and (6) low- or high-context culture.

Keywords: Cultural difference · User characteristics · User interface design

1 Introduction

The recent proliferation of the internet and digital products makes it necessary to think about user interface (UI) design considering the many users worldwide. It is important to understand various users' characteristics within the context of various preferences, values, and cultural backgrounds, and to localize UI design appropriately or find standard solutions among several users. Especially in recent years, there has been an emphasis on an approach that focuses on users' subjective experience (Design for user experience (UX)), rather than just eliminating usability drawbacks for efficient and effective UI operation. The quality of users' subjective experience is highly dependent on the values of each user. It is necessary to design UI with a good understanding of the cultural background underlying the user's values to achieve a better UX.

According to Petrie et al. [1], approaches to developing accessible and usable designs vary based on each cultural context. In terms of design characteristics in different cultures, Cyr and Trevor-Smith [2] and Doi and Murata [3] investigated differences in web design characteristics in several countries to examine the design direction of localized

R. S. Goonetilleke et al. (Eds.): AHFE 2021, LNNS 273, pp. 391–396, 2021.
https://doi.org/10.1007/978-3-030-80713-9_49

websites. The results showed that websites in these countries were reported to have some statistically significant differences in terms of several characteristics: Barber and Badre [4] and Marcus and Gould [5] reported some features of UI design across countries with different cultural characteristics, such as Alexander et al. [6, 7] who tried to develop cross-cultural web design guidelines, based on the discovery of cultural differences between several countries; Zhou and DeSantis [8], Fang and Rau [9], and Doi and Murata [10] conducted a usability evaluation of websites in countries with different cultural backgrounds and compared the results. These studies show that usability drawbacks may differ from cultural differences.

Thus, different countries, culturally, have different design preferences and web design characteristics. To design with cultural differences in mind, the designers should examine how cultural differences relate to user characteristics during UI operations. The present study aimed to understand the differences in user characteristics related to UI design from the perspective of cultural differences and examine how to thereby approach UI design. This study mainly focused on the cultural differences between the United States and Japan to examine user characteristics based on cultural differences. Previous studies have shown that values, attitudes, personalities, and behavioral patterns differ greatly in terms of the cultural differences between countries and regions. In Japan and the U.S., the focus of this study, various cultural differences have been reported. This study summarized previous studies and examined how these differences in user characteristics based on cultural differences between Japan and the U.S. may relate to UI design.

2 Cultural Differences Between the U.S. and Japan

From past surveys, it is clear that there are differences in values, attitudes, personalities, and behavioral patterns based on various cultural differences between countries and regions. It is also said that there are various cultural differences between Japan and the U.S. As mentioned above, these cultural differences need to be taken into account when designing UI. Here we focus on Japan and the U.S. and list the cultural differences that are thought to affect UI interaction (Table 1).

Table 1. Summary of tendency of cultural differences in values, attitudes, way of thinking, and communication style between the United States and Japan.

	United States	Japan
(1)	Short-term orientation	Long-term orientation
(2)	Individualism	Totalism
(3)	Control over the environment	Acceptance of a given environment
(4)	Preferring change	Risk-averse and seeking for stability
(5)	Emphasizing the value of time	Emphasizing the value of human relation
(6)	Low-context communication	High-context communication

2.1 Long-Term or Short-Term Orientation

According to Hofstede [11], Japanese people tend to have a long-term orientation, while Americans tend to have a short-term orientation. Long-term orientation means that people tend to think in the long run and that there is more social pressure for patience, frugality, and continuity. Conversely, a short-term orientation means that people tend to think in terms of immediate results, and social pressure for consumption tends to be stronger. For example, in Japan, which tends to have a strong long-term orientation, it is said that there is a strong hierarchical relationship based on age and status, that long-term results are sought, and that patience is considered a virtue. On the other hand, in the U.S., where short-term orientation is relatively strong, people are said to be meritocratic, seek short-term results, and value upward mobility.

2.2 Individualism or Totalism

It is said that the United States tends to be more individualistic than Japan [11–13]. In individualistic cultures, the interests of the individual take precedence over the interests of the group. Individual interests take precedence over group interests, and the ties between individuals are loose, and each individual's differences tend to be respected. In contrast, Japan is less individualistic and more totalistic than the United States [11]. In totalism, the interests of the group are given priority, even at the expense of personal convenience.

In the context of individualism, people tend to be diplomatic, and their duties tend to be more important than their human relationships in business. Individual opinions are respected, and each individual's abilities determine employment and promotion in business. They also tend to place value on self-actualization. On the other hand, in the case of totalism, many people are introverted, and human relations tend to be more important than work. Group harmony is respected more than individual opinions, and group harmony is important for hiring and promotion in business. They also tend to value social harmony and consensus.

2.3 Control Over the Environment, or Acceptance of a Given Environment

According to Kohls [12], Americans believe that they can change their environment to achieve their goals, and they find value in controlling their environment. On the other hand, Japanese people tend to accept the given environment, partly because they consider patience a virtue, as described in Sect. 2.1. It is said that even in extremely harsh working environments and conditions, people do not consider it acceptable to disturb the harmony with their surroundings and tend to accept it quietly [14].

2.4 Preferring Change or Risk-Averse and Seeking for Stability

According to Hofstede [11], Japan has one of the most "highly uncertainty-averse cultures" globally. In such a culture, people tend to avoid risk, seek for stability, and not decide on a course of action unless they are sure that the risk is as low as possible. Because they believe that uncertainty can be avoided by increasing predictability, they have many customary rules that they prefer to follow in their daily lives. In contrast, the U.S. has a relatively low uncertainty-avoidance tendency [11] and is said to find value in "change" [12]. It tends to encourage people to take risks to succeed, tackle things in ways that are different from conventional practices, and keep trying even when they fail.

2.5 Emphasizing the Value of Time or Emphasizing the Value of Human Relation

Americans tend to ascribe a high value to time and do not like to waste time [12]. They see the value in increasing productivity without wasting time. They also prefer to get things done on time rather than build deep human relationships [12]. On the other hand, the Japanese are punctual and value getting things done on schedule [15], but tend to place less importance on avoiding time-wasting and productivity. In some cases, Japanese people work in an unproductive manner, placing more importance on human relations and group harmony in the workplace than on working efficiently according to each individual's duties and abilities [14]. As an example of the emphasis on human relations rather than productivity, it is said that in Japanese companies there are many unnecessary meetings and overtime work to maintain harmony within the group. Besides, it is said that in business decision-making, not only the context of the job but also personal relationships formed outside of work hours often have an impact [15].

2.6 Communication Style

According to Hall [16], there is a cultural difference in communication style between high-context and low-context cultures. In high-context cultures, non-verbalized communication is more important than linguistic expressions and is not always explained in a strict language. It is not enough for the receiver of information to take the sender's intentions literally as they are worded. It is also necessary to infer the sender's intentions and emotions, including information other than words. That is, the onus is on the receiver of the information and requires the ability to understand the sender's intentions.

On the other hand, in low-context cultures, linguistic communication is important, and rigorous, specific, and clear verbal explanations are preferred. The receiver of information does not need to infer the sender's intentions but receives the information just as it is worded. The sender of information must convey the information concisely and concretely to convey everything in verbal information. The United States is classified as a low-context culture, while Japan is classified as a high-context culture.

3 Differences in Users' Characteristics Related to UI Design Between the U.S. And Japan

The cultural differences described so far are expected to affect the way we think and act toward UI. Indeed, it has been shown that the characteristics of the appearance and structure of web design are different between Japanese and American websites [3], and the tendency of usability problems is also different [10]. In this study, we examine below how the cultural differences described so far affect the user characteristics that should be taken into account when designing UI.

3.1 Characteristics of American Users

Americans tend to value their time and seek short-term results. They are not afraid of change, and since many products are returnable, they are quick to give up on products that are difficult to use or unsatisfactory. In a culture where individual opinions are respected and self-fulfillment is important, people do not put up with products that are difficult to use or unsatisfactory but rather, point out the discovered problems. In other words, in UI design, care should be taken to avoid dissatisfaction based on first impressions and initial ease of use.

Also, Americans prefer to control their environment rather than just follow the environment they are placed in. Thus, they tend to prefer to "search" for information on their own rather than be "guided" when operating websites. Indeed, American websites often place a search box in a prominent place on the top page [3].

Besides, it is difficult for users to read designers' intentions if the UI is difficult to understand because of the low-context communication culture. It is necessary to think of a simple screen that is easy to understand and clearly and concisely verbalized in the UI design.

3.2 Characteristics of Japanese Users

The Japanese tend to be long-term oriented and do not seek short-term results, and patience is seen as a virtue. Designers need to think about how to improve the value to users in the long-term. Japanese people tend to accept the environment they are given and do not like to raise their voices that disturbs their surroundings' harmony. Also, because of the high-context communication culture, even if the user interface is somewhat difficult to understand, the user may be able to guess the designer's intentions and continue to operate it. For these reasons, users are unlikely to openly raise their voices about UI that is difficult to use. Because of this culture, there is a tendency for users to speak less when they "think-aloud" in usability testing [17].

The Japanese tend to accept their environment and prefer to be guided by a website's design rather than search for information. For this reason, many Japanese websites have a large number of links on the top page [3].

Japan is an uncertainty-averse culture, and people tend to avoid making decisions until they are sure that the risk is low. For this reason, when making a decision, they prefer to gather a quantity of reassuring information to reduce uncertainty. Therefore, they seem to feel more secure with products that have more functions than necessary and with advertisements that provide more information than just a simple sales proposition.

4 Conclusion

In this study, we introduced six cultural differences between Japan and the United States that have emerged from previous research: (1) long-term orientation, (2) individualism or totalism, (3) control of the environment or acceptance of a given environment, (4) preferring change or being risk-averse and seeking for stability, (5) emphasizing the value of time or the value of human relation, and (6) low- and high-context culture. As a point of view for examining UI design considering cultural differences, differences in user characteristics related to UI design between Japan and the U.S. were examined, and how to approach UI design considering cultural differences was examined.

References

1. Petrie, H., Weber, G., Darzentas, J.: Designing for accessibility. In: Bernhaupt, R., Dalvi, G., Joshi, A., Balkrishan, D.K., O'Neill, J., Winckler, M. (eds.) INTERACT 2017. LNCS, vol. 10516, pp. 387–390. Springer, Cham (2017). https://doi.org/10.1007/978-3-319-68059-0_36
2. Cyr, D., Trevor-Smith, H.: Localization of web design: an empirical comparison of German, Japanese, and United States web site characteristics. J. Am. Soc. Inform. Sci. Technol. **55**(13), 1199–1208 (2004)
3. Doi, T., Murata, A.: Cross-cultural analysis of top page design among Brazilian, Chinese, Japanese and United States web sites. In: Goossens, R.H.M., Murata, A. (eds.) AHFE 2019. AISC, vol. 970, pp. 609–620. Springer, Cham (2020). https://doi.org/10.1007/978-3-030-20145-6_61
4. Barber, W., Badre, A.N.: Culturability: the merging of culture and usability. In: Proceedings of 4th Conference on Human Factors and the Web (1998)
5. Marcus, A., Gould, E.W.: Cultural dimensions and global web user-interface design. Interactions **7**(4), 33–46 (2000)
6. Alexander, R., Murray, D., Thompson, N.: Cross-cultural web usability model. In: Bouguettaya, A., et al. (eds.) WISE 2017. LNCS, vol. 10570, pp. 75–89. Springer, Cham (2017). https://doi.org/10.1007/978-3-319-68786-5_6
7. Alexander, R., Murray, D., Thompson, N.: Cross-cultural web design guidelines. In: Proceedings of the 14th International Web for All Conference on The Future of Accessible Work, 10 p. ACM (2017)
8. Zhou, Q., DeSantis, R.: Usability issues in city tourism website design: a content analysis. In: 2005 IEEE International Professional Communication Conference Proceedings, pp. 789–796 (2005)
9. Fang, X., Rau, P.P.: Culture differences in design of portal sites. Ergonomics **46**(1–3), 242–254 (2003)
10. Doi, T., Murata, A.: Comparative analysis of website usability between United States and Japan. In: Karwowski, W., Goonetilleke, R.S., Xiong, S., Goossens, R.H.M., Murata, A. (eds.) AHFE 2020. AISC, vol. 1215, pp. 501–507. Springer, Cham (2020). https://doi.org/10.1007/978-3-030-51549-2_66
11. Hofstede, G., Hofstede, G.J., Minkov, M.: Cultures and Organizations: Software of the Mind. McGraw-Hill, New York (2010)
12. Kohls, L.R.: The values of Americans live by, Meridian House International (1984)
13. Nisbett, R.: The Geography of Thought: How Asians and Westerners Think Different-ly…and Why. Free Press (2004)
14. Murata, A., Doi, T.: Attitude toward long working hours from the perspectives of cross-cultural difference of viewing things and scarcity-slack relation. In: Karwowski, W., Goonetilleke, R.S., Xiong, S., Goossens, R.H.M., Murata, A. (eds.) AHFE 2020. AISC, vol. 1215, pp. 508–514. Springer, Cham (2020). https://doi.org/10.1007/978-3-030-51549-2_67
15. Meyer, E.: The culture map–Breaking through the invisible boundaries of global business, PublicAffairs (2014)
16. Hall, E.T.: Beyond Culture. Doubleday, New York (1976)
17. UX alliance: Cultural differences in usability testing around the world. http://www4.uxalliance.com/wp-content/uploads/2014/06/Cultural-differences-in-usability-testing_0.pdf

Mechanism of Improving Performance by Expressing Human Service Employees' Positive Emotion

Noriko Okabe(✉)

Center for Economic Growth Strategy, Yokohama National University, 79-4, Tokiwadai, Hodogaya-ku, Yokohama 240-8501, Japan

Abstract. This study aims to examine how human service employees' positive and negative emotions affect their task performance. Affective delivery, presenting positive emotions in the service interaction, is used as a positive emotion, and emotional exhaustion is an employees' stressful negative emotion. The author tested hypotheses by using a multiple mediation model. The variables used in this study are first measured using descriptive statistics and correlation analysis. The author verified direct and mediator effects between the variables used in this study by path analysis of covariance-based structural equation modeling (SEM). The result shows that employees who practice positive emotional expression in service interaction can improve job performance and increase job satisfaction. Understanding this mechanism could also help prevent stress and emotional exhaustion among employees in the workplace.

Keywords: Human service employees · Emotions · Emotional Exhaustion · Job performance · Job satisfaction · Airline

1 Introduction

In the aviation industry, many traditional airlines have reduced costs and changed employment relationships due to deregulation and the intensified competition. Such an environment allowed LCC carriers (Low-Cost Carriers) to enter the market. Further, the airlines in the world have continued to operate with as few people as possible. In other words, the professional role required of modern flight attendants has gradually changed from the state it was in the past few decades [1, 2].

Role ambiguity occurs when an employee lacks sufficient information to perform a job and does not know what to do [3]. Therefore, modern flight attendants may perceive role ambiguity in the workplace. Besides, employee perceived role ambiguity reduces employee job performance and job satisfaction [4, 5].

In contrast, previous studies have shown that if a company is market-oriented, its performance will improve, and results would be obtained [6]. Ultimately, market-oriented companies and organizations' profitability will improve [7]. Additionally, employees' confidence in the company and their job commitments would increase [8, 9]. Notably,

R. S. Goonetilleke et al. (Eds.): AHFE 2021, LNNS 273, pp. 397–404, 2021.
https://doi.org/10.1007/978-3-030-80713-9_50

many sales representatives consider meeting customer needs imperative for the customer service provider to help a company succeed.

For service organizations, individual service employees mainly implement market orientation. In organizations that provide many customer services, individual service employees practice their basic principles as their duties. A key determinant of customer satisfaction with a service is the customer-employee interaction that individual service employees practice in providing the service. Thus, customer orientation is a factor composed at the personal level, and it is a central component due to which service companies continue to be market-oriented [6].

This study uses recent social psychology research to explain how customer service employees' positive emotions and negative stress emotions affect job performance. Furthermore, it utilizes customer service employee data to validate a customer service employee's job satisfaction result.

2 Literature Review and Hypotheses Construction

2.1 Relationship Between Role Ambiguity, Job Performance, and Job Satisfaction

Many researchers have conducted studies on the relationship between role ambiguity, job satisfaction, and job performance [10, 11]. For example, a classical Hawthorne study [12] concluded that productive workers were the most satisfied workers. Locke [13] believed that high performance resulted in employees feeling job satisfaction. On the other hand, some researchers said that the employees were satisfied only when their performance was rewarded correctly [14]. However, other researchers have shown that job satisfaction is multifaceted [15] and that different job factors lead to job satisfaction and job performance in different people [16]. Thus, for various reasons, the relationship between job satisfaction and job performance has been repeatedly studied. However, in the end, the relationship between role ambiguity, job satisfaction, and job performance varies from person to person at the professional level. Therefore, individual differences should affect job satisfaction [17].

2.2 The Basic Mediation Model

The central idea in the mediator model is that the effects of a stimulus (e.g., role conflict) on behavior (e.g., job performance) are mediated by various transformation processes internal (e.g., emotion and stress) to the organism. Figure 1 shows this essential causal chain involved in mediation. This basic model assumes a three-variable system such that two causal paths feed into the outcome variable (e.g., job satisfaction): the direct impact of the independent variable (path c) and the impact of the mediator (path b). There is also a path from the independent variable to the mediator (path a) [18].

Hypotheses 1: The "role ambiguity" perceived by service employees has a negative relationship (direct effect) that reduces "job performance."

Hypotheses 2: The "role ambiguity" perceived by service employees has a negative relationship (direct effect) with the "affective delivery," which refers to employees' expressing positive emotion in service interactions.

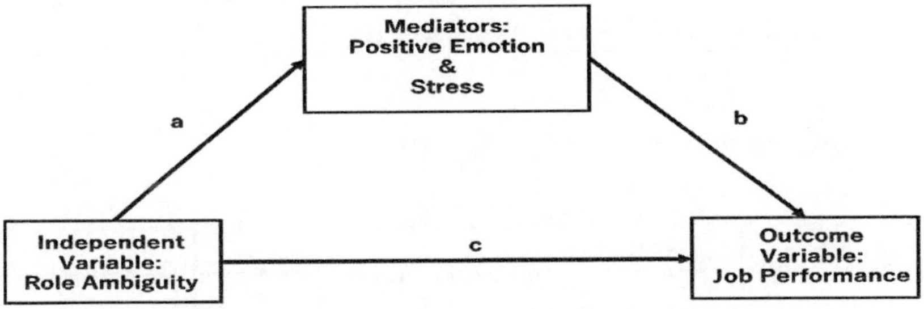

Fig. 1. The basic mediation model tested in the study

2.3 Emotions in the Organization: "Affective Delivery" Practiced by Service Employees

"Affective delivery" refers to express positive emotions in a service interaction to satisfy customers. Human service employees often practice an affective delivery, for example, smiling and showing friendliness. Conversely, employees may not always feel positive, but they may control their emotions to show only positive emotions in their duties [19]. When service employees practice affective delivery, customers impress that they have received good service, which improves customer satisfaction [20]. Thus, "affective delivery" influences the service employee's behavior and increases job performance.

Hypotheses 3: The "affective delivery" expressed by service employees has a positive relationship (direct effect) with "job performance."

Customers satisfied with the employees' excellent service are more likely to be repeaters [21]. Affective delivery practices can improve customers' purchasing behavior and enhance job performance [22].

Hypotheses 4: The "affective delivery" expressed by service employees has a mediating effect that transforms the negative direct relationship between "role ambiguity" and "job performance" into a positive one.

2.4 Emotional Exhaustion

Emotional exhaustion refers to a state of chronic physical and mental fatigue due to excessive work and stress [23]. The employees who perceive role ambiguity can also experience emotional exhaustion. The employees experiencing emotional exhaustion due to extreme physical and psychological stress may have reduced job performance (Fig. 2).

Hypotheses 5: The "role ambiguity" perceived by service employees has a positive relationship (direct effect) with "emotional exhaustion."

Hypotheses 6: The "emotional exhaustion" perceived by service employees has a negative relationship (direct effect) with "job performance".

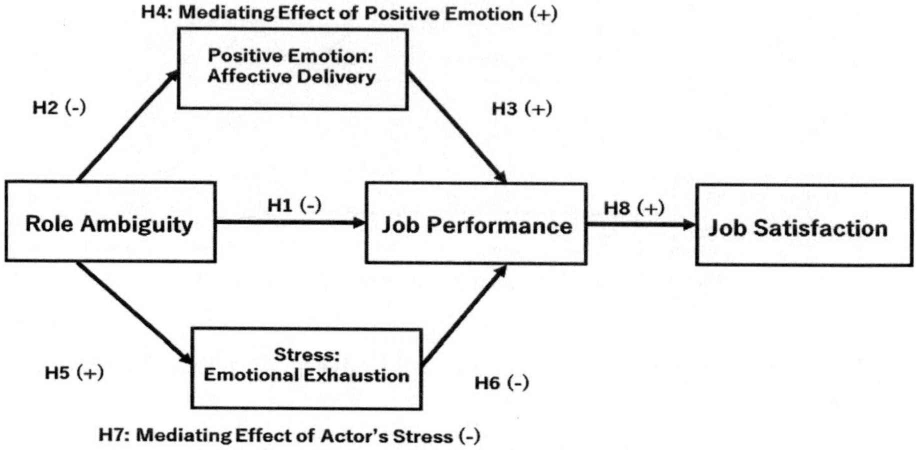

Fig. 2. The multiple mediation model tested in the study

Hypotheses 7: The "emotional exhaustion" perceived by service employees has a mediating effect; nonetheless, it does not transform the negative direct relationship between "role ambiguity" and "job performance" into a positive one.

Hypotheses 8: The "job performance" of service employees positively affects "job satisfaction."

3 Method

In this study, the author distributed a five-point Likert-type questionnaire to 500 flight attendants working for European airlines and received a valid response from 414 (82.8%). In testing the hypotheses, the author first measured the variables used in this study using descriptive statistics and correlation. Table 1 shows the descriptive statistics of the variables used in this study, the correlation analysis, and the Cronbach's alpha coefficient. Next, the author verified the direct effect between the variables and the mediator effect by path analysis in covariance-based structural equation modeling (SEM).

4 Results

4.1 Direct Effects

Regarding Hypothesis 1, the correlation analysis found a significant negative correlation between role ambiguity and job performance ($r = -.29, p < .01$). The results of structural equation modeling (SEM) also found a negative path coefficient ($r = -.23$, $p < .01$) between role ambiguity and job performance. Hypothesis 1, which predicted a negative relationship (direct effect) between role ambiguity and job performance, was supported. The goodness-of-fit summary of the SEM model shows $\chi^2/\text{df} = 2.467$, CFI = .953, GFI = .922, TLI = .943, RMSEA = .060.

Regarding Hypothesis 2, the correlation analysis found a significant negative correlation between role ambiguity and affective delivery ($r = -.20, p < .01$). The results of structural equation modeling (SEM) also found a negative path coefficient ($r = -.23, p < .01$) between role ambiguity and affective delivery. Hypothesis 2, which predicted a negative relationship (direct effect) between role ambiguity and affective delivery, was supported.

Regarding hypothesis 3, the correlation analysis found a significant positive correlation between affective delivery and job performance ($r = .36, p < .01$). The results of structural equation modeling (SEM) also found a positive path coefficient ($r = .40, p < .01$) between affective delivery and job performance. Hypothesis 3, which predicted the positive relation between affective delivery and job performance, was supported.

Table 1. Descriptive statistics and correlations

	Variable	Mean	s.d.	α[4]	Correlations 1	2	3	4	5	6	7	8
1	Gender [1]	.74	.44									
2	Tenure [2]	4.11	1.51		.12**							
3	Age [3]	3.98	.79		.02	.76***						
4	Role Ambiguity	1.63	.47	.84	-.12**	-.14***	.12**					
5	Job Performance	4.24	.52	.82	.09*	.12**	.08*	-.29***				
6	Job Satisfaction	4.08	.71	.83	.05	.14***	.11**	-.14***	.10**			
7	Affective Delivery	4.50	.53	.82	-.10**	.14***	.22***	-.20***	.36***	.04		
8	Emotional Exhaustion	3.45	.87	.91	-.16***	-.03	.01	.25***	-.17***	-.21***	-.06	

Note: *** p < .01, ** p < .05, * p < .10. N = 414.

[1] Gender: coded as Male = 0, Female = 1.

[2] Tenure: coded as 1 = 0 ~ 5 years, 2 = 6 ~ 10 years, 3 = 11 ~ 15 years, 4 = 16 ~ 20 years, 5 = 21 ~ 25 years, 6 = 26 ~ 30 years, 7 = more than 30 years.

[3] Age: coded as 1 = less than 20, 2 = 21 ~ 30, 3 = 31 ~ 40, 4 = 41 ~ 50, 5 = 51 ~ 60, 6 = more than 60

[4] Reliability represents Cronbach Alpha coefficients.

4.2 Mediating Effects

Regarding hypothesis 4, Fig. 3 shows that "affective delivery" is a mediator between role ambiguity and job performance. At the time, the negative direct effect of role ambiguity on job performance was transformed positively by the moderating effect of "affective delivery." Thus, hypothesis 4, which predicted: "affective delivery" expressed by service employees as having a mediating impact and transforms the negative direct relation between "role ambiguity" and "job performance" into a positive one, was supported.

Regarding hypothesis 5, the correlation analysis found a significantly positive correlation ($r = .25, p < .01$) between role ambiguity and emotional exhaustion. The results

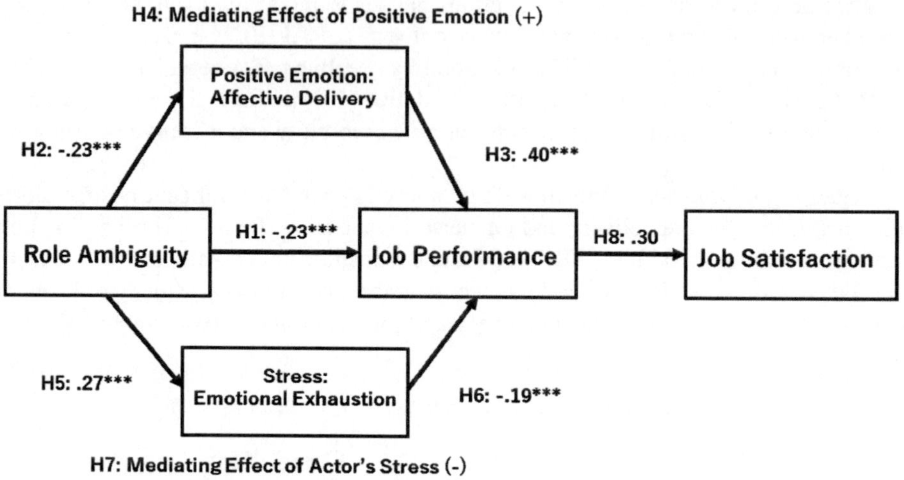

Note: *** p < .01, ** p < .05, * p < .10.
χ^2/df= 2.467, CFI=.894, GFI=.922, TLI=.943, RMSEA=.060.
Path coefficients are described as standardized coefficients

Fig. 3. Result of the multi mediation model

of structural equation modeling (SEM) also found a positive path coefficient ($r = .27$, $p < .01$) between role ambiguity and emotional exhaustion, and hypothesis 5, which predicted a positive relationship between role ambiguity and emotional exhaustion, was supported.

Regarding hypothesis 6, the correlation analysis found a significantly negative correlation ($r = -.17, p < .01$) was found between emotional exhaustion and job performance. The results of structural equation modeling (SEM) also found a negative path coefficient ($r = -.19$, $p < .01$) between emotional exhaustion and job performance, and hypothesis 6, which predicted a negative relationship between emotional exhaustion and job performance, was supported.

Regarding hypothesis 7, Fig. 3 shows that emotional exhaustion is a mediator between role ambiguity and job performance. The negative direct effect of role ambiguity on job performance was not transformed positively by the mediating effect due to emotional exhaustion. Thus, hypothesis 7, which predicted that "emotional exhaustion" perceived by service employees is having a mediating impact; nonetheless, it does not transform the negative direct relationship between role ambiguity and job performance into a positive one, was supported.

Regarding hypothesis 8, the correlation analysis found a significantly positive correlation ($r = .10$, $p < .05$) was found between job performance and job satisfaction. The results of structural equation modeling (SEM) also found a positive coefficient ($r = .30$, $p < .05$) between job performance and job satisfaction, and hypothesis 8, which predicted the positive relationship between job performance and job satisfaction, was supported.

5 Discussion and Implication

"Affective delivery," or expressing positive emotions in service interactions, helps satisfy customers. Research has shown that affective delivery has a positive effect on service interaction with customers. This attitude includes smiling and friendliness that is positively associated with essential customer outcomes, such as the intention to return, intent to recommend a service to others, and perception of overall service quality [19–21].

In this study, "affective delivery," presenting a positive emotional expression during the service encounter, was used as the "positive emotion," and "emotional exhaustion" as the "stressful negative emotion" of human service employees such as flight attendants.

This study's findings show that service employees who practice "affective delivery" tend to have better job performance and job satisfaction, even if they perceive role ambiguity in the workplace. In contrast, service employees who perceive stress and emotional exhaustion due to role ambiguity tend to have reduced job performance.

Finally, understanding the mechanism by which expressing positive emotion improves job performance and increases job satisfaction helps service employees prevent stress and emotional exhaustion.

Acknowledgment. JSPS KAKENHI supported this work, Grant Number JP19K13800.

References

1. Bergman, A., Gillberg, G.: The cabin crew blues. Middle-aged cabin attendants and their working conditions. Nordic J. Work. Life Stud. **5**, 23 (2015)
2. Okabe, N.: Creating of customer loyalty by cabin crew a study of the relation between emotional labor and job performance. Transp. Res. Procedia. **25**, 149–164 (2017)
3. Katz, D., Kahn, R.: The Social Psychology of Organiations. Wiley, New York (1978)
4. Rizzo, J., House, R., Lirtzman, S.: Role conflict and ambiguity in complex organizations. Adm. Sci. Q. **15**, 150 (1970)
5. Tubre, T., Collins, J.: Jackson and Schuler (1985) revisited: a meta-analysis of the relationships between role ambiguity, role conflict, and job performance. J. Manag. **26**, 155–169 (2000)
6. Brown, T., Mowen, J., Donavan, D., Licata, J.: The customer orientation of service workers: personality trait effects on self-and supervisor performance ratings. J. Mark. Res. **39**, 110–119 (2002)
7. Narver, J., Slater, S.: The effect of a market orientation on business profitability. J. Mark. **54**, 20 (1990)
8. Deshpandé, R., Farley, J., Webster, F.: Corporate culture, customer orientation, and innovativeness in Japanese firms: a quadrad analysis. J. Mark. **57**, 23–37 (1993)
9. Jaworski, B., Kohli, A.: Market orientation: antecedents and consequences. J. Mark. **57**, 53 (1993)
10. Abramis, D.: Work role ambiguity, job satisfaction, and job performance: meta-analyses and review. Psychol. Rep. **75**, 1411–1433 (1994)
11. Jackson, S., Schuler, R.: A meta-analysis and conceptual critique of research on role ambiguity and role conflict in work settings. Organ. Behav. Hum. Decis. Process. **36**, 16–78 (1985)
12. Dodd, P.: Roethlisberger, F.J., Dickson, W.J.: Management and the Worker, pp. xxiv, 615. Harvard University Press, Cambridge (1939)

13. Locke, E.: Job satisfaction and job performance: a theoretical analysis. Organ. Behav. Hum. Perform. **5**, 484–500 (1970)
14. Cherrington, D., Reitz, H., Scott, W.: Effects of contingent and noncontingent reward on the relationship between satisfaction and task performance. J. Appl. Psychol. **55**, 531–536 (1971)
15. Smith, P.: The Measurement of Satisfaction in Work and Retirement: A Strategy for the Study of Attitudes. Rand McNally, Chicago (1969)
16. Hackman, J., Lawler, E.: Employee reactions to job characteristics. J. Appl. Psychol. **55**, 259–286 (1971)
17. Gould, S., Hawkins, B.: Organizational career stage as a moderator of the satisfaction-performance relationship. Acad. Manag. J. **21**, 434–450 (1978)
18. Baron, R., Kenny, D.: The moderator–mediator variable distinction in social psychological research: conceptual, strategic, and statistical considerations. J. Pers. Soc. Psychol. **51**, 1173–1182 (1986)
19. Grandey, A.: When "the show must go on": surface acting and deep acting as determinants of emotional exhaustion and peer-rated service delivery. Acad. Manag. J. **46**, 86–96 (2003)
20. Pugh, S.: Service with a smile: emotional contagion in the service encounter. Acad. Manag. J. **44**, 1018–1027 (2001)
21. Tsai, W.: Determinants and consequences of employee displayed positive emotions. J. Manag. **27**, 497–512 (2001)
22. Totterdell, P., Holman, D.: Emotion regulation in customer service roles: testing a model of emotional labor. J. Occup. Health Psychol. **8**, 55–73 (2003)
23. Maslach, C.: Burnout: The Cost of Caring. ISHK, Los Altos (2015)

Weakness of Real Estate Collateral Valuation Policy in Changed Financial World

Jukka Rantala[1(✉)], Atte Rantanen[2], Maria Yllikäinen[2], and Timo Holopainen[2]

[1] Turku University of Applied Sciences, Joukahaisenkatu 3, 20520 Turku, Finland
jukka.rantala@turkuamk.fi
[2] Turku University of Applied Sciences, Turku, Finland

Abstract. The change in the real estate market has been rapid, partly due o mortgages, partly due to banking regulation and tightening construction control because of climate change. The importance of the appraisal expertise and correct collateral value related to mortgage lending has been further emphasized as a means of risk management. However, mortgage valuation, collateral valuation and calculation models related to mortgages have not evolved with regulation, because of which new market risks are emerging in the real estate market, the combined effect of which may lead to strong market reactions.

Keywords: Real estate · Collateral · Valuation

1 Introduction

The housing market is deeply connected with the mortgage market. The property to be purchased is almost always utilized as collateral for the loan to be applied for, regardless of whether the property to be acquired is to be utilized for the personal use or for the investment purposes. The value of the collateral or the property valuation method is not regulated by law in Finland, which leads to inequality from the perspective of lending.

This article examines the Finnish mortgages collateral value practices typically utilized by the financial institutions, like first-time buyer privileges, regime guarantee, collateral, authentic estate appraisal practice, and the tax deductibility of the loan. The article also introduces European Union member states practices and seeks to initiate a discussion on more suitable practices by presenting an alternative to the current approach.

2 Property Valuation

Property valuation has a consequential impact on access to finance, and current legislation does not guide the appraisal process. Acquiring your own house is one of the greatest investments in human life, and currently, an estimate of the surmised value of a property for financing can be made without proven qualifications in the field of real estate evaluation. The resulting estimate is predicated on statistical market value during

R. S. Goonetilleke et al. (Eds.): AHFE 2021, LNNS 273, pp. 405–412, 2021.
https://doi.org/10.1007/978-3-030-80713-9_51

a period and does not consider seasonal variations, property characteristics, or longer-term development plans of the region. Real estate valuation based on market value has received a lot of criticism partly due to deficient methodological criteria and due to evaluators own subjective part.

3 Real Estate Agents

In Finland, an authorized real estate agent does not need to know specific criteria's for evaluating a home in practice and understand how to evaluate a home. Regional differences are mostly identified, but the structural condition, localization, or energy efficiency between real estate, are not necessarily reflected in the prices. This poses its own risks to the housing market, especially in situations, where the owner runs into financial difficulties. Valuation expertise has not been considered as a key risk stabilizer, but the value of housing has been largely based on market knowledge.

Real Estate is a licensed activity that is supervised by the Finnish authorities. A real estate company must register in the real estate register. In a real estate agency, half of the real estate agents must have passed a a formal real estate exam LKV, and they can be called a legal real estate agent LKV. The LKV test has remained very similar for decades and its main goal has been to know the real estate-related legislation during the trial period. In recent years, parts which deal with good real estate practice have been introduced to the test. However, the test does not actually measure the ability of a prospective real estate agent in real estate appraisal. The person who has passed the test has the right to assess the transaction value of the property and issue an appraisal book on the value of the property. Finland uses international valuation standards and the related IVS 400, which deals with the determination of the value of the property being valued [1].

4 Finnish Housing Market

In the early 2000s, there was rapid household indebtedness across Europe. In the Finnish housing market, prices rose by 84% in relations to consumer prices. For this reason, the loan ceiling was taken into use in Finland [2]. After this time, the loan ceiling has largely been canceled due to the overheating of the Finnish housing market. The purpose of these measures has been to prevent mortgages from overheating. A U.S. study found that mortgage lending is just one way banks get involved in the mortgage business. Banks may have mortgages as part of, or under other names, loans that comply with regulation. Alongside these, credit limits and loans have been brought in with the help of shadow banks, increasing, in turn, interest rates [3].

The development of house prices has not taken into account the intensification of housing in large growth centers, the investor activity and the possibility of price bubbles. The aging of the population and increasing regulation of banks have contributed to the change in mortgage lending. In the financial markets, the number of shadow banks is increasing. In a study conducted in Finland, the variation in the total interest rate on a loan among five banks was 0.04 per cent [4]. This also shows that there is little competition. The purpose of the loan ceiling has been to increase stability, when, it enables the shadow banking market. The impact of Basel IV on the German capital requirement is as high as 26%, which is likely to contribute to government funding, as this does not require capital investment, in their study, Rantala et al. [5] highlighted the rise of shadow banks as part of funding guarantee deficits.

The loan ceiling was introduced in Finland in 2016, in 2020 the number of new mortgage borrowers was marginal. At the moment, in Finnish housing market new home buyer has to fill a 20% to 25% security gap. To fill this gap, home buyers and brokers have used creativity in the market. In the United States, for example, the tightening of credit in the housing market has turned lending into a creative measure. A U.S. study found that mortgage lending is just one way banks get involved in the mortgage business. Banks may have mortgages as part of, or under other names, loans that comply with regulation. In line with the banking advantage, banks have reduced their exposure relative to mortgage lending. Alongside these, credit limits and loans have been brought in with the help of shadow banks [6].

5 Practices Outside of Finland

A study executed at EU level showed that the size of the banks, has a particular impact, including on their credit supply to its customers. The effect of foreign capital on credit growth varies by type of loan, and in mortgages the effect of bank size was negative relative to the mortgage [7]. When banks apply for a lower risk of lending from shadow banks, social benefits do not compensate for the deficit, Basel IV seems to have the greatest impact on people's economies [8] A study in the UK found that Basel II contributes to fixed cost increases in mortgage lending, with all smaller lenders being disadvantaged in retail mortgages relative to the needs of large banks [9].

In a study conducted in Saka, the separation of Spain, Greece and Ireland from countries where mortgage lending is growing much faster than house prices, which means that reliably past the previous periods, housing price bubbles. Finland is the only Scandinavian country where a significant increase in the number of mortgage loans, is observed [10]. This can be partly attributed to the shadow banks [5].

The entry of the shadow banks to the mortgage market introduced a high potential interest rate behavior. Mortgage market is less dynamic than real estate where the solvency loan is the norm. Mortgage regulation causes an improvement in the position of large international banks in financial markets [10]. U.S. study in which the role of shadow banks in mortgage lending increased interest rates as shadow mortgage lenders weaken the banks' lending channel and thus change housing tips, employment and wage sensitivity to monetary policy [11].

The share of fast-moving shadow banks in financing is currently 55% in the euro area and the role of shadow banks as financiers of the real economy is still rising in 2019. Using AI and blockchain thinking, commercial banks seek to produce collateral valuation, but there are problems with the definition [12].

Heterogeneity and thinness in the housing market affect the definition of fixed value, which can also affect the relative proportion of property sold by a creditor or a homeowner. An extensive study by Pennington-Cross revealed the phenomenon that the value of an apartment sold by a creditor is 22 percent inferior to that of an apartment sold by a private seller [13].

6 Value of Real Estate

The value of real estate is demand in relation to the utility function [14]. The value can be rational cost value, return value, balance sheet value or market / trade value. The difficulty related to the value is, that the benchmark is only in relation of being individual. Arbitrage opportunities arise when homogeneous offers differ from each other by the juxtaposition of home buyers and home sellers. This is not an issue in an efficient market [13]. In Germany, the mortgage market impact on the bank's operations has been affected specifically by the Basel IV capital requirement. Banks providing mortgages have declined in the German market and banks have focused on providing municipalities and private customers with other loans that are clearly visible in marginal operations. The impact of regulation on loan costs has become clear. Gordon's eternity problem with the formula becomes when the parties in the environment trade with each other for amounts that increase the value of real estate throughout the area. The measure is also possible in the other direction, so identifying market failures from the average is challenging [15].

In Finland, the statistics on real estate purchase prices are based on the geographical division provided by the postal codes. However, the regionalization of postal codes does not reflect the actual factors affecting geographic pricing. The interaction between geographic area and price on a property has been described by Culley [16]. In the geographical distribution of the real estate market for sub-markets such as office space, a method for determining price and geographical area based on the analysis of empirical data has been proposed as a method [17].

In practice, real estate agents' estimates of real estate values are mainly based on the definition of a statistical market value, and the consideration of the specific features of the property being assessed has generally received little attention. However, the features and characteristics of the properties have evolved tremendously in recent decades. The properties of the properties have been affected by the emphasis on ecology and the ongoing climate change, which has affected the thermal insulation of structures, the implementation of air conditioning, water and sewage systems and heating systems.

A typical problem is the pricing of the difference in real estate caused by technical development between new and old buildings, deviating from the statistical definition. In the case of exceptionally large changes in construction methods and regulations, the statistical treatment of average prices results in an increase in the variance created around the average price, thus increasing the significance of the pricing error contained in the statistical average price. Typically, the difference can be seen in the property as increased investment costs caused by both modern technical equipment and new implementations. Repayment of increased construction costs for new properties is most likely due to reduced energy demand and, in some cases, the sale of surplus energy production. Similar difference can be seen with older properties, where the value of the repair does not always go as intended or in full directly to the value of the property. There are cases where the correction has only affected the liquidity of the property, at the time of sale. Another example is when only about half of the cost of extensive repairs to water and sewage pipelines has changed to the value of the dwelling [18].

7 Trends and Practices in Finnish Housing Market

One of the future driving forces for change in Finland may be tax deductibility of the loan interest actualized from the previously acquired home. The right to deduct interest has been linked to the mortgage, and at the same time the government has sought to regulate mortgage properties such as loan repayment periods and other conditions such as collateral to loan ratio. In Finland's neighboring country, Sweden, the mortgage repayment is linked to the ratio of the value of the home to the amount of the mortgage, so that when the amount of the mortgage falls below a certain limit, the loan does not have to be shortened at all. Typically, Finns take the loan they need to buy their home as a whole from one bank, and it is rare for a loan for an apartment to come from different sources. However, after a transitional period, the tax deduction will now disappear completely in 2023.

In Finland, apartment buildings occupied by owner-occupied dwellings are typically housing companies, where each apartment corresponds to a certain amount of shares, which creates the right to live in a certain apartment. At the same time, this creates the responsibility to pay for the maintenance and repair costs related to the apartment. Thus, the sale of an apartment building is in fact a sale of shares in a limited company. As a result of the ease of the transaction, the sale of apartments in a housing company is perceived to be simpler, the number of apartments in the companies is large and the construction methods and implementations of the buildings are well-established, which means that the apartments are very similar. A major change in construction methods has taken place since the generalization of prefabricated construction in the 1970s.

The large number of apartment buildings and a certain degree of periodic similarity make it easier to determine the values of the apartments in question, as the volume of statistical data and reasonably small individual variations reduce the error in the statistical average price. The number of dwellings also contributes to the sale time of individual dwellings, ie liquidity and thus also to the perceived collateral value.

Changed building regulations due to the climate change are now also affecting the technical properties of apartment buildings, but they are mainly a result of improved energy tightness and increased insulation thickness due to the energy efficiency of the buildings compared to the previous construction method. In the case of detached houses, the diversity of houses and very different structural and technical characteristics both between buildings of the same era and between different eras make the use of statistical data challenging. The lack or absence of reference material increases the difficulty of determining the price of the property and at the same time affects both the liquidity of the property, and the determination of the fair value of the collateral.

A housing company is like any independent company with statutory obligations. A housing company is usually set up by a founding builder who is the sole shareholder to finance the renovation project, so that the loan is taken over by the company being built. When a construction company sells the apartments of a completed apartment building, ie a series of shares, the shares sold are subject to a share of the housing company's loans, ie the new shareholder only from those loans through its own shares. According to the law, in connection with the sale of shares, the buyer must be informed of the sale price of the apartment, the price of the corporate debt and the share of the apartment in the loans.

Depending on the method of calculation of the banks, if the housing company loan is taken into account in the calculation of the maximum collateral which may result in a situation where the housing company loan on the apartment prevents the mortgage from being granted to the buyer. However, the buyer has to obtain (buy) an additional guarantee that allows him to get a mortgage up to 90% of the debt-free price of the property.

The ability of traditional banks to respond to changes in collateral values caused by house prices has traditionally been very schematic, and the regulation of collateral valuation following banking regulation has further weakened banks' ability to respond to rapid market changes. The increased rigidity of collateral values underlying the mortgage market will further accelerate the accelerated fall in rural housing prices, especially in the housing sub-market, for example as a result of rural-urban migration. The market value, which is almost purely based on demand, reflects the market value of the short-term market and, at the same time, the development of the short-term collateral value, however, the maturity of mortgages is several decades. The rigid guidance of mortgage lending related to the state real estate market, together with the lack of expertise in assessing the individual features and technical characteristics of real estate properties, will play a growing role in increasing the uncertainty in the general housing market an especially in the housing sub-markets. At the market level, the lack of assessment expertise can be reflected, for example, in the increase in mass-produced and highly standardized housing solutions such as "house packages".

8 Changes in the Housing Market

The Finnish housing market is now facing several different forces for change, and the know-how requirements related to real estate brokerage have become an obvious shortfall. If, because of banking regulation, some traditional banks in Finland also end up withdrawing from mortgage financing or decide to limit lending to the housing market, the most likely consequence would be a change in the housing financing model so that purchase financing would consist of different financing sources in the future. This change may be accelerated in part by the simultaneous removal of the tax deduction on mortgage interest, as well as the general rise in interest rates. If the change is linked to the inclusion of new sources of funding and international shadow banks in mortgage lending, as well as problems with the valuation of housing and the associated collateral valuation, or even the deterioration of collateral value, it is likely that there will be varying degrees of market disruption based on temporary shortfall of lending.

The major trends underlying the real estate market affecting mortgage lending are also significantly affected at EU level, as well as a result of banking regulation, so in practice it is possible to choose to influence the implementation of collateral calculation as the most effective means of taking into account the locality of the housing market. The housing market in different countries is not identical due to climatic factors alone, not to mention the housing culture, in which case special attention should be paid to the valuation of housing in the local housing market and the collateral calculation methods used.

9 Proposed Solutions

To ensure market stability, a component that increases the stability of the collateral value of a dwelling, typically the part based on the trend of sub-market prices, should be added to the definition of the collateral value. The move would also protect the homeowner and buyer from temporary market disruptions and remove some of the short-term risk arising from the rigidity of collateral regulation, which is likely to accelerate negative price risk. At the same time, the trend would indicate the direction of possible price development in the region in the future, when the mortgage bank would be able to anticipate, for example, a long-term decline in demand in the sub-market, individualization of the maximum collateral value. Part of a possible negative change in the market price can always be compensated because of the reduction in the loan ratio included in the loan repayment plan, in which case the position of the bank that granted the mortgage is also secured.

There is only one way to increase real estate appraisal skills and understanding of the overall operation of the market, and that is to increase the skills requirements. Traditionally, those already working in the field may oppose this thinking, drawing on the skills already learned in the field, but the accelerating change in the industry and the critical position of assessment skills as a whole require development and maintenance of skills by individuals in the field. In practice, the change in the industry's expertise can be implemented in the future with a slow change in the industry, which will increase the qualification requirements and training of new real estate agents entering the industry.

References

1. IVSC-2020-FINAL-suomeksi.pdf. (www.skayry.fi)
2. Kiiras, J., Tammilehto, S.: Kiinteistökehitys, Kiinteistömedia Oy
3. Bank of Finland, Financial Report 2013 and European Commission Marrez, Peter Potumach (2013)
4. Taloustaito 2020, Veronmaksajain keskuliitto r.y.
5. Rantala, J., Yllikäinen, M., Holopainen, T. Competences in expert sales. In: Kantola, J., Nazir, S., Salminen, V. (eds.) Advances in Human Factors, Business Management and Leadership. AHFE 2020. Advances in Intelligent Systems and Computing, vol. 1209. Springer, Cham. https://doi.org/10.1007/978-3-030-50791-6_16
6. Morris-Levenson, J., Sarama, R., Ungerer, C.: Does Tighter Bank Regulation Affect Mortgage Originations? (2017). https://paprers.ssrn.com
7. Kouretas, G.P., Pawłowska, M., Szafrański, G.: Market structure and credit procyclicality: Lessons from loan markets in theEuropean Union banking sectors, Elsevier (2020)
8. Demary, M., Voightländer, M.: The Impact of Basel IV on Real Estate Financing, IW-Policy 18/2020 (2020)
9. Benetton, M., Eckley, P., Kirwin, L., Latsi, G.: Capital reguirements and mortgage pricing: Evidence from Basel II. Elsevier (2020)
10. Darvey et al. (2020)
11. Zink, D.: The Effects of Shadow Banking on the Transmission of Monetary Policy The Effects of Shadow Banking on the Transmission of Monetary Policy by David Zink :: SSRN (2020)
12. Linné, M.R.: Artificial intelligence, avms and blockchain: Implications for valuation. Apparaisal Institute (2018)

13. Pennington-Cross, A.: The Value of Foreclosed Property. FRB of St. Louis Working Paper No. 2004–022A
14. Virtanen, P.: Kiinteistöjen arviointikirja. Suomen kiinteistöarviointiyhdistys (1991)
15. Kiiras, J., Tammilehto, S.: Kiinteistökehitys: Käsitteet, markkinat, prosessit, talous ja riskit. Markkinointi, rakennuttaminen ja suunnittelu. Tuotteistus, muuntojousto ja kaavoitusyhteistyö (1. p.). Helsinki: Kiinteistöalan Kustannus (2014)
16. Culley, J., Bailey, L., Postila, M.: Prime London housing – drivers and submarkets, Conference paper at ERES 2014 Conference in Bucharest, Romania. (2014). (https://eres.architexturez.net/doc/oai-eres-id-eres2014-81)
17. Dunse – Leishman – Watkins, : Classifying office submarkets. J. Property Investment Finance **19**(3), 236–250 (2001)
18. https://www.op-media.fi/asunnon-osto/nain-putkiremontti-vaikuttaa-asunnon-arvoon/vii ttaus 20.1.2021

Micro Loans to Over-Indebtedness, Causes and Consequences, Perspective on Youth Spending

Jukka Rantala[1](✉), Henri Untinen[2], Maria Yllikäinen[2], and Timo Holopainen[2]

[1] Turku University of Applied Sciences, Joukahaisenkatu 3, 20520 Turku, Finland
jukka.rantala@turkuamk.fi
[2] Turku University of Applied Sciences, Turku, Finland

Abstract. The research includes both the creation of a model included in the teaching for the needs of teaching economics, as well as the testing of the created model with the help of students, and the development of activities as the project progresses. The goal of validating the model is to bring the student into a learning event at different ages and lower the learning threshold. The ease and attractiveness of e-shopping brings constant temptations to a young person's life, making understanding how to use money a vital basic skill.

Keywords: Micropayments · Spending · Digitalization

1 Introduction

The digital environment enables children and young adults to use money in a easy and fast manner, therefore forming their view and understanding toward value of money. Change created by the development of technology and online access, children and young adults have become part of the consumer society in much earlier age than generations before. Consumption is associated with the growth of the identity of children and young adults, as well as it is associated with the construction of the socio-economic space of parents. The understanding of the value of money in relation to reality has become blurred for children and young adults, as reflected in young adults' attitudes towards long-term savings.

This article presents a study that sought to find out how young people and children saw money in in-depth interviews with more than 10 young consumers and 4 groups during the study also observing children's attitudes towards money in general. The interviews highlight the perspectives of young consumers on the causes and effects of a lack of understanding of the values of money.

The article also presents a teaching method to prevent the negative effects associated with a lack of understanding of money. Utilizing the learning method of learning, high school-level young consumers educate basic-level consumers related to economic understanding and value for money.

R. S. Goonetilleke et al. (Eds.): AHFE 2021, LNNS 273, pp. 413–420, 2021.
https://doi.org/10.1007/978-3-030-80713-9_52

2 Consumption as Part of Socio-economic Status

Consumption is a priority in a socio-economic position that has become intrinsically valued, given its own position vis-à-vis the Community, and which tends to create value and future prospects. Young people and children are directly exposed to different markets and exposed to rapid spending. Consumer research has taken a more on ethical and environmental issues than on morality. Young people and children are increasingly confronted with the values that consumption produces through games, media, education and the home. They create an idea of what kind of clothing is part of the agency in relation to the environment. The economic revolution has brought with it rapid consumer credit and shadow banks into consumer culture [1]. Most often, they focus on teaching interest rate risks, rather than the need to take credit for consumption. Consumption has become part of the emphasis on ego and living standards.

Mobile payments, quick levers, one-time credit cards and technical requirements in the life of a child and young person have increased unobtrusive spending. Rapid digitalization and the acceleration of consumption have placed young people and children in a new part of consumer society. Today, children have to deal with the economy even before they reach adulthood [2].

Consumption has become part of the emphasis on socio-economic status and its impact on the ability of the whole person to assess themselves in relation to the environment. It has become part of the construction of identity and the functioning of the social environment. Through consumption, the aim is to find and increase one's position in relation to others. In the current worldview, consumption maintains social relations and places a person in his socioeconomic position. Consumer culture creates the need for parents to buy as gifts, toys, and technical equipment that a child wants to adapt to their peer group, while in developing countries, children's toys are seen as an increase in value relative to the environment; In a parent, a child's consumption raises the child's position to success so that the parent feels successful as an educator [3]. Childhood conditions in later life can predispose to impulsive and short-sighted consumption habits [4]. The history of the UN Convention on the Rights of the Child concerns the child's ability to "want". Young people and children are more vulnerable consumers if their financial skills and financial management are not up to date [5].

Online stores are part of the hidden advertising and enabler of social media. Social media and various free applications constantly collect user data and target marketing correctly. Indeed, mobile platforms have enabled and diversified the concept of e-commerce [6]. When children and young people get reflections from the online store, it affects the general consumption behavior of the family. In family purchasing decisions, children play an important role directly and indirectly in purchasing decisions [7]. In ecommerce, purchasing decisions were influenced by ease of shopping, social impact and facilitation [6]. According to a (2003) reaching by Thomas & Lang, young people search for goods on the Internet because they are more individual and thus more exclusive. The drop in per capita purchases of more expensive goods increases as the possibility of a quick loan and low monthly payments allow for higher consumption and thus the importance of learning financial skills increases [8].

3 On Micropayments About Consumption Habits

Micropayments are fast payments that are discreetly made on mobile platforms in general or as small purchases. The goal of purchasing mobile games can be internal or external to the game. The goal is to achieve a faster and better position in the community in the game. The game has become one of the big social medians. In the world of youth and children, trading platforms are social media, and their learned standard of living connects youth and children globally. Regardless of the starting point, people should always strive for an accepted position in relation to the environment, even virtually. In the virtual world, children and young people are learning to use micropayments, leading to a habit of rapid consumption. Adults justify small purchases by improving the child's socioeconomic status and ethical consumption relative to the good. In reality, the use of money is thus consumption-centric without buzzing. With the help of mobile phones, small purchases have increased the total number of bills and thus the use of children and young people's money in customs on the subject [3] Problems arise if he does not understand budgeting, i.e. how to control spending relative to the money spent [2].

Online stores are part of the hidden advertising and enabler of social media. Social media and various free applications constantly collect user data and target marketing correctly. Indeed, mobile platforms have enabled and diversified the concept of e-commerce [6]. When children and young people get reflections from the online store, it affects the general consumption behavior of the family. Rantala et al. (2020) study also shows the importance of online stores in general as places to shop and their position in relation to the economy. In family purchasing decisions, children play an important role in purchasing decisions directly and indirectly [7]. In e-commerce, the purchasing decision was influenced by the ease of shopping, the social impact and the facilitation of conditions [6]. According to a 2003 study by Thomas & Lang, young people search for goods on the Internet because they are more individual and thus more exclusive. The drop in per capita purchases will increase as the possibility of quick loans and low monthly payments allow for higher consumption and thus the importance of learning financial skills will increase [8].

With games being their own social world where everyone spends time, there is a need for the child to spend time playing. It can be said that games are the television of modern society. The experiences and opportunities it brings are shared. Within games Children engage in a changing economy in which the child needs to donate goods, money or services to create and strengthen social relationships [3]. The purpose of games is to attach the player to the social environment so that when the player stops playing, his social environment also disappears [9]. Online social platforms and gaming platforms not only sell products, but also create an environment where children and young people can participate in sharing economics in everyday life [3]./renewal, improving competitiveness, tailoring characters or games, or giving potential gifts [8]. With small payments, the player improves his own position in relation to others and the low prices blind the player to the use of money, so the player does not even notice the cash flow. There may also be virtual betting in games, as in many countries betting on money is criminalized and virtual objects and currency are used instead of money, allowing the gaming company to make a profit on the proceeds [9].

4 Interview

Based on the picture describing the change in the expenditure of young people and children, we made precise criteria for the thematic interview of young people and children. The study found that it was very transparent, with no pressure or challenges in the current money debate. The aim was to obtain a research result on how young people and children experience money and consumption. The emphasis in the study was on the name of your own under-18s. The young person and the child were given the freedom to talk about money, and the research was deepened with additional questions from the young people and thus the young person was allowed to bring an open view to the interviewer. At the same time, the interviewees were allowed to ask themselves about the money planned for things and deepen their own contribution. Discussions also led to insights and information mapping. No time pressures or goals were set for the interviewees, but the situation of discussion and interaction was justified, which made it easier to talk about and reflect on money in everyday life from an early age. This allowed for an understanding of the whole and an open conversation with the young person without boundary conditions.

Based on the thematic interview, it can be argued that the use of money by young people and children, regardless of cultural identity, is highly consumption oriented. The respondents' view of the purpose of money was precisely consumption. Everyone was basically asked what you think about money. Many young people even think that money is an important part of spending, and the younger people went, the more common the answer became, "money makes everything nice". This all-nice thing was a garment, toy or other consumer good or thing.

In an everyday encounter from the mouth of a pre-school child, the word money became "BMW is the secret of happiness". The child had thus reflected the value of money in relation to the parent's goals. This also came to the fore in the spending of young people's money. The focus of young people on what they get with money came largely from the atmosphere and goals of the home. About how money and the purpose of money had been taught in the home. The older the respondents went, the more influences came from outside the home, but entry-level morale was tied to the home, like "I'm saving money on a new phone now because it's released in the first month and I have it first, the other mom said you wouldn't save for independence".

The results of the interview also clearly show the impact of social media on consumption habits. Interviewees included living standards ideas through social media. In addition, the attitude towards the environment decided what type and size of things the young person was willing to spend money on. Just like home values, ecological change shifted consumption only to different products. On the other hand, many young people did not consider the money spent on games or other micropayments to be consumption. "It was just such a small, nice addition." Together with a few interviewees, we counted these "little ones" into one whole, which helped young people and children understand the amount of consumption. However, the reason for the behavior to continue was that "you understand that my position in the game will not weaken and I will not achieve my goal." Among the interviewees, the players considered the use of micropayments reasonable in relation to the goal. In other social media, pastimes saw glamor as a way of life and luxury as a more important consumption goal. Likewise, more expensive products and glamor were considered ethical and ecological and thus an acceptable way to consume. Consumption itself was life.

Many socioeconomically weaker participants had the desire and understanding to learn how to use money if parents had actively sought to teach it to them. For young people and children whose money was not discussed at home or whose parents were already in a weak financial situation themselves and were unable to manage the finances, it was difficult for the interviewees to see their own chance of succeeding in managing the finances. The interview revealed more that the level of home education in itself did not affect the young person's opportunities, but rather the use of money and talking about it openly.

On the contrary, the financially better off children and young people who were talked about money had only worries about their other peers. On the other hand, the typical economic situation at home was seemingly good at home, but there was not much talk about money for key consumer and even financial difficulties. This group saw money as an enabler of the socio-economic status. As an opportunity to be in a better position in relation to the possible. Those young people, typically talking about money, saw money again as a possible thirst and future focused. All the young people are ready to learn from money in the end when they realized their own potential. Motivational learning and storytelling from the interviewer helped here.

Interviewees see saving more as goals for higher consumption than owning their own home. Saving was seen as central to the pursuit of the greater cause. Many people think that taking out a loan is not a bad decision in relation to a better position in their own environment. A small loan is only part of a normal economy. This was reflected especially in young people where the use of money in the home environment was consumption-oriented and parents had instalments.

Social media and cleaved prices for ads appear as cultures of opportunity and a normal part of achieving a goal. Some young people understand that hard interest-bearing loans should not be given, but banks accept the existence of consumer credit as part of the economy and loans for financial management. Many young people are unable to borrow car finance, mobile phone payments, billing and other financial arrangements, but are large financial arrangements because mobile phones can be of a certain type to maintain their own socio-economic status relative to other young people.

The interviews also revealed the difficulty of talking about money. Talking about money was originally considered a personal matter. This was also inside them, where money was talked about at home. Young people and children, in their own words, talk openly to each other about money, but not with adults. Adults don't reportedly want to. Through sharing the interviewer's experiences, young people and children are able to open up their own views on the issue. The most insightful insights made through learning and discussion happened precisely through reflection. Indeed, the young people interviewed felt able to talk about money and financial openness to experience experiential expertise. Indeed, young people and children felt that they were taught and talked about too little about money. Children and young people also felt that education in economics and money was too theory-based and materially old-fashioned. Indeed, the young people who had been talked about the money had taken the initiative to seek information about the money and the possibilities for saving.

Part of the research in the interviews was conducted in a group. In this way, the young people in the group were allowed to reflect their own skills and knowledge in relation to each other. Young people found it easier to approach money in terms of the experience and understanding of another young person. Opportunities in relation to one's own possibilities thus arose as an insight into the experiences of another young person. The discussions were deepened mainly by the solution of key questions from the interviewer who helped the young people to advance their knowledge skills. Based on these discussions, the theory was created and strengthened by learning by teaching.

5 Functional Teaching of Money

Research and theory highlight the acquisition of children's and young people's knowledge of the socio-economic environment. The young person and the child compare themselves and their opportunities in relation to other people in the socio-economic environment. This particular feature is utilized in the teaching of finance. In this way, the learner is able to compare their own possibility and become better understood when unnecessary identity scripts brought by age groups are left out. It can be said that this creates more topical and reflective learning situations for young people and children. The Fig. 1 below describes how to teach how to understand and invest the value of money.

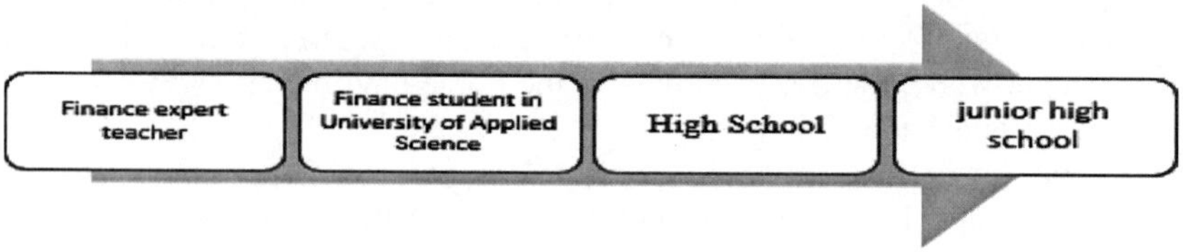

Fig. 1. Finance teaching model

The Fig. 1 shows how teaching becomes a process that takes advantage of the growth of free will and identity, providing an opportunity for natural communication between student and teacher. In this way, the teacher is not in itself a greater authority but part of the community, which makes it easier for the student to understand the diverse purpose of funding through examples. Teaching is thus not too learning-centered but interactive. Teaching must not be teacher-led, in which case it is not possible for the student to develop into learning, but the situation becomes performance-based. [12] In this way, the student can question what has been learned so that he or she gathers extensive knowledge and understanding of finance. In the next step, the economics student teaches the subject to high school students. At this point, the teaching student creates value for themselves and can apply the knowledge to their environment. A high school student will have a better understanding of learning when there is likely to be a smaller difference between the pupil and the student using the scripts brought by the socioeconomic environment and age group. On the other hand, a high school student gathers to question teaching and teaches high school how to use value for money. In secondary school, it is easier to position oneself in relation to a high school student and his or her socioeconomic class.

Figure 2 shows the learning model used in this study. The model serves as a general frame of reference for learning by teaching method which is used during the project.

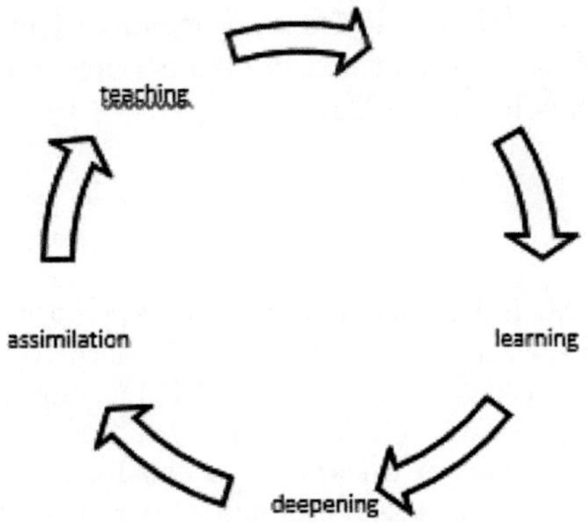

Fig. 2. Learning cycle

The figure above demonstrates how a student deepens what he or she has learned so that he or she can teach knowledge further. After learning, continuous learning is a must-have in order to be able to teach with current knowledge. By deepening the learning-after, the student is able to answer questions about the teaching situation. After theoretical reference development, the student's ability to equate knowledge with that around him or her helps both the teacher and the learning student to observe and construct their learning. In the last stage, the student is able to apply what he or she has learned to the environment, this stage occurs when the student realizes that he or she is an expert in relation to what he or she is teaching. The child's analog reasoning is based on observations in which the child assimilates new information onto the old data structure [11].

6 Summary

The change in young people's and children's consumption behavior and understanding of saving is rapid and globalized. The pursuit of luxury and the globalization of consumption have also brought about a rapid change in household indebtedness and thus in the national economy. Increasing socio-economic status because of the glorification of money has made managing the economy more consumer centric. Outdated textbooks in schools and a lack of material largely contributed to the loss of interest in the motivation to learn to save and invest.

The aim of this research and related project is to create and develop a model and learning method for teaching economics to children and young people. A key approach to the method is to lower the learning threshold at all levels of education involved in this research and teaching project. A method in which young people teach young people lowers the threshold for interaction and exchange of ideas and promotes learning. At the heart of the implementation is a teaching method in which young people teach young people, which lowers the threshold for interaction and promotes exchange of ideas and learning. Based on the material produced by the practical part to be implemented, it is possible to further study the possibilities and development of the model and the reproducibility of the created model in different environments. The project, which includes a practical teaching project, was launched in collaboration with the local chamber of commerce and its members in early 2021.

References

1. Rantala, J., Yllikäinen, M., Holopainen, T.: Competences in expert sales. In: Kantola, J., Nazir, S., Salminen, V. (eds.) Advances in Human Factors, Business Management and Leadership. AHFE 2020. Advances in Intelligent Systems and Computing, vol. 1209. Springer, Cham (2020). https://doi.org/10.1007/978-3-030-50791-6_16
2. Luukkanen, L., Raijas, A.: Nuorten näkemyksiä taloudellisesta osaamisesta (2012)
3. Ruckenstein, M.: Lapsuus ja talous 2013 Helsinki: Gaudeamus OY (2013)
4. Mittal, C., Griskevicius, V.: Sense of control under uncertainty depends on people's childhood environment: a life history theory approach 2014
5. Martikainen, L.: Mitä se sulle kuuluu? Nuorten elämän valinnat puntarissa. Artikkeli 5 Peura-Kapanen, L. 2015. Nuorten talouden hallinta. EU: United Pree Glopal 2015 (2015)
6. Bucko, J., Kakalejčík, L., Ferencová, M.: Online shopping: Factors that affect consumer purchasing behaviour (2018)
7. Thomson, E.S., Laing, A.W., McKee, L.: Family purchase decision making: exploring child influence behaviour (2007)
8. Pankki, S.: Talouslukutaito 2020-luvulla.. Artikkeli 18 Peltonen T 2018. Mikä ihmeen talouslukutaito? 57 (2018)
9. Määttä, M., Nuotilta, E.: Opas peli liiketoimintaan Oulu 2016: Oulun ammattikorkeakoulu (2016)
10. Mäyrä, F., Karvinen, J., Ermi, L.: Pelaajabarometri 2015. Tampere: Tampereen Yliopisto (2015)
11. Nurmi, J.-E., Ahonen, T., Lyytinen, H., Lyytinen, P., Pulkkinen, L., Ruoppila, I.: Ihmisen Psykologi. Helsinki: WSOY oppimateriaalit (2006)
12. Lahdes, E.: Peruskoulun uusi didaktiikka Helsinki; Kustannusosakeyhtiö Otava (1997)

A Comparison on the Development Mode of Traditional and Emerging Cultural Innovation - A Case Analysis of Electronic Sports-League of Legends

Xinyao Huang and Wei Ding(✉)

School of Art Design and Media, East China University of Science and Technology, Shanghai, China
heyyo.hxy@icloud.com

Abstract. In recent years, the cultural and creative industry(CCI) in mainland China has been developing rapidly. At present, CCI focus more on traditional and regional culture. However, with the improvement of cultural industry policies, the future development of CCI should not only realize platform and technology upgrading, but also realize content upgrading. This paper selects e-sports culture as the research object. Adopts the research methods of the comparative study, from the "Image" "Star" "Object" "Field" four dimensions analysis the differences between traditional and emerging CCI, this research mainly take League of Legends (LOL) as an example, compared to traditional CCI. In culture mining, IP creation, product extension, marketing methods and other aspects separately, compare the similarities and differences between two kinds of CCI.

Keywords: Electronic-sports culture · Traditional culture and innovation · IP · League of Legends (LOL)

1 Background

Recent years, China's CCI industries had developed rapidly. From various surrounding areas to the game industry, from historical culture to corporate culture, IP exploitation has gradually formed a set of mature system. It is inevitable that the public will turn to cultural consumption, and the value of design will have more space to play in the field of CCI [1].

Therefore, this paper selects e-sports culture as the research object. It is worth mentioning that the current development of the entire CCI industry also advocates grafting cutting-edge cultural and creative industry resources, realizing the coordinated development of cultural and creative design, supply chain management and service operation, thus forming the basic model of integrated development of cultural and creative IP field.

R. S. Goonetilleke et al. (Eds.): AHFE 2021, LNNS 273, pp. 421–426, 2021.
https://doi.org/10.1007/978-3-030-80713-9_53

2 Comparative Study

The contrast point of this paper is the "Image", "Star", "Object" and "Field" created in cultural creation design. Image-making refers to digging the connotation of culture, keenly sensing the cultural vein of the whole city, and finding some impressive visual language among the seemingly vast city and cultural materials [2]. Star-making refers to creating a fitting IP. The core is to activate cultural or historical resources so that IP can better serve CCI industries. Object-making, relatively easy to understand, is to create a physical or virtual cultural products. It is the real communication bridge between cultural creation and our users. Field-making, is the last step of the text and design, is to create consumer scenarios. Including online and offline, scenic spots or some service contact, triggered by the user purchase behavior, it is also the most crucial step in CCI market. A really good design should be good with scene matching products, planning a very complete interactive chain, make the product and the business community.

3 A Comparative Study of Traditional and Emerging CCI in Four Dimensions

3.1 Image-Making

Traditional cultural creation is mainly based on a relatively concrete culture. To make a traditional cultural creation first is to study and analyze the culture it relies on. By studying the local culture, we can make use of some local specialties or intangible cultural heritage skills to recreate.

Country revitalization project of Tongji university, for example, through the local investigation and research, we found "Chongming rice" and "Pole Play" the two breakthrough point [3]. One is a local specialty, the other one is a local intangible. We make the "Rice milk series" food and cosmetics packaging, and refining pole drama characters and complete a set of VI design. Through the design of this set of visual scheme, expand the influence of Chongming rice culture and Pole Play (Fig. 1).

Fig. 1. Cultural and creative products of Chongming rural revitalization

There is no culture on this savage planet, because people have created it, it exists. From the perspective of culture portrait, the development time of e-sports in China is not long. However, from the perspective of the development breadth of e-sports culture, there is no doubt that no other culture can match the speed of it. E-sports culture can also be divided into two categories:

First, there are the empty heroes and story lines in e-sports. Take LOL for example, Riot Games created a summoner's canyon world. He created a complete game background, and these stories can make you feel that there is a certain interaction between characters and players in the game universe. The characters in these games like real people flesh and blood.

Second, the theme of e-sports is people, and the core of e-sports culture is players [4]. LOL can go so far, not by a person or a team, but by the joint efforts of countless people who love LOL and love e-sports. This love is worthy of respect. As users said, because of love, so alliance.

Traditional image-making relies more on the existing specialties of the region and city. But the image of e-sports is to construct a parallel world with flesh and blood, love and hate. E-sports culture not only refers to the love and hate of the heroes created, but also more about the personal character and charm of the players shown in it.

3.2 Star-Making

We have entered an era where IP empowers industries. For example, the series of "The Glory of Time" of Xinhua Net created five IP images of mythical animals through the study of Jingdezhen porcelain culture in Jiangxi Province. Each mythical animal represents a unique personality with its own story and a complete vein of a certain kind of ceramics. By means of multimedia video exhibitions and cultural and creative derivative products, the young consumer groups can better pay attention to, know, understand and perceive porcelain. This IP makes the traditional porcelain making technology, traditional products and traditional culture re-enter our life through new forms of display and new IP (Fig. 2).

Fig. 2. The Glory of Time IP creation

E-sports IP can be divided into two categories:One is a virtual character's IP build: mention LOL, for example, I might be the first reaction is to Galen's "man live,the tower exist". The hero's skills and personality have become the game's star IP.

The other one is a player star IP, mention LPL, had to mention IG the teams, 2018 IG win the S8 LOL finals champion. This is the first in the history of the China division champion, is also broke the monopoly of South Korea's Europe North America countries, during the game attracted many new young people pay attention to the LOL. We this generation of young people learn from LOL is to constantly learn lessons, continuous progress, and constantly tamp the foundation, fill up the short board.

The IP building of traditional culture relies on cultural mining, finding the angle through the characteristics of culture, so as to create an IP tailored to local characteristics. In e-sports culture, IP not only represents the quality and value of the game itself, but also the combination and sublimation of the spiritual core of the game and cultural elements. It is the foundation for a game to maintain long-term vitality. Riot Game has talked more than once about turning LOL into a Marvel IP, and they're definitely working on it [5].

3.3 Object-Making

The products of traditional culture can also be divided into two kinds. The first kind is local specialty. For example, Songxi, Fujian Province is the producing area of sugarcane, so we can make use of the advantages of sugarcane origin to build the centenary sugarcane into a local characteristic brand. The second is the cultural extension products, such as Dunhuang Museum, which uses the precious treasures left in the Mogao Grottoes to design a series of derivative products through re-creation. Such as bookmarks, brushes, iron clasps, refrigerator magnets and so on (Fig. 3).

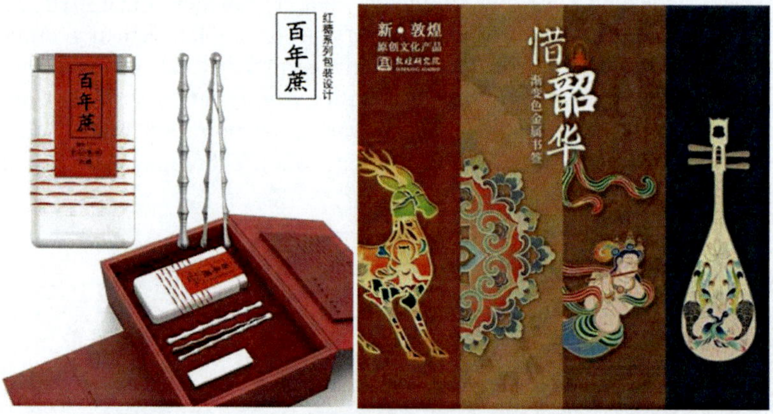

Fig. 3. Songxi and Dunhuang cultural derivative design

E-sports is a special industry, to say his virtual than entity, therefore, e-sports creation can also be divided into two kinds: the first is a virtual product, such as customized game skin, LOL launched every year on the basis of the hero and his story background architecture design of the skin [6]. The second is the actual product. In the narrow sense, it is about the handmade games. With the popularity of e-sports, more offline brands have begun to have joint names with LOL, LV, Nike and other brands have caught up with this trend, and attracted the attention of countless e-sports fans.

Traditional cultural and creative products mainly rely on physical output. They attract the attention of tourists and achieve the effect of cultural promotion and brand building through products. On the contrary, the products of e-sports tend to be virtual, but in recent years, both online and offline have been involved. Take the League of Legends Professional League (LPL) in mainland China as an example, the income from live matches, game peripheral goods, broadcasting and other aspects jointly builds the commercial influence of League of Legends in China and even the world.

3.4 Field-Making

Field-making is to create sales contact points for cultural and creative design, in which users are the ultimate goal of cultural and creative design, and commercial profit is the direct goal of field-making. The business model of CCI design mainly relies on the platform of vertical integration of large cultural tourism industry, such as the Palace Museum, city museums and so on [7]. They create a scene in the scenic spot, the tourists just enjoy the charm of traditional culture, and to provide a platform to buy for everyone. Secondly, relying on platforms such as e-commerce and we-media, traditional cultural creativity promotes local culture through web celebrity effect, resonates design and operation, and builds cultural brands and communities (Fig. 4).

Fig. 4. Cultural and creative shop in the Forbidden City scenic spot

The "field-making" created by e-sports literature is mainly to provide players with a specific buying atmosphere. Combing LOL immersive experience, LOL provide to each player is not only the experience of the game, but also a comprehensive ecological system, the hero alliance will event into a scene type stage model. At the same time start the star plan, in order to attract a large number of e-sports fans, form a different club "Fan culture" [8]. Under the influence of the top competition, the E-sports ecology, club ecology exclusive to the LOL have been gradually formed.

In terms of business model, traditional cultural and creative marketing is relatively simple. It forms a complete closed business loop by integrating supply chain and community new retail operations. On the contrary, the marketing of e-sports cultural creation runs through the entire game-related ecosystem. It's a game, it's an e-sport, it's an Asian

Games event. Like other sports such as football and basketball, Suning and other enterprises sponsor the completion of competition operation and surrounding construction, forming an industrial chain. It is also an important source of income for enterprises and an inseparable part of enterprises in the new era.

4 Conclusion

To sum up, traditional cultural and creative design is based on cultural insight, accurate market positioning and design ability to promote the development of tourism and local industry. At the same time to spread a unique regional cultural image to the public, to achieve the upgrading of large-scale cultural tourism products, and to embody the charm of the city. E-sports cultural creation is a product created in the whole ecosystem with games as the main body and players, competitions and surrounding areas as the extension. This paper mainly takes LOL as an example, compares traditional cultural creation, analysis in cultural mining, IP creation, product extension, marketing methods and other aspects, and compares the similarities and differences between the two kinds of cultural creation. Along with the development of e-sports and radiating surface expands, e-sports culture has also been gradually find the breakthrough point of new transmission relationship between the traditional culture and online games. The rise of online games has greatly changed people's ideas and status, providing a virtual way of spreading traditional culture.

E-sports culture, although it is not an art in itself, but with very strong artistic elements, and provides the design more opportunities, the image of e-sports in the public mind is also slowly changing. Game is not only bringing a moment of relaxation, but wisdom and motivation of the eternal. Every love is worthy of respect, and the future of e-sports cultural creation is foreseeable.

References

1. Li, Z.: Analysis of the Spread of E-sports Games from the Perspective of Heizinha's "Game Theory" (in Chinese) (2020)
2. Tian, Y.: Optimizing Digital Architecture Pattern to Stimulate the Vitality of Cultural and Creative Industry (in Chinese)" (2020)
3. Liu, Y.: Research on the Current Situation and Protection of "Carry-pole Opera" in Xincai, Henan Province from the Perspective of Intangible Cultural Heritage (in Chinese) (2017)
4. Florian, B., Philipp, B., Günter, W., et al.: Motivational Profiling of League of Legends Players. **11**, 1307-1307 (2020)
5. Unity Technologies; Riot Games Chooses to Build Next Games in League of Legends Franchise on Unity (2020)
6. Park, H., Seo, G.: A study on game character design based on mythical motif - focused on league of legends. **20**(12), 2329–2340 (2019)
7. Jin, Y.: Research on the Creative Industrialization Strategy of Traditional Culture in web celebrity Economic Era -- Taking the Cultural Innovation of the Imperial Palace as an Example (in Chinese) (2020)
8. Zhang, Z.: Discussion on the Contemporary Path of Traditional Culture Cultivating Cultural Confidence Based on "web celebrity Cultural Creation" (in Chinese) (2020)

Understanding the Value Rankings of Chinese Middle Class

Wenhua Li and Jiaying Huang(✉)

Guangzhou Academy of Fine Arts, Guangzhou 510000, China

Abstract. This study explored the shared values among the emerging Chinese middle class with List of Value instrument and identifies four of the most important shared values: sense of security, sense of accomplishment, self-respect and well respected. The demographic differences of the Chinese middle class on value priorities were compared. The impact of dominant values on lifestyles and consumption attitudes was studied as well. The dominant value rankings and demographic differences on value preference can be a theoretical reference for designers and marketers.

Keywords: Value · Middle class · Lifestyle

1 Introduction

China has a high cultural context with a long historical accumulation of Confucian culture [1]. Certain values, beliefs, religions, and culture form the collectivistic characteristics of Chinese people, which have been stable for a long time. With economic and cultural globalization occurring in the past 100 years, the value system in China has become a melting pot, mixing value systems from Chinese and western, traditional and modern culture [2]. The uneven development of the two forces leads to ever-changing dominant social values. Eastern culture, individualism and collectivism, Confucius traditions make China unique in marketing and consumer behavior studies. Values are essential in understanding consumers because of its influences on consumer interests and decision making. Therefore, this study intends to examine the dominant values shared among the emerging Chinese middle class.

The studies on value priorities of Chinese consumers are worthy. Furthermore, given that value priorities and lifestyle preferences have significant impact on consumption decisions, understanding the shared values, lifestyles, and desires of Chinese middle class can help marketers precisely position their marketing strategies for potential consumers and obtain in-depth consumer insights.

2 Theoretical Frameworks

Three key theories are adopted in this study: Kahle and Goff Timmer's List of Value [3], Hofstede's five dimensions of national culture [4] and Schwartz's theory of basic values [5].

R. S. Goonetilleke et al. (Eds.): AHFE 2021, LNNS 273, pp. 427–434, 2021.
https://doi.org/10.1007/978-3-030-80713-9_54

List of Value (LOV for abbreviation) is an important value measurement instrument in culture and consumer studies. Several cross-cultural studies used the LOV instrument to rank the importance of values. Results indicate that differences exist in value rankings among different countries. To use LOV instrument in this study enables the researcher to measure value priorities and compare the results with the value rankings of people in different countries.

The cultural dimension of collectivism/ individualism from the theory of Hofstede's five dimensions of national culture [4] will be adopted in this study as supplementary theoretical guide, to explain the value ranking results related to collectivistic/ individualistic cultures in this study.

The value system proposed by Schwartz contains ten universal values from four motivational directions: self-enhancement, openness to change, conservation and self-transcendence [6]. The four motivational directions of values and the characteristics of each direction will be used to explain the possible reasons for value ranking results in this study.

3 Research Method

This study examined the dominant values shared among the emerging Chinese middle class with List of Value instrument (LOV for abbreviation) [3]. The research method of questionnaire was adopted in this study. The questionnaire contains three parts. The first part is about demographic information, including career, educational and family background, and financial situation. The second part includes value ranking; it uses LOV instrument to rank the nine values according to their importance to the respondents. The third part contains lifestyle statements, to collect psychographic information from the respondents. All the questions related to attitudes used a five-point Likert scale to indicate the extent to which an individual would agree with the statements. The language used in the questionnaire is Chinese. The English version of this questionnaire was translated by a Ph.D. candidate student and double-checked by a professional translator. To test the clarity and total comprehensiveness of the questions, a pilot survey was launched to 100 respondents with snowball sampling strategy through the online survey platform. Respondents in the pilot study sent feedback to the researcher, thus helping improve the holistic performance of the questionnaire.

A professional online sampling database was adopted in this research. Potential respondents in this database who met the criteria received an email inviting them to join the online survey. stratified random sampling method was used according to certain stratification criteria, including city of residence, age, income, and educational background. The respondents were randomly selected from Tier-1 cities in China, including Beijing, Shanghai, Guangzhou, and Shenzhen, which are the four largest cities with the largest population base and highest income level and GDP proportions. The age of the respondents ranged from 25 years to 40 years. Their personal monthly income level was above 5,000 CNY (about 770 U.S Dollar). The educational background of respondents should be at least college level. These well-educated middle-class respondents are usually salaried employees who work as white collar, senior managers, IT programmers, or engineers in large companies. The respondents are completely anonymous to the researcher. In total 753 invitation requests were sent with 400 responses with valid completed questionnaires. The online survey platform carefully controlled the distribution of the quota sample volume. Data was processed through SPSS software package version 22. The sample was composed of 400 respondents, 50% men and 50% women, 35.8% aged between 26–30 and 64.2% aged between 31–40. We purposely chose to survey only in this age category because they are more likely to be the emerging middle class. In China, university students usually graduate at the age of 24. All the respondents hold educational certificate of college or above. 69.3% of the respondents have bachelor degree. The respondents are equally distributed in four Tier-1 cities.

4 Results and Discussions

4.1 The Value Rankings of Chinese Middle Class

This study uses the LOV instrument to identify the value rankings in urban China to determine if significant differences are present among the emerging middle class. Nine values were listed in the question, and respondents were instructed to rate the values according to the importance in their daily life. For the rating, 1 is the most important and 9 is the least important. The mean score for each respondent on the nine values was calculated, and the score for each value was divided by their overall mean to indicate the relative importance of the value for the respondents. Value ranking is more important than the mean scores because value priorities, the relative importance of the different values, can guide people's behavior and attitudes, not the importance of any one value [6]. The importance rankings of LOV are shown in Table 1.

Table 1. Important value rankings in everyday life

Important values in everyday life	Means	Standard deviation	Rankings
Sense of security	3.86	2.479	1
Sense of accomplishment	4.21	2.482	2
Self-respect	4.36	2.351	3
Well respected	4.71	2.380	4
Sense of belonging	4.87	2.446	5
Self-fulfillment	4.96	2.371	6
Enjoyment	5.31	2.543	7
Warm relationship	5.47	2.276	8
Excitement	7.25	2.360	9
Valid Number: 400			

Note: Scale: 1 = most important; 9 = least important

The middle-class respondents in this study endorse sense of security, accomplishment, and self-respect as three most important values. The bottom 3 important values are enjoyment, warm-relationship and excitement. Sense of security is selected as the most important value by middle-class respondents in urban China. Chinese people show a strong need for sense of security, and they work hard to acquire enough security. In this survey, 86.8% of the respondents feel secure in their current life. According to Maslow's hierarchy of needs, sense of security is the fundamental need of human being [7]. The respondents endorsed sense of accomplishment as the second most important value in life. Self-respect is ranked as the third most important value in life by Chinese respondents. Self-respect is an individualistic value, which implies personal sense of control and one's dignity, and being well respected is a collectivistic value because it is about the perceptions of others [8, 9]. Advertisements with value theme of well-respected usually depict people being admired and respected by others. Values of self-respect and being well respected are about the maintenance of "face". Saving face means maintaining a good image of the self. Losing face means losing dignity, whereas giving face means paying respect to other people. In China, saving and giving face are important in social relations. The value of being well respected is related to the sense of belonging, which means being accepted and respected in social groups. Sense of belonging is found across cultures. Belongingness to social groups and positions in the social groups likewise help individuals mark their social status. Sense of belonging is one of the most important human motivations. A strong sense of insecurity arises if people feel they do not belong. Self-fulfillment refers to achieve personal dreams, mainly endorsed by young professionals [3]. It shows little difference from sense of accomplishment. Young people desires for personal success, which can be social recognized success, so called "accomplishment", or fulfilling personal dreams, so called "self-fulfillment". Hedonism values, including enjoyment and excitement are ranked as the least important values by middle class respondents, which implies that striving for success, being recognized in social groups, keeping harmonious social relations are more important than personal pleasures.

The scores of value ranking among gender categories are similar without significant differences, except for the mean score of sense of accomplishment. The top 3 most important values for male and female respondents are the same but the ranking is different. Male respondents rank sense of accomplishment as the second most important value, while female respondents endorse self-respect as the second most important value. Mass media in China usually shape the images of "successful man" instead of "successful women", which may form a psychological hint that man should value success over woman. According to Kahle's research in U.S., people who endorse sense of accomplishment are middle-aged successful, well-educated men [3]. Female respondents care about personal dignity and are eager to receive respect from society. This result could be explained as the awakening of female self-perception. Furthermore, female respondents endorse sense of belonging over male respondents. This phenomenon is also identified in the U.S. [3].

The mean scores of value ranking among age categories show few significant differences. The respondents above age 30 show a stronger desire for the values of accomplishment and well-respected by others, while younger respondents show a stronger need for self-respect and sense of belonging.

The ranking differences of shared values among city of residence categories are in the values of sense of security, accomplishment, sense of belonging, self-fulfillment, and warm relationship. Respondents in Guangzhou and Shenzhen share similar value preferences. They express their strongest desire on accomplishment and their weakest interest on enjoyment and excitement. Respondents in Shanghai and Beijing consider sense of security as the most important value, followed by self-respect and accomplishment. Respondents in Beijing value being well respected more than people in the other three cities.

The ranking differences of shared values among education level categories are significant. Respondents with college degrees prefer collectivistic values. These respondents consider security and being accepted and respected by groups very important. The respondents with master degree endorse sense of accomplishment as the most important value. The respondents with doctoral degrees are likely to take individualistic values as more important. These respondents rate self-fulfillment, self-respect, and accomplishment as the most important values in life. After receiving higher education and cross-cultural knowledge, higher educated people have the chance to reach the top need of self-actualization, which is identified in Maslow's hierarchy of needs [7].

The characteristics of professions influence the preferred values. The ranking differences of shared values among profession categories are significant. Self-employed individuals, professional executives, teachers, and engineers place sense of accomplishment as the most important value in life. Their work is usually result-oriented, which could explain their strong desire for accomplishment. White collar respondents, middle-level managers, and officers rate sense of security as the most important value. Their jobs are usually stable without too many changes. Senior managers, programmers, and IT practitioners consider self-respect as the most important value in life. Salespersons desire for self-fulfillment most.

4.2 Impact of Dominant Values on Lifestyles of the Middle Class

The emerging Chinese middle class is not similar to the middle class in western countries because of different government policies. The social safety net, medical insurance system, and public education services are closely related to the daily life of the middle class. To some extent, middle-class people are vulnerable to financial freedom issues. Without financial stability and freedom, maintaining a decent middle-class life will be difficult. For the new Chinese middle class, maintaining their present social status is not easy. By contrast, compared with the middle class sticking to their current life, upper-class people pay more attention to the future. They feel more confident in their financial situation and are willing to accept high-quality life and invest in stocks, real estate, or financial products. Meanwhile, lower-class people consider more about the current situation to survive in the fast-changing society. They prefer saving, which provides the safest solution and promise for the future. The World Bank China Poverty Assessment [10] finds that Chinese families save a considerable portion of their income even when they are near-poor. The middle class is stuck between caring about both security needs and investment needs. The middle class is saving for house property, medical expenses, educational expenses for children, and retired living. Hence, the middle class encounter difficulty in freewheeling consumption. They desire high-quality life, but worry about the current status quo at the same time. For instance, 86% of the middle-class respondents in this study express interests in financial and investment activities. Many middle class are cautious on their daily spending. To some extent, the middle class still comprise price-sensitive consumers who would be attracted to sales promotions.

Credit card consumption has gained popularity since 2000 in China. Many older consumers are cautious about credit card usage, whereas younger consumers who are heavily influenced by American culture use credit cards more frequently. Around 455 million credit cards were issued as of 2015 in China [11]. Zhou [12] believes that credit cards are the key stimulus for consumer spending and China's economic growth. The Chinese middle class usually use credit cards for convenience rather than borrow money for advanced consumption. For them, "saving first, spending second" is still applicable.

This study purposely selected the new middle class in urban China as the respondents. Among the respondents, 85% agree that daily consumption reflect social status. Furthermore, 81.1% of the respondents believe that fortune and social status are the best criteria to evaluate individual success. The value of accomplishment stimulates strong materialistic desires and conspicuous consumption. Schor [13] argues that middle-class consumers are eager to upscale their lifestyles, practice conspicuous consumption, and compete to acquire status goods. She proposes a new status, which is the disconnection between consumer desires and incomes. A common phenomenon in urban China is that a woman may purchase a Prada handbag that costs her salary of two months.

Advertisements with the theme of accomplishment are tempting for the middle class. These advertisements depict products as necessities for upper-class lifestyles. Given that middle-class people are eager for recognition by social groups and realize upward social mobility, the fast way is to consume and imitate the lifestyle of the upper class. In fact, the main sources and knowledge of upper-class lifestyles are from mass media and advertising [14]. Through imitating fashion, home decorations, and leisure activities of the middle and upper classes depicted in movies, TV series, and advertisements, consumers strengthen their values of accomplishment and social status.

The Chinese middle class have a strong will to become the leading and fashionable consumers around the world [15]. An increasing number of Chinese consumers can afford luxury goods and the latest high-tech products. They show strong interests in high-end brands. According to Bain and Company [16], the new Chinese middle class has become more sophisticated and knowledgeable about luxury. They are familiar with global pricing and overseas channels for purchase. Buying overseas has likewise been a trend for years. Although, middle-class consumers indulge themselves with luxury goods, they compare global prices and choose the best deals.

A trend identified in this study is that the Chinese middle class values hedonic values over pragmatic values. They like hedonic advertisements more than informational advertisements. Chinese consumers used to be considered as pragmatic consumers who care for functional needs and best cost performance. However, 81% of the middle-class respondents agree that enjoying life is very important to them. Furthermore, 83.3% of the respondents pursue the life of high quality.

5 Conclusion

The dominant value rankings and demographic differences on value preference are examined in this study. The most important values shared by the Chinese middle class are sense of security, sense of accomplishment, self-respect, well respected, and sense of belonging. A trend of dominant social values from collectivism to individualism is identified. Both male and female respondents show endorsements on individualistic values, such as accomplishment and self-respect. The younger middle class endorses individualistic values over collectivistic values, whereas the older middle class endorses collectivistic values over individualistic values. The respondents with a college degree endorse values related to fundamental needs. The respondents with higher education levels are likely to pursue material and spiritual success, instead of security needs.

The impact of dominant values on lifestyles and consumption attitudes is examined, which shows that the Chinese middle class interests in status and conspicuous consumption. This finding is consistent with the values of accomplishment and social status. Moreover, hedonic consumption has become popular among the Chinese middle class. Enjoying life is considered important for the Chinese middle class, a perspective that is different from tradition. The dominant value rankings and demographic differences on value preference can then be a theoretical reference for designers and advertisers. Paying close attention to culture and value changes will be useful in making precise marketing strategies.

References

1. De Mooij, M.: Global marketing and advertising. Sage Publications (2013)
2. Cheng, H.: Toward an understanding of cultural values manifest in advertising: A content analysis of Chinese television commercials in 1990 and 1995. J. Mass Commun. Quart. **74**(4), 773–796 (1997)
3. Kahle, L.R., Kennedy, P.: Using the list of values (LOV) to understand consumers. J. Consum. Mark. **6**(3), 5–12 (1989)
4. Hofstede, G.: Culture's consequences: International differences in work-related values, **5**. Sage (1984)
5. Schwartz, S.H.: Universals in the content and structure of values: theoretical advances and empirical tests in 20 countries. Adv. Exp. Soc. Psychol. **25**, 1–65 (1992)
6. Schwartz, S. H.: An overview of the Schwartz theory of basic values. *Online readings in Psychology and Culture*, 2(1), (2012)
7. Maslow, A.: Maslow's hierarchy of needs. Salenger Incorporated (1987)
8. Homer, P.M., Kahle, L.R.: A structural equation test of the value-attitude-behavior hierarchy. J. Pers. Soc. Psychol. **54**(4), 638 (1988)
9. Kahle, L.R., Rose, G., Shoham, A.: Findings of LOV throughout the world, and other evidence of cross-national consumer psychographics: introduction. J. Euromarketing **8**(1–2), 1–13 (2000)
10. World Bank China Poverty Assessment: China - From poor areas to poor people: China's evolving poverty reduction agenda - an assessment of poverty and inequality (2009). http://documents.worldbank.org/curated/en/816851468219918783/China-From-poor-areas-to-poor-people-Chinas-evolving-poverty-reduction-agenda-an-assessment-of-poverty-and-inequality;jsessionid=W3AyYgusimg4THQaAbrLXgJ3. Accessed 28 Oct 2020

11. Qin, F.: The credit card industry report 2015 (2015). http://news.51credit.com/woaika/10762564.shtml. Accessed 20 Oct 2020
12. Zhou, F.: Credit cards are key for stimulating China's economic growth (2013). http://www.kapronasia.com/china-payments-research/credit-cards-are-key-for-stimulating-china-s-economic-growth.html. Accessed 28 Oct 2020
13. Schor, J.: What's wrong with consumer society. Consuming desires: Consumption, culture, and the pursuit of happiness, 37–50 (1999)
14. Schiffman, L.G., Wisenblit, J.: Consumer behavior. 11th Edition. Pearson (2014)
15. Kharas, H., Gertz, G.: The new global middle class: a cross-over from West to East. Wolfensohn Center for Development at Brookings, 1–14 (2010)
16. Bain and Company: Mainland China's luxury spending continued its decline in 2015, however, emerging signs signal a reversal in 2016 (2016). http://www.bain.com/about/press/press-releases/China-Luxury-Report-2016-press-release.aspx. Accessed 28 Oct 2020

The Concept, Development, Evolution and Practice of Poverty Alleviation Design

Jie Zhou(✉), Wei Ding, Yuyao He, Yiran Zhang, Yisha Wang, and Xinyi Yu

East China University of Science and Technology, 130 Meilong Road, Xuhui District, Shanghai, China

Abstract. Based on the background of East Asian culture, history and economy, this paper summarizes the concept and motivation of design poverty alleviation. Based on the logic and mode of China's design poverty alleviation action, it shows rich design poverty alleviation practice cases. From the perspective of sharing design cultural resources and sustainable development, this paper proposes effective ways for design to improve the lifestyle and economy of poor areas. Hope to be able to achieve balanced development of regional economy in the future to make a contribution to design.

Keywords: Design poverty alleviation · The concept, development, evolution · Transformation and upgrading

1 Introduction

The year of 2020 is the end of China's fight against poverty. The road of poverty alleviation in China can be divided into the following four main stages: Firstly, the broad concept of poverty alleviation was put forward under the planned economic system from 1949 to 1977. Secondly, driven by structural reform, China carried out a large-scale poverty alleviation from 1978 to 1985. The third is the development-oriented poverty alleviation during the period of rapid economic growth from 1986 to 2000. The fourth is the development-oriented poverty alleviation in the process of building a moderately prosperous society in an all-round way proposed since 2001 [1, 2].

According to the focus of poverty alleviation, it can be summarized as development-oriented poverty alleviation and targeted poverty alleviation [3]. The former relies on economic growth to promote poverty reduction, which is agriculture-driven, industrial-driven and state-led, planned and targeted development-oriented poverty alleviation practices in rural areas. According to Multidimensional Poverty Theory proposed by Sen, income is the direct dominant factor of poverty, but is not the cause of poverty [4].

In the latter stage, poverty alleviation cannot be achieved by relying only on economic growth, so "targeted poverty alleviation" emerges as the times require. Targeted poverty alleviation is an accumulation of policies in the process of poverty alleviation, rather than a substitute. At the same time, the original poverty alleviation measures will continue to play a role.

R. S. Goonetilleke et al. (Eds.): AHFE 2021, LNNS 273, pp. 435–441, 2021.
https://doi.org/10.1007/978-3-030-80713-9_55

Asia is in the midst of the world's fastest economic and social transformation. Rapid urbanization and industrialization have brought increasing mobility. At the same time, developing countries in Asia have gradually developed from poverty and underdevelopment in the past decades. Infrastructure, public services and social governance are far from being able to cope with the various crises caused by the rapid economic and social transformation. From the perspective of East Asia, the focus of this research on the design for poverty alleviation is not to increase income, but to provide the path of industrial transformation through design. Intellectual support is the premise of poverty alleviation, and then income can be boosted by providing blood transfusion to poor areas and their own hematopoietic capacity. From the perspective of how to get rid of poverty, this paper focuses on providing a general framework and effective strategies that can be duplicated, promoted and achieved, highlighting how design, as a purposeful activity, provides intellectual support for poverty-stricken areas through service and system design.

2 Research Status

From the perspective of discipline, domestic research on design poverty alleviation mainly focuses on agricultural economy, architecture, Party history and Party construction. Since the theory of design poverty alleviation comes from rural poverty alleviation work and later targeted poverty alleviation policy, until 2018, the Ministry of industry and information technology further put forward the concept of design poverty alleviation, combining design activities with poverty alleviation direction, design poverty alleviation can be divided into the following centralized categories: Macro evaluation mechanism. Case Study. This kind of works focuses on regional characteristics, but its universal applicability needs to be considered. The design of poverty alleviation projects has also been concerned in the international community. Many countries are constantly trying to find ways suitable for their own countries. Generally speaking, the design of poverty alleviation projects in the international community are mostly from the perspective of environmental protection, personal safety and social vulnerable groups At the same time, we can find that the poverty alleviation projects in various countries are carried out in combination with their own national conditions and characteristics. Allen J. Scott. Cultural-products industries and urban economic development (2004) [5] and other experts elaborated the design of poverty alleviation from the perspective of market mechanism; Ann Markusen (2006) [6] and other experts mainly analyzed urban development and political policy; Karenjit Clare (2013) elaborated how to use and transform regional resources [7]. Foreign research on design poverty alleviation mainly focuses on the renewal of industrialization and post industrialization era, based on the perspective of urban planning.

Domestic research on poverty alleviation by design started late. We can learn from foreign experience. At the same time, it should be in line with the national conditions. At present, we need more industrial upgrading and transformation as a foothold to carry out poverty alleviation by design. The domestic practices of design for poverty alleviation are mainly reflected in the following aspects: first, it emphasizes the concept of "co-creation" of designers and villagers in the process of design for poverty alleviation [8]. The second is to emphasize that design provides sustainable support for the industry, deeply excavate the cultural characteristics of ethnic minorities, and make further design innovation and empowerment. The third is to emphasize industrial upgrading and transformation through design, and inte-grate design into regional industrial development plan. For example, Ding Wei and Cheng Jianxin put forward the concept of building a county by design, taking the county as a unit, driving the integration and upgrading of the primary, secondary and tertiary industries through design empowerment, and on this basis, they further put forward ten models of building a county by design, which has a good guiding effect on the theory and practice of design poverty alleviation [9].

3 The Concept and Development of Design for Poverty Alleviation [10]

With the development of poverty alleviation in breadth and depth, higher requirements are put forward for poverty alleviation. In 2013, the concept of "targeted poverty alleviation" was first introduced. Two years later, the Implementation Plan for the Establishment of a Targeted Poverty Alleviation Mechanism was formulated, and nationwide targeted poverty alleviation was launched in an all-round way.

In sorting out the historical context of the development of China's poverty alleviation policies, the most basic experience is to formulate policies and measures in accordance with the national conditions in different historical stages, and find a path of poverty alleviation and development with Chinese characteristics. Focusing on severe poverty-stricken areas and aiming at improving product quality, promoting residents' living conditions, developing the characteristic cultural industries of rural areas and upgrading characteristic advantageous industries, Design for Poverty Alleviation fully simulates mobilizes the enthusiasm, initiative and creativity of organizations in the design sector and enterprises in the industry to share design concepts, provide design solutions, cultivate design talents and stimulate the internal power for development for rural areas. In April 2018, the United Nations Industrial Development Organization, the Ministry of industry and information technology of the people's Republic of China and China Industrial Design Association jointly launched the "design for poverty alleviation initiative". The Ministry of Industry and Information Technology released a three-year action plan for poverty alleviation (2018–2020) on August 13, 2018. The plan also put forward a number of major actions such as regional brand creation, implantation of scientific and technological achievements, talent training can assign, intangible reengineering activation, natural studies education, beautiful rural culture, rural community construction, cultivation of characteristic industry, love for special groups of people and low-end industry transformation, to enhance the level of product design in poverty-stricken areas and improve people's life quality in poor areas.

4 The Feasibility Principle of Design for Poverty Alleviation

Design is a way of thinking and a set of concept system committed to positively changing the world. It is also a methodology and a feasible system for innovative exploration.

Design thinking aims at constantly improving people's quality of life, and carries out creative design and practice in accordance with cultural ways and methods. Due to the characteristics of its methods and thinking, design is destined to play an important role in poverty alleviation. In the practice of design for poverty alleviation, we should do a more systematic and in-depth research and exploration in the following aspects, and integrate resources to help fulfill the expectation of design for poverty alleviation design.

Adjust one's measures to local conditions. It is necessary to conduct a more detailed investigation and excavation of the implementation sites, and build a collaborative promotion system of disciplines and industries in different industries, such as design, sociology, ethnology, geography, so as to give full play to the role of all parties. The protection and utilization of regional cultural creative products is not only a matter of economy, spatial commodity or aesthetic taste, but also the interpretation and expression of regional context [11].

Intelligence input. We should establish the relationship between resource-advantaged areas and poverty-stricken areas. Design talents working in advanced areas can inject their own design resources and design thinking into poverty alleviation destinations through design projects and activities. At the same time, they can also bring the special products of poverty-stricken areas to advanced areas for publicity and promotion. In recent years, with the continuous development of China's design industry, a large number of design industries and talents have emerged in the Yangtze River Delta, Pearl River Delta and other places. However, in the more extensive central and western regions, there is a lack of entrepreneurial design talents and corresponding mechanism guarantee. Due to the numerous ethnic groups and cultural diversity, central and western regions often put forward higher requirements for cultural development and design for poverty alleviation. Therefore, it is necessary to continuously organize design forces to conduct in-depth research in poor areas, and provide corresponding design solutions for their specific design needs. Some characteristic products in poor areas do not have competitive advantages in physical design, sales channels, advertising and marketing, resulting in meager profits. Designers can help local poor people to increase their income by redesigning packaging and formulating feasible marketing plans.

Extensive collaboration. It is advised to strengthen the integration and coordination of a region's resources. Core industries will play a role in promoting the development and upgrading of the whole region. And it is also important to enhance the collaboration between different institutions and departments, give full play to the role of design enterprises, universities and research institutions to guide local governments and relevant units to fulfill their social responsibilities, and to work with leading enterprises, e-commerce platforms, logistics and other fields.

5 A Case Study: Design for Poverty Alleviation to Help Regional Brand Building

Songxi County is located in the north of Fujian Province, with many hills, beautiful environment, inconvenient transportation, low agricultural output value and weak industrial foundation. It is a key county for poverty alleviation and development at the provincial level. Green tea, porcelain and sword casting are local specialties, but the green tea market is mature and competitive, while porcelain and sword casting are difficult to become the pillar industries of the region due to their small scale and low output. It is urgent for Songxi to find a way suitable for its own development and walk out of poverty under the condition of a lack of abundant products. Through in-depth investigation, we have learned that there is a rare agricultural species in Songxi, namely the sugarcane field with a history of 292 years, which has the longest ratoon life of sugarcane in the world, and is known as the Centennial sugarcane. According to the Institute of Medicine of Shanghai Academy of Sciences, the brown sugar produced by "Centennial sugarcane" has strong resistance to stress. Its extract is rich in 28 kinds of nutrients, such as calcium, iron, riboflavin and nicotinic acid, and contains five times more folic acid than common brown sugar. The rare polysaccharide components in the Centennial sugarcane have many special functions, such as anti-Alzheimer's disease and cancer prevention. They are not only of great significance in agriculture, but also have broad prospects in food, health care products, disease prevention and drug development.

It is not realistic for a poor county to invest heavily in mature industries such as tea and porcelain. It is necessary to find a unique product of Songxi and use it to drive the regional brand. Songxi County Government has established a close cooperation with East China University of Science and Technology and Shanghai Institute of Medicine. The following are what we have done from the perspective of design for poverty alleviation.

The regional brand of Songxi can be established in the form of Cultural Tourism Festival firstly. The local county government used to be the co-organizer of Guling science conference. From the perspective of resource integration, the science conference can be become a part of the tourism festival. The theme of this conference can be developed from the perspective of combining design and science around the development of sugarcane in the past century.

The second is to carry out design competition themed by Songxi. The theme design for Songxi was created by taking advantage of talents from colleges and universities as well as design institutions in advanced areas. Songxi originally had some regional characteristic agricultural products, but because of its lack of competitive advantage in appearance design, sales channels, advertising and marketing, the profit was meager. Designers can help local poor people to increase their income by redesigning packaging and formulating feasible marketing plans.

Thirdly, it is necessary to build a space for the centralized display of Songxi products. Green tea, brown sugar and other bundles appear in this exhibition and sales space, jointly creating Songxi regional brand of a thousand years of Songxi sugarcane. It only focuses on the construction of offline entities but also established flagship stores to promote the sales with the help of new internet media.

The above measures of poverty alleviation have improved Songxi City's visibility, and significantly increased the turnover of sugar and other related products. From the traditional appearance design, such as product packaging, design extension to the creation of categories and brands, updated production and sales models in related industries have produced innovative changes in the design field, and provided strong support for the transformation and upgrading of regional industries and the development of design services industry.

Design for poverty alleviation is a systematic project, in which we should grasp the key links and carry out active and diversified cooperation. Specifically, the first is the role of the government to actively guides and optimizes the industrial policy; the second is the industrial positioning: docking with high and new technology and releasing the industrial vitality; the third is the characteristic resources: digging deep into the characteristic resources and building the regional brand; the fourth is the function construction: emphasizing incubation and transformation, and supporting innovation and entrepreneurship; the fifth is the carrier construction: integrating into the economic transformation and reshaping the regional image; the sixth is t industrial synergy: improving the industrial chain and giving full play to the synergy effect; the seventh is new marketing: diversified investment and financing, new retail formats; the eighth is industry-city integration: urban-rural integration to form a creative new area.

6 The Future Development Direction of Design for Poverty Alleviation

Under the strategy of targeted poverty alleviation, design for poverty alleviation will be deeply integrated with Rural Revitalization Strategy in the future. At present, some pillar industries serving at the county level can more accurately serve specific villages. The rural specialties need to be specifically mine to create characteristic industrial villages, enabling each village to have its own characteristic industry or cultural specialty. explore its own development path suited to its local conditions and achieve a positive and stable development, further promoting the economic development of the county. Secondly, extensive services will be provided for farmers in the future so as to more accurately increase their benefits, stimulate their initiative and creativity, generate a sustainable power for economic growth in rural areas and making these areas wealthier.

Finally, in the future, it will serve the industrial development of less developed areas in a more interconnected manner. The platform of regional support projects is jointly built with the linkage of online and offline, and the design industry, enterprises and regional economic transformation and development are thought together, so as to explore more directions and possibilities for the development of design and industry, and provide strong support for further rural revitalization plan at the theoretical level.

References

1. Chuliang, L.: Economic growth, inequality and poverty in rural china. Econ. Res. J. **2**, 15–27 (2012)
2. Shen, Y., Zhan, P., Li, S.: China's rural multidimensional poverty under the evolution of poverty alleviation policies. Econ. Perspect. **7**, 53–66 (2018)
3. Report of Boao Forum for Asia (2019)
4. Sen, A.: Famines as failures of exchange entitlements. Econ. Polit. Weekly. 11, 1273, 1275, 1277, 1279–1280 (1976)
5. Scott, A.J.: Cultural—products Industries and urban economic develepment: prospects for growth and market contestation in global context. Urban Affairs Review. **4**, 39 (2004)
6. Markusen, A.: Urban development and the politics of a creative class: evidence from the study of artists. Environ. Plann. A. **10**, 1921–1940 (2006)

7. The essential role of place within the creative industries: boundaries, networks and play. Cities. **34**, 52—57 (2013)
8. Ying, J.T.P.: Community and network based design and social innovation. Hunan University (2012)
9. Ding, W., Zhang, Z., Lai, H.: Design and transformation: the development path and the ten models of the design strategy of the country plan —shanghai-yangtze river delta industrial design project outsourcing services platforme design county planning as an example. Design. **7**, 118–120 (2014)
10. Ben, Z., Disha, S.: to define the concept of designing poverty alleviation in three dimensions. Art Life – J. Fuzhou Univ. **2**, 18–22 (2019)
11. Qiao, Z., Ma, J., Huang, J.: Identification of regional cultural and creative elementsand brand remolding in rural tourism. Packag. Eng. **40**, 95–103 (2019)

Towards Better Working Conditions for Visually Impaired: A Pilot Study on Occupational Risk Assessment for Visually Impaired Massage Workers in China

Linghong Li(✉)

Soochow University, Suzhou 215123, Jiangsu, China

Abstract. According to the China Disabled Persons' Federation (CDPF) statistics report in 2016, the Chinese visually impaired population is around 17.31 million, in which 23.5% are under age 30. The vast majority of visually impaired in China are engaged in massage job occupation. The working hours of them are usually long hours per day, and all the work is done by using both hands continuously. The visually impaired massage workers are regularly exposed to risks of occupational injuries and work-related hands' musculoskeletal disorders (WMSDs). With the aim of assessing the occupational risks of visually impaired massage workers, this study was conducted in one city region of China. The Cornell Musculoskeletal Discomfort Questionnaire (CMDQ) was used. This paper provides recommendations according to ergonomics for mitigating the risks identified and better working conditions for visually impaired by fully considering the working environment in China.

Keywords: Visually impaired massage workers · Occupational risk · Cornell Musculoskeletal Discomfort Questionnaire · Hand

1 Introduction

The International Classification of Diseases 11 (2018) classifies vision impairment into two groups, distance and near presenting vision impairment [1]. Vision impairment is commonly referred as blindness and low vision. According to the World Health Organization (WHO) statistics report in 2010, China has the largest visually impaired population in the world [2]. According to the China Disabled Persons' Federation (CDPF) statistics report in 2016, the Chinese visually impaired population is around 17.31 million, in which 23.5% are under age 30 [3].

In China, the job occupations that the visually impaired can be recruited are extremely limited. Massage work is one of the few jobs by which the visually impaired can make a living in China. The overwhelming majority visually impaired are engaged in massage occupation in China. Most of the visually impaired are engaged in massage professions since young age. The China Disabled Persons' Federation (CDPF) has also vigorously carried out massage training for the visually impaired. The massage work is a strenuous

R. S. Goonetilleke et al. (Eds.): AHFE 2021, LNNS 273, pp. 442–448, 2021.
https://doi.org/10.1007/978-3-030-80713-9_56

task. The visually impaired massage workers must stay indoors and use both hands for continuous work all day (Fig. 1 and Fig. 2). The hands' musculoskeletal is hard to get enough rest.

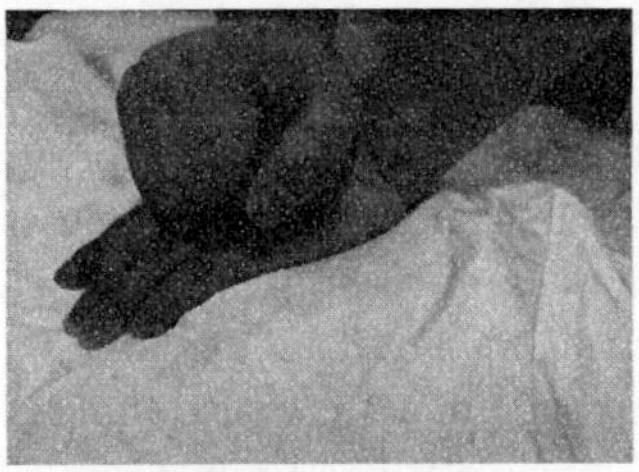

Fig. 1. Hands in use I

Fig. 2. Hands in use II

However, massage is very popular in China. In addition to the visually impaired, there are a lot of people with normal vision are engaged in massage work. To compete, quite many visually impaired masseurs have to work even more harder, such as work overtime. They often work continuously from 9 a.m. or 10 a.m. till midnight, and work on weekends and holidays. Thus, they prolong the time of frequent use of hands every day.

In addition, once the visually impaired masseurs have been engaged in massage work, even if they feel the severe disorder to their hands, they cannot easily switch jobs in China. Only extremely limited number of them can switch to another job. For the most visually impaired massage workers in China, massage job is an occupation they take from young to old, no matter how hard it is. In contrast, if the masseurs with normal vision feel disorder to hands, they may easily switch to other jobs.

Therefore, the problem of hands' musculoskeletal injuries is far more common among the visually impaired massage workers than among masseurs with normal vision in China. Due to excessive exertion for a long time, the thumbs of the visually impaired masseurs are in varying degrees of bending and deformation (Fig. 3).

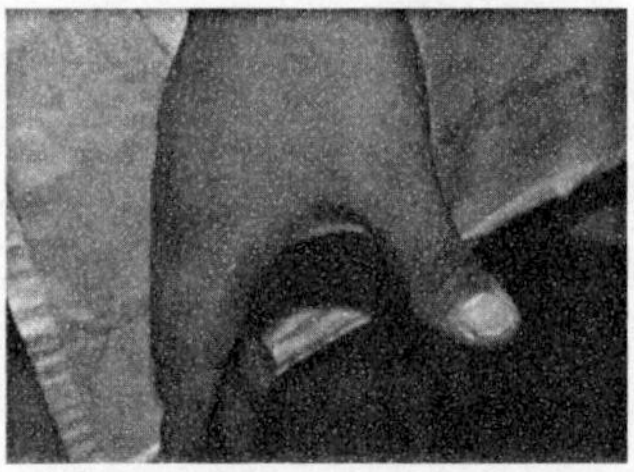

Fig. 3. One visually impaired masseur's finger joints deform after years of work.

2 Objectives of the Study

The objectives of the study are to assess the occupational risks of visually impaired massage workers, and to provides recommendations for mitigating the risks identified and better working conditions for visually impaired by fully considering the working environment in China.

3 Methodology

To determine the factors affecting hands' musculoskeletal disorders for the visually impaired massage workers, the Cornell Musculoskeletal Disorder Questionnaire (CDMQ) was distributed to 40 visually impaired masseurs working in small scale massage shops located in Shanghai. The questionnaire was read out and explained question by question in Chinese by researchers, and the researchers fill in the questionnaire instead of the visually impaired masseurs. The participants consisted of 10 females and 30 males with age ranging from 23 to 38 years (Mean = 25.1 yrs). The working hours and working years were collected.

The Cornell Musculoskeletal Disorder Questionnaire (CDMQ) used in this paper was English version (Fig. 4 and Fig. 5) [4]. In this study, because the respondents who spoke Chinese did not fill in the questionnaire themselves, so the English version did not cause obstacles to the questionnaire survey. Meanwhile, the English version of CDMQ has a positive effect on the presentation of this study.

The hands act as most essential "work tools" of the visually impaired massage workers. As a pilot study, this paper studied the musculoskeletal disorder of both hands. No other part of the body was studied in this paper.

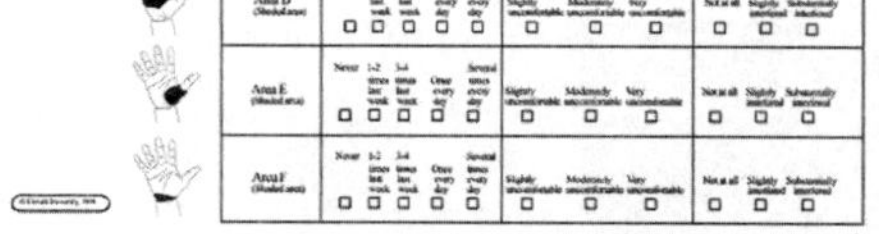

Fig. 4. The Cornell Musculoskeletal Disorder Questionnaire (CDMQ) for right hand used in this paper.

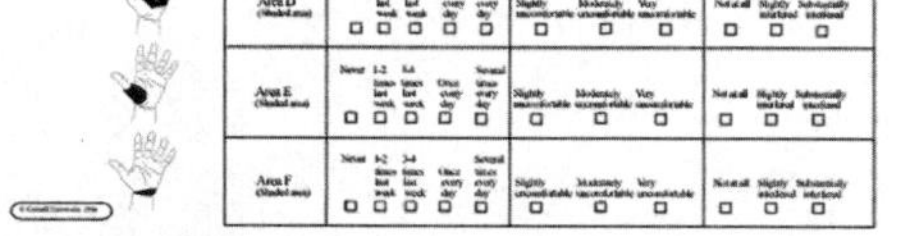

Fig. 5. The Cornell Musculoskeletal Disorder Questionnaire (CDMQ) for left hand used in this paper.

In the study, there are 40 CDMQ questionnaires for the right hand and 40 CDMQ questionnaires for the left hands were collected. A total of 80 questionnaires were collected.

In addition to the Cornell Musculoskeletal Disorder Questionnaire (CDMQ), the basic demographic information, the working related information and the general health information were also collected. The self-perceived health questions were required to answer and to give the specific situation if any.

4 Results

The characteristics of study participants are presented in Table 1. There were 40 visually impaired massage workers were studied, who worked in small scale massage shops located in Shanghai. The participants consisted of females (25%) and males (75%) with age ranging from 23 to 38 years with mean age 25.1 yrs. Overall, they did not suffer from illnesses, but more than half of them felt unhealthy (57.5%). The Table 1. gave the demographic and the general health information of the visually impaired massage workers under this study.

Table 1. Descriptive characteristics statistics of the 40 visually impaired massage workers

Variable	% or Mean (N = 40)
Gender	
Female	25%
Male	75%
Age (mean)	25.1 yrs
General health	
Illnesses (mean, total)	0,0
Self-perceived health (mean, total)	57.5,23

The working related statistics of the 40 visually impaired massage workers under this study are presented in Table 2. Compared with the daily regular eight-hour working in China, the 40 visually impaired massage workers had significant higher daily working hours (Mean = 13.35 h), and the daily working hours with hands were more than 8 h (Mean = 11.35 h). The break time in working per day was 2.075 h. 92.5% of them need to work on weekends, which means 7-day working with hands per week. The working posture was a typical standing posture with high daily standing hours to work (Mean = 8.75 h). The sitting hours during working included lunch and super time (Mean = 2.275 h). The other posture during working was extremely rare (Mean = 0.325 h), which meant they hardly had time to relax the muscles and bones during working.

The Cornell musculoskeletal discomfort scores for right hand are presented in Table 3. The Cornell musculoskeletal discomfort scores for left hand are presented in Table 4. The scores were collected by weighting the rating scores (Never = 0, 1–2 times/week = 1.5, 3–4 times/week = 3.5, Every day = 5, Several times every day = 10).

Comparison of CMDQ scores between right hand and left hand showed that there is no significant difference in scores bwtween both hands from Area A to Area F. The highest discomfort score for right hand was Area C (score = 315). The highest discomfort score for left hand was Area C (score = 320). Area C was thumb. The second highest score for right hand was Area A (score = 280.5). The second highest score for left hand was Area A (score = 275.5). Area A consisted of index finger, middle finger, and half of the ring finger (beside the middle finger). The discomfort score for right hand of Area B was 253. The discomfort score for left hand of Area B was 250. Area B consisted of

Table 2. Working related statistics of the 40 visually impaired massage workers

Variable	% or Mean (N = 40)
Work experience, yrs (mean)	2.8
Working hours	
Daily working hours (mean)	13.35
Daily working hours with hands (mean)	11.35
Overtime per day, hrs (mean)	5.35
Overtime per day with hands, hrs (mean)	3.35
Working on weekends (mean, total)	92.5%,37
No work on weekends (mean, total)	7.5%,3
Time spent on other jobs, hrs (mean)	0
Break time in working per day, hrs (mean)	2.075
Working posture and hours	
Daily standing hours (mean)	8.75
Daily sitting hours (mean)	2.275
Daily other posture, hrs (mean)	0.325

Table 3. The Cornell musculoskeletal discomfort scores for right hand

Right hand	Discomfort score	Percentage of a single item total score (%)
Area A	280.5	70.13
Area B	253	63.25
Area C	315	78.75
Area D	199.5	49.88
Area E	198	49.5
Area F	182.5	45.63

half of the ring finger (beside the pinkie finger) and pinkie. Area B was the third highest score. It identified the most serious discomfort problems of both hands were the fingers.

Table 4. The Cornell musculoskeletal discomfort scores for left hand

Left hand	Discomfort score	Percentage of a single item total score (%)
Area A	275.5	68.88
Area B	250	62.5
Area C	320	80
Area D	198	49.5
Area E	208.5	52.13
Area F	179.5	44.88

5 Discussion and Conclusion

The Cornell Musculoskeletal Disorder Questionnaire (CMDQ) scores are not affected by gender and employment status [5]. In summary, the study provides novel evidence that in persistent massage working contribute to hands' impairments of the visually impaired massage workers regardless of gender. In this study, the mean working experience of the subjects was 2.8 years and the mean age was 25.1 yrs. The unhealthy feeling of them (57.5%) did not obviously relate to the long years working or old age. The identified similarities in their work were that over 8 h working per day, lack sufficient hand rest time, overtime work, and over 8 h daily standing hours (Mean = 8.75 h). Fatigue caused by long-term overtime work, which reduced their feeling of self-perceived health. The thumbs were the most severely injured parts (78.75%, 80%), and the rest fingers were injured but less (70.13%, 63.25%, 68.88%, 62.5%). It should be noted that these proportions were all higher than 50%. This means highest potential of occupational risks of visually impaired massage workers. These occupational risks include pain, strain, bending and deformation of the finger joints, and constant muscle tension and pain, and the risk of causing more disabilities in the long run.

Musculoskeletal disease is the leading cause of disability worldwide. Further measures should be implemented to mitigate the occupational risks identified of the visually impaired massage workers.

In this study, musculoskeletal discomforts and symptoms related to visually impaired massage workers' work are largely due to excessive exertion and long-term use of hands without enough rest. The working environment and the social factors in China should be taken into account. Therefore, according to ergonomics, (1) improve working conditions, implement ergonomic interventions in the workplace, design ergonomic chairs that can be switched between sitting and lying at any time for the workplace, to increase the random break time and relieve fatigue, allow visually impaired massage workers to lie down and rest at any time no customers, (2) ensure no more than 8-h daily workload, shifts are suggested to carry out to meet more hours workload, (3) design hand massage equipment for the visually impaired massage workers, set minimum daily use time, help them to relax hand muscles and reduce injury risks, the equipment should be adapted to both the workplace and the home environment.

In future studies, musculoskeletal studies on the sitting and standing positions of visually impaired massage workers will be added. Findings from this study would inform the next stage of research.

Acknowledgments. We would like to thank all the visually impaired masseurs as participants and the volunteer research assistants, for this study would not have been possible without their participation.

References

1. Blindness and Vision Impairment (WHO). https://www.who.int/news-room/fact-sheets/detail/blindness-and-visual-impairment
2. Visual Impairment and Blindness 2010 (WHO). https://www.who.int/blindness/data_maps/VIFACTSHEETGLODAT2010full.pdf
3. The Visually Impaired Population in China Is About 17.31 Million, Young People Under 30 Account for 23.5%. https://china.huanqiu.com/article
4. Cornell Musculoskeletal Discomfort Questionnaires (CMDQ). http://www.ergo.human.cornell.edu/ahmsquest.html
5. Borres, R.D., Lim, J.U., Robielos, R.A., Pacaña, M.J.: Musculoskeletal disorders among orthodontists: risk factors and ergonomic intervention. In: Bagnara, S., Tartaglia, R., Albolino, S., Alexander, T., Fujita, Y. (eds.) IEA 2018. AISC, vol. 820, pp. 709–733. Springer, Cham (2019). https://doi.org/10.1007/978-3-319-96083-8_87

Cognitive Biases in Game Momentum, Winning Strategy, and Jinx in Baseball

Atsuo Murata(✉)

Department of Intelligent Mechanical Systems, Okayama University, 3-1-1 Tsushimanaka, Kita-Ward, Okayama, Japan
murata@iims.sys.okayama-u.ac.jp

Abstract. This study investigated game momentum, strategy, and jinx related to baseball games. Based on such investigations, cognitive biases underlying such phenomenon were extracted to show the irrationality and biased thinking in baseball games. We discussed countermeasures for removing cognitive biases in baseball games in order to view, plan, and manage a baseball game rationally, scientifically, and without cognitive biases so that higher achievements are acquired in a game.

Keywords: Game momentum · Strategy · Jinx · Cognitive bias · Irrationality

1 Introduction

In Nippon Professional Baseball (NPB) and Major League Baseball (MLB), a lot of game strategies or tactics seem to exist and widespread among commentators, coaches, and players. In other words, it is believed by baseball commentators, coaches, and players that such strategies or tactics are scientifically valid and effective to acquire higher achievements in a game. Therefore, a lot of teams in NPB and MLB are affected by such strategies or tactics in managing and planning a game. These strategies or tactics seem to be generally accepted by baseball commentators and coaches. However, it seems that there is no rationale to judge that the generally accepted strategies or tactics are reasonable and rational ones.

Our biased mind of random sequences that is believed to occur in basketball games is called as the myth of hot hand [1, 2, 3, 4, 5, 6, 7, 8]. The hot hand myth represents our misunderstood thinking that the percentage shoot success of a particular player increases due to the momentum shift of the game. However, this was scientifically and statistically proved to be wrong. We mistakenly use the phrase "momentum shift." This represents one of our irrational minds (cognitive biases). As well as the myth of hot hand, there are many cognitive biases in baseball games. The following phenomenon are examples of cognitive biases in game momentum or game strategy.

R. S. Goonetilleke et al. (Eds.): AHFE 2021, LNNS 273, pp. 449–456, 2021.
https://doi.org/10.1007/978-3-030-80713-9_57

(a) Giving a lead-off batter a walk turns a tide of game worse than permitting a base hit,
(b) A pinch after a chance,
(c) Bases loaded with no outs scores few, and
(d) Mistakes in offenses turn a tide of game worse and give impetus to opponents.

The aim of this study was to explore cognitive biases in baseball games from the following three perspectives: (1) game momentum, (2) winning strategy or tactics, and (3) jinx.

First, we briefly summarized what cognitive biases are. We tend to get trapped into a cognitive bias such as availability bias or representativeness bias. For example, we generally tend to be affected by gambler's fallacy as follows. When a batter with a batting average more than .300 goes down at bat two times in a row, a baseball commentator states that it is time for this batter to make a hit. This is apparently wrong, and a batting average more than .300 merely means that independent of former at bat, the probability of a batter making a hit is about 1/3 (.300). This phenomenon represents an irrational way of thinking and corresponds to a cognitive bias.

Next, cognitive biases [9, 10, 11] in understanding game momentum were discussed to examine whether well-recognized game momentum such as "a pinch after a chance", "an error or mistake induces game momentum shift", and "a fine play induces game momentum shift." We also examined whether the well-recognized winning strategy such as "sacrifice bunt is an effective tactic", "Wait one pitch at a three-nothing situation", "a left-handed pitcher for a left-handed hitter is effective." is actually rational or not. Finally, a cognitive bias in a jinx in baseball such as "Seventh inning is a lucky one" or "the situation of bases loaded with no outcounts does not lead to large scores." was discussed.

In this manner, this study attempted to demonstrate that there are many irrational ways of thinking, that is, cognitive biases in many situations of baseball such as game planning and management. Based on such discussion, countermeasures for removing cognitive biases in baseball games in order to view, plan, and manage the game rationally, scientifically, and without cognitive biases so that higher achievements are effectively and certainly acquired in a game.

2 Cognitive Biases [9, 10, 11]

Heuristics, such as availability, representativeness, or affect, cause biases, including confirmation biases, anchoring and adjustment, hindsight, availability bias and conjunction fallacy. An event easily imaginable is more available than an event that is difficult to imagine. For example, the availability of the vividness of imaginable events biases our perception of the occurrence frequency of similar events. This might lead to wrong decision making on the frequency of such events. Managers predict a salesperson's performance based on an established category of salespeople. This corresponds to the representativeness heuristic. While this heuristic offers a proper approximation of salespeople in some cases, it can induce a biased understanding of salespeople and lead to serious errors in other cases. Such errors include ignorance of base rate or insensitivity to sample size. People naturally tend to seek information that confirms their expectations and hypotheses, even though information disconfirming their expectations and hypotheses is more useful. This induces a biased recognition of causality and leads to serious errors.

Not only heuristics, but also overconfidence and framing causes a variety of biases. Bounded awareness prevents one from focusing on useful, observable and relevant information. Due to such bounded awareness, it is valid to assume that we occasionally cannot behave rationally. Moreover, it is assumed that our bounded awareness and uncertain (risky) situations form the basis of heuristics, overconfidence and framing.

3 Cognitive Biases in Baseball Game

The data in the following subsections (Sect. 3.1–3.3) were derived from Katoh and Yamasaki [12] who used data in 2007 of NPB.

3.1 Game Momentum

3.1.1 A Walk to a Lead-off Batter Leads to Bad Game Momentum

Baseball commentators and coaches frequently state that a walk to a lead-off batter makes game momentum bad. If this is the case, the percentage of loss or mean runs allowed at the corresponding inning must be larger for the walk to a lead-off batter than for the hit to a lead-off batter. The percentage of loss and mean runs allowed did not differ am among the walk and the hit: 39% and 0.83 for the walk and 40.5% and 0.83 for the hit. Moreover, the percentage of gain or mean runs gained at the next inning were compared between the walk and the hit to a lead-off batter. No significant difference of the percentage of gain or mean runs gained at the next inning was detected between the two situations. In short, no game momentum was observed as a result of statistical comparisons.

3.1.2 Opportunity Brings Tough Time

The percentage of loss or mean runs allowed at the next inning were compared between the two situations: when the advantage of the opportunity was taken and when it was not taken. The percentage of loss or mean runs allowed at the next inning for the former were 26% and 0.49, respectively. The same indices for the latter were 26% and 0.49, respectively. These statistics show that no such game momentum is observed.

3.1.3 An Offensive Error Leads to Bad Game Momentum

We calculated the percentage of loss and mean runs allowed after the inning in which an offensive error occurred. Those data at the next inning (defense) were 22.8% and 0.436, respectively. These values were not different significantly from the mean of all innings. The percentage of gain (score) and mean runs at the next, next inning (offense) were 26.2% and 0.476, respectively. These values were also not different from the mean of all innings. These statistics are not affirmative to the game momentum "An offensive error leads to bad game momentum."

3.1.4 Long Defensive Time Leads to Bad Game Momentum

The duration of defensive inning was classified by the number of pitches in each inning as follows: (1) short: 10 or less, (2) medium: between 11 and 20, and (3) long: more than 20. Aa a result of calculating the percentage innings scored and the mean score, these data did not differ significantly among three categories (1)-(3). It seems that long defensive time does not necessarily lead to bad game momentum as baseball commentators express.

3.1.5 Summary

Based on the statistics in Sect. 3.1.1–3.1.4, it seems that generally accepted game momentum is not observed at all. Availability bias, which forces people to judge based on few experiences, ease of recollection, and ambiguity in memory, must have caused such illusion in interpretation of game momentum.

3.2 Game Strategy

3.2.1 Frequent use of Stolen Bases and Hit-And-Run is an Effective Strategy

The percentage score and the mean score were compared between the case where a stolen base was tried and that where a stolen base was not tried. The percentage score and the mean score were (41%, 0.87) and (37.3%, 0.74), respectively. It must be noted that the percentage score and the mean score were different between the success and the failure of a stolen base ((50.2%, 0.98) and (14%, 0.31)).

The percentage score and the mean score were compared between the case where a hit-and-run was tried and that where a hit-and-run base was not tried. The percentage score and the mean score were (40.5%, 0.86) and (43.3%, 0.84), respectively. The percentage score and the mean score were different between the success and the failure of a hit-and-run ((72.1%, 1.53) and (28.6%, 0.48)).

Although it seems that the hit-and-run is an effective strategy so long as it is successful. However, we must also pay attention to the failure of a hit-and-run. Taking into account the failure of a hit-and-run, this strategy does not seem to be effective as coaches or commentators state.

3.2.2 Sacrifice Bunt is an Effective Strategy

As a result of comparing the percentage score and the mean score between the case where a sacrifice bunt was tried and that where a sacrifice bunt was not tried, the percentage score and the mean score were (40.6%, 0.89) and (40.4%, 0.76), respectively.

The percentage score and the mean score were different between the success and the failure of a stolen base ((44.0%, 0.82) and (28.4%, 0.55)). It seems that a sacrifice bunt is an effective strategy so long as it is successful. However, taking into account the failure of a sacrifice bunt, that is, the risk of a sacrifice bunt, this strategy does not seem to be effective as recognized by coaches or commentators.

3.2.3 A Left-handed Batter for a Right-handed Pitcher and a Right-handed Batter for a Left-handed Pitcher

For a right-handed pitcher, a right-handed and a left-handed batter batted in as a pinch hitter 696 and 1,075 times, respectively. In case of a left-handed pitcher, a right-handed and a left-handed batter batted in as a pinch hitter 961 and 76 times, respectively. It seems, from these data, that coaches pay more importance on the match of a right-handed batter with a left-handed pitcher.

The batting averages of left-handed hitters were .272 and .283 for a left-handed and a right-handed pitcher, respectively. The batting averages of right-handed hitters were .277 and .264 for a left-handed and a right-handed pitcher, respectively. These data show that coaches must consider the low batting average of a right-handed batter for a right-handed pitcher, which means that a right-handed hitter should be replaced with a left-handed hitter when a right-handed pitcher is on the mound. In this manner, the game strategy "a left-handed hitter for a right-handed pitcher" should be more enhanced. Coaches seem to be too cautious about the strategy "a right-handed batter for a left-handed pitcher."

3.2.4 Summary

Based on the statistics in Sect. 3.2.1–3.2.3 it seems that generally accepted game strategy is not effective as coaches or baseball commentators think. This must represent status quo bias in baseball strategies. Coaches or baseball commentators irrationally follow a precedent or follow suit with a generally accepted precedent without analyzing statistics appropriately.

3.3 Jinx

3.3.1 Home Team Gains Ground at Extra Innings

The data (win, loss, draw, percentage win) of the home team were (404, 365, 0, 0.525) except for extra innings and (28, 31, 18, 0.475) for extra innings. The data definitely demonstrates that there is no home team advantage.

3.3.2 A Walk-off Game Makes a Team Ride on the Crest of Winning more Games

The percentages win at the next game were 28 wins (50%) and 28 losses (50%) when the team won an extra inning game. When the team lost an extra inning game, the corresponding data were 26 wins (46%) and 30 losses (54%), and 1 draw. This data does not prove that a walk-off game makes a team ride on the crest of winning steaks. This can be further confirmed by the following data. When there was no extra inning, the percentages win at the next game were 433 (54%) wins and 371 losses (46%) when a team won the game. When the team lost a game, the corresponding data were 372 wins (54%) and 435 losses (46%), respectively. The data does not show the validity of the jinx "A walk-off game makes a team ride on the crest of winning more games."

3.3.3 Situation of Bases Loaded with no Outs Does not Necessarily Lead to Large Scores

Baseball commentators often mention that the situation of bases loaded with no outs does not necessarily lead to a score. We will show whether the jinx is true or not. For the no-out situation, the percentage score and the mean score were (26.4%, 0.49), (40.7%, 0.86), and (84.5%, 2.40) for the situations of no runner on the base, the runner on the 1st base, and the bases loaded, respectively. For the one-out situation, the percentage score and the mean score were (15.7%, 0.27), (27.5%, 0.56), and (66.4%, 1.64) for the situations of no runner on the base, the runner on the 1st base, and the bases loaded, respectively. For the two-out situation, the percentage score and the mean score were (6.8%, 0.11), (13.4%, 0.25), and (32.8%, 0.78) for the situations of no runner on the base, the runner on the 1st base, and the bases loaded, respectively. These data apparently show that the situation of bases loaded with no outs has high likelihood to lead to a score than other situations. Therefore, the jinx does not seem to be true.

3.3.4 Luck Seven (7th Inning is a Lucky One that Leads to more Scores)

Announcers often mentions the lucky 7th inning. Is this jinx true? The following statistics were used to investigate the validity of this jinx. The percentage score and the mean score at 6th and 7th inning in NPB were (26%, 0.53) and (24%, 0.45), respectively. The corresponding data at 6th and 7th inning in MLB were (26.5%, 0.49) and (24%, 0.41), respectively. The data definitely show that there is no advantage of 7th inning over other innings.

3.3.5 Summary

Section 3.3.1–3.3.4 seem to show that generally accepted jinx is not true as announcers or baseball commentators state. Such a jinx must stem from availability bias, under the effect of which people tend to judge based on few experiences, ease of recollection, and ambiguity in memory.

4 Discussion

Baseball umpires are more reluctant to call the fourth ball. People tend to avoid behaving and committing an error (error of commission) rather than missing an opportunity (omission). While we are blamed for committing an error, we are rarely blamed for not acting. This is an irrational act of umpires called omission bias. Umpires, due to omission bias, do not want to issue a walk or strikeout, because they want to let batters determine the outcome by prolonging the at-bat.

The curse of Billy's goat or curse of Bambino in a famous jinx in MLB. The longer Chicago Cubs and Boston Red Sox are away from their world championship, the more often the reason is attributed to the curse. Such a curse stems from a self-attribution bias. When got trapped into this bias, people attribute a success to their skill and competence and a failure to a luck. People tend to attribute the cause that they don't want to explain to a luck.

Myth of hot hand insists that streaks of success can predict future performance. This myth predicts that recent performance directly influences immediate future performance. In short, a team or player currently riding the wave fare better in the next game than one who is not. However, there was no evidence that the streaks had any carryover effect and streaks could predict future performance. This is one of cognitive biases called representativeness bias. What players did on their previous free throws didn't affect what they did on the next free throw. In short, streaks of success were not an accurate predicter of future performance.

Why people believe in "game momentum" in spite of no evidence of hot hand is discussed. People generally ascribe patterns to events. Gamblers fallacy means that gamblers on losing streaks mistakenly think they keep gambling, hoping that their luck would balance out. People are not good at thinking rationally based on probability (risk). Therefore, it is difficult to understand luck or chance (randomness) causes streaks for both the best and the worst players. People cannot understand that the true performance of teams or players can be measured best in large samples and that small samples ruled by randomness are unreliable.

Based on such discussion, we unconsciously experience cognitive biases in game momentum, game strategy, or jinx described in Sect. 3.

5 Conclusions

As demonstrated in this paper, there exist many cognitive biases in game momentum, game strategy, and jinx in baseball. Such cognitive biases hinder establishing a rational and effective game strategy or management. Such biases must be removed based on accurate statistics in order to establish an effective game strategy and management and increase the winning probability as high as possible. The strategy or game momentum spoken by coaches or baseball commentators should be carefully verified using available baseball statistics. Unbiased and rational game strategy would lead to an effective game management with a high probability of winning the game.

References

1. Gilovich, T., Vallon, R., Tversky, A.: The hot hand in basketball: on the misinterpretation of random sequences. Cogn. Psychol. **17**, 295–314 (1985)
2. Gilovich, T.: Judgmental biases in the world of sports. In: Straub, W.F., Williams, J.M. (eds.) Cognitive Sports Psychology. Sports Science Associates, New York (1984)
3. Koehler, J.J., Conley, C.A.: The 'Hot Hand' myth in professional basketball. J. Sports Exerc. Psychol. **25**, 253–259 (2003)
4. Camerer, C.F.: Does the basketball market believe in the 'Hot Hand'? Am. Econ. Rev. **79**, 1257–1261 (1989)
5. Ayton, P., Fischer, I.: The hot hand fallacy and the gambler's fallacy: two faces of subjective randomness? Mem. Cogn. **32**, 1369–1378 (2004)
6. Brown, W.O., Sauer, R.D.: Does the basketball market believe in the 'Hot Hand'? Am. Econ. Rev. **83**, 1377–1386 (1993)
7. Burns, B.D., Corpus, B.: Randomness and induction from streaks: 'Gambler's fallacy' vs. 'Hot Hand'. Psychon. Bull. Rev. **11**, 179–184 (2004)

8. Moskowitz, T., Wertheim, L.J.: Scorecasting-The Hidden Influences Behind How Sports are Played and Games are Won. Three Rivers Press, New York (2011)
9. Murata, A.: Cross-cultural difference of Free Agency (FA) system between MLB and NPB (Nippon Professional Baseball). In: Kantola, J., Nazir, S., Barath, T. (eds.) AHFE 2018. AISC, vol. 783, pp. 668–680. Springer, Cham (2018)
10. Murata, A.: Cognitive biases and distorted decision making that prevent rational and efficient sports management - cross-cultural difference between MLB and NPB. In: Goossens, R.H.M., Murata, A. (eds.) AHFE 2019. AISC, vol. 970, pp. 561–571. Springer, Cham (2019)
11. Lewis, M.: The Undoing Project: A Friendship That Changed Our Minds. W.W. Norton & Co Inc., New York (2017)
12. Katoh, H., Yamasaki, T.: Illusion in Baseball, Data Stadium, in Japanese (2008)

Cultural Preparation for Digital Transformation of Industrial Organizations: A Multi-case Exploration of Socio-technical Systems

Aurangzeab Butt[1](✉), Faisal Imran[1], Jussi Kantola[2], and Petri Helo[1]

[1] School of Technology and Innovation, University of Vaasa, Yliopistonranta 10, 65101 Vaasa, Finland
{Aurangzeb.Butt,Faisal.Imran,Petri.Helo}@uva.fi

[2] Mechanical Engineering, University of Turku, Joukahaisenkatu 3-5, 20520 Turku, Finland
Jussi.Kantola@utu.fi

Abstract. Digital technology adoption is not new; however, digital transformation in industrial organization is recognized as a contemporary phenomenon. Extant literature, however, offers limited guidance on how the socio-technical system could culturally evolve during the digital transformation. Our limited research focused on the exploration of cultural preparedness in the traditional large-scale organizations with a global footprint. With a multi-case study research design, we collected data from three industrial organizations, and perform data analysis with a grounded theory approach. Our results are a collected set of values, assumptions, and artifacts pertinent to the cultural preparedness of industrial organizations.

Keywords: Digital transformation · Culture · Values · Assumption · Case study · Industrial organization

1 Introduction

Advancement in digital technologies underline the unprecedented integration of cyber and physical worlds [1]. Most commonly cited digital technology options fuel the digitizing and digitalizing trends e.g. advanced algorithms and artificial intelligence-aspired data science, machine learning, self-learning systems, AI engineering and industrialized AI, API gateways, cybersecurity mesh, industrial automations, cloud computing and 5G-enabled distributed cloud for industrial/ smart factory applications, SaaS to PaaS, industrial internet of things (IIoT), robotic process automations (RPA), cognitive science, internet of behaviors (IoB) etc. [2, 3]. Only considered as limitation, however, technology is bringing multiplex opportunities for businesses. Not only businesses; excited by plethora options and evolution speed, the predominant focus of research community remained to digital technology adoptions [4, 32]. This excitement to capitalize on digital technology advancements has led to tiring efforts on digitizing businesses at an unprecedented pace [5]. Unambiguously and arguably technology is transforming value chains, business operations, and the whole competitive spectrum [4, 6].

R. S. Goonetilleke et al. (Eds.): AHFE 2021, LNNS 273, pp. 457–463, 2021.
https://doi.org/10.1007/978-3-030-80713-9_58

Certainly, digital transformation is a socio-technical phenomenon [1, 29]. Social embeddedness of the deployed technologies impacts value generation and business transactions [7]. Because large changes, like digital transformation, consequence in prudent design updates to the social and technical systems of an organization [8]. These updates involve organizations' goals, people, infrastructure, processes, technology, and their culture [9, 32]. Socio-technical perspective therefore strives to delineate best-fit between aspired technological adoptions and the people [10, 11]. Bringing socio-technical perspective to digital transformation enables collective attention to organizations' structures, people in these structures, the strategic goal-setting and the delivery of goals (as tasks), and the technology as an enabler for desired business performance [12, 13]. Enabling congruence in socio-technical system drive cultural preparedness for digital future of its people, structures, and the tasks [14, 32].

An organization' culture covers all aspects of its "learned response" to business challenges and value integration activities of its people [15]. Culture operationalizes macro and micro level organizational mechanisms of change [16], thus it holds regulating role during transformations [17]. Organizational culture furnishes a system of accepted meanings for the employees to interpret a change situation and the response actions [18]. Nurturing organization's culture enables innovativeness, flexibility & agility, engagement, superior financial performance, and hence sustained competitive advantage [19]. Whereas stagnated culture becomes a liability, a source of resistance and increased inertia, and leads to loss-making business [19]. For example, in Kodak's (the film company) sluggish cultural response to digital filming technologies led to loss of market share, 80% decline in its workforce, and stalled the futureproofing of the company [20]. Despite such recorded cases in literature, limited literature on cultural preparedness for digital transformation is a known dilemma [6, 12].

In the current research, our focus is on "how" [21] of the cultural changes can foster the digital transformation of industrial organizations [22, 23, 32]. Our aim is not to predict a universal model of cultural change however to explore its emergence [23] within a single context of digital transformation. Accordingly, this study attempts to answer the main research questions: ***How industrial organizations prepare their culture for digital transformation?*** Herein this research excludes devising cultural diagnostics tool [24]. Rather discusses the cultural preparedness for organizational *values*, *assumptions*, and *artifacts* [15] that propel the digital transformation.

2 Methodology

Culture ties distinctive characteristics of an organizational system; each organization has its own culture. While the social mechanisms of change embedded in the context relay cultural preparedness digitally transforming organization, a contemporary phenomenon. We deployed qualitative case study which is an accepted research approach to investigate a phenomenon embedded in its context wherein the phenomenon and the context are inseparable [25]. We interviewed digital transformation experts and business executives of three (3) multinational industrial organizations with rich history and collective revenue of over eighteen (18) billion euro. For the results presented in this short paper, we included the analysis of sixteen (16) semi-structured interviews from the case companies (see

Table 1). All these interviews were conducted in 2019 and during Q1 of 2020 (prior to the pandemic).

Table 1. Research data context

Research Data origin	Case_Alpha	Case_Bravo	Case_ Charlie
Business focus	Offshore & Onshore Energy OEM	Wood processing and Energy production	HVAC & Electric Power OEM
Footprint & revenue	Global; multi-billion €	Global; multi-billion €	Global; multi-billion €
History	+120 years history as engineering & manufacturing; Nordic company	+20 years history as process & manufacturing; Nordic company	+80 years old engineering & manufacturing; Nordic company
Interviewee profiles	• VP Digital Transformation • VP Open Innovation • Director Digital Culture • Director business development • GM Business OD • GM Open Innovation	• VP IT & Digital Transformation • VP SCM • Director SCM • Director IT (business unit) • Director IT & Digitalization (business unit)	• VP/Head of a Business-unit • VP Services Business • Director Enterprise Architecture • Director IOT Platform Architecture • Director/ Head of Digital Product Management

We performed grounded theory-based data analysis of the data [26]. The data analysis led to a compilation of organizational culture *values*, *assumptions*, and *artifacts* [15] the leaders in our case organizations are prioritizing to focus for digital tranformation. Interview transcriptions and data analysis results are captured in NVivo.

3 Results and Discussion

All the three cases have had history of successfully futureproofing their businesses since decades. Our initial analysis guides that the stated cultural preparedness in the strategy (published) materials appear quite similar, however, there were quite some peculiarities in practice [19]. Still, there were few common however implicit values revealed during the data analysis process. These commonalities could have been the consequent of successful survival over several decades in a similar industry i.e. engineering, manufacturing, and process. All our case companies value code of conduct based on: ***precision***; ***accuracy***; and, ***authentic knowledge***. This led to open communication culture where people have built-in assumptions on what matter an individual can speak to get appreciated, and in which matters not to participate. In most parts of these organizations, the foremost reason for technology adoption is given to ***operational efficiency*** – their business model evolved at considerably slow, however, at a consistent rate. It is noted that ***disruptive ideas*** for technology adoption are appreciated when communicated from higher-ups in the ***business hierarchy***. However, it is accepted that the adoption of technologies for improving technical efficiency in lieu of cost effectiveness [27] are promoted by individuals or a team with authentic knowledge. Another, commonality in these cultures is assumptions related to customer reach and collaboration. Customer collaboration is meant to be the discretion of sales and marketing team.

Our case companies are undertaking digital technology adoption seriously with program-like initiatives on corporate, business division or functional teams' level. All the three companies approach digital transformation as a big opportunity to bring customers and business operations closer, not just a business imperative. Our analysis led to three common purposes as the targeted outcomes of digital transformation: ***agility***, ***customer-centricity***, and ***collaboration***. We note these three outcomes are ubiquitous as global trends of digital transformation across all industries, however, with our collected data couldn't specify the cause of this commonality.

Our data analysis results show that all three case companies are approaching cultural transformation in a discernable manner. The Case Alpha, deployed cultural preparation by updating values on corporation level with a top down approach. The cultural preparedness in Case_Bravo is taken a top-down route however with focus on developing new artifacts mainly the work procedures, modern systems & tools, and ways to manage development and operational projects. In Case_Charlie, the responsibility of cultural change predominantly remained with individual business units and support functions, herein the top leadership encouraged the middle management to take responsibility for enabling cultural-drive in their respective business lines.

Hereunder, *Fig.* 1 presents the collective focus cultural preparedness learned from the interview experts of our three cases.

Organizational values are enduring beliefs that guides the preferable mode of conduct its people apply to get things done [16]. Living both in implicit and explicit fabric of organizational culture, the values are acceptable behaviors in pursuit of "*a roadmap for future action*" [15]. Our case organization are very much focusing to build & reinforce values of experimentation, problem sharing, team learning, together, transparency & openness, as well as user validation to be first step of developing new products. All these values promote service mindset. To be at service-of-customers drives towards agility and promote beyond formal-hierarchy collaborations to integrate expert knowledge, and seek for externals' involvement. It is not typical in traditional formalized organizations. These values originate from deeply held assumptions validated during the organizational experimenting and learning in response to environmental challenges [23].

Cultural values in their manifestation are explicitly discussed while assumptions remain as tacit believes to interpret day-to-day situations and response behaviors of people as individuals and in a group [17]. Discussed in the introduction section, prior focus on digitizing and digitalization alone has led to job insecurity due to assumptions that automation and artificial intelligence is competing with human competence and workforce. With the insecurity about jobs in a precision-liking culture, people tend to assume that they do not have "*good enough idea*" to share and collaborate on with other

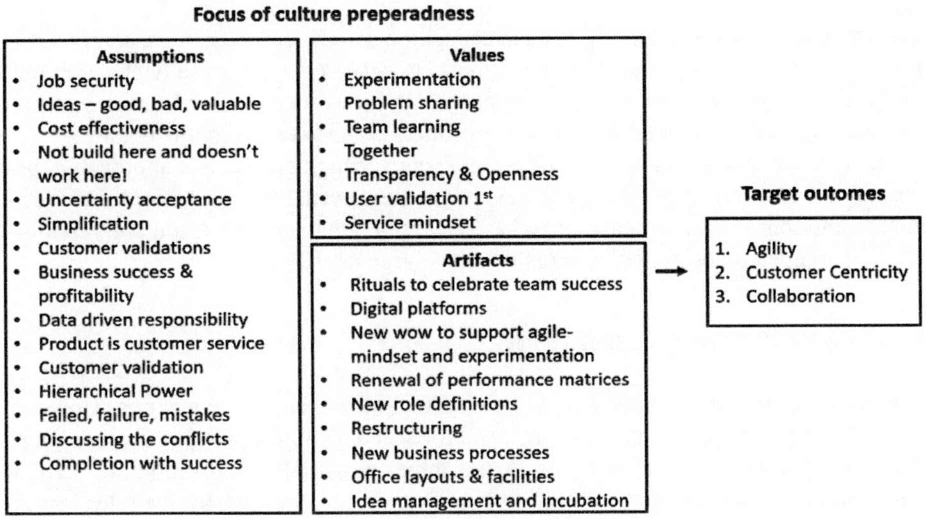

Fig. 1. Focus of cultural preparedness in the case industrial organizations.

colleagues. One expert directed their focus on correcting this assumption by involving people while "*an idea is going through different phases*". Ideation and implementation of ideas involves "*a lot of uncertainty*". By involving people during different phases help to accept uncertainties and discussion of the conflicts. The conflicts are between either different ideas (customer/business problem can be solved by multiple ways) or idea implementation approach. In both cases, the leaders have strong role [30, 31] in updating the assumption that "*conflicts are necessarily bad*" rather constructive conflicts can simplify the solution and its implementation tasks. This shall include validating that "*we make sure it's actually something the customer wants*". Another, relevant assumption the leaders in our case organizations are targeting is business success versus profitability. The cost-conscious culture inherits that successful business are profitable because of cost efficient operations. Such related assumptions trigger most economical responses in within formal hierarchies. While integrating tasks done with cost on top agenda though result in low-cost solutions, which however offer low customer value; result is loss of business and low profitability. Another related assumption in the adoption of new innovations to the business operations: if the tools, systems, and way of working is not based on the existing processes then those are thought to not align to work with our old processes and tools.

Most apparent layer of organization culture is formed from numerous artifacts. The artifacts are workable solutions tested over the course of group actions. Artifacts covers all e.g. dress code, tools, work procedures, methods, business processes, rituals of reward and punishment, IT platforms, as well as the governance structures. In their seminal work, Schein [15] remarked that the first things a new entrant/employee in organization observes and feels are its artifacts. Our exploratory research reveals that case companies are working on developing rituals for celebration, renewal of performance matrices, new role descriptions, office layouts & facilities upgradations, renewing way-of- working for agile-mindset and experimentation, idea management and incubation hubs, new business processes, and restructuring the organizations. However, expect in case Alpha, we could not specify a sizable restructuring to support digital transformation. Meanwhile, all the case organizations deployed new structures functions and teams) in this regard.

Insofar, analyzed data analysis does not specify which company's approach is best working, neither it is purpose herein. Result however reveal that targeted cultural values and assumptions – what to cease and what to instill – remained with leadership-assumed role and competencies to deploy transformative urgency [28].

4 Conclusions and Limitations

Although digital transformation is the new normal for business organization. However, cultural preparedness is response to the need for speed for futureproofing business organizations. Our limited data analysis of the three industrial organizations resulted in compilation of assumptions, values, and artifacts being focused by their leadership. The findings are promising and need further investigation. The future extension of their research will include more data from the case organizations as well we shall explicate on the role of leadership competences and needed role in preparing organizational culture. Alongside the importance of formalization, hierarchy, and integration of organizational structures shall be discussion.

References

1. Sony, M., Naik, S.: Industry 4.0 integration with socio-technical systems theory: a systematic review and proposed theoretical model. Technol. Soc. **61** (2020)
2. Deloitte: Tech trends 2021 (2020)
3. Gartner: Top strategic technology trends for 2021 (2020)
4. Porter, M.E., Heppelmann, J.E.: How smart, connected products are transforming companies. Harv. Bus. Rev. **93**(10), 96–114 (2015)
5. Verhoef, P.C., et al.: Digital transformation: A multidisciplinary reflection and research agenda. J. Bus. Res. **122**, 889–901 (2021)
6. Vial, G.: Understanding digital transformation: a review and a research agenda. J. Strateg. Inf. Syst. **28**(2), 118–144 (2019)
7. Dacin, M.T., Ventresca, M.J., Beal, B.D.: The embeddedness of organizations: dialogue and directions. J. Manage. **25**(3), 317–356 (1999)
8. Pasmore, W., Winby, S., Mohrman, S.A., Vanasse, R.: Reflections: sociotechnical systems design and organization change. J. Chang. Manag. **19**(2), 67–85 (2019)
9. Davis, M.C., Challenger, R., Jayewardene, D.N.W., Clegg, C.W.: Advancing socio-technical systems thinking: a call for bravery. Appl. Ergon. **45**(2), 171–180 (2014)
10. Mitki, Y., Shani, A.B.R., Greenbaum, B.E.: Developing new capabilities: a longitudinal study of sociotechnical system redesign. J. Chang. Manag. **19**(3), 167–182 (2019)
11. Trist, E., Baumforth, K.: Some social and psychological consequences of the longwall method of coal getting. Hum. Relat. **4**(1), 7–9 (1951)
12. Hartl, E.: A characterization of culture change in the context of digital transformation. Twenty-fifth Am. Conf. Inf. Syst. 1–10 (2019)
13. Lyytinen, K., Newman, M.: Explaining information systems change: a punctuated socio-technical change model. Eur. J. Inf. Syst. **17**, 589–613 (2008)

14. Kane, G.C., Palmer, D., Nguyen, A.P., Kiron, D., Buckley, N.: Aligning the organization for its digital future. MIT Sloan Manag. Rev. Deloitte Univ. Press, **58180**, 1–29 (2016)
15. Schein, E.H.: Organizational Culture. Am. Psychol. **45**(2), 109–119 (1990)
16. Arz, C.: Mechanisms of organizational culture for fostering corporate entrepreneurship: a systematic review and research agenda. J. Enterprising Cult. **24**(4), 361–409 (2017)
17. Schein, E.H.: The Corporate Culture Survival Guide (2009)
18. Pettigrew, A.M.: On studying organizational cultures. Adm. Sci. Q. (1979)
19. Barney, J.B.: Organizational culture: can it be a source of sustained competitive advantage? Acad. Manag. Rev. **11**(3), 656–665 (1986)
20. Lucas, H.C., Goh, J.M.: Disruptive technology: how kodak missed the digital photography revolution. J. Strateg. Inf. Syst. **18**(1), 46–55 (2009)
21. Hedström, P., Ylikoski, P.: Causal mechanisms in the social sciences. Annu. Rev. Sociol. **36**, 49–67 (2010)
22. Canato, A., Ravasi, D., Phillips, N.: Coerced practice implementation in cases of low cultural fit: cultural change and practice adaptation during the implementation of six sigma at 3M. Acad. Manag. J. **56**(6), 1724–1753 (2013)
23. Giorgi, S., Lockwood, C., Glynn, M.A.: The many faces of culture: making sense of 30 years of research on culture in organization studies. Acad. Manag. Ann. **9**(1), 1–54 (2015)
24. Cameron, K.S., Quinn, R.E.: Diagnosing and chaning organisational culture, 3rd edn. Jossey-Bass, San Francisco (2011)
25. Yin, R.K.: Applications of case study research. Appl. Soc. Res. Methods Ser. (2013)
26. Gioia, D.A., Corley, K.G., Hamilton, A.L.: Seeking qualitative rigor in inductive research: notes on the gioia methodology. Organ. Res. Methods **16**(1), 15–31 (2013)
27. Shuen, A., Feiler, P.F., Teece, D.J.: Dynamic capabilities in the upstream oil and gas sector: managing next generation competition. Energy Strateg. Rev. **3**, 5–13 (2014)
28. Schein, E.: Organizational Culture and Leadership. San Diego. California, USA: Jossey-Bass (2017)
29. Imran, F., Kantola, J.: Review of industry 4.0 in the light of sociotechnical system theory and competence-based view: a future research agenda for the evolute approach. Adv. Intell. Syst. Comput. **783**, 118–128 (2018)
30. Imran, F., Shahzad, K., Butt, A., Kantola, J.: Leadership competencies for digital transformation: evidence from multiple cases. In: Kantola, J.I., Nazir, S., Salminen, V. (eds.) AHFE 2020. AISC, vol. 1209, pp. 81–87. Springer, Cham (2020). https://doi.org/10.1007/978-3-030-50791-6_11
31. Imran, F., Kantola, J.I.: "A Co-evolute Approach to Analyze the Competencies of Sales Personnel of Banking Sector of Pakistan", in Advances in Human Factors, pp. 125–136. Springer, Business Management and Leadership, Los Angeles, California, USA (2017)
32. Imran, F., Shahzad, K., Butt, A., Kantola, K.: Digital transformation of industrial organizations: toward an integrated framework. J. Chang. Manag. 1–29 (2021)

Study Abroad in the Philippines and Canada by Japanese Undergraduate Students: A Comparative Mixed Methods Study

Chihiro Tajima[1](✉) and Michael D. Fetters[2]

[1] Faculty of Intercultural Communication, Gakushuin Women's College, 3-20-1 Toyama, Shinjyuku-ku, Tokyo, Japan
chihiro.tajima@gakushuin.ac.jp

[2] Mixed Methods Program and Department of Family Medicine, University of Michigan, 1018 Fuller Street, Ann Arbor, Michigan 48104-1213, USA
mfetters@umich.edu

Abstract. The aim of this research is to compare differences in language learning and affective factor outcomes, and to contrast the experiences of students in two types of study abroad destinations—one where English is used as the lingua franca, a common language used mostly by non-native speakers of English, and the other where English is used mostly by native speakers of English. To achieve this aim, this research compares two previously published mixed methods evaluation studies, one of 21 Japanese university students who studied in the Philippines and another of 29 Japanese university students who studied in Canada. Outcome measures and qualitative experiences data addressed language proficiency gains and changes in attitudes and affect such as motivation, willingness to communicate, and language anxiety. An integrated mixed methods analysis highlights similarities and differences of the mixed data findings between the two study abroad destinations.

Keywords: Study Abroad · Outcomes · Canada · Philippines · Mixed Methods Research

1 Introduction

As reported by the Japan Ministry of Education, Japanese university students studying abroad in traditional destinations such as the US and Canada have been decreasing while undergraduates studying abroad in untraditional destinations such as the Philippines have been increasing [1]. However, research comparing differences in learning outcomes and experiences for these two types of study abroad (SA) contexts are lacking. To fill this gap, this research aims to compare outcomes and experiences of university students on short-term SA in the non-English speaking country of the Philippines, and a traditional SA destination in the English-speaking country of Canada.

This research represents "stage three" of a body of research. In "stage one", the outcomes of 21 Japanese undergraduate students from short- term SA in the Philippines

R. S. Goonetilleke et al. (Eds.): AHFE 2021, LNNS 273, pp. 464–469, 2021.
https://doi.org/10.1007/978-3-030-80713-9_59

were reported based on a mixed methods design incorporating both quantitative (QUAN) and qualitative (QUAL) data [2]. "Stage two" involved reporting of QUAN outcomes of 29 Japanese university students after short-term SA in Canada [3], as well as a mixed methods case study design conducted by the additional integration with QUAL data [4]. As stage three, this analysis examines findings from stages one and two to clarify the SA outcomes and experiences for the differing contexts of the Philippines and Canada.

2 Methods

2.1 Design, Research Purpose, and Questions

The study utilizes a pre- and post-experience comparative mixed methods design. The purpose is to compare and contrast second language (L2) proficiency test scores from survey data collected at pre- and post-short-term SA, as well as post-SA interview data from students in the two destinations. This study had four research questions: (1) Does short-term SA in the Philippines and Canada improve learners' L2 proficiency similarly? (2) Does short-term SA in the Philippines and Canada develop learners' affective factors of L2 willingness to communicate (WTC), L2 motivation, and L2 anxiety similarly? (3) Based on post-SA semi-structured interview data, how do the SA experiences compare between the Philippines and Canada? (4) What can be understood about L2 learning in the different SA contexts by integrating the QUAN and QUAL data?

2.2 Situating Two Study Abroad Contexts

Philippines. For five weeks in the city of Bacold, 21 female Japanese students from a single university in Tokyo studied English from Monday through Friday based on four hours a day of one-to-one instruction and two hours a day of classroom instruction. While speaking in their second language of English, Filipino instructors provided instruction to SA students. Participating students stayed in a dormitory located next to the language school with many SA Korean students. Participants used English in common areas. The participants visited two orphanages and other activities organized by the program.

Canada. For four weeks in the Province of Alberta and British Columbia, 29 female Japanese university students from two universities in Tokyo studied English Monday through Friday for four to six hours a day. Mainly native speakers of English taught them in classroom settings. All students participated in a homestay. Students who chose to do so could participate in activities and trips organized by the program, e.g., visiting museums and going on a ski trip.

2.3 Mixed Data Collection

All participants provided written informed consent to participate. For both SA destinations, QUAN and QUAL data were collected with the same timing, pre-departure and post-return. Table 1 lists the four measurement instruments used to collect QUAN data at

Table 1. Data collected at pre-departure and post-return for the Philippines and Canada study abroad programs

Measurement Instruments
1. English proficiency test (Computerized Assessment System of English Communication: CASEC)
2. Willingness to communicate questionnaire
3. Language anxiety questionnaire
4. Motivation questionnaire

pre-departure and at post-return. For details of the instruments used, see the previously published article of "stage one" of this program of research [2].

QUAL data for this research consisted of the interviews collected from all participants' post-return. Each interview lasted approximately 20 to 30 min. Interview guide questions were developed to match the four QUAN domains listed in Table 1. The interviews were conducted in Japanese, transcribed, then translated by the author to English for dissemination. The interview guide is available upon request.

3 Results

Research question one was to investigate whether the short-term SA in the Philippines and Canada improved learners' L2 proficiency similarly. As shown in Table 2, both groups made gains in language proficiency. Students going to the Philippines had lower baseline proficiency, but made greater gains than Canada-bound students.

Research question two concerned whether short-term SA in the Philippines and Canada develop learners' affective factors of L2 WTC, L2 motivation, and L2 anxiety similarly. Tables 3, 4, and 5 show that both groups made improvements in all domains. As also shown in these tables, participants in the Philippines had greater gains for all of the affective factors. In terms of L2 motivation, the Philippines group had higher baseline and higher post-SA scores (Table 3). For L2 WTC, the Philippines group had lower baseline scores and higher post-SA scores (Table 4). For L2 anxiety, participants going to the Philippines had higher baseline scores and greater reduction in post-SA scores than those who went to Canada.

Research question three was intended to compare differences qualitatively between learners in the Philippines and Canada. For both groups, a variety of speaking practices outside of the classroom was an important explanation for gains in English as shown in Table 2. As in Tables 3, 4, and 5, students in the Philippines group strongly valued individualized feedback during one-to-one instruction, and greatly benefited from seeing active and natural L2 use in the lingua franca environment.

Research question four sought an overarching understanding about L2 learning in the two different SA contexts by integrating the QUAN and QUAL data. For both groups, the measured increase in proficiency and decrease in L2 anxiety derives from opportunities to speak English outside of class, though one-on-one lessons in the Philippines provided notable benefit. Higher gains in scoring on WTC and motivation to speak English for SA in the Philippines were largely attributable to having non-native speakers as role models for SA in the L2 lingua franca setting.

Table 2. English Proficiency: Pre- and Post- Comparison of Language Proficiency and Experiences of Japanese Female Undergraduate Students for Study Abroad in the Philippines and Canada

Study abroad country	QUAN English Proficiency Score				QUAL Interviews	Mixed Methods (MM) Meta-inferences (Interpretation of QUAN and QUAL)
	Pre-Study Abroad Mean (SD)	Post-Study Abroad Mean (SD)	Change	T score P value r score	Post-Study Abroad	
Philip-pines	442.57 (75.62)	483.62 (76.51)	+41.05	*T(20) = -3.23, p<.01, r =.34*	"I studied really hard every day…I spoke a lot of English with Koreans in the dorm." "Every day was meaningful… Each day was filled with a lot of content." "I read aloud a lot, and my teachers told me I was speaking faster."	Gains in proficiency were not solely due to one-on-one English instruction. Students used English as a common language. Feedback received during one-on-one sessions was engaging.
Canada	537.86 (79.07)	554.83 (69.14)	+16.97	*T(28)= -2.15, p<.05, r =.14*	"My English improved. I couldn't think of what to say [before]. But I had to speak even if I didn't want to, so, I am able to speak much more than before." "To improve English, it's important to talk to the host family." "I feel happy when I can communicate with foreigners. So, I want to study and improve my English."	Gains in proficiency were largely due to opportunities to practice English outside of classroom settings, e.g., homestay. Positive attitudes towards English maybe another contribution to increased communication and proficiency gains.
Interpre-tations	QUAN: Both groups made gains in language proficiency. Students going to the Philippines had lower English proficiency scores, but made greater gains than students going to Canada.				QUAL: Speaking outside of class was an important experience for participants. For the Philippines group, individual feedback from the instructor was a factor.	MM: Increased proficiency derives from opportunities to speak English more outside of class, with additional benefit of one-on-one teaching sessions in the Philippines.

Table 3. Second Language Motivation: Pre- and Post- Comparison of Language Proficiency and Experiences of Japanese Female Undergraduate Students for Study Abroad in the Philippines and Canada

Study abroad country	QUAN Second Language Motivation				QUAL Interviews	Mixed Methods (MM) Meta-inferences (Interpretation of QUAN and QUAL)
	Pre-Study Abroad Mean (SD)	Post-Study Abroad Mean (SD)	Change	*t* score *p* value *r* score	Post-Study Abroad	
Philip-pines	.69 (.16)	.83 (.10)	+.14	t(20) = -5.06, p < .001 r = .56	"The best part of the trip was my one-on-one lessons… It was so much fun…The teacher was friendly so we talked about many things." "Teachers spoke Tagalog as their native language, but they were all educated to use English as a second language. Teachers had studied English very hard." "Not just teachers, but many people on the island were bilinguals. Many people were taught in English on this island."	QUAL responses showed reasons for the improvements in motivation. Seeing the Filipino instructors as bilinguals motivated students. Many Filipinos studied English and were bilinguals - this surprised the SA students.
Canada	.67 (.16)	.73 (.12)	+.06	t(28) = -3.09, p < .01 *r* =.25	"I want to study abroad again. It was great to be in an environment where I could speak in English. There were things that I couldn't have learned [in Japan]." "I want to work in the hospitality industry. So, either in Japan or in another country, there will be foreign customers, and I want to communicate with various people. I like English."	QUAL responses explained the major reasons for gains in motivation. Speaking opportunities and engagement in English communication motivated them. Some were eager to study abroad again and continue with English learning.
Interpre-tations	QUAN: Motivation increased for both groups. The Philippines group had higher baseline motivation, and greater scores post-SA.				QUAL: Communication opportunities were important for motivation. Being in a lingua franca context was particularly useful as noted in the Philippine group.	MM: Higher gains in motivation to speak English appear attributable to having non-native speakers as role models in the Philippines.

Table 4. Second Language Willingness to Communicate: Pre- and Post- Comparison of Language Proficiency and Experiences of Japanese Female Undergraduate Students for Study Abroad in the Philippines and Canada

	QUAN L2 Willingness to Communicate				QUAL Interviews	Mixed Methods (MM) Meta-inferences (Interpretation of QUAN and QUAL)
Study Abroad Coun-try	Pre-Study Abroad Mean (SD)	Post-Study Abroad Mean (SD)	Change	*t* score *p* value *r* score	Post-Study Abroad	
Philip-pines	.46 (.04)	.66 (.04)	+.20	*t(20) = -6.08, p<.01, r =.65*	"I am used to people starting conversations with me in English… At shops, I am not surprised, and I can just talk." "At first, I didn't want to make mistakes, so I wasn't speaking much. But, later, I didn't care about mis-takes. Since there were a lot of one-on-one lessons, I began to speak a lot." "I used to never start a conversation. Also, if I didn't understand, I never asked. But I could ask now…I don't care about mistakes anymore."	Having more opportunities to speak in English in authentic communications with non-na-tive speakers of English made the SA students less con-cerned about making mis-takes.
Canada	.48 (.16)	.57 (.16)	+.09	*t(28) = -2.61, p<.01, r =.20*	"Even in Japan in Japanese, if I needed to search something, I searched it myself. Now, I don't know if my communication skills are higher, but I ask others." "Just a little bit, but I am able to initiate conversation now." "I am just a bit more cheerful now. I feel like I am letting my feelings come out. I want to express them more openly. I think it's better to say these things. I want to be more communicative."	Students were more willing to initiate conversations after SA and confident and posi-tive about communicating in English. Their higher self-perceptions of their commu-nication skills may have re-sulted in higher levels of will-ingness to communicate.
Inter-preta-tions	QUAN: Willingness to communicate increased for both groups. However, the Philippines group had lower scoring at baseline and more gains as evidenced by higher scores.				QUAL: Both groups talked about opportunities to speak English out of class as important experience. The Philippine group also talked about the im-portance of one-to-one instruction.	MM: Authentic communica-tion opportunities explain greater WTC. Observing Fili-pino non-native L2 role mod-els reduced concern for mis-takes.

Table 5. Second Language Anxiety: Pre- and Post- Comparison of Language Proficiency and Experiences of Japanese Female Undergraduate Students for Study Abroad in the Philippines and Canada

	QUAN L2 Language Anxiety				QUAL Interviews	Mixed Methods (MM) Meta-inferences (Interpretation of QUAN and QUAL)
Study Abroad Country	Pre-Study Abroad Mean (SD)	Post-Study Abroad Mean (SD)	Change	*t* score *p* value *r* score	Post-Study Abroad	
Philip-pines	.75 (.26)	.57 (.28)	-.18	*t(20) = 3.30, p<.01, r =.35*	"My one-to-one teacher tried hard to understand my English, so I wasn't nervous anymore." "[in the one-on-one lessons] My teacher worked on my pronunciation a lot... Everyone on the street was so friendly there." "In Japan, I hated speaking in English…I hated talking to non-Japanese…But, I got used to it there, and really had fun talking to people from other countries."	Decrease in learner anxiety was largely due to one-to-one lessons. Students were highly engaged and had fun in the Philippines communicating in English.
Canada	.68 (.27)	.54 (.26)	-0.14	*t*(28) = -3.14, *p*<.01, *r* =.26	"I am so shy so I can't speak loud, my voice be-comes shaky, and I speak so fast... But I am able to speak [better] now so I feel good." "When foreigners are lost in Japan, I was asked [for directions], but I really couldn't understand so, I said 'sorry' and ran away. I used to have an image of foreigners as scary. But I feel like I can initiate conversations now."	Learners were less anxious to speak, but did not give rea-sons for the change. Decease in learner anxiety may have been due to becoming used to speaking with non-Japanese speakers.
Interpre-tations	QUAN: Both groups experienced reduced language anxiety. Students going to the Philippines had higher baseline scores. These students experienced greater reduction in their anxiety scores at post-SA.				QUAL: Speaking practice contributed to reducing language anxiety for both groups. For the Philip-pines group, one-to-one lessons were important experiences.	MM: Language anxiety de-creased in both countries due to experience speaking Eng-lish, but reduced L2 anxiety for students in Philippines re-lated especially to one-on-one lessons.

4 Conclusion

This research shows that short-term SA in a non-traditional destination like the Philippines and a traditional destination like Canada effectively improved L2 proficiency, as well as affective factors such as L2 WTC, L2 motivation, and lowered L2 anxiety. Use of mixed methods joint display analysis [5] as seen in Tables 2, 3, 4 and 5 revealed that the Philippine group had higher gains on outcomes for all 4 measurements. In addition, by comparing the interview data from both SA destinations, one-on-one instruction, instruction by a second language speakers of English, and living in a lingua franca SA environment catalyzed improvement of affective factors for the SA participants in the Philippines. These novel findings illustrating the benefits of SA in a lingua franca context have not been reported previously. Methodologically this highlights the value of using a mixed methods research approach. As for study limitations, all participants in the Philippines and Canada were female. Incremental advances in WTC and reduced L2 anxiety in the Philippines appear attributable to living in a lingua franca SA environment, but could also be influenced by the one-on-one instruction. The SA period in the Phillipines was slightly longer. Caution should be used in extrapolating the results.

Acknowledgments. This paper is the final stage of a research project to clarify the differences of short-term SA programs in traditional and non-English speaking SA environments. This research was funded by Grants-in-Aid for Scientific Research, Research (C), the Ministry of Education, Culture, Sports, Science and Technology Japan. Grant number: 16K02984. The authors would like to thank Dr. Paul Joyce at Kindai University, Japan, for providing support for statistical analysis throughout all stages of this program of research.

References

1. Ministry of Education, Culture, Sports, Science and Technology (MEXT) Report (2020). https://www.mext.go.jp/content/20200421-mxt_gakushi02-100001342_1.pdf
2. Tajima, C.: Language training in the philippines: measuring and exploring learner experiences. In: Nazir, S., Teperi, A.-M., Polak-Sopińska, A. (eds.) AHFE 2018. AISC, vol. 785, pp. 430–442. Springer, Cham (2019). https://doi.org/10.1007/978-3-319-93882-0_41
3. Tajima, C.: Language and cultural learning in a short-term study abroad: an investigation of japanese sojourners in Canada. In: Goossens, R.H.M., Murata, A. (eds.) AHFE 2019. AISC, vol. 970, pp. 572–577. Springer, Cham (2020). https://doi.org/10.1007/978-3-030-20145-6_57
4. Tajima, C.: A mixed methods case study: japanese sojourners' intercultural awareness. In: Karwowski, W., Goonetilleke, R., Xiong, S., Goossens, R.S., Murata, A. (eds.) AHFE 2020. AISC, vol. 1215, pp. 515–521. Springer, Cham (2020). https://doi.org/10.1007/978-3-030-51549-2_68
5. Fetters, M.: Mixed Methods Research Workbook. Thousand Oaks, Ca, SAGE (2020)

Cultural Mediations Between Branding and Lifestyles: A Case Study Based Model for the Articulation of Cultural Strategies and Urban Tribes

Nelson Pinheiro Gomes[1(✉)] and William Afonso Cantú[1,2]

[1] CEAUL, Centro de Estudos Anglísticos da Universidade de Lisboa, Faculdade de Letras, Alameda da Universidade, Lisbon, Portugal
nelsonpinheiro@campus.ul.pt

[2] IPAM, Instituto Português de Administração e Marketing, Quinta do Bom Nome, Estr. da Correia 54, Lisbon, Portugal

Abstract. This paper aims to cross concepts and analytical approaches in culture, strategy and urban tribes to provide a process that can diagnose brand communication pieces and larger campaigns within a specific cultural landscape or larger contexts. With a mix of qualitative methods, this articulated critical approach allows for the understanding of branding and strategic communication objects. Within a scope of cultural analysis, it is possible to activate a textual reading that highlights both visible and invisible webs of meaning. These impact the way we read brand texts and can be made to inspire urban tribes and activate identification mechanisms that inhibit collective mindsets. This cultural diagnosis process of brand objects can generate important strategic insights for brand communication and management.

Keywords: Cultural analysis · Strategy · Branding · Urban Tribes · Cool

1 Introduction

This research aims to contribute to the conceptual-methodological debate between approaches that cross strategy and cultural analysis in terms of a strategic cultural analysis and management. We highlight the proposals of Cultural Branding [1], Cultural Strategy [2] and Tribal Marketing [3] to see possible contributions of the latest to the first two.

This is a diagnosis process and model to generate strategic guidelines. We start the model with a review and mapping of methods, like semiotics, to understand the analyzed cultural objects and their contexts. This is followed by an articulation with social groups and the shared tribal link and bonds that objects may enclosure and manifest. This model enables a solid diagnosis of brand communication and allows us to understand (i) the construction and signs of brand cultural expressions and their possible associations with urban tribes and (ii) deconstruct the applied cultural formula. With this, we gain strategic

R. S. Goonetilleke et al. (Eds.): AHFE 2021, LNNS 273, pp. 470–476, 2021.
https://doi.org/10.1007/978-3-030-80713-9_60

insights and generate better future solutions. To test the analytical component of our model, we will apply it to a case study. Namely, an Instagram object of Nintendo Switch, in an in-depth hermeneutical analysis of the mix between brand DNA and lifestyles DNA.

2 Strategic Cultural Analysis for Brand Communication

In 2019, Gomes articulated conceptual and methodological approaches of cultural branding and strategy with that of tribal marketing. This suggested that applied cultural analysis at the strategic level is gaining attention. We can see this mainly through the works of Douglas Holt, Douglas Cameron, Bernard Cova and Véronique Cova [1–3]. These contextualized the proposal for a strategic management of culture with said approaches. The proposed methodological roadmap can be applied in the development of communication strategies and brand management, considering their relationship with the various audiences and an attention to emerging changes in collective mentalities [4]. We take this research as a starting point for our literary review and discussion. Here, cultural analysis practices - from a critical perspective to field work or semiotic based hermeneutical readings - take the form of an instrument that unveils sociocultural dimensions, dynamics and shared meanings capable of generating strategic insights. The context for this work starts with cultural branding [1] and strategy [2] and is followed by tribal marketing [3].

In 2004, Holt was already looking into cultural branding and the importance of looking at culture as a source. The objective of this perspective is to create iconic brands. According to Holt, "they have distinctive and favorable associations, they generate buzz, and they have core consumers with deep emotional attachments" [1, p.35] but these are to be understood as the result of successful mythmaking [1]. This way, we harness myth to "commercial purposes" [5, p. 101]. "Customers buy the product to experience these stories" [1, p. 36], so the main objective is to connect with narratives and meanings, an identification process that is promoted by communication instruments. For this, the author underlines the importance of having the right cultural expression: "the particular cultural contents of the brand's myth and the particular expression of these contents in the communication. [...] detail the brand's stakes in the transformation of culture and society and the particular cultural expressions the brand uses to achieve these transformations" [1, p. 36–37].

In terms of cultural strategies, Holt and Cameron underline the concept of cultural innovation as "a brand that delivers an innovative cultural expression" [2, p. 173]. The authors largely elaborate on this idea, suggesting that cultural innovations have the right ideology (a concept) expressed via cultural codes (a composition of elements that allow a correct interpretation of meanings) that build a myth, which is built upon source materials (subcultures, media myths and brand assets) that delve outside a mainstream view [2]. There is a link between this idea of cultural innovation and emerging socio-cultural trends. As the authors underline, "the engine of cultural innovation is historical change in society that is significant enough to destabilize the category´s cultural orthodoxy, creating latent demand for new cultural expressions" [2, p. 185]. Schroeder adds, "brands are not only mediators of cultural meaning; they themselves have become ideological referents that shape cultural rituals, economic activities and social norms" [6, p. 1523]. As representations of change, these expressions act as signals of socio-cultural movements and trends. Thus, cultural innovations can also be understood as a creative/cool signal, the translation of socio-cultural changes and ideas emerging in the construction and dissemination of certain objects (artifacts, practices, and representations). Cultural innovations, resulting from an innovative cultural expression, are narratives and objects that reveal disruptions and changes in socio-cultural orthodoxy [2], an understanding close to the concept of "cool" [7–9]. To build cultural innovations and a cultural strategy, Holt and Cameron underline six steps: "Map the Category's Cultural Orthodoxy"; "Identify the Social disruption that can Dislodge the Orthodoxy"; "Unearth the Ideological Opportunity"; "Cull Appropriate Source Material"; "Apply Cultural Tactics"; "Craft the Cultural Strategy" [2, p. 196–199]. In our diagnosis process, it is important to consider the cultural orthodoxy and mainly the social disruption, the change. This will give us an insight into the socio-cultural context and the nature of the emerging elements. Also, understanding the source material will give us an insight into the used cultural codes, the construction of the cultural innovation (via a cultural formula), and how the elements were articulated to construct a narrative. We use a mainly hermeneutical approach because, as O'Reilly states, "given the symbolic dimensions of culture, and given that branding is a symbolic enterprise, a discursive-analytical treatment of branding as discursive practice would open to way to a critical appraisal of the relationships between business and culture" [10, p. 585].

Taking the highlighted role of identities and subcultures, and adding to this conceptual-methodological roadmap, it is also important to consider the role of tribal marketing. Again, we follow the review and articulation already established by Gomes [4]. The perspective of Torelli [11] that an iconic brand symbolizes an abstract image that can be valued by a subculture opens this discussion. But we underline the concept of tribe as a gathering of a heterogeneous group of people connected by a strong link, a shared passion/interest [3]. Or, as Canniford suggests, a group based on emotions that have a diffused structure, "who, through the features of a hybrid, affectual, performative and changeable tribal network, enter into productive, democratic and symbiotic dialogue with market offerings. In so doing, they foster and nurture linking value" [12, p.70]. In this sense, our analytical process must identify elements that may call out for urban tribes. The cultural expression construction can take place in articulation with the tribes or with tribes in mind, to generate identification processes and facilitate connections. Tribes can be an active agent in the construction of objects, including communication.

3 Analytical Model and Methodological Process

This analytical process is built upon a model that allows us to define socio-cultural signals as cool [8] phenomena and as cultural innovations [2]. Its objective is to understand the hidden cultural formula. It encompasses (i) the analytical perspective of cool signals proposed by Rohde [7] and Dragt [13] in the description and decoding of cool characteristics and trends associations; (ii) semiotics and ethnographic practices to understand cultural objects and contexts as well as social groupings; followed by (iii) an understanding of the shared tribal links [3] and bonds that objects enclosure and manifest. This work highlights the diagnosis case of brand communication. Registration and analysis should follow the described process:

1. *Title, the illustrative visual element and source(s) for additional information.* Following the scheme of Rohde [7] and Dragt [13], after presenting a title, the analyst should add a visual element that illustrates the signal along with sources and links that provide additional information and the date of the analysis.
2. *Description of the main signal characteristics and context.* Again, following the scheme of Rohde [7], the analyst must highlight the main characteristics of the signal, its visible elements, physical and functional attributes, creation date, production and display contexts, among others that can give a clear description of the object. In a communication piece it may also be relevant to highlight participants.
3. *Connotative and denotative analysis, with the possible identification of myth(s).* In terms of cultural branding [1], it is important to be on the lookout for myths. For this specific exercise, we follow the analytical procedure of Barthes [14, 15] and Volli [16]. We start with a denotative reading (the more descriptive, immediate and almost dictionary like meanings) of the signs (considering the signifier and the signified), followed by a connotative reading (further associations) [14, 16, 17]. After this, we unveil the possible myth that inhabits the narrative. As Barthes states, "it is constructed from a semiological chain which existed before it: it is a second-order semiological system" [15, p. 113]. Myths provide a more complex reading and a wider network of association on top of the first reading, which only becomes a new signifier. There is a parallel between this conception of myth and the advertising sign proposed by Volli, as a process of mythical construction where there is a direct connection between the (denotative) sign and the brand/product [16].
4. *Interviews, inquiries and focus groups on the object.* Only when pertinent, the analyst may choose to conduct interviews, inquiries or focus groups regarding the brand/product and the reception of the communication object.
5. *Cool DNA analysis.* The analyst must go through the characteristics of cool, like Rohde [7] suggests, to define the signal/object as cool or not. The Trends and Culture Management Lab proposes a definition of cool, having followed an extensive literary research of the concept [8], as something "relevant", "viral", "current", "irreverent", "instigating", and with a proposal of "discontinuity" [8].
6. *Articulations with urban lifestyles and tribes.* Following our literary research on the topics [3], this specific exercise entails identifying if the narrative calls out to one or more tribes by (i) identifying elements that may address identities and shared passion/interests and (ii) finding the links between these elements.
7. *Definition of the formula and cultural innovation based on emerging mindsets.* The analyst defines if there is a cultural innovation by identifying the change in the cultural orthodoxy that the object addresses and the proposed solution [2]. We review the cultural codes [2] of this object, and we take into consideration the cultural formula, the way in which the advert was constructed and the elements present in the composition. We follow here the revision made by Gomes [18] of the Marlboro case study developed by Holt and Cameron [2]: we cannot just copy and paste cultural elements in communication pieces, "it is necessary to generate the right context (characters, scripts, dialogues, music, among others) capable of managing and articulating cultural codes […] and give rise to cultural expressions that meet the expectations and emerging mindsets of consumers" [18, p. 63]. So, the focus is on the formula presentation and articulation of cultural elements.

8. *Strategic Insights.* Based on the results of the former exercises (from 1 to 7), the analyst tries to understand (i) the creative germ(s) behind the object that may be translated to others, (ii) the main strategic topic and (iii) the major socio-cultural changes being addressed.
9. *Articulation with identify[ ]ed socio-cultural trends.* The final exercise, based mainly on the results of exercises 5, 7 and 8, is to associate the signal/object with an already identified socio-cultural trend(s) from the analyst's chosen trend map/bank.

4 Case Study: Nintendo Switch, Mario Kart Live

1. "Mario Kart Live: the veil between the physical and the digital".
   Visual Element (video) in www.instagram.com/p/CF2U1wdDTu5, and additional information in mklive.nintendo.com. The analysis took place on the 31st December 2020.
2. The object is a promotional trailer video for the "Mario Kart Live", a game for the Nintendo Switch console. Following the former Mario Kart(s), this game provides a physical kart (with Mario or Luigi) equipped with a camera and cardboard gates that can be placed in one or more divisions of a person's home. When we play the game on Nintendo Switch, the physical car reacts to our commands and the race starts in the physical space and the console. The game generates other characters, and the physical space is transformed in the screen adding visual elements of the game.
3. From a car to house elements (sofa, table, closets), plants, cardboard objects, and game elements like bananas, shells, bullets, among symbolic characters, all compose a list of signs that provide denotative readings. As a whole, the connotative reading is one where Mario related elements mix with our personal physical reality: "Mario Kart/World can be anywhere, including your living room; Mario world is your world".
4. Following other games that blend the physical and the digital world, this is another cool signal that shatters that division of realities. Also, considering the current context of pandemic and social confinement, it provides means to transform our home into a new space adding to (i) the important ways in which we are transforming our relationship with our personal spaces and (ii) the way we define home and its dynamics.
5. There is a connection to gamers, although it is not exclusive and can enroll new players in family/friends' dynamics and Nintendo lovers in general. It highlights a Super Mario symbolic world, elements we recognize (images and sounds) from the former games, and it appeals to a competitive spirit, among other gaming elements.
6. The trailer is created to have a descriptive and explanatory nature. Narrative takes place at two levels enacted simultaneously: (i) it explains how to set the game in the space and console and how play it - a simple how to assemble and do it; (ii) also, the trailer takes you to the Mario Kart symbolic world, starting with Super Mario to take you further into race sets and their elements, among other characters. In general, the formula is assembled by mixing the physical and game elements so that we see the transformation of our home into a new world filled with the characters/elements of Mario Kart in an increasingly immersive story.

7. Strategic Insights. Considering the increasing time we spend in the digital world, there is much to do in terms of articulating the physical reality with the digital one. It is more than providing augmented reality, it is about changing spaces, functionalities and adding experiences taking the familiar spaces to new places: create realities!
8. Considering the Trend Map of the Trends and Culture Management Lab [19], an academic and scientific project, we can associate this signal mainly to the "Anchored Narratives" and "Lifestyle Redesign" macro trends.

5 Final Considerations

The proposed model articulates concepts, practices and perspectives from cultural analysis, cultural branding and strategy, and tribal marketing into an approach of strategic cultural analysis and management. The model has a mainly hermeneutical process that calls to semiotics readings and also the reading of creative/cool signals that comes from Trend Studies. It allows us to go deep into the meanings of a signal, in this case a strategic communication object, to understand its cultural context and to generate strategic insights. It can be applied as a diagnosis tool in the final stage of approval of a communication piece; or it can be a benchmark exercise for new objects.

References

1. Holt, D.: How Brands Become Icons: The Principles of Cultural Branding. Harvard Business School Press, Boston (2004)
2. Holt, D., Cameron, D.: Cultural Strategy - Using Innovative Ideologies to Build Breakthrough Brands. Oxford Press, Oxford (2010)
3. Cova, B., Cova, V.: Tribal Marketing: The tribalization of society and its impact on the conduct of marketing. Eur. J. Mark. **36**(5/6), 595–620 (2002)
4. Gomes, N.: Gestão da Cultura: estratégia e articulações no âmbito do branding e das tribos urbanas. e-revista Logo. **8**(2), 25–47 (2019)
5. Cayla, J., Arnould, E.: a cultural approach to branding in the global marketplace. J. Int. Mark. **16**(4), 86–112 (2008)
6. Schroeder, J.: Corporate branding in perspective: a typology. Eur. J. Mark. **51**(9/10), 1522–1529 (2017)
7. Rohde, C.: Serious Trendwatching. Fontys University of Applied Sciences and Science of the Time, Tilburg (2011)
8. Trends and Culture Management Lab: A natureza do Cool, um desafio académico. https://creativecultures.letras.ulisboa.pt/index.php/gtc/
9. Gomes, N., Cohen, S., Flores, A.: Estudos de Tendências: Contributo para uma abordagem de análise e gestão da cultura. Moda Palavra. **11**(22), 50–81 (2018)
10. O'Reilly, D.: Cultural Brands/Branding Culture. J. Mark. Manag. **21**, 573–588 (2005)
11. Torelli, C.J.: Globalization, Culture, and Branding: How to Leverage Cultural Equity for Building Iconic Brands in the Era of Globalization. Palgrave Macmillan, New York (2013)
12. Canniford, R.: A typology of consumption communities. Res. Consum. Behavior. **13**, 57–75 (2011)
13. Dragt, E.: How to Research Trends: workbook. BIS Publishers, Amsterdam (2018)
14. Barthes, R.: Elements of Semiology. Hill and Wand, New York (1986)
15. Barthes, R.: Mythologies. Noonday, New York (1991)

16. Volli, U.: Semiótica da publicidade: a criação do texto publicitário. Edições 70, Lisboa (2003)
17. Oswald, L.: Marketing Semiotics: Signs. Strategies and Brand Value. Oxford University Press, Oxford (2012)
18. Gomes, N.: A análise cultural e o estudo de tendências na comunicação e gestão de marcas: Estudo de caso publicitário em contexto semiótico. International J. Mark. Commun. New Media. (7), 56–79 (2020)
19. Trends and Culture Management Lab: Tendências Socioculturais - Um Mapa de Macro e Micro Tendências (2020). http://creativecultures.letras.ulisboa.pt/index.php/gtc-reports/

Correlations Between Inspections, Maintenance Errors, and Accidents

Toshiyuki Wakimizu[1,2](✉), Atsuo Murata[2], Toshihisa Doi[2], Yutaka Yoshida[1], and Keisuke Fukuda[1]

[1] West Japan Railway Company, Safety Research Institute, 2-12 Matsuzaki-cyou 1-chome, Abeno-ku, Osaka 545-0053, Japan
{toshiyuki-wakimizu,yutaka-yoshida,keisuke-fukuda}@westjr.co.jp
[2] Department of Intelligent Mechanical Systems, Graduate School of Natural Science and Technology, Okayama University, 3-1-1, Tsushimanaka, Kita-ward, Okayama 700-8530, Japan
murata@iims.sys.okayama-u.ac.jp, tdoi@okayama-u.ac.jp

Abstract. Factors common to past accidents and incidents were identified by analyzing and summarizing background factors related to accidents and incidents caused by vehicle and equipment inspection and maintenance errors. As a result, measures for improving railway safety were classified. Reports on railway and aircraft accidents and serious incidents in Japan were analyzed, and their background factors were identified. Then, these factors were classified using the m-SHEL model. The results indicated that work environment-related factors were observed in 80% of the cases.

Keywords: Accident · Incident · Railways · Background factors · m-SHEL

1 Introduction

On August 8th, 2015, a West Japan Railway Company's (JR West) eight-car Shinkansen was traveling from Shin-Osaka to Kagoshima-Chuo Station at approximately 295km per hour through the Shiroumaru Tunnel. Then, the side closure plate installed on the left corner of the second car fell off and collided with the left side of the third car and injured passengers. The increased torque on the bolt attaching to the side closure plate, which did not have the specified torque value, caused this accident. As a result, the bolt fell off because of the moving train's vibration. After conducting tests, the Japan Transport Safety Board (JTSB) reported that maintenance errors when installing the closure plate caused the accident. Inspection workers did not identify this fault, even though regular inspections had been conducted before the accident [1].

Regular inspections and maintenance ensure the safety of public transportation, including railways. JR West is conducting inspections of Shinkansen cars every two days, regular inspections every 30 days or every 30,000 km, bogie inspections every 18 months or every 600,000 km, and general inspections every 36 months, or 1,200,000 km. Accidents might happen if there are specific errors during these inspections. It is the essential task of every worker to reduce human errors and conduct accurate inspections.

R. S. Goonetilleke et al. (Eds.): AHFE 2021, LNNS 273, pp. 477–482, 2021.
https://doi.org/10.1007/978-3-030-80713-9_61

This study identified factors common to past accidents and incidents by analyzing and summarizing background factors resulting from equipment maintenance and vehicle inspection errors and clarified measures to improve railway safety.

2 Method

The present study analyzed ten cases [3–11] based on reports by JTSB. These cases were accidents and serious incidents caused by human errors during inspections or maintaining railway vehicles or large aircraft. Regular inspections are conducted on railway vehicles and aircraft based on the distance they have traveled or the number of days. Much of this work and inspections are conducted by humans, which might result in human errors. Human errors are caused by individual workers and by different background factors. This study used the m-SHEL model [2] for analyzing background factors in human errors (Fig. 1). The m-SHEL model classifies human error factors into five categories; Management, Software, Hardware, Environment, and Live-ware, or the people that work and related others. People are at the center of the model, and problems occur when there is disharmony between these people and their surroundings.

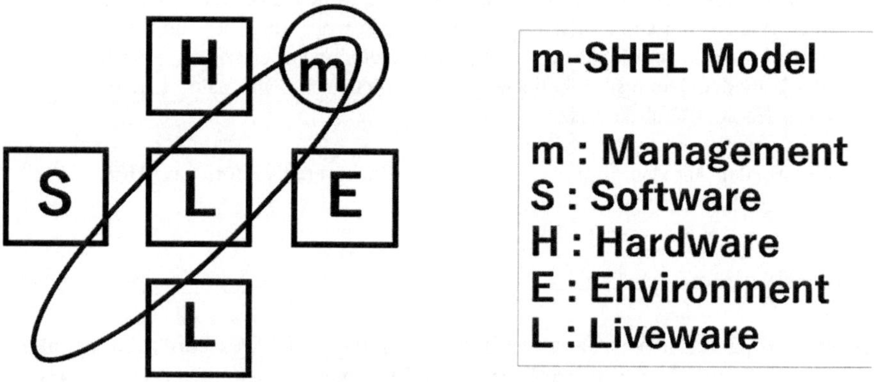

Fig. 1. m-SHEL model.

The “L” at the center of the m-SHEL model is the person conducting inspections or maintenance. To extract the background factors, two workers with on-site work experience conducted an analysis using the “Why-Why” analysis (or the 5 Why method).

Figure 2 shows the results of analyzing faulty vehicle parts that were not detected during inspections, which caused the personal injury accident because a side closure plate separated (JTSB, RI2016-5). We repeated “Why?” for the events described in the report and visualized their background factors. The last item at the end of a branch was classified into the m-SHEL model.

3 Results

Table 1 shows 15 background factors of ten cases that were classified according to the m-SHEL model and the percentage that each factor was observed. “L1: insufficient

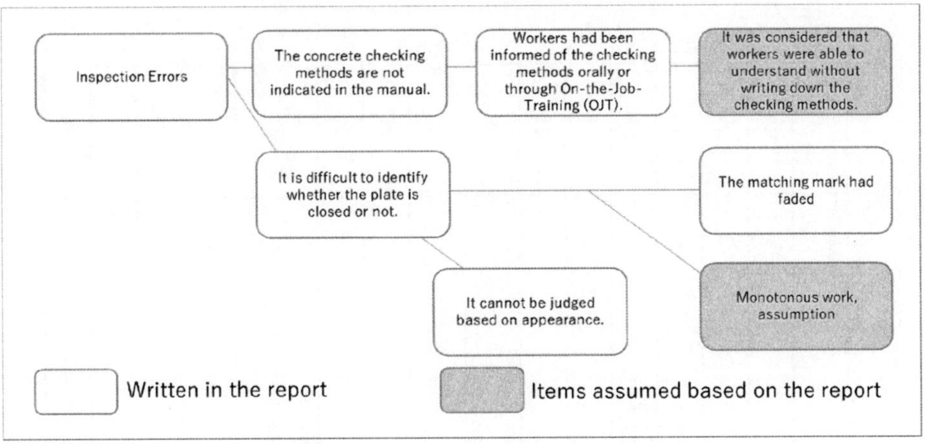

Fig. 2. An example of "Why-Why" analysis

understanding and experience" was observed in five cases because of factors related to the person making errors, including "Not understanding why the rule was made", "Not understanding the characteristics of electric wires or flux, or points to note when using them", and "Doing the work for the first time", among others.

"E4: problems in the work environment" was observed in eight cases caused by factors related to the dark and small space that is difficult to check or work, among others. Inspection and maintenance work of railway vehicles and aircraft is sometimes conducted in small and dark places, including under inspection covers, which could have caused such problems (As shown in Fig. 3).

"L-H: structural problems," including "vehicles of the same model sometimes have different structures" or "clear signs are not posted," was detected in six cases, and "L-S: problems in manuals and checklists", including "there are no manuals" or "checkpoints are unclear", was also observed in six cases.

Regarding management-related problems, "m4: call for attention" was detected in two cases, including attending too closely to the items that were identified as priority checking items based on previous cases and not attending to surrounding items, which resulted in overlooking faults.

4 Discussion

The results indicated eight cases included background factors related to the work environment. The darkness in the work area was observed in five cases, and locations difficult to work or check were observed in three cases. Several areas of a large structure, such as railway vehicles, cannot be properly illuminated for inspection and maintenance work by only using general lighting. Ensuring good visibility is an important factor for successful visual inspection. More careful work or local lighting is required to ensure visibility.

The background factors of the two cases was a call for attention to similar cases causing inspection errors. Notification methods should be improved when adding priority checkpoints to prevent such errors.

Table 1. Background factors of ten cases.

Background factors			Examples of case analysis	The number of cases	The percentage of observations
m	m1	Safety culture	Not investigating the causes of problems and not taking countermeasures	3	30%
	m2	Role division problems	The worker tightening the torque was not decided.	2	20%
	m3	Risk management	Inadequate fail-safe designs. Not understanding working conditions and problems.	5	50%
	m4	Call for attention	Only concentrate on prioritized items for checking and not attend to surrounding items	2	20%
	m5	Workers' management	OJT by inexperienced workers (inadequate training)	2	20%
L-S		Problems in manuals and checklists	Too many items and items not fitting the current condition	6	60%
L-H		Structural problems	No signs. Vehicles of the same model have different structures. Impossible to judge whether the closure plate is closed or not by the appearance	6	60%
L-E	E1	Rare work	Work conducted only once a year Unusual situations	3	30%
	E2	Monotonous work	Repeating the same job many times	1	10%
	E3	Time pressure	Completing inspections within approximately 10 minutes before the opposite train passes	3	30%
	E4	Problems in the work environment	The work area is dark and difficult to see. The location is difficult to work on or check. The work area is narrow and difficult to work on. It is difficult to concentrate because of working with new employees.	8	80%
L	L1	Insufficient understanding and experience	Not understanding of why a rule was made. Not understanding the object of a job or the materials to use.	5	50%
	L2	Reduced attention	Sleepiness, fatigue, other reasons	1	10%
	L3	Assumptions	Assuming something is ok because there were no problems in the past.	4	40%
L-L		Communication problems	Cannot cast doubt on what the boss says. Communication is insufficient for confirmation.	4	40%

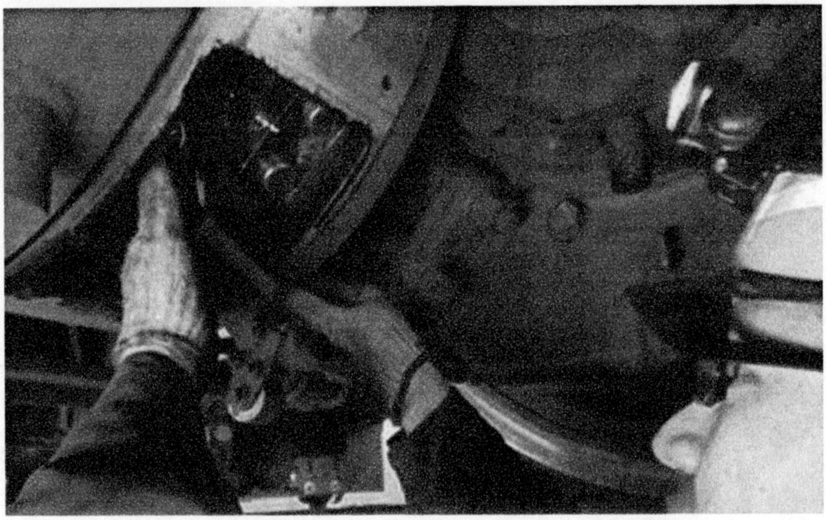

Fig. 3. An example of working in small and dark space

We plan to contribute to further improving railway safety by increasing the number of cases that we analyze.

References

1. The Japan Transport Safety Board: Railway accident investigation report. A personal injury accident between Ogura Station and Hakata Station on the Sanyo Shinkansen, West Japan Railway Company (in Japanese). RI2016-5, pp. 1–35 (2016)
2. Murata, A.: Science of Human Error. Nikkan Kogyo Shinbun, Ltd. Japan (2008)
3. The Japan Transport Safety Board: Railway serious incident investigation report. A serious incident related to vehicle troubles at Shirakibaru Station on Tenjin Omuta Line, West Japan Railway Company (in Japanese). RI2019-2, pp. 1–33 (2019)
4. The Japan Transport Safety Board: Railway accident investigation report. A derailment accident on Kitairie signal station premises on Muroran Line, Japan Freight Railway Company (in Japanese). RA2018-5, pp. 1–60 (2018)
5. The Japan Transport Safety Board: Railway accident investigation report. A train fire between Tsukuda Station and Iwamoto Station on Joetsu Line, East Japan Railway Company (in Japanese), RA2014-8, pp. 1–26 (2014)
6. The Japan Transport Safety Board: Railway serious incident investigation report. A serious railway incident between Tsunetoyo signal station and Kamiatsunai Station on Nemuro Line, Hokkaido Railway Company (in Japanese). RI2014-3, pp. 1–42 (2014)
7. The Japan Transport Safety Board: Aviation accident investigation report. Engine fire during a take-off run. Korean Air (in Japanese). AA2018-5, pp. 1–77 (2018)
8. The Japan Transport Safety Board: Aviation accident investigation report. China Airlines. B18616 (in Japanese), AA2009-7, pp. 1–97 (2009)
9. The Japan Transport Safety Board: Railway serious incident investigation report. A serious incident at Miyazakijingu Station on Nippou Line, Kyushu Railway Company (in Japanese). RI2009-4, pp. 1–39 (2009)

10. The Japan Transport Safety Board: Railway accident investigation report. A train collision at Bifuka Station on Souya Line, Hokkaido Railway Company (in Japanese). RA2006-1, 13p (2006)
11. The Japan Transport Safety Board: Railway accident investigation report. Train fire at Awakawaguchi Station on Dosan Line (in Japanese). RA2004-4, pp. 1–24 (2004)

Globalization, Cultural Pluralism and the Space of the Human "Borderless Career" World

Agnieszka Cybal-Michalska(✉)

Adam Mickiewicz University, ul. Wieniawskiego 1, 61-712 Poznan, Poland
agnieszka-cm@wp.pl

Abstract. The postmodern, ambiguous and ambivalent social and cultural reality, undergoing never-ending fluctuations, leads to a radical social change expressed as permanent self-creation. The driving force behind it is the desire for equilibrium on the one hand and conflict and tension in the social structure on the other. The multidimensional global changes which affect the modern world and touch nearly all aspects of social existence seem unavoidable, and their complexity hard to grasp. This world's globalist tendencies have deprived today's society of the stability which was taken for granted in the modern industrial society era. All the same, the dialectic character of the globalization phenomenon determines the ability of societies to anticipate hazards and handle them on the individual, social, and global level. New opportunities and new possibilities arise on micro and macro scale, accompanied by anxieties and fears which form the outlines of the society of the future. The strictly global discourse on contextual theory of reality, determinants of change in the social system, and the situation of the modern man who lives in a random global structure and is made to find their place in the constantly changing reality, is one of the most important ones for the discussion of the phenomenon of "borderless careers". The qualitative aspects of the evolution of relations between globality and locality, society and the individual, and their links, are not without consequence for the multicontextual changes in the labor market, which present new challenges to employees. This vision implies a completely new perception of reality, the methods of interpreting the world and the quality of opinions on the situation of the modern man – including man as the manager of their own career. This vision implies completely new ways of interpreting the world in which "a career makes a career" – they are problems of "a borderless career", "a varied career", "a post-corporate career" as a *novum*, of sorts, in the planning of a broadly understood career, the rise in importance of having a career and the ability to plan, manage and monitor one's own career in the perspective of one's whole life. The most important to an individual – internal – meaning of a career must be juxtaposed with its external characteristics. The co-existence of cognitive, normative, emotional and behavioral components is accompanied by the objective of achieving things that are desirable for the individual in their career. Engaging in a career is characterized by the development of personal career objectives, attachment and identification with these objectives and, finally, becoming engaged in them. Changes in the world of "borderless" careers shine a new light on the problem of individual proactivity – an individual's proactive behaviors in their career. A distinguishing feature of proactive career planning, development and management is the awareness of being an agent who takes steps in the direction they themselves desire, and shapes the surrounding reality by initiating changes.

R. S. Goonetilleke et al. (Eds.): AHFE 2021, LNNS 273, pp. 483–494, 2021.
https://doi.org/10.1007/978-3-030-80713-9_62

Proactivity understood as an individual's attitude is revealed in the processes of striving for goals and objectives, which in effect turns the individual into the creator of reality rather than a mere forecaster. Proactive persons will explore their environment in search for opportunities for change, think outside the limitations of a situation and handle it in such a way as to mine it for benefits resulting from changing it. Career development in the world of "borderless careers", in the reality of a globalizing world, affords to an individual a multitude of ways to create themselves and foster their inherent potential.

Keywords: Globalization · Career · Professional identity · Boundaryless career

1 Introduction

The transformations of the contemporary world and the attempts aimed at characterizing the broad context of processes related to globalization lead us to seek answers to questions concerning the origins and the scope of the concept of globalization. The notion of "globalization" is one of the terms bearing ambiguity and is sometimes interpreted in various ways in the literature. The following theoretical considerations on what the concept of globalization denotes and connotes present an attempt to bring this commonly used and ambiguous term closer to the reader by reviewing the most important definitions and aiming to identify some of the most frequently occurring elements of meaning to define the phenomenon of globalization that interests us.

2 Origin and Scope of the Concept of Globalization

The issue of globalization as a description of a common denominator for the peculiarity of processes taking place in the world has been the focus of interest of various scientific disciplines (philosophy, sociology, political science, cultural studies, pedagogy, economics) quite recently, so it does not have an established and rich tradition. The first sociological article in which the term "globalization" appeared in its title was the text by R. Robertson (1985), entitled *The Relativization of Societies: Modern Religion and Globalization* [1]. It is only in the 1990s that we notice - as M. Waters states - the globalization of the use of this concept, which becomes ubiquitous and may constitute a basic category of humanist discourse and as a less controversial term, it replaces the concept of "post-modernism" to describe the transformations of the contemporary world [2]. The term has become a part of the "global consciousness", a concept capturing "global" aspects. Although the adjective "global" was in use much earlier (to denote the "world-wide range" and "totality"), it is symptomatic of contemporary empirical interest in globalization to focus on describing global interdependencies [3].

The starting point for considering the broad spectrum of the issue under discussion is the recognition of globalization as a phenomenon. In those terms, globalization is understood as the fact that "can be observed, perceived with the senses, as something unique" [4], "occurring in some field" [5] or fields. Considering globalization only as a process, i.e., a course of causally related, successive changes constituting stages of development, transformation of something [6], only partially brings this phenomenon closer. An inherent feature of considerations concerning the globalization is to pay attention to the structure and function of this phenomenon. Thus, in the analyses of many authors it will be possible to see a qualitative diversity of approaches to globalization, in process-related, structural and functional terms, interpreted as a phenomenon that plays specific roles, contributing to the image of the complexity of the contemporary world. Moreover, as R. Robertson points out, the diversity of "responses" to globalization has an impact on the quality, direction and outcome of this process, allowing us to conclude that the shape of the "global field" is to a large extent "dependent on ourselves" [7].

According to M. Albrow, globalization refers to all those problems by which the nations of the world are integrated into a single world society, the global society [8]. The direction of similarly understood transformations which cover the entire globe, is illustrated by R. Robertson, claiming that globalization is a set of processes that create one common world [9]. The position of R. Robertson, referred to as "voluntaristic theory", is based on the belief that individuals, societies, "systems" of societies as well as humanity as a whole should be treated as a single, coherent analytical framework. The avoidance of reductionism, with its overtly functionalist, utilitarian and materialist forms, is an equally important issue in this regard. "The global field", as a whole, is a socio-cultural "system", which on the one hand is the result of "world compression" but on the other hand - makes cultures, civilizations, ethnic groups, national societies, "intra-" and "cross-" national movements and organizations independent [10]. Processes coupled with the phenomenon of globalization are often treated as diffusions of cultural patterns and referred to as westernization, macdonaldization (G. Ritzer) or Americanization, which is emphasized in their analyses by H.P. Martin and H. Schumann, when they write that America admittedly is not everything, but without America everything has been nothing so far [11]. M. Golka adds that nowadays the world watches the mythical America, and America shows the mythologized, crafted world [12], contributing to the development of widely understood consumption. The views presented are linked by the attempt to define globalization as a phenomenon or process aiming at the creation of a homogeneous world and a single common society - the world society. From the perspective of the approach discussed, all transformations of the contemporary world are recognized as cultural or civilization influences, covering the area of the entire globe. It is difficult to answer unequivocally whether globalization is simply a process of homogenization, since the forces of fragmentation and hybridization are equally strong. More convincing is the interpretation of globalization that emphasizes its heterogeneous nature [13].

3 Globalization as a Phenomenon Arising from the Process of Diversification and Cultural Pluralism in the Contemporary World

Out of opposition to conventional theories of social modernization (and in particular their "western-centrism") and lack of interest in civilization and cultural diversity, the perspective of recognizing the world as a whole evolving in "globally" suggested directions was born. According to M. Golka, globalization has not created a homogeneous world and it is not clear whether this vision will ever materialize. An echo of this opinion can be found in Featherstone's statement claiming that the current process of globalization, which contradicts earlier expectations of an increasingly homogeneous world, leads to an increasing sensitivity to difference, is a consequence of the fact that the flows of information, knowledge, money, goods, people and ideas have intensified to such an extent that the sense of spatial distances that separated and isolated people from the need to take into account all the other social actors that make up humanity has been destroyed. As a consequence, we are all in each other's backyard [14]. Globalization as a phenomenon arising from the process of diversification and cultural pluralism in the contemporary world is situated in the context of L. Roniger's considerations. According to the author, globalization is both the diffusion of specific models of economic development, growth, market development and the corresponding adaptation or rejection of cultural patterns of westernization [15]. In this context, globalization should rather be understood as a worldwide web of interdependencies, affecting individual societies and states, so that they are a part of a certain entirety [16]. The best-known representatives of the approach under discussion: A. Giddens, A. McGrew and P. Streeten, argue that globalization seen as interdependence, interaction, intensification of relations between states is an expression of perceiving the world as a web of connections and only in this context can globalization be treated holistically and its civilization-related implications indicated. Underlying A. Giddens' considerations is the assumption that globalization is the intensification of social relations with a worldwide scope, which connects different localities in such a way that local events are shaped by events occurring many thousand miles away and influence them themselves [2]. In specifying the scope of this concept, it should be recognized, following A. McGrew's reasoning, that globalization consists in the multiplicity of connections and interactions of states and societies that make up the current world system [17]. Consequently, as L. Roniger points out, globalization is characterized by: "transnationalization of cultural patterns, continentalization of economic exchanges, regional transnationalization and the growing role of localism" [15]. P. Streeten also draws attention to the aspect of deepening world connections in almost all spheres of the contemporary socio-cultural, economic and political life, defining globalization as intensification of economic, political and cultural relations across borders [18].

At the root of the above statements, there is no resolution as to whether the term globalization refers to global consequences or global undertakings. The answer to this question is provided by Z. Bauman, emphasizing that the ubiquitous concept of globalization most often refers to global effects, still unintended and unforeseen rather than global initiatives and undertakings. A consequence of this view is the recognition of globalization as a largely uncontrollable, spontaneous and also irreversible process, where it is simultaneously difficult to determine the state of globalization in the contemporary world, whose fate depends largely on chance at the level of global activities, dependencies and interests. Perspectives of viewing the globalizing world are therefore difficult to grasp due to their dynamic and always in the process of becoming nature. This view finds contemporary confirmation in the deliberations of Z. Bauman, according to whom the notion of globalization conveys the indeterminate, unpredictable and autonomous character of the world and its affairs, the absence of the center, the absence of the desktop, the team of directors, the board office. Globalization is another name for the "new world disorder" [19] and refers directly to G.H. von Wright's "anonymous forces"; forces operating in the void, in a foggy, sinuous and untraversable "no man's land", stretching beyond the reach of anyone's capacity for concrete planning and action [20].

The above considerations indicate that the nature of the concept of globalization is complex and that it is difficult to determine its basic meaning. However, it can be concluded that the problematic scope of this term has the character of a historical and social construction taking into account the temporal dynamics of many socio-cultural and civilization processes that constitute the image of the contemporary world, which indicates the relative novelty of the concept of globalization.

4 The World of "Infinite Careers"

The contemporary globalizing society in the world of rapid economic fluctuations tries to respond adequately to the inevitability of the updated fourth wave in economics, to which J. G. Maree and Z. Pollard pay special attention [21]. Contemporary processes of globalization of the global economy, its reorganization and restructuring, foster considerations on the specific nature and predominance of the global economic transformation [22, 23]. Undoubtedly, a constituent element of the phenomenon of economic globalization and in particular the development of free market economy, are the changes in the occupational environment, the structure of work, the perception of work and also in the sphere of characteristics, meanings and values attributed to work. It is difficult to overestimate the significance of these changes for the quality of career construction, career progression and the modification of its tailored paths.

The world of careers is a world of numerous micro-transitions. In the "portfolio" of micro-change, the time between them is shortened - after a period of stability, re-separation occurs more and more quickly. Moreover, multiple career transitions can occur simultaneously [24]. The conceptualization of a qualitatively new approach to the issue of career as a "property" of an individual, points to the multidimensional nature of the contemporary discourse, which combines the implications of interdisciplinary dialogue and creates the need to review theoretical reflections on the ways of understanding the career and the conditions and determinants of its formation. There is no doubt that active coping in the reality undergoing permanent change requires adaptation to the constantly changing context of individual career creation, while the new way of thinking about it means, as W. Lanthaler has put it, being a manager of knowledge in your own case [25] and constructing a specific career "portfolio". In career capital, as opposed to social or cultural capital, special attention is paid to personality as: a) a commercial value being the object of potential transactions for other forms of capital; b) reproducible, i.e. modified value from the point of view of preconceived assumptions; c) deliberately constructed value developed by the subject taking into account temporal dynamics [26]. Hence the conclusion that the contribution of the theory of development in the recognition and interpretation of the peculiarities in the human life cycle and personality development in relation to the definition of the cognitive field of human career development (whether it refers to the whole, the most important part or most of the components of the object of cognition), as well as the phenomenon of investment and renewal of career capital is difficult to overestimate. The immanent property of career capital renewal is the continuous development.

The image of an individual as a causal agent is an important theoretical construction. This problem is outlined by Herr (1992), stating that it is individuals who are capable of creating careers. Careers do not exist, just like professions or jobs [27]. This specific, as K. Obuchowski notes, shift of orientation of an individual from external conditions of being to internal conditions [28] inclines to consider the career in connection with the individual as an individual being whose property is the individual career [29]. At this point it is important to recall a fragment of Collin and Watts' (1996) discussion in which the authors assume the need to re-evaluate career thinking. They state that it is necessary to focus more on career as a subjective construction of the individual rather than on career as an objective construction [27]. Accordingly, the subject develops a career on the basis of perceptions and attitudes towards it, which means, as Patton and McMahon (1999) emphasize, that a career is a pattern of influences that coexist in an individual's life [27]. This view represents the individualistic tendency of the individual (ambition, sense of empowerment, motivation to act), which finds its legitimacy in economic theories that promote investment in human resource potential within organizations [30]. This view is a framework for thinking about careers as a "property" of an individual, taking into account individual career choices, individual career planning strategies or individual stages of career development. For the evaluation of this position, a critical aspect is to emphasize that while individuals, to a large extent, exercise control over their careers, their management must also take into account the stream of intra-organizational experience that forms the mechanisms of the career system. Recognizing a career as "a property of an individual" (Y. Baruch - 2004, A. Bańka - 2005), is based on an individualistic assumption about the unique quality of each person's career, as it is the accumulation by an individual of a series of unique workplaces, jobs, positions and professional experiences [31] and the responsibility of the subject for constructing his/her career. In the foregoing approach to the career, several varieties of its recognition can be distinguished, which is classified by A. Bańka, indicating the selected distinctive criteria of career: the criterion of promotion, the criterion of profession, the criterion of stability, the criterion of practicing work. The contemporary approach to the definition of career is not limited to aspects related to the practice of promotions, the possession of a specific profession, the satisfactory professional situation of an individual or the stability of the internal links of the content of the profession practiced. A much broader scope of what is denoted and connoted by the notion of "career" contains in addition to the purely professional situation, also the level of *psychological well-being*, understood as the absence of tensions disorganizing the activity of the individual, economic and social prosperity, and a successful family situation [32]. Contemporary approaches to careers also emphasize the importance of activities that are not directly related to work, such as leisure activities, recreation, education and family roles which are associated with employment [33]. This less restrictive approach to defining the "career" emphasizes the importance of constructing (rather than choosing) a career for designing the quality of life [34]. A career understood in such a way, as D. T. Hall puts it, means a sequence of experiences of an individual (understood as shaping of internal processes of an individual, such as: ambitions and aspirations, satisfaction, self-image, attitudes towards work under the influence of changing roles) [35] related to the professional role fulfilled by him/her, constituting his/her professional life history. As A. S. King (1999) states, the psychological strength of one's own identity, insight into oneself and endurance in pursuit of career goals represent the main components of motivation and "commitment" understood as engagement in the career as well as cooperation, cohesiveness in an organization [36]. A single person, as an individual, assigns specific individual meanings to selected elements of reality thanks to his/her ability to read cultural codes, creates his/her own individual record and has a history ahead of him/her. He/she constructs his/her own life, and thus his/her personal career by identifying (assigning meanings to) his/her own professional behavior and numerous workplace experiences [37], also assigning meaning to the context in which these experiences occur. The assumption of responsibility by an individual for seeking the meaning of his/her role in the practiced profession, the source of which Ch. Handy, author of the work entitled *The Age of Paradox*, attributes to: direction (the individual's feeling of acting for a legitimate cause), continuity (the subject's belief in the survival and continuation of the products of his/her work) and connectivity (participation in a community with which one identifies oneself and co-creates), is a lifelong process and a lifelong learning task [38]. The British scholar adds that the meaning will come to those who develop their own sense of direction, continuity, and connectivity [39] in what they practice professionally.

The broad view of *the* career as a "property of an individual", the cognitive attitude dictates looking for a set of elements constituting this position. This seems to be necessary to determine the scope of the properties of constructing a career "portfolio" within the theoretical tradition chosen. An important element of the distinguished definitions of "career" is their clear *subjective, personal* shape. The career is always attached to a specific individual, it is a state of his/her possession and this is what gives it its specific individual meaning. We do not turn to emphasize the practiced profession (e.g.: I have a profession…) but "being" a representative of the practiced profession (e.g.: I am…). The occupation is only the context within which one's own career develops, which is shared by everyone who works or even (as in the case of the unemployed), seeks work [40]. In this sense, each subjective career has a unique nature. Based on these assumptions, D. T. Hall emphasizes that the career is an arrangement, a sequence of work-related experiences of a particular individual, it is a unique arrangement, imposed by autonomous choices [41]. This is not surprising, since individual dreams, desires, longings and images constitute one's own, most personal model of a life ideal [42]. Another element in the subjective approach is the attention paid to the existence of *determinants in developing the* individual's *career.* Both objective and subjective factors, considered in isolation, do not exhaust the scope of the concept in question. The holistic approach to the issue of career development takes into account both objective elements (e.g., responsibilities, positions, activities, roles, career decisions) and subjective elements (values, aspirations, attitudes, expectations, needs, orientations, emotional and feeling aspect of work experiences). An individual managing his/her career can change either objective elements of the career development environment (e.g.: change of job) or subjective elements (e.g.: change of expectations). Moreover, in the case of a career similarly developed and revealed by individuals, systematic changes occur in both the objective events associated with labor market dynamics and the subjective reactions of the subject to these events

[40]. The subjective perception and feeling of the career means that what is a career for one individual is a minor distinction for another one [43]. The career process, as D. T. Hall recognizes, is therefore the resultant of two dimensions: what is observable (the objective dimension) and what is not observable (the subjective dimension), which are closely interrelated [44].

Adaptation to an amorphous environment takes place through the practice of learning the new context in which one participates, also contributing to its change. The power of influence on the current situation, or on the social environment, is of tailored nature and depends on the subject's inclination to take active measures that indirectly trigger these changes in the environment. Contemporary studies on school quality should take into account its broadly defined pro-developmental dimension referring in essence to the level of what T. S. Bateman and J. M. Crant have called the shaping of individuals' proactivity. Initiality, as a distinguishing component of proactive behavior, understood as the ability to initiate action and gather resources and support for the change process whose essence is not narrowed down to the initiation of change, but extends to the feature of involvement in the process of achieving the goal of bringing change to completion [45], seems too important at the educational level to be omitted. The space of education fosters and increases the chances of assigning meaning to the subject's right decision for the future. A temporal orientation towards the future will allow the individual to better focus on the "choice" rather than on the "fate" or "randomness". This thinking is closer to the promotion of autonomous causal subjectivity than to adaptation to existing conditions. A path of subjective activity is facilitated by an educational impact that is oriented towards openness to new possibilities and situations rather than uncritically anchoring traditional homogeneous assumptions and rules. The proactive behavior, as intentional actions by a subject, has been the focus of interest of Z. King, (2004); R. A. Noe (1996); C. Orpen (1994). The studies allowed to distinguish two groups of proactive behavior components which can be described as: cognitive components and behavioral components [46]. The main characteristic feature of proactivity is undertaking the initiative to change the environment, which means that the individual has the ability to shape the environment to an extent that exceeds the ability of the environment to shape behavior [47]. The pro-development approach as a cognitive practice and implemented in a continuous way, recognized as a turn towards the development of a proactive personality, will be the "foundation" for the constitution of the following qualities: seeking change, seeing opportunities, creating situations, taking initiative, taking action [48]. The way we think about proactivity as a personality feature and proactivity as an attitude of commitment resulting from contextual conditions, needs and circumstances has been significantly affected by the views of T. Bateman and J. M. Crant. According to the authors, proactive individuals are distinguished by seven interrelated characteristics. Emphasizing the personalistic dimension in the culture of school reality and embarking on the path of (self-)education means concentrating on the creation of the following qualities: seeking opportunities for change, setting effective and change-oriented goals, anticipating problems and taking countermeasures, searching for ways to achieve goals, embarking on the path of action with the awareness of risk and assuming responsibility, persistence in the pursuit of the goal and achieving the goal, demonstrating achievements and implementing changes by affecting the environment [49]. An important aspect of the proactive status and conditions, a personality feature highlighted by E. W. Morrison and C. C. Phelps is the "responsible commitment". This construction is defined as a meaningful effort by a subject targeted at negating *the status quo* in order to introduce functional changes in the scope of the tasks performed [50]. Similarly, the concept of Frese, Kring, Soose and Zempel (1996) emphasizes the problem of personal initiative of the subject. It defines proactivity as the behavior oriented towards taking the initiative (doing something when no one tells them to and when the role does not require it; long-term focus, anticipation of future problems or opportunities), and persistence (overcoming limitations so that changes can occur) [51].

While looking at the issue of career development as a problem in the "boundaryless" career world, the emphasis is placed on the fact that the responsibility for shaping careers is shifting from the organization to the individual and the basic feature of the post-organizational era is the orientation towards knowledge which, being appreciated on the labor market, is becoming the basis for the subject's mobility within the career [52]. The classification of careers of international itinerants developed by M. Banai and H. Harry can be considered as a continuation of the foregoing typology. In the world of "boundaryless" careers, eight types of managers can be distinguished. The career models proposed by the authors are not mutually exclusive which means that a subject in a career can belong to several types within the systematization. The self-managed career of international itinerants can be systematized by indicating: a) "failed" expatriates from international business organizations, b) managers with exceptional expertise, c) cosmopolitan professionals, d) individualists, e) returning citizens and f) innovation seekers [53].

"Failed expatriates" are international itinerants who have chosen to settle abroad as a result of "failure" in relation to their employing organization but not necessarily in relation to the development of their career prospects. It is not uncommon for people who choose autonomy and independence to orient themselves towards the style of an international itinerant. The ability to freely construct a career (which a traditional employer is unlikely to offer) is the driving force behind action in the career. *Exceptional professionals* are international itinerants who have learned their craft well. Constantly updating their knowledge and developing their talents in an international market where they can sell their knowledge, they move from one contract to another, doing work for the company that won a tender for the project. *Cosmopolitan professionals* are international itinerants with professional knowledge, skills and abilities that are updated in a cross-cultural environment. They highly appreciate the opportunity to work in an international environment and career mobility. Emphasizing the role of early socialization, it is stressed that in these people the lifestyle they lead during childhood, adolescence or early adulthood often allows them to multiply their circles of identification with foreigners or other international itinerants rather than with their compatriots [54]. *Individualists* are international itinerants with the professional knowledge and skills to perform a specific task or project. Each project is treated as a stopover "for a while", after which they continue their journey. The difference between individualists and exceptional professionals is that in their case "self-direction" (the employer does not guide them, but only gives instructions and specifies the time and budget) is more important than expertise. *Returning compatriots* are international itinerants who, as valuable workers, decided to

return to their homeland. These are usually students and managers who have gone abroad (usually from a developing country to a developed country) to acquire education, plan their prospects and develop their careers. However, it happens (this practice is prevalent in South-East Asia and China) that when they return to their home country they have a lower salary (with the same qualifications and experience) than expatriates simply because they are compatriots. *Innovation seekers* as international itinerants do not feel a close bond with their homeland. Despite having formal attributes of being assigned to their country (having a passport and a citizenship) they feel part of a larger community and even a world community. The global village allows them to seek employment in the global labor market. It is not uncommon that having a partner of a different race, religion or nationality, on the one hand, opens the "wandering" to other communities and on the other hand, cuts them off from the community in which they grew up [55]. To a certain extent, the international career of itinerants is like a product that one has to keep investing in and keeping on the market. Career self-management is about making sense of the multitude of opportunities found in the world.

5 Conclusion

Summing up, it should be stressed that the vision of the world of constant fluctuation, ideas that undermine the existence of "once and for all" career competence, emphasize the need to focus on the issue of "proactivity". The development of career competence is a kind of postulate concerning the human condition, which is not without significance for the shape of the social order. It is a continuous process of the subject acquiring new skills and improving existing ones. The multiplicity, fragmentation, variability and complexity of forms of social life organization influence changes in the perception of career development and overcoming the tension between the experience of the past and the opportunities of the future. The individual as a conscious creator of his/her own biography participates in the process of "investing" in the career capital.

References

1. Robertson, R.: The relativization of societies: modern religion and globalization. In: Robbins, T., Shepherd, W., McBride, J. (ed.): Cults, Culture and the Law, Chico (1985)
2. Kempny, M.: Globalization. In: Encyclopaedia of Sociology volume I, Warsaw, p. 242 (1998)
3. Robertson, R.: Globalization. Social Theory and Global Culture, London, p. 89 (1992)
4. Wierzbicka, E. (ed.): Dictionary of Contemporary Polish, Volume II, Warsaw, p.655 (1998)
5. Sobol, E. (ed.): Small Dictionary of Polish, Warsaw, p.1275 (2000)
6. Wierzbicka, E. (ed.): Dictionary of Contemporary Polish, Volume II, Warsaw, p. 148 (1998)
7. Robertson, R.: Globalization. Social Theory and Global Culture, London, p. 161 (1992)
8. Kempny, M.: Globalization. In: Encyclopaedia of Sociology, volume I, Warsaw, p. 242 (1998)
9. Ibidem
10. Robertson, R.: Globalization. Social Theory and Global Culture, London, p. 61 (1992)
11. Martin, H.P., Schumann, H.: The Globalization Trap, Wrocław, p. 258 (1999)
12. Golka, M.: Civilisation. Europe. Globalization, Poznań, p. 136 (1999)
13. Barker, C.: Television, pp. 38–39. Globalization and Cultural Identities, Philadelphia (1999)
14. Kahn J.S.: Culture, Multiculture, Postculture, London (1995), pp.126–128

15. Starosta, P.: Globalization and the New Communitarianism. In: Culture and Society, No 3/2000, p. 48
16. Golka, M.: Culture in the global space. In: Blok, Z. (ed.) Social Problems of Globalization, Poznan, p. 79 (2001)
17. Golka, M.: Civilisation. Europe. Globalization. Poznań, p.114 (1999)
18. Liberska, B.: Globalization - causal forces and mechanisms. In: Liberska, B., Globalization. Mechanisms and Challenges, Warsaw, p. 17 (2002)
19. Bauman, Z.: Globalization, Warsaw, p. 71 (2000)
20. Ibid., pp. 72–73
21. Maree, J.G., Pollard, Z.: The impact of life design counselling on adolescents. J. Psychol. Afr. **19**(2), 169–175 (2009)
22. Cybal-Michalska, A.: Tożsamość młodzieży w perspektywie globalnego świata. Studium socjopedagogiczne, Poznań (2006)
23. Cybal-Michalska, A.: Młodzież akademicka a kariera zawodowa, Kraków (2013)
24. Mayrhofer, W., Iellatchitch, A.: Rites, right? the value of rites of passage for dealing with today's career transitions. Career Dev. Int., (1), 58 (2005)
25. Lanthaler, W., Zugmann, J.: Action "I". A new way of thinking about career, Warsaw (2000)
26. Kobasa, S.C., Maddi, S.R., Kahn, S.: cited after: Bańka A.: Psychological career advice, Poznań, p. 85 (2007)
27. Patton, W., McMahon, M.: Career Development and Systems Theory. Connecting Theory and Practice, Rotterdam, p. 2 (2006)
28. Obuchowski, K.: An intentional man, or how to be yourself, Poznań, p.62 (2000)
29. Bańka, A.: Motivation of achievements, Poznań-Warsaw, pp.8–9 (2005)
30. Rosenbaum, J.E.: Organization career systems and employee misperceptions. In: Arthur, M.B., Hall, D.T., Lawrence, B.S.: Handbook of career theory, Cambridge, p. 330. (2004)
31. Bańka, A.: Proactivity and modes of self-regulation, Poznań-Warsaw, p. 23 (2005)
32. Bańka, A.: Motivation of achievements, Poznań-Warsaw, p. 8 (2005)
33. Bańka, A.: Proactivity and modes of self-regulation, Poznań-Warsaw, p. 26 (2005)
34. Maree, J.G.: Brief overview of the advancement of post-modern approaches to career counselling. J. Psychol. Afr. **20**(3), 362 (2010)
35. Miś, A.: Career shaping in organizations. Król, H., Ludwiczyński, A.: Human Resources Management, Warsaw, p. 477 (2006)
36. Adekola, B.: Career planning and career management as correlates for career development and job satisfaction. a case study of Nigerian bank employees. Aust. J. Bus. Manag. Res. 2, p.104 (2011)
37. Maree, J.G.: Brief overview of the advancement of post-modern approaches to career counselling. J. Psychol. Afr. **20**(3), 363 (2010)
38. Piotrowska, D.: Career Management, Introduction, Warsaw, p. 11 (2006)
39. Biolos, J.: Career models of the 21st century. In: Career Management, Warsaw, p. 30 (2006)
40. Bańka, A.: Proactivity and modes of self-regulation, Poznań-Warsaw, p. 25 (2005)
41. Miś, A.: Career shaping in organizations. In: Król, H., Ludwiczyński, A., Human Resources Management, Warsaw, p. 478 (2006)
42. Bauman, Z.: Career. Four sociological sketches, Warsaw, p. 18 (1960)
43. Worach-Kardas, H., cited after: Szymański, M.: Career paths of UAM sociology students, Warsaw, p. 80 (2010)
44. Miś, A.: Career shaping in organizations. In: Król, H., Ludwiczyński, A.: Human Resources Management, Warsaw, p. 478 (2006)
45. Bańka, A.: Proactivity and modes of self-regulation, Poznań-Warsaw, pp. 8–9 (2005)
46. De Vos, A., De Clippeleer, I., Dewilde, T.: Proactive career behaviours and career success during the early career. J. Occup. Organ. Psychol. **82**, 763 (2009)

47. Bańka, A.: Proactivity and modes of self-regulation, Poznań-Warsaw, p. 8 (2005)
48. Bańka, A.: Proactivity and modes of self-regulation, Poznań-Warsaw, p. 12 (2005)
49. Bańka, A.: Proactivity and modes of self-regulation, Poznań-Warsaw, pp.9–11 (2005)
50. Bańka, A.: Proactivity and modes of self-regulation, Poznań-Warsaw, pp.13–14 (2005)
51. Parker, S.K, Turner, N., Williams, H.M.: Modelling the antecedents of proactive behaviour at work. J. Appl. Psychol. (3), pp.636–637 (2006)
52. Mayrhofer, W., Iellatchitch, A.: Rites, right? the value of rites of passage for dealing with today's career transitions. Career Dev. Int. (1), pp. 56–57 (2005)
53. Banai, M., Harry, W.: Boundaryless Global Careers. Int. Itinerant Int. Stud. Manage. Organ. (3) (2004)
54. Banai, M., Harry, W.: Boundaryless global careers. Int. Itinerant Int. Stud. Manage. Organ. (3), pp. 102–103 (2004)
55. Banai, M., Harry, W.: Boundaryless global careers. Int. Itinerant Int. Stud. Manage. Organ. 3, pp. 101–110 (2004)

Trade Gravity Models for the Factors Affecting Foreign Trade in the Political-Administrative Regions of Chile

Manuel Ayala[1], Hanns de la Fuente-Mella[2(✉)], Víctor Leiva[1], and Ana María Vallina-Hernández[2]

[1] Facultad de Ingeniería, Pontificia Universidad Católica de Valparaíso, Avenida Brasil 2241, Valparaíso, Chile
manuel.ayala.b@mail.pucv.cl

[2] Facultad de Ciencias Económicas y Administrativas, Pontificia Universidad Católica de Valparaíso, Avenida Brasil 2830, Valparaíso, Chile
{hanns.delafuente,ana.vallina}@pucv.cl

Abstract. Chile is recognized as a highly centralized country, the literature in general and organizations such as the OECD identify Chile with a strong economic dependence on a few regions, where in addition, the OECD itself points out that this centralization and concentration limits Chile's capacity for increase. The main objective of the following research is to analyze the degree of incidence of macroeconomic, social, territorial and political variables that influence the foreign trade of the political-administrative regions of Chile with the World. Evaluating the impact of trade liberalization policies on the development of productive activity in the political-administrative regions of Chile. Regarding the methodology, a formulation of a commercial gravity model will be carried out applying the analysis methodology of panel data, considering for this the regionalized import and export figures, extracted from the customs databases, comprised between the years 2010 and 2019 and information from different national and foreign data sources published as open sources. Thus, it seeks to obtain a regional econometric analysis instrument that identifies the main variables that influence the flow of regional trade in Chile, and consequently measure the degree of territorial deconcentrating achieved from the international trade of Chile's regions as a percentage.

Keywords: International trade · Latin America · Gravity model · Econometric modeling

1 Introduction

At present, Chile and its trade policy are proclaimed as the country with the greatest access to part of world trade (88% of World GDP), data from the Organization for Economic Cooperation and Development (OECD) for 2019, record 29 Free Trade Agreements (FTA), facilitating access to 65 world economies with a total of more than 4.9 billion people. This commercial dynamism made Chile stand out that same year as

R. S. Goonetilleke et al. (Eds.): AHFE 2021, LNNS 273, pp. 495–503, 2021.
https://doi.org/10.1007/978-3-030-80713-9_63

the largest exporter of various products, for example: fresh plums, sea urchins, algae, salmon fillet, iodine and lithium, coniferous cellulose and copper cathodes. Evidence of a diversity of products beyond traditional mining [1], also sensing a productive export dynamic throughout the national territory.

Chile is recognized as a highly centralized country, the literature in general [2–5] and organizations such as the OECD identify Chile as having a strong economic dependence on a few regions, where the OECD itself also points out that this centralization and concentration limits Chile's ability to growth [6].

Chile has an economic model open to the world that favors trade and investment [7], this country economic model has generated in recent decades a growing network of trade agreements and free trade agreements with countries and trade blocs throughout the world, with a total of 29 trade agreements in force [8], and according to the Organization for Economic Cooperation and Development (OECD) Chile has the most powerful network of Free Trade agreements in the world, with access to more than 60% of world GDP [9]. Adding to this, Chile presents a diversity of its exports oriented to Goods and Services with increasing dynamism and according to the Ministry of Foreign Affairs it is one of the keys to the success of our country.

The commercial model implemented in Chile translates into a high degree of dependence on the economy of international trade, showing the dynamism of exports of goods and services through GDP, in 1960 Chilean exports represented 13% of national GDP and in the year 2019 corresponded to 28%.

Considering the export variety of Chile and taking the year 2019 as an example, the productive mining sector corresponded to 51% of the country's total exports, the rest is mainly made up of fruits and fruits, sea products, forestry and their derivatives, viticulture, meats edible oils, cereals, compost and other foods [10].

The 10 main exports made in 2019 and corresponding to 56.7% of all national exports, copper products stand out in general with 45.4% of exports, followed by the export of forestry production 3.6%, of salmon with 2.9% and fruit products with 2.0% of exports. The same document indicates that the main means of outbound transport for national production is by sea, moving 96.5% of cargo and the main destination countries are China with 31.3%, United States 14.9% and Japan 8.7%. Regarding imports from Chile, they mainly involve fuels such as oil, liquefied natural gas and coal, imports of cars, mobile phones and bovine meat also stand out.

The 10 main products imported by Chile and which correspond to 22.4% of total purchases abroad, mainly marked by fuels such as oil, liquefied natural gas and coal, representing 13.9%, then the main import of non-fuel products, cars stand out, representing 5.2%, cell phones 2.0%, and bovine meats with 1.5%. From the same yearbook, it is highlighted that the means of transport for accessing merchandise from abroad is by sea 88.1%, with origins mainly from the American continent 33.3%, from Asia and Europe 17.9%.

All this commercial flow to and from Chile with various economies in the world, has generated economic growth that has been highlighted by different organizations specialized in finance, in this context the World Bank explains about Chile: "Chile has been one of the economies of more rapid growth in Latin America in recent decades, which has allowed the country to significantly reduce poverty. However, more than 30% of the population is economically vulnerable and income inequality remains high" [11].

In addition to the above, Chile is among the most centralized countries in the OECD and in South America [12–14], territorial inequality in Chile has been growing, especially since the 19th century. Its main manifestation is the concentration of the population in the Metropolitan Region [15].

The inequality of income and spatial economic distribution of these throughout the national territory has been a constant occupation of the State of Chile, which for many years has tried to decentralize the administration in processes of powers and resources from a central government to the administrations sub-nacionales o unidades más pequeñas, entregándoles autonomía en la toma de decisiones y poder decisorio local acerca de materias delimitadas por la misma autoridad central, consecuente con lo anterior, Chile está suscrito junto a otros 192 países al programa de la ONU 2015–2030 "Agenda 2030 para el Desarrollo Sostenible", esta agenda entre sus lineamientos de acción plantea que "La dimensión territorial es clave en las trayectorias de desarrollo de los países, es el eje de la organización político - administrativa del Estado y la base material de la actividad económica" [16].

Uniting the country policies that Chile has planted, an economic model opens to the world and decentralization as country development, it seems interesting to evaluate the joint impact of these State policies. [2] affirm that authors such as [17, 18] have pointed out that trade liberalization policies favor territorial deconcentrating, [2] state "because the Export-oriented companies that depend on imports for the purchase of inputs will not have incentives to locate in the national center, where their chains are weak and they face diseconomies of agglomeration. The commercial opening would therefore lead to a deconcentrating process that reduces the primacy and generates more balanced urban systems". In accordance with the above, the idea of analyzing the effect of the commercial liberalization applied by the State and the effects of the economic model in the national territory is proposed, investigating the commercial flow of each of the political-administrative regions of the country. For this, it is proposed to use a trade gravity model that characterizes the impact of Chile's trade policies at the level of Chile's regions and the territorial impact that it could have stimulated.

Thus, Chile has become one of the most active countries in international trade agreement policies, [19] identifies the identity of Chile since its founding has had what he calls "multilateral intuition", the author that has developed a set of principles and values that have facilitated an involvement in the international community and have guided our multilateral action in different contexts, has participated in negotiations and persists in the search for convergence with other countries to generate dynamics that allow multidimensional advancement.

In an inclusive and globalized social context, the trade agreements made by Chile as a whole, based on social connections that provide dynamism and commercial liberation, play a role in human development, poverty reduction, the reduction of inequalities and inclusion. social.

"International trade does not take place in a vacuum, but the possibilities of taking full advantage of its benefits depend not only on the internal policies adopted by a country, but very fundamentally on those adopted by the countries with which it trades" [20].

This open policy has resulted in a commercial liberalization, expanding and facilitating transactions of goods and services with the whole world. According to information from the Undersecretary of International Foreign Relations, the trade opening strategy has evolved from the unilateral reduction of tariffs, towards a wide network of bilateral, as well as plurilateral and multilateral trade agreements, thus establishing a model of growth and development. country based on imports of goods and services and at the same time, an increase in foreign investment in the national territory and the investment of Chilean companies abroad.

Considering the unilateral opening of Chile, a sample of the actions taken is the MFN (Most Favored Nation) tariff, which has remained constant since 2003 at 6%, this has allowed to build the basis of international trade opening. For a small country like Chile, unilateral opening contributes to a more adequate resource allocation and consequently, maximizes the general welfare of the entire country [21–23]. In the same way, Chile's trade policy does not discriminate, there are no protectionist policies that restrict trade, not applying, for example, import licenses, nor subsidies for the export of agricultural products.

The Bilateral and Plulateral agreements signed and updated over time by Chile have complemented the unilateral opening and covered the aspects in which it cannot resolve or intervene, this considers preferential Free Trade Agreements, first in In the 1990s, free trade agreements were achieved with geographically close and Spanish-speaking countries, such as Mexico, Argentina, Bolivia, Venezuela, Colombia, Ecuador, Peru, to later reach agreements with Mercosur, as well as geographically distant economies and different languages, such as Canada, United States, Block P4, Japan, China, Turkey, Australia, Hong Kong, Vietnam, among others.

For Chile and its trade policy, the fact of belonging to plurilateral trade blocs allows achieving substantial trade flow results in shorter terms than is feasible at the level of multilateral treaties. In the Multilateral context, Chile is one of the founding members of the World Trade Organization (WTO for its acronym in English), belongs to the Asia-Pacific Economic Cooperation Forum (APEC) and a member of the Organization for Cooperation and Development. Economic (OECD). This multinational context in which Chile is inserted requires reliable external and internal policies, given globalization, regimes are assumed that must be guarantees of peace, development and cooperation, in addition to being subject to rules and agreements established by multinational organizations.

One aspect to consider in Chile's trade policy is its ease of trade integration, it is considered that successful trade integration begins with the ability of a country to move goods across borders reliably, quickly and at low costs [24], given Chile's territorial disposition and active participation in a multilateral context, allows commercial ease, considering that in 2014 an agreement was signed between member countries of the Mundial Trade Organization, which advances in trade facilitation, and on the other hand [25] based on the guidelines of the Asia-Pacific Economic Cooperation Forum, identifying three general aspects for success in trade integration: physical connectivity, institutional connectivity and connectivity between towns.

The main objective of the following research is to analyze the degree of incidence of macroeconomic, social, territorial and political variables that influence the foreign trade of the political-administrative regions of Chile with the World. Evaluating the impact of trade liberalization policies on the development of productive activity in the political-administrative regions of Chile.

The proposed research raises the following specific objectives: to formulate a foreign trade gravity model that identifies Chile's regional trade relations with the World; to empirically validate the macroeconomic, social, territorial and political variables that are statistically significant in the foreign trade of the Chilean regions; to characterize the effect of the trade agreements applied by the State of Chile in the foreign trade flow of the regions; measure the impact represented by the geographical distance of Chile's regions on the flow of international trade; relate the foreign trade flow and the existence of maritime ports in the regions of Chile; estimate spatial migrations and population redistribution in the political-administrative regions of the country based on regional trade flows and their effect on territorial deconcentrating.

To comply with these proposed objectives, a commercial gravity model will be formulated applying the analysis methodology in panel data, considering for this the regionalized import and export figures, extracted from Customs databases, comprised between the years 2010 and 2019 and information from different national and foreign data sources published as open sources.

2 Methodology

Regarding gravity models, these have been widely used in the literature to predict international trade flows [26–32], this type of methodology has normally been used for trade flows between countries, although there are few investigations carried out at the Latin American and regional level, this work is expected to have an analytical instrument that quantifies the flows of foreign trade and the possible impact on the regions of Chile, contributing to the economic decentralization of the country, a consequence of the economic opening in force for more than two decades in the country.

Since the works of Jan Timbergen [33–35], economic science uses the gravity model, which is an analytical tool for the study and prediction of trade flow. This econometric model is based on the analogy from a model of theoretical physics and classical mechanics to economics, particularly econometrics, using Isaac Newton's Law of Universal Gravitation, as a trade model that predicts flows. of trade between two economies as gravitational forces in direct function of their mass (GDP, economic income of the country or its population) and inversely proportional to the square of the distance between them.

Anderson [27], made one of the first theoretical explanations to the gravity equation for trade flows based on economic theory, showing that it can be derived from the properties of expenditure models in a context of differentiated goods, using the assumption of Armington [36], differentiating unique goods by country of origin. From the theoretical definition proposed by Anderson, a series of theoretical approaches to economic gravity models followed one another based on international trade specifications, relying on the summary of works that have contributed to the theoretical foundation applied to trade models, carried out by [37], a brief summary of the methodological milestones of the theoretical evolution of the model is presented. [38], based his approach on models of monopolistic competition, [23], they developed their theory of the gravity model on assumptions of product differentiation and economies of scale, [39] showed that different theories of standard trade converge on the general gravity equation and [40] specify the multilateral resistance to coefficients, which helps to interpret the results from the trade gravity equation, regarding the adequacy and improvement of the gravity equation as a modeling tool for trade flows, over time econometric specifications have been made to the gravity equation and the inclusion of new variables such as, [41].

The gravity equation applied to international trade is documented in summaries and guides of organizations such as the WTO, which cover theoretical aspects of the model, analysis and techniques of application of the model, mentioning some [42].

On the other hand, in the Latin American context, the application of the gravity model according to [24], has increased in recent years the articles that use the gravity model to explain the export behavior of Latin American countries, these include Chile in an indirect and significant way, but there are no specific publications on Chile's foreign trade applying a gravity model.

From the literature review carried out considering the last 10 years, the following works that include Chile and apply a gravitational trade model are recognized: [43] evaluate the effects of trade facilitation among a set of countries including Chile, estimating the main variables associated with trade facilitation and the pattern of trade, estimating a gravitational model for the period 2003–2006 using Panel data with fixed effects. [44] analyze the effect of internationalization and technology in the Mercosur block relative to trade, estimating the impact of direct foreign investment and the internal variables of the countries through a gravity model, applying a joint regression model for the period 1992–2008 by estimation of Ordinary Least Squares (OLS). [45] analyzes the exporting and regionalist behavior as an enhancer of the industrialization of Latin American countries during the period 1996–2005, estimating the influence of certain variables that facilitate the Latin American bilateral trade flow, applying a gravitational model with an OLS estimation. [46] estimate variables that characterize bilateral trade between Latin American economies, using a triple severity model indexed and estimated by OLS with fixed effects over time. Consider the period 1990–2014 for 20 countries.

It is estimated [24] as the first specific study for Chile that applies a gravity model, this analyzes the effects of trade facilitation on Chilean exports using a gravitational model expanded by pool and data static panel effects with fixed, random and dynamic effects. Among the results obtained by the applied gravitational model, geographic distance, common border, common language, and real exchange rate are not significant in the model. The study suggests increasing the size of the ports and number of ports in Chile to reduce freight costs, access more adequate logistics chains according to products, also considering that the ports of southern Chile have better maritime accessibility conditions, being able to attract more number of shipping companies to the southern areas of the country.

Then, [47] study the Chilean production of molybdenum, influence and export behavior through a gravitational model expanded to export behavior according to the Chilean reality and the world molybdenum market, the results of the model indicate that the geographic distance variable positively influences the flow of trade as well as the model makes up the direct relationship between the GDP of the importing country and the export, and that the common border variable presents a negative coefficient, not contributing to trade.

3 Conclusions

The main results of the research will be able to respond with the proposed objectives, in addition, it seeks to obtain a regional econometric analysis instrument that identifies the main variables that influence the flow of regional trade in Chile, and thus measure the degree of territorial deconcentrating achieved from the international trade of the regions, to generate an instrument that is useful for regional governments in decision-making.

To meet the objectives of the study, the volume of bilateral international trade of the Chilean regions will be modeled with the countries that generate the flow of exporting and importing trade in a period of time, for this an econometric analysis is proposed through a gravitational model of trade that contains the main variables that explain trade, it is intended to estimate the effect of these variables on the volume of trade in the regions of Chile. The panel data method will be used as an analysis tool, considering that the gravity models are based on the exporter and importer data characterized in time and these are generically represented in a panel data structure.

References

1. SUBREI: Beneficios Tratados de Libre Comercio. Subsecretaria de relaciones Internacionales, Chile (2020). https://www.subrei.gob.cl/landings/beneficios
2. Atienza, M., Aroca, P.: Concentración y crecimiento en chile: una relación negativa ignorada. EURE Santiago **38**, 257–277 (2012)
3. Frigolett, H., para el Desarrollo, P.C.T.: Economías regionales en Chile: desigualdad y heterogeneidad. Santiago Chile RIMISP (2013)
4. PNUD: Nueva publicación del Programa de las Naciones Unidas para el Desarrollo en 2018 (2018)
5. Dazarola, G.: Descentralización en Chile Avances y temas pendientes (2019). https://obtienearchivo.bcn.cl/obtienearchivo?id=repositorio/10221/27720/1/BCN_Estado_Descentralizacion_Chile_2019pdf
6. OECD: Territorial Reviews: Chile 2009 | READ online. OECD iLibrary (2009). https://read.oecd-ilibrary.org/urban-rural-and-regional-development/oecd-territorial-reviews-chile-2009_9789264060791
7. SUBREI: Acuerdos Comerciales Vigentes. Subsecretaria de relaciones Internacionales, Chile, October, 2020. https://www.subrei.gob.cl/acuerdos-comerciales/acuerdos-comerciales-vigentes
8. SUBREI: Beneficios Tratados de Libre Comercio. Subsecretaria de relaciones Internacionales, Chile, October, 2020. https://www.subrei.gob.cl/landings/beneficios
9. MINREL: Chile en el Exterior. Ministerio de Relaciones Exteriores, Chile, October, 2020. http://chile.gob.cl/chile/economia
10. de Aduanas, S.N.: Anuarios Estadísticos de Comercio Exterior. aduana (2019). http://www.aduana.cl/anuarios-estadisticos-de-comercio-exterior/aduana/2018-12-14/113928.html
11. WTO: Examen de Políticas Comeciales, Informe de Chile (Examen de políticas comerciales No. 15–5206; p. 23) (2015). https://www.wto.org/spanish/tratop_s/tpr_s/tp415_s.htm

12. Frigolett, H., para el Desarrollo, P.C.T.: Economías regionales en Chile: Desigualdad y heterogeneidad. Santiago de Chile: RIMISP (2013)
13. Dazarola, G.: Descentralización en Chile Avances y temas pendientes (SUP: 121007). Biblioteca del Congreso Nacional de Chile (2019). https://obtienearchivo.bcn.cl/obtienearchivo?id=repositorio/10221/27720/1/BCN_Estado_Descentralizacion_Chile_2019_def.pdf
14. von Baer vL, H., Bravo, N.: Desarrollo territorial colaborativo: Descentralizando poder, competencias y recursos (2019)
15. Aroca, P.: Desigualdades regionales en Chile. Foreign affairs: Latinoamérica, p. 9 (2009)
16. PNUD: Nueva publicación del Programa de las Naciones Unidas para el Desarrollo en 2018. Programa de las Naciones Unidas Para el Desarrollo (2018). https://www.desiguales.org/regiones
17. Krugman, P., Elizondo, R.L.: Trade policy and the third world metropolis. J. Dev. Econ. **49**(1), 137–150 (1996). https://doi.org/10.1016/0304-3878(95)00055-0
18. Hanson, G.H.: Regional adjustment to trade liberalization. Reg. Sci. Urban Econ. **28**(4), 419–444 (1998). https://doi.org/10.1016/S0166-0462(98)00006-4
19. Somavía, J., Oyarce, P.: Chile actor del sistema multilateral. Una tradición nacional. Santiago, Academia Diplomática de Chile Andrés Bello (2018)
20. Sáez, S., Valdés, J.G.: Chile y su política comercial lateral. Revista de la CEPAL (1999)
21. Duran, J., Finot, A., LaFleur, M.: Análisis de la apertura comercial sobre el bienestar de los hogares: una aplicación para Chile 1999–2006. ECLAC, United Nations, December 2010
22. Bernhofen, D.M., Brown, J.C.: A direct test of the theory of comparative advantage: the case of Japan. J. Polit. Econ. **112**(1), 48–67 (2004)
23. Helpman, E., Krugman, P.R.: Market structure and foreign trade: Increasing returns, imperfect competition, and the international economy. MIT press, Cambridge (1985)
24. Fuenzalida-O'Shee, D., Valenzuela-Klagges, B., Corvalán-Quiroz, A.: La facilitación comercial y sus efectos en el comercio bilateral chileno de 2006 a 2014. Revista de la CEPAL **2018**(124), 172–191 (2018). https://doi.org/10.18356/630ad7fa-es
25. Shepherd, B.: Did APEC's trade facilitation action plans deliver the goods? J. Asian Econ. **43**, 1–11 (2016). https://doi.org/10.1016/j.asieco.2016.01.003
26. Head, K.: Gravity for beginners. University of British Columbia, 2053 (2003)
27. Anderson, J.E.: A theoretical foundation for the gravity equation. Am. Econ. Rev. **69**(1), 106–116 (1979)
28. Helpman, E., Melitz, M., Rubinstein, Y.: Estimating trade flows: trading partners and trading volumes. Q. J. Econ. **123**(2), 441–487 (2008). https://doi.org/10.1162/qjec.2008.123.2.441
29. Kepaptsoglou, K., Karlaftis, M.G., Tsamboulas, D.: The gravity model specification for modeling international trade flows and free trade agreement effects: a 10-year review of empirical studies. Econ. J. **3**(1), 1–13 (2010). https://doi.org/10.2174/1874919401003010001, ~!2009-07-09~!2010-01-28~!2010-04-22~!
30. Paz, A., de la Fuente-Mella, H., Singh, A., Conover, R., Monteiro, H.: Highway expenditures and associated customer satisfaction: a case study. Math. Probl. Eng. **2016**(4630492), 1–9 (2016)
31. Coughenour, C., Paz, A.: Hanns de la fuente-mella and ashok: multinomial logistic regression to estimate and predict perceptions of bicycle and transportation infrastructure in a sprawling metropolitan area. J. Public Health **38**(4), 401–408 (2016)
32. de la Fuente-Mella, H., Fuentes, J.L.R., Leiva, V.: Econometric modeling of productivity and technical efficiency in the Chilean manufacturing industry. **139**, 105793 (2020).https://doi.org/10.1016/j.cie.2019.04.006
33. Tinbergen, J.: An analysis of world trade flows. Shap. World Econ. **3**, 1–117 (1962)
34. Pöyhönen, P.: A tentative model for the volume of trade between countries. Weltwirtschaftliches Arch. **90**, 93–100 (1963)

35. Linnemann, H.: An econometric study of international trade flows. North-Holland Pub. Co (1966)
36. Armington, P.S.: Una teoría de la demanda de productos distinguiéndolos según el lugar de producción. Staff Papers **16**(1), 159–178 (1969). https://doi.org/10.2307/3866403
37. Martínez-Zarzoso, I., Cantavella, M., Guerrero, J.: Estimación y aplicaciones de una ecuación de gravedad para el comercio atlántico de la Unión Europea. Información Comercial Española, ICE: Revista de economía, No. 806, 2003 (Ejemplar dedicado a: Relaciones económicas entre la Unión Europea y Latinoamérica), pp. 23–32 (2020). ISSN 0019-977X
38. Bergstrand, J.H.: The gravity equation in international trade: some microeconomic foundations and empirical evidence. Rev. Econ. Stat. **67**(3), 474–481 (1985). https://doi.org/10.2307/1925976
39. Deardorff, A.: Determinants of bilateral trade: does gravity work in a neoclassical world? En The regionalization of the world economy, pp. 7–32. University of Chicago Press (1998)
40. Anderson, J.E., van Wincoop, E.: Gravity with gravitas: a solution to the border puzzle. Am. Econ. Rev. **93**(1), 170–192 (2003). https://doi.org/10.1257/000282803321455214
41. Matyas, L.: Proper econometric specification of the gravity model. World Econ. **20**(3), 363–368 (1997). https://doi.org/10.1111/1467-9701.00074
42. Yotov, Y.V., Piermartini, R., Monteiro, J.-A., Larch, M.: An Advanced Guide to Trade Policy Analysis: The Structural Gravity Model. WTO (2016). https://doi.org/10.30875/abc0167e-en
43. de Souza, M.J.P., Burnquist, H.L.: Impactos da facilitação de comércio: evidências do modelo gravitacional. Rev. Econ. Soc. Rural. **49**(4), 909–940 (2011). https://doi.org/10.1590/S0103-20032011000400005
44. Alvarez, I., Fischer, B., Natera, J.M.: MERCOSUR: tendencias de internacionalización y capacidades tecnológicas (2013). https://repositorio.cepal.org//handle/11362/11570
45. Klagges, B.V., Brito, L.E.: Regionalismo latinoamericano y comercio bilateral. Revista Pilquen Sección Ciencias Sociales **18**(2), 1–11 (2015)
46. Vallina-Hernández, A.M., Martinez, P., Gonzalez, C., Fuentes, R., de la Fuente-Mella, H.: Gravity models for Latin American economies. In: Kantola, J.I., Nazir, S., Salminen, V. (eds.) AHFE 2020. AISC, vol. 1209, pp. 495–501. Springer, Cham (2020). https://doi.org/10.1007/978-3-030-50791-6_63
47. Donoso, R., Klagges, B.V., Bubert, A.S., Pavlov, V.M., Klagges, I.V.: Producción chilena de molibdeno: influencia en el mercado mundial y su comportamiento exportador (2007–2016). Revista de Economía del Rosario **23**(1), 109–148 149–172 (2020). https://doi.org/10.12804/revistas.urosario.edu.co/economia/a.8629

Natural Color System Quantization Design of Economy Class Seat Driven by Perceptual Imagery

Jianghua Xu[1] and Shuangle Ding[2(✉)]

[1] Shanghai University of Engineering Science, Shanghai 201620, China
[2] Nanchang Hangkong University, Nanchang 330063, China

Abstract. Objective. In order to improve the color comfort of economy class seats and reduce the dependence of seat color design scheme on designers' sensory experience, the perceptual demand is quantified to obtain a color design scheme more in line with users' perceptual needs. **Method.** A design method combining kansei engineering and NCS color quantization system is proposed. Cluster analysis and gray correlation analysis are used to screen out the optimal color design scheme. Firstly, the semantic difference method (SD method) was used to construct the economy class color image vocabulary and the mapping model of a single color. Secondly, the NCS color system is used to quantify and classify each color. Then, the image value of each group of color design schemes is obtained according to the relationship between color design schemes and color image positioning vocabulary. Finally, based on MATLAB grey correlation analysis, the correlation degree analysis is carried out on the image evaluation matrix of color design scheme, and the correlation coefficient and correlation degree of each color design scheme's evaluation on passenger image are obtained, so as to obtain the priority of color design scheme of economy class seat. **Conclusion.** By quantifying the emotional value of passengers, this method prioritizes the final color design scheme and obtains the color design scheme of economy class seats that better meets the emotional needs of passengers, which provides a good theoretical basis for the follow-up research.

Keywords: Perceptual image · Economy class seat · Comfort · NCS color quantification · Gray correlation analysis

1 Introduction

With the rapid development of the civil aviation transportation industry, aircraft cabin comfort has gradually received more attention. The cabin comfort is affected by many factors such as the size of the cabin space, the color of the cabin facilities, the size of the leg space, the comfort of the seats, and the cabin environment noise. The cabin interior

Shanghai Art and Science Planning Project 2020 General Project "Innovation Research on Design Method of Civil Large Aircraft Cabin Space and Cabin Facilities" (YB2020F02).

R. S. Goonetilleke et al. (Eds.): AHFE 2021, LNNS 273, pp. 504–511, 2021.
https://doi.org/10.1007/978-3-030-80713-9_64

and facility color design are the main components of the cabin comfort of civil aviation aircraft [1]. Color is one of the influencing factors of passengers' psychological comfort, and its brightness, chroma, and chromaticity configuration have a significant impact on the human space perception system and cabin comfort [2]. Colors are transmitted to the brain through the human visual system and produce perception of things. In order to reduce operating costs, some low-cost airlines continue to increase the number of seats in economy class and reduce the distance between seats, resulting in a sharp decline in the comfort of economy class. Based on this, this article quantifies the passenger's perception of the color image of economy class seats from the perspective of perceptual engineering, and uses cluster analysis and gray theory to further refine color design schemes with higher comfort to meet passengers' perceptual needs to a greater extent and improve cabin comfort.

2 Color Quantification Method System

Color quantification refers to the premise that the color performance is consistent with the human perception, that is, the combination of "visual", "physical", and "conscious", and is guided by human needs to achieve the consistency of color design and human sensory needs mapping results. The economy class seat of a civil aviation aircraft is not only an intermediary for passengers to obtain the flight experience, but also the subject of passenger perception, the existence of "consciousness" and the "phenomenon" itself [3]. Combining "awareness" with "phenomenon" can maximize the comfort of economy class.

2.1 Semantic Difference

The Semantic Difference Method (SD Method) is a method based on user surveys to understand the real perceptual needs of users, and is a commonly used research method in Kansei Engineering. Firstly, the representative color design plan of economy class seats and the collected perceptual image vocabulary are used to construct a semantic difference scale; secondly, a mapping model between the color design plan and the semantic difference scale is established, and the sensibility of each color design plan is obtained using the questionnaire Evaluation, and finally get the perception space of perceptual image of color design of economy class seat.

2.2 NCS Color Quantization System

A survey conducted by the American Popular Color Research Center shows that when people choose a dazzling array of products, they only need 7 s to determine whether they are interested in these products, and in this short 7 s, color plays a role in 67% [4]. It can be seen that color has a great influence on the perceptual value of products. The aircraft cabin facilities are composed of seats, luggage compartments, ceilings, etc. The difference in color combination also affects people's mood changes. Therefore, the color design of the cabin is particularly important.

The color of the economy class has many effects on passengers. The color, contrast, and brightness of the color itself have different degrees of influence on the cabin's sense of space, passengers' mood and seat comfort [5]. This article uses the NCS color system to systematically quantify the three elements of color, and obtain a visual color design plan. It mainly includes 6 reference colors: white (W), black (S), yellow (Y), red (R), blue (B), and green (G). The hue, lightness, Saturation, which allows each color to be quantitatively described and analyzed in the NCS system [6].

2.3 Grey Correlation Method

Gray system theory analyzes the gray correlation degree between various factors, and finds the correlation relationship between the factors through a certain method. Therefore, the gray correlation degree analysis can measure the development and change trend of a system and conduct dynamic analysis [7]. This paper uses MATLAB analysis software to perform gray correlation analysis on the color design scheme of economy class seats obtained by the SD method, obtain the degree of relevance and sort, and finally obtain the optimal solution of the color design scheme. The modeling method of grey relational system is as follows:

According to the SD method, the semantic difference component table is obtained. After the subjects are scored and evaluated, a decision matrix is established with the collected data:

$$V = \left[v_i(j)\right]_{p\times q} \tag{1}$$

In the formula: p – the number of color design schemes.

q – he number of image words.

$V_i(j)$ – the evaluation value of the i-th color design scheme for the j-th image vocabulary, I = 1, 2, ..., p, j = 1, 2, ..., q.

(1) Standardization of decision matrix

For the image vocabulary matched by each color design scheme, the higher the evaluation value of each image, the better, and it is standardized by the following formula:

$$x_i(j) = \frac{v_i(j) - \min_{1\le t\le p}\left[v_t(j)\right]}{\max_{1\le t\le p}\left[v_t(j)\right] - \min_{1\le t\le p}\left[v_t(j)\right]} \tag{2}$$

The standardized decision matrix is $X = \left[x_i(j)\right]_{p\times q}$

(2) Determine the ideal solution

There are p image vocabularies, and the ideal solution for the maximum evaluation value of q color design schemes by users is:

$$x^* = \left[x^*(1)x^*(2)\cdots x^*(p)\right] \tag{3}$$

Among them:$x^*(j) = \max\left[x_i(j)\right]$

(3) Calculate the gray correlation coefficient
According to the ideal solution obtained in the previous step, calculate the correlation coefficient of the i-th color design scheme to the ideal solution on the j-th image vocabulary, as follows:

$$\mu\left(x^*(j), x_i(j)\right) = \frac{\min_i \min_j |x^*(j) - x_i(j)| + \beta \max_i \max_j |x^*(j) - x_i(j)|}{|x^*(j) - x_i(j)| + \beta \max_i \max_j |x^*(j) - x_i(j)|} \tag{4}$$

β-- resolution coefficient, used to reduce the influence of distortion caused by the maximum absolute value difference, and to improve the significance of the difference between the correlation coefficients, generally $\beta = 0.5$.

(4) Calculate the degree of gray correlation
Suppose the weight vector of m image vocabulary is $M = (m_1, m_2, \ldots, m_z)$, according to the gray correlation coefficient equation, the gray correlation degree of the i-th color design scheme to the ideal solution is:

$$\mu\left(X^*, X_i\right) = \sum_{j=1}^{q} m_j \cdot \mu\left(x^*(j), x_i(j)\right) \tag{5}$$

The result obtained is the optimal ranking order of the popularity of the color design scheme of economy class seats [7].

3 NCS Color Quantification Design of Aircraft Economy Class Seats

3.1 Analysis on the Location of Color Image Vocabulary of Economy Class Seats

Each airline has its own visual characteristics and target passenger groups, and the color design of aircraft cabin interiors and facilities should meet passenger preferences as much as possible. Through the collection of perceptual image vocabulary and positioning analysis, the color image positioning vocabulary can be obtained, so as to meet the perceptual needs of passengers to the greatest extent.

Collect 230 color image vocabularies from the Internet, periodicals, magazine advertisements and literature, and select expert groups and investigative team members who have experience in flying and engaged in the design industry, of which 5 experts form an expert group to initially screen and remove irrelevant color images Vocabulary: A survey team composed of designers, economy class passengers, crew members, etc., conducted an experimental investigation on the color image vocabulary selected, and selected color image vocabulary with a frequency of more than 60%. The final 10 sets of color image vocabulary are: crowded-loose, bright-simple, tedious-concise, vulgar-technological, irritable-calm, depressive-relaxed, cold-warm, Tired-relaxed, hyperactive-gentle, sleepy-awake.

After multiple rounds of analysis and selection by the expert group and survey group, 10 groups of color image vocabulary were clustered and analyzed. This paper selects the K-means clustering algorithm based on partition and takes the color comfort of economy class seats as a measurement standard, and finally obtains 3 sets of representative economy class seat color image positioning vocabulary, as shown in Table 1 below.

Table 1. Vocabulary of color image of economy class seat

Cabin environment	Psychological state	Physiological state
Crowded-relaxed	Depressed-relaxed	Sleepy-sober

3.2 Construction and Classification of Sample Library of Color Design Schemes for Economy Class Seats

In order to better meet the psychological needs of passengers, 15 passengers and 15 designers were selected for perceptual testing. Among them, passengers and designers have experience of flying in economy class. According to the three sets of color image positioning vocabulary in Table 1 above, 10 relevant color combinations are selected on the NCS color circle, a total of 30, and the number of colors selected by each passenger for each set of image vocabulary shall not be less than 1 and no more than 10, a total of 427, as shown in Fig. 1 below.

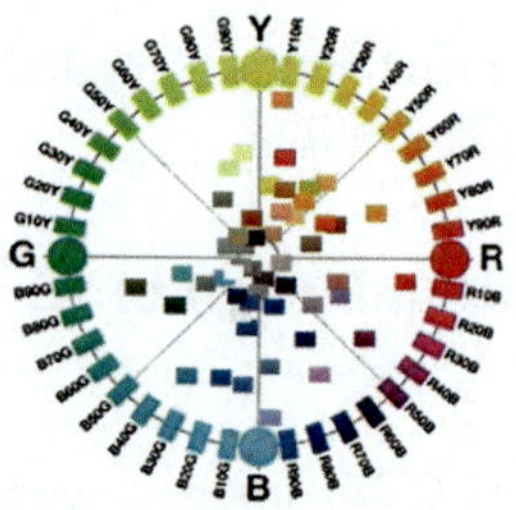

Fig. 1. Schematic diagram of NCS color design scheme sample base

Invite 5 design-related experts to discuss in groups and use the KJ method for selection. The 3 sets of color image vocabulary correspond to 9 sets of NCS color design schemes for aircraft economy class seats, as shown in Fig. 2 below.

Fig. 2. Color image positioning vocabulary corresponds to NCS color design scheme

3.3 Grey Relational Sorting of Color Design Schemes of Economy Class Seats

(1) Iteratively optimize the color design of the economy class seats of the aircraft to better meet the intended needs of target passengers
The 9 color design scheme samples and 3 pairs of color image semantics were combined to form a likert 5-level scale. The evaluation value is set between 1 and 5, and the perceptual evaluation scale of the subject's color design scheme is expressed according to the numerical value.
(2) Perform average processing on the collected scores to form the decision matrix shown in Table 2 below.

Table 2. Image evaluation decision matrix of color design scheme

Sample	Image score		
	Crowded-relaxed	Depressed-relaxed	Sleepy-sober
Group 1	4.35	3.26	3.17
Group 2	3.82	3.97	4.01
Group 3	3.14	3.05	3.92
Group 4	3.56	4.01	3.91
Group 5	2.99	2.59	2.98
Group 6	3.52	3.34	3.47
Group 7	4.15	4.23	3.64
Group 8	3.81	3.92	3.67
Group 9	3.50	3.08	3.11

(3) Grey Relational Analysis of Color Design Scheme
The above Table 2 color design scheme image evaluation decision matrix is systematically modeled, and the popularity of 9 groups of color design schemes is analyzed. According to the Matlab algorithm, the gray correlation coefficient is ranked as: Group4 > Group6 > Group5 > Group9 > Group7 > Group1 > Group2 > Group3 > Group8. According to the sorting results of the color design schemes, it can be seen that Group4, Group5, and Group6 have a higher degree of correlation with the evaluation values of the passenger user's images. Where Group4 for the best color scheme.

4 Application of Color Design for Economy Class Seats of Widebody Passenger Aircraft CR929

CR929 is a long-range wide-body passenger aircraft jointly developed by China and Russia. It is China's first large passenger aircraft after the narrow-body medium-sized passenger aircraft C919. The color design of CR929 economy class seats should fully tap the potential needs of passengers and enhance passengers' satisfaction with the comfort of the aircraft cabin [8].

According to the test flight regulations, adjust the color matching of the aircraft cabin and facilities, and use the optical illusion to increase the sense of space in the economy class [9]. The CR929 cabin interior and lighting system uses white as the main color, and the optical illusion brings passengers a sense of space extension. The light color of Group4 is used as the main color block of the seat, which complements and integrates with the cabin environment, as shown in Fig. 3 below. Using emerald as a secondary color and reducing the purity can effectively relieve the mood of passengers and improve the comfort of passengers on long-distance flights, as shown in Fig. 4 below.

Fig. 3. Color evolution and matching

Fig. 4. Overall rendering of the cabin

5 Conclusion

This article combines perception and rationality to discuss the color design of passenger seats in economy class of long-distance large-scale civil aviation aircraft, which makes up for the lack of quantitative color in perceptual engineering, and perfects the lack of passenger perceptual needs in the qualitative analysis process, so as to obtain more precise satisfaction. The color design scheme for passengers' perceptual needs provides method support for the subsequent color design of economy class seats. The research also has some shortcomings. There are many influencing factors in perceptual testing, and there is a certain degree of randomness, which affects the analysis results.

References

1. Vink, P.: Aircraft Interior Comfort and Design, pp. 24--50. CRC Press Inc., Bocaraton (2013)
2. Zhong, Y.: Research on structure optimization and modeling design of civil aviation seat. Shenyang Univ. Aeronaut. Astronaut. **23**(15), 13–14 (2015)
3. Tang, L.: "body-consciousness" in design. Packag. Eng. **40**(20), 1–8 (2019)
4. Lin, C.: Product color orientation design method based on user experience. Packag. Eng. **40**(22), 46–49 (2019)
5. Chu, J., Zhang, M., Wang, L., Wang, J., Ju, Y.: Research on color and material design of aircraft cabin seats based on perceptual image. Mech. Design **37**(02), 126–130 (2020)
6. Zhang, W.: Comparative analysis of colors under different conditions using the NCS color system. Public Lit. (19), 141 (2013)
7. Sun, Z.: Multi-image optimization of product color scheme based on gray correlation analysis. Mech. Design **32**(01), 120–122 (2015)
8. De Crescenzio, F., Bagassi, S., Asfaux, S., Lawson, N.: Human centred design and evaluation of cabin interiors for business jet aircraft in virtual reality. Int. J. Interact Design Manuf. (IJIDeM) **13**(2), 761–772 (2019). https://doi.org/10.1007/s12008-019-00565-8
9. Aviation Industry Standard of the People's Republic of China. HB 7045-2014, Design Requirements for Interior Decoration of Civil Aircraft Cabins, pp. 13–14(2014)

Innovation in Value Chain in the Medical Tourism Industry in Tijuana, Baja California

Alma Laura Bonilla-Hernández[1(✉)] and Rodolfo Martínez-Gutiérrez[2]

[1] Advance Health Medical Center SA de CV, Diego Rivera 2339-101 A Zona Urbana Rio, 22010 Tijuana, Baja California, Mexico
direccion.medica@ssdmedicalcenter.com

[2] Tecnológico Nacional de México/IT de Tijuana, Calzada Del Tecnológico S/N, Fraccionamiento Tomas Aquino, 22414 Tijuana, Baja California, México
rodolfo.martinez@tectijuana.edu.mx

Abstract. According to Patients Beyond Borders (2012), Mexico takes the second place in Medical Tourism, due to its geographical location and the dynamics of services between Mexico and the United States, making it the busiest border in the world.

The Tijuana-San Diego border region has been characterized as a development pole for the health traveler's medical services industry sector, with the added value of enjoying the variety of tourism and recreation options in combination with high quality and low cost medical care.

ProMéxico (2014) refers that Mexico in the border cities is a world center of health care that offers a complete range of specialties and procedures that compete directly with those offered in other developed countries, taking in consideration the quality, doctor's professionalism and in the first place, the price.

As part of the emerging actions that the medical tourism industry in Tijuana has developed, we find the world class value chain that goes from transportation, highly qualified health professionals, the hotel and restaurant industry, hospitals, laboratories and cabinets, pharmacies and suppliers of supplies and equipment, which together with strategic alliances with the Medical Tourism Cluster and the support of the government with its public policy in this area, allow us to offer comprehensive services for the smooth crossing of the border.

The results of this research are based on a systemic methodology with the purpose of articulating all sectors of the Medical Tourism industry in Tijuana to develop databases and indicators, as well as applicable local, regional and international regulations, the catalog and statistics of specialized service teams with supply and demand as a promotion strategy at the national and international level for the patients.

This scientific research study is based on the comparative academic information between countries that lead this type of service and productive chains mentioned before.

Keywords: Medical tourism · Innovation · Value chain

R. S. Goonetilleke et al. (Eds.): AHFE 2021, LNNS 273, pp. 512–517, 2021.
https://doi.org/10.1007/978-3-030-80713-9_65

1 Introduction

Patients from countries around the world are exercising increasing degrees of autonomy over their health care options by obtaining information from sources other than their regular health care providers and, in some cases, by electing to pursue care alternatives outside their domestic medical system [1].

It is important to determine the difference between health tourism and medical tourism. Health tourism covers all forms in which patients travel to other countries in search of an improvement in health. [2] [mentioned by Arias et al., 2012] Almost the same, Ross [2] [mentioned by Arias et al., 2012] argues that health tourism occurs when people travel from their place of residence for health reasons. Health tourism is a fortunate concept that describes the phenomenology associated with travel, which for health or wellness reasons people perform abroad or outside the place where they live. The international definition of health tourism says that it is the process in which a person travels outside his place of residence with the aim of receiving health or wellness services. The concept of medical tourism is more useful for distinguishing the realization of certain medical-surgical interventions.

Mexico's medical tourism industry represents for about 5.6% of the world, with an income of between $4.1 and $4.5 billion dollar, with the economic spill of $2.7 billion to the health sector and between $1.4 and $1.8 billion dollar to the tourism industry [3].

Many countries around the world have opened their health systems for greater cross-border patient circulation. High costs of surgeries and long waiting lists, new medical technologies and skills in destination countries, along with lower transportation costs and internet marketing have played an important role [2]. Mexico is the second most preferable destination as medical tourism place, attracting over one million foreign patients annually from the United States; many of these medical tourists are of Hispanic origin, mostly from the states of California, Arizona and Texas. Today, "tourism on its present international scale could not occur without the existence of a large and sophisticated 'industry' which enables people to be tourists." For the medical tourist experience, the majority of people also require accommodation, transportation, entertainment, and other essential support [4]. In the study market opportunities and health prospects in Mexico [3] Delloite referred to that Health Tourism in Mexico has grown at an average annual rate of 33.7% between 2013 and 2018, hoping that by 2023 it will reach a value of $10 billion dollar. Currently, the Mexican government has made efforts to develop nine clusters, including Tijuana's Baja Health Cluster, made up of doctors, hospitals, auxiliary services such as laboratories and radiology services, as well as hotels, restaurants and transportation.

Health is an important element to be taken into account when planning for cross-border tourism. Infrastructure and health services are to be integrated with constant attention to the traveler. The concept is based on tourism recreation and leisure spas. Among establishments offering such services are health spas, meditation centers, hospitals, etc. Different types of health care systems are likely to produce different configurations of demand for medical tourism, which influences the range of policy instruments available to governments and other actors seeking to influence decision-making and behavior within their particular context [1].

As economic positive influences, medical tourism generates rising incomes (exports), investments in modern equipment and technology conducting to less pollution, saving energy and protection of natural resources (in many situations the prime factor for the medical recovery or rehabilitation in medical tourism). Cultural and social positive impact focus on inter cultural changes, tolerance or preservation of local authenticity. Generating new possibilities of employment, the medical tourism offers the opportunity of getting back home the brains emigrated years ago in west developed countries. Oriented to research and permanent education, medical tourism develops the entire society in the spirit of acceptance [5]. Mexico must seek to increase its productivity through knowledge, based on integration of global value chains and take steps to advance the construction of an innovation system consistent with a more active role of the private sector, and strengthen stronger industrial academic ties [6] (World Bank, 2006).

A cluster can be defined as the initiative to consolidate the supply of services and/or goods competitively in a geographical region, articulating the socio-product organization of the different public, private and knowledge-generating actors through links of cooperation and competence in the long term. One of the ways to strengthen the competitiveness of health tourism for export is by forming clusters [7]. The Ministry of Tourism in México, refers that the main destinations for medical tourism are: Tijuana, Mexicali, Ensenada, Rosarito, Tecate, Ciudad Juárez, Navojoa, Hermosillo, Reynosa, Matamoros, Nuevo Laredo, México City, Monterrey, Cancún, Guadalajara, San Luis Potosí, Puebla and Querétaro [6].

Because of its Latin American nature and being the benchmark in the range of savings, it brings for patients from developed countries such as the United States, the Health Tourism Cluster in Medellin needs to be highlighted, as it presents a number of important actors that generate value in the productive efficiency of the resources used and the amount of products or services developed Fig. 1 [6].

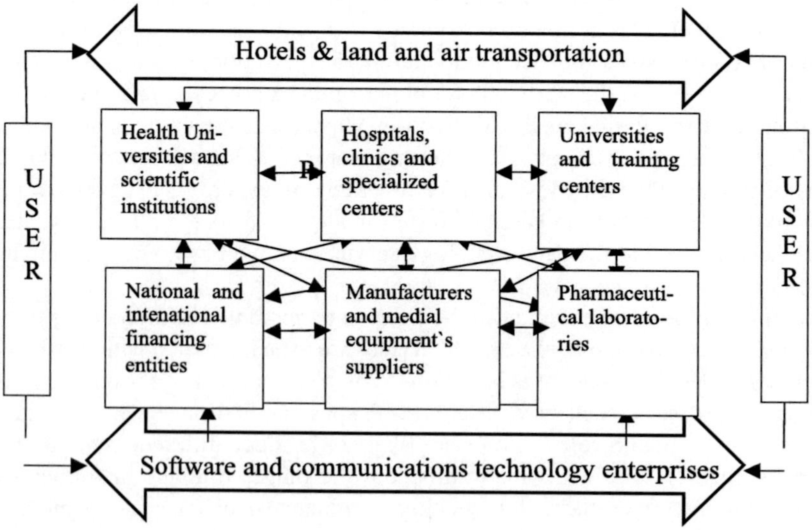

Fig. 1. Source: own elaboration [2021]. Information from Casolco [2015].

Colombia is part of the top five wellness destinations in Latin America along with Mexico, Argentina, Brazil and Puerto Rico. Since 2007, Colombia's Ministry of Trade industry and tourism has been developing the Productive Transformation Program, focused on strengthening health tourism as a world-class sector; also, within the framework of public policy, it is established as a sectoral policy to positioned Colombia as a world-class health tourism destination [7].

In Tijuana, Baja California, México. Postgraduate research has generated contributions to the state of the art in research methodologies on the frontier of innovation, for the benefit of society, such as the case of the Fifth Systemic Helix (FSH) in Spanish known as Quinta Hélice Sistémica (QHS), based on focus group techniques for sector integration studies and evaluation of public policies [8–10].

The influence of the leadership of SMB in the hospital sector reflects the success of the image of quality services and care in Tijuana, Baja California, Mexico [11].

In order to understanding about the Tijuana´s successful in Medical Tourism, this investigation question is: What is the value chain in medical tourism industry in Tijuana, Baja California?

2 Methodology

To test these investigation question, a qualitative research approach was utilized. Two in depth interviews were carried out on October, 2020, which were recorder and transcribed. The first one, was made on to the Baja Health Cluster´s Director and the second was carried out to the Health and Wellness Tourism ex coordinator of the Baja California Tourism Ministry.

This is a qualitative approach in phenomenological design and pretended to know, in depth, how the value chain that leads to leadership in medical tourism to the city of Tijuana was put together.

3 Results

In order to answer our research question: What is the value chain in medical tourism industry in Tijuana, Baja California? Below are the results obtained from the two in-depth interviews conducted, which will be shown according to the chronological sequence of the creation of the value chain.

2010 is the beginning year in the history of clusters in Baja California, Mexico, which have been accompanied since its creation of a promotion and advertising company. Baja California is the leader in health tourism in Mexico, so in 2015 public policy is developed to formalize the exercise of health tourism.

The initial strategy was to improve the arrival infrastructure of the city, providing to the visitors with an affordable ecosystem, offering high-level health services with quality and dignified treatment of the patient, as well as certified and bilingual staff.

The project was based on models of Colombia, Spain and Chile and is tropicalized to the State, because there are different micro regions, each with different strengths and experience in the care of health tourism.

Public policy is designed based on experience, not the promotion of a tourist destination, hence the concept "*The Baja Experience*", linking the entire value chain, strengthening from the perspective of each sector its needs and integrating them into a single project to achieve the unique experience for the visitor.

For the integration of each actor in the value chain, a recognition in health and well-being is designed, issued by the Secretary of Economy of Baja California, which is granted to service providers that fulfill four parameters: minimum experience of two years providing the service, bilingual receptionist, compliance with the corresponding regulations in infrastructure and facilities and certified staff in their specialty.

Baja Health cluster is a private initiative agency that works hand in hand with the Government of Baja California in the safe and responsible promotion of Health Tourism, a situation that made it a creditor to the European Secretariat for Cluster Analysis (ESCA).

One of the great innovations in the health tourism value chain in Baja California is the design of a digital platform in which all providers that fulfill the requirements in regulations are integrated, which is promoted in the United States and Canada with the aim of the visitor conducting the search for reliable and accredited services, in addition to finding recommendations and packages of all services at the best price and with the security of care with a certified staff.

The challenge for 2021 is the post-pandemic COVID 19 economic revival whose first objective is to support and certify that service providers fulfill with all established anti-COVID measures, to give certainty and safety to the visitor of health tourism, and start with an outward promotion through social media, offering our reactivation with all anti-COVID prevention measures.

Once the U.S.-Canada border is re-established, a tour will be held to 21 consulates in Nevada, Arizona, California and Southern Canada in collaboration with the Baja California government to promote health tourism with the entire value chain ecosystem.

4 Conclusions

Health tourism is an industry with an annual growth rate between 8 and 10% and an economic spill for Mexico between $1.4 and $1.8 billion dollar [3].

It is important to mention that part of the success of Tijuana's industry leadership is due to joint work with the Baja California government and the cluster's alliance with the promotion and marketing company.

In the research study Market Opportunities and Perspectives of Health Tourism in Mexico, a cluster maturation scheme is proposed to realize assessments and roadmaps that contribute to the professionalization and development of the industry, whose objective is to achieve phase four, where the cluster generates synergies, activities are innovative, support regional development and anticipate the Market [3], finding an area of opportunity in the creation and development of scientific and academic research to support the growth and leadership of the Baja Health cluster.

Mexico publishes 1% of the world's production in tourism, contributing academic bodies by 21.5% [12].

The proposal derived from this scientific research is precisely to integrate a research observatory for innovation and development that collaborates together with educational institutions hand in hand with their academic bodies.

References

1. Béland, D., Zarzeczny, A.: Medical tourism and national health care systems: an institutionalist research agenda. Glob. Health **14**, 68 (2018). https://doi.org/10.1186/s12992-018-0387-0
2. Arias, F., Caraballo, A., Matos, R.:. Health tourism: concept, history, development and current state of global market. Rev. Clío Am. **6**(11) (2012). ISSN-e 2389-7848, ISSN 1909-941X
3. Delloite: Oportunidades de Mercado y Perspectivas de Salud en México (2019). https://www2.deloitte.com/mx/es/pages/real-estate/articles/turismo-de-salud-en-mexico.html
4. Contreras, T.C.: An Approach to Medical Tourism on Mexico's Northern Border. Eurasia Border Review, vol 6. Slavic-Eurasian Research Center, Hokkaido University. (2015). https://doi.org/10.14943/ebr.6.1.45
5. Florenta, L., Tigu, G.: Medical tourism market trends - an exploratory research. In: Proceedings of the International Conference on Business Excellence, vol. 11. The Bucharest University of Economic Studies, Bucharest (2017). https://doi.org/10.1515/picbe-2017-0114
6. Casolco, S., Mateos-Espejel, L., Ochoa-Cáceres, H.: Hacia la creación de un clúster de turismo médico en la ciudad de Puebla: un análisis de variables estratégicas. In: XIX Congreso Internacional de Investigación en Ciencias Administrativas 21 al 24 de Abril de 2015. Universidad Juárez del Estado de Durango (2015)
7. Causado-Rodríguez, E., Mojica-Cueto, A., Charris-Fontanilla, A.: Clúster de turismo de salud en Colombia: referentes para la competitividad. Duazary **15**(3), 307–323 (2018). https://doi.org/10.21676/2389783X
8. Martínez, R.: Fifth Systemic Helix (FHS). A method to evaluate the international competitiveness of electronics sector in Baja California, Mexico (2012). Quinta hélice sistémica (QHS). Un método para evaluar la competitividad internacional del sector electrónico en Baja California, México (2012). Invest. Adm. (110), 34–48 (2012). ISSN 1870–6614. https://www.redalyc.org/articulo.oa?id=4560/456045338003. Accessed 28 Apr 2020
9. Martínez Gutiérrez, R.: Fifth Systemic Helix (FSH), a model for the development of public policies. Cooper. Desarrollo **20**(101) (2012). https://revistas.ucc.edu.co/index.php/co/article/view/17
10. Martínez-Gutierrez, R.: Methodology of the fifth systemic helix for the development of public sector policies. J. Competit. Stud. **22**(3–4), 147+ (2014). https://www.questia.com/library/journal/1G1-427666157/methodology-of-the-fifth-systemic-helix-for-the-development
11. Bonilla-Hernández, A.L.: Accountability in assertive decisions making. In: Kantola, J.I., Nazir, Salman, Salminen, Vesa (eds.) AHFE 2020. AISC, vol. 1209, pp. 143–148. Springer, Cham (2020). https://doi.org/10.1007/978-3-030-50791-6_18
12. Chávez, R., Bravo, M., Andrade, E., Cornejo, J., Espinoza, R.: Contribución científica de los cuerpos académicos en turismo de México. Rev. Iberoamericana Turismo **9** (2019). https://doi.org/10.2436/20.8070.01.144

Author Index

R. S. Goonetilleke et al. (Eds.): AHFE 2021, LNNS 273, pp. 519–521, 2021.
https://doi.org/10.1007/978-3-030-80713-9